Photovoltaic Engineering Handbook

Photovoltaic Engineering Handbook

France Lasnier and Tony Gan Ang

Asian Institute of Technology, Bangkok, Thailand

Adam Hilger

Adam Hilger, Bristol and New York

British Library Cataloguing in Publication Data

Lasnier, France
Photovoltaic engineering handbook.
1. Solar energy. Photovoltaic conversion
I. Title II. Ang, Tony Gan
621.31244

ISBN 0-85274-311-4

US Library of Congress Cataloging-in-Publication Data

Lasnier, France.
Photovoltaic engineering handbook/France Lasnier and Tony Gan Ang.
568 p. 24 cm.
Includes bibliographical references.
ISBN 0-85274-311-4
1. Photovoltaic power generation. 2. Solar cells. 3. Solar batteries.
I. Gan Ang, Tony. II. Title.
TK2960.L27 1990
621.31′244—dc20 90-4286

Consultant Editor: **Dr R H Taylor,** CEGB

Published under the Adam Hilger imprint by IOP Publishing Ltd
Techno House, Redcliffe Way, Bristol BS1 6NX, England
335 East 45th Street, New York, NY 10017-3483, USA

Typeset by MS Filmsetting Limited, Frome, Somerset
Printed in Great Britain by Galliard (Printers) Ltd, Great Yarmouth

To Rose, Anita, our sisters, brothers, parents and grandparents

Contents

Preface

This book is intended to be a comprehensive introduction to photovoltaic (PV) technology and systems engineering. Six years of teaching and research experience at the Asian Institute of Technology (AIT)—a regional multicultural post-graduate technological institute located near Bangkok, Thailand—have contributed to this achievement.

This book has been elaborated within the regional socio-economic scenery which frames the industrial development of South and East Asian economies, is aimed at assisting the growth of PV industry, and the dissemination of PV generators in rural areas in particular. It provides the necessary technical knowledge which is a preliminary condition for acquisition of this technology by less developed countries.

Through a theoretical, practical and multi-disciplinary approach, it provides a broad understanding of PV industrial options, system-designing methods, economic appraisal as well as components field trial procedures and characterization methods.

Owing to its multi-disciplinary approach, the authors believe this book will meet the need of a broad range of readers in post-graduate institutions and technical colleges, and will serve as a reference book in electronics and electrical courses as well as in energy economics and planning and energy technology related fields of study.

The materials contained in this book encompass the overall spectrum of PV technology starting from the solar-cell-manufacturing processes to the analysis of the entire system of energy production. It is basically divided into seven parts, i.e. technological processes, the PV generator, systems engineering, PV characterization and testing methods, PV sizing procedure, economic analysis and instrumentation. The order adopted is aimed at a progressive and global understanding of photovoltaics and each part constitutes specifically a perfect and independent learning entity.

Acknowledgment should be made to the world community of researchers whose different publications and sharing of findings in international and regional forums contributed to make the bibliographical sections of this book well supported and densely informative. References and bibliographies are listed at the end of each chapter for further in-depth research work.

Although many individuals have participated in one way or another, by question or comment, to the quality of the present edition, a special word of thanks has to be addressed to AIT students who helped the authors to find the right focus of this book, i.e. to answer the regional needs in photovoltaics.

France Lasnier
Tony Gan Ang
June 1989

Part I

Technological Processes

1 Solar Cell Technologies

1.1 INTRODUCTION

It has been over 150 years since Becquerel in 1839 first discovered that a photovoltage was developed when light was directed onto one of the electrodes in an electrolyte solution. Adams and Day were the first to observe the photovoltaic (PV) effect in a solid (selenium) in 1877. Early solid-state researchers including Lange, Grondahl and Schottky did pioneering work on selenium and cuprous oxide PV cells. However, it was not until 1954 that scientific literature published results on the use of the PV effect in energy conversion processes.

The wide spectrum of semiconductor-related technologies encompasses the field of solar cells (see figure 1.5) and all manufacturing processes described below are derived from and enjoy the latest advances made in microelectronics. From this point of view, PV converters belong to the so-called high-technology field and any improvement in that field contributes to the progress in semiconductor technology as a whole.

The silicon solar-cell-manufacturing process involves basically three stages.

(i) Material preparation and shaping.
(ii) Cell processing.
(iii) Cell interconnection and encapsulation.

The conventional technological process, the flow chart of which is presented in figure 1.1, relies conventionally on the high-purity semiconductor-grade silicon (SG-Si) available from electronics industry wastes. In this process, quartzite as a raw material is reduced into metallurgical-grade silicon (MG-Si), the purity of which lies in the 98–99% range. MG-Si is purified to qualify as the so-called SG-Si or electronic-grade silicon (EG-Si). SG-Si is then melted in a crucible and pulled out to form a single-crystal silicon ingot by using the traditional Czochralski (CZ) technique (figure 1.2). Ingots are cut into wafers which are etched to remove the damage induced during the slicing operation. Wafers are then processed into a complete solar cell

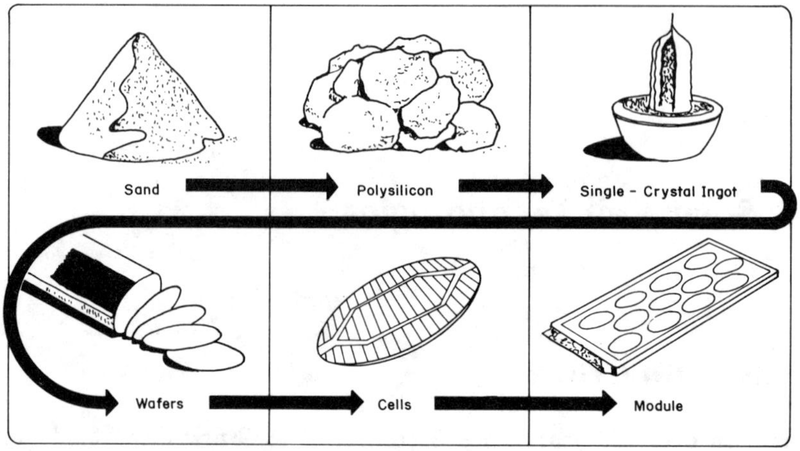

Figure 1.1 Solar-cell-manufacturing process.

(figure 1.3), i.e. junction diffusion, contact elaboration and anti-reflection coating. Finally, cells are interconnected to achieve a practically useful voltage and encapsulated into a module by using a lamination technique.

In 1982, silicon material contributed nearly 50% of the total cost of a PV module whereas cell processing and module assembly contributed respectively 30% and 20% of the total cost breakdown. Since that time, critical paths in the silicon technology process have been investigated intensively, leading to major technical breakthrough and substantial cost reduction as well as the definition of various specific silicon routes for solar cell fabrication (figures 1.4 and 1.5).

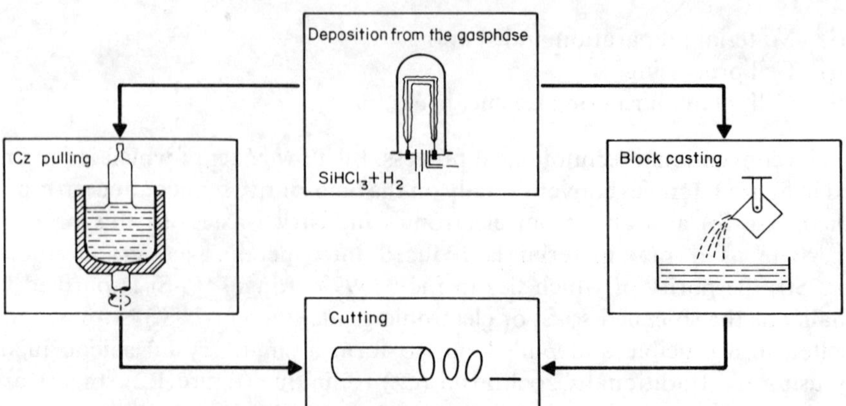

Figure 1.2 Production and slicing of silicon ingots.

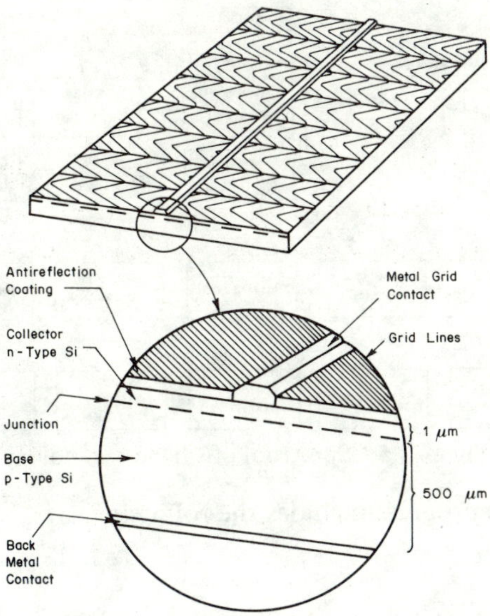

Figure 1.3 A typical single-crystal silicon p–n junction solar cell.

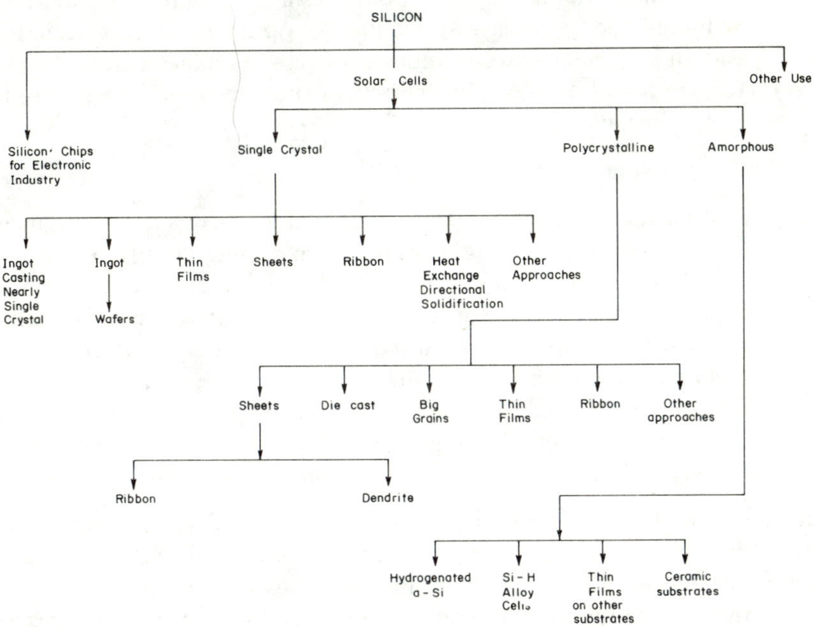

Figure 1.4 Silicon and its various routes of utilization, particularly the solar-cell route.

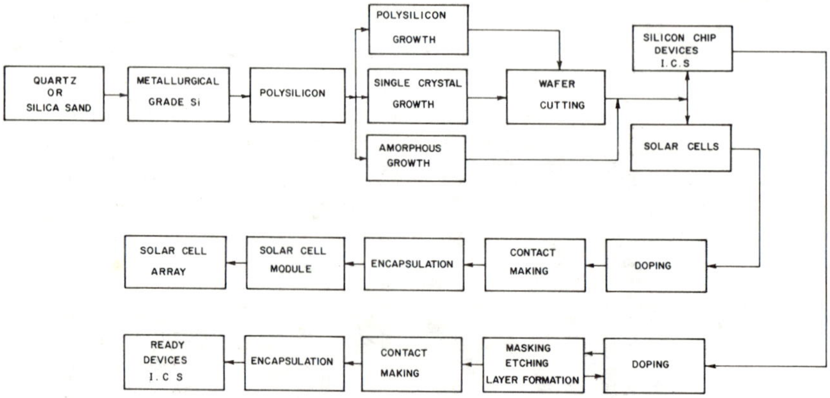

Figure 1.5 Spectrum of silicon technology.

A review of recent progress includes the following.

(i) Material.

(*a*) Preparation and definition of 'solar-grade' silicon (SOG-Si) material as a low-cost and solar-specific alternative to the expensive semiconductor-grade material which will soon be depleted.

(*b*) Development of the polycrystalline technology route based on SOG-Si as an industrial alternative to the single-crystal technology and including the ingot-casting technique, optimized slicing techniques and, in particular, the continuous process to replace batch processes.

(*c*) Research and development efforts in sheet technology (ribbon, foil, thick films, etc).

(ii) Cell processing.

(*a*) Surface texturization to decrease optical losses in single-crystal cells.

(*b*) Cold-junction processing (CJP) in which doping impurities are introduced by ion or molecular implantation or evaporated or spray deposited on the wafer surface. They are then driven into the material to form a p–n junction by pulsed annealing (laser or electron beam); this process is amenable to full automation.

(*c*) Screen printing and to a lesser extent electroplating for deposition of contacts. Research is being carried out to replace noble materials (silver) by low-cost alternative materials (aluminium).

Moreover, substantial cost reductions have been achieved through process automation, and it is widely recognized that the industrial dimension is of paramount importance for the economic use of PV converters.

A further cost reduction scenario implies more fundamental changes and, in the year 2000, amorphous silicon (a-Si) (or other material) thin-film technology is anticipated to emerge as a key technology for future low-cost mass production of solar cells. In the last few years, remarkable progress has been seen in both physics and technology development of hydrogenated amorphous silicon (a-Si:H) solar cells which in a very short period has moved through all the phases from discovery to industrial production. Nowadays a-Si solar cells have a considerable market size particularly in the consumer electronics field such as pocket-sized calculators, wrist watches and battery rechargers. Although a-Si technical readiness has still to be proven to enter the power production sector, the basics of the process are reviewed in the present chapter.

1.2 LARGE-SCALE PRODUCTION OF SILICON BASE MATERIAL

1.2.1 Conventional technology process

1.2.1.1 Quartzite to metallurgical-grade silicon

After oxygen, silicon is the most abundant element in the Earth's crust. Geological processes have provided different ways of concentrating silicon dioxide in vast deposits all over the world. Quartzite rock and special strata of quartz sand have been the most widely used raw materials for the production of elemental silicon. Their inexpensive mining requires little energy and has negligible environmental impact. The quality ranges of either genuine or refined silicon dioxide have been computed in table 1.1. The preference or necessity to use a certain quality depends largely on the type of reduction and refining process sequence and possibly the geographical

Table 1.1 Different types of SiO_2 sources.

Mining product	Upgrading	Nature of particles	Impurities (wt.%)
Quartzite rock	Standard (technical grade)	Chunks	≈1
Special silicates	Chemical conversion	Powder	<0.3
Quartz sand	Standard (technical grade)	Grains	<0.2
Selected quartz sands, pegmatites	Chemomechanical treatment	Grains	<0.05
Special quartz veins, rock crystal	Standard	Nuggets	<0.03

situation. Prices of quartz product can be 50 times more expensive when obtained from rock crystal (US\$0.2 kg^{-1} in 1979) as opposed to selected quartz sand.

Present commercial extraction processes use the crystalline form of silicon dioxide, quartzite, with 90% or more silica (SiO_2) rather than sand whose major constituent is also silica. Sand with its high amount of impurities is too costly to process and therefore does not make an economic choice as a raw material. The quartzite is reduced in large arc furnaces (figure 1.6) by carbon to produce elemental silicon according to the following chemical reaction:

$$SiO_2(s) + 2C(s) \rightarrow Si(l) + 2CO(g).$$

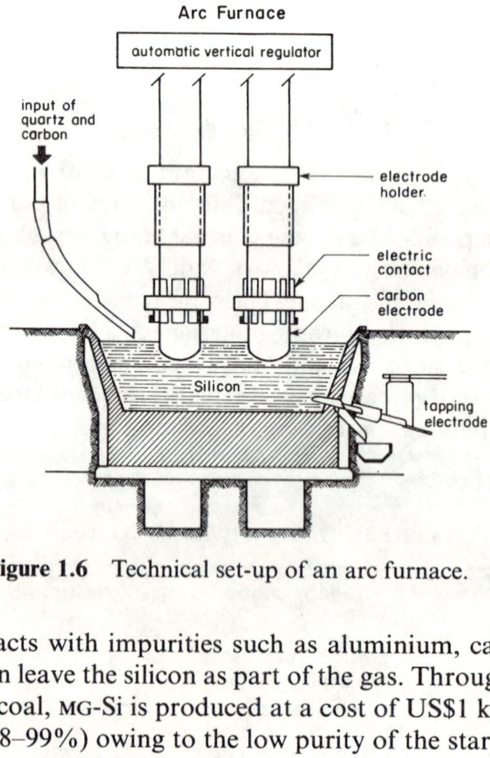

Figure 1.6 Technical set-up of an arc furnace.

The vapour reacts with impurities such as aluminium, carbon and magnesium which then leave the silicon as part of the gas. Through the reduction of quartzite with coal, MG-Si is produced at a cost of US\$1 kg^{-1}. The purity of MG-Si is low (98–99%) owing to the low purity of the starting materials.

1.2.1.2 Metallurgical-grade silicon to electronic-grade silicon

The Siemens C process is at present the standard method for purifying MG-Si to EG-Si. The silicon produced by this method is 99.9999% pure (electronic-grade) polycrystalline form.

$$Si + 3HCl \rightarrow SiHCl_3 + H_2$$
$$SiHCl_3 + H_2 \rightarrow Si + 3HCl.$$

In this process, trichlorosilane ($SiHCl_3$) is first obtained. A bed of fine MG-Si particles is fluidized and chlorinated with hydrochloric acid in the presence of a copper catalyst to induce the reaction. The low purity of the MG-Si results in impure $SiHCl_3$ which must then undergo subsequent fractional distillation to reduce impurities by a factor of 10^4–10^5. A chemical vapour deposition (CVD) method is subsequently used to isolate the pure silicon from the electronic-grade $SiHCl_3$. Vaporized $SiHCl_3$ is decomposed and reduced with hydrogen at about 1000 °C, resulting in the deposition of silicon precipitates on the hot surface of a bridge in the shape of an inverted U. The bridge is made of slim silicon rods and has been heated in a reactor by passing an electric current through it. By this process, six polycrystalline silicon rods of length 1 m and diameter 12 cm can be produced simultaneously.

The price increase from US\$1 kg^{-1} for MG-Si to more than US\$70 kg^{-1} for EG-Si reflects the costliness and energy-intensiveness of the Siemens C

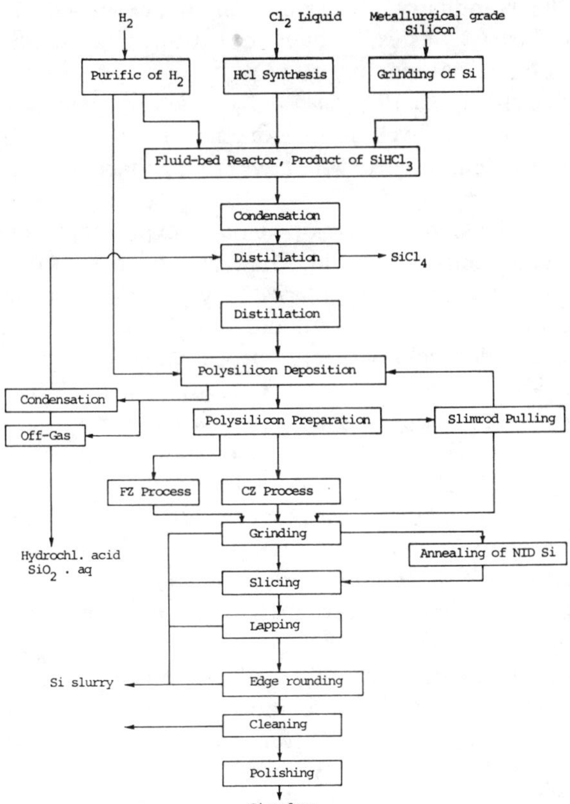

Figure 1.7 Purification of MG-Si to make silicon wafers.

process, the most energy intensive being the last step, the deposition of silicon from the gas phase (150 kWh (kg Si)$^{-1}$). In addition, the deposition process is characterized by a low yield (26%), a low throughput and a discontinuous operation (figure 1.7). Analysis of the present module costs show that the manufacturing of the silicon substrate alone amounts to nearly 50% of the total cost per peak watt with the remainder being costs of cell fabrication and packaging.

1.2.2 Low-cost silicon alternative production processes

The present worldwide production of crude silicon (Eastern European countries excluded) is roughly identical by volume with the annual world production rates of copper and zinc. Only 0.5% of it is used in the semiconductor industry. Roughly 5% goes into silicon production whereas the remaining is consumed by steel and aluminium industries in the form of alloys or 'silicon metal'.

Until now the PV industry has relied upon an inexpensive raw material, i.e. electronics industry wastes, which constitutes a deposit of about 400 t year^{-1}, used to manufacture 50 MW_p year^{-1}. This raw material capacity will be sufficient up to 1992 or 1995. However, owing to the restricted nature of this deposit as well as the forecasted growth of the PV industry itself, alternative technologies will have to produce a large quantity of necessary silicon.

Different ways have been developed or re-explored to manufacture an alternative low-cost silicon material. The target of these worldwide efforts is to provide a bulk material characterized by the following features: abundance of resources, low-cost manufacturing (US$10 kg^{-1}), impurity optimization (solar grade) and low-energy processing.

Figures 1.8 and 1.9 present flow charts of the different technological processes investigated. Figure 1.8 shows a low-purity approach, i.e. all those processes which start with MG-Si or at a corresponding impurity level. Figure 1.9 characterizes a high-purity approach based on an upgraded silicon dioxide (SiO_2) starting material. Moreover two types of process sequence can be distinguished: whether they are relying upon silicon dioxide (SiO_2) as a primary silicon source or volatile compounds as secondary silicon sources. All volatile products have one property in common that makes them very attractive, namely the ease of refining, e.g. by fractionated distillation or desorption. Among the volatile halides, silicon tetrafluoride and chloride are the most economically interesting compounds. Monosilane (SiH_4) can be considered as an at least tertiary silicon source and characterized largely as an intermediate product. These various processes produce intermediate qualities of silicon (figure 1.10) which aim at further refining to reach the 'solar-grade' quality, i.e. a solar cell feedstock material qualified for the manufacturing of solar cells with a high conversion efficiency (12%).

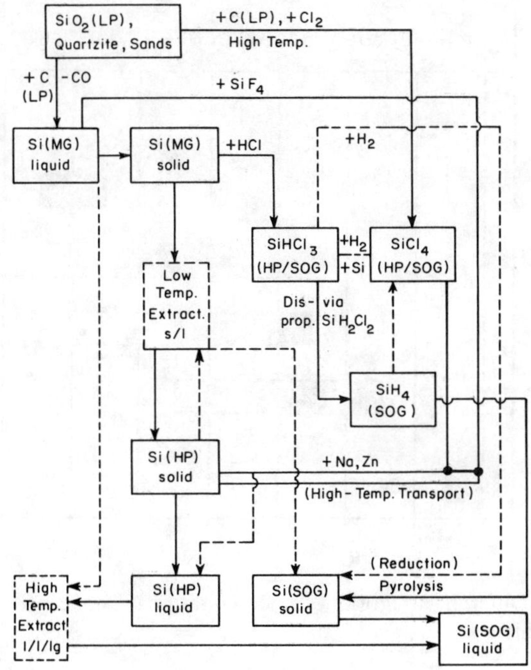

Figure 1.8 Low-purity approaches to large-scale silicon production: LP, low purity; HP, high purity; SOG, solar grade.

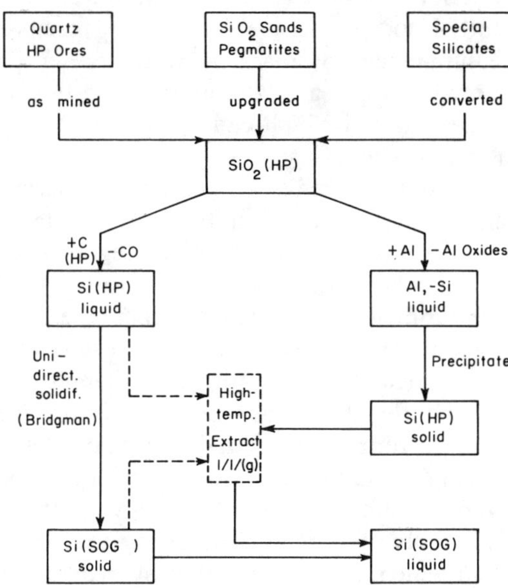

Figure 1.9 High-purity approaches to large-scale silicon production: HP, high purity.

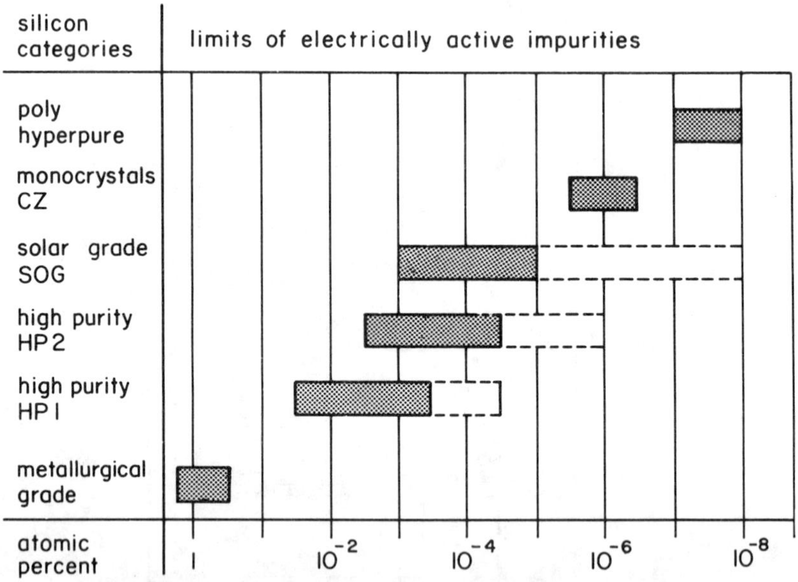

Figure 1.10 Different categories of silicon quality: ---, concentrations of fast-diffusing metals.

1.2.2.1 The 'solar-grade' silicon concept

The SOG-Si concept comes out of the need for the PV industry to develop a low-cost silicon material to replace the EG-Si material.

The development of SOG-Si came about as the result of the apparent insensitivity of medium-efficiency solar cells to the presence of some typical impurities; for example, copper and aluminium in n-type and phosphorus and copper in p-type material which can be tolerated in excess of 1 ppm without major cell degradation.

SOG-Si material has not to be seen as a second-rate material but as a specific material for efficient solar cell production. Furthermore it is more accurate to speak of solar-grade 'synthesis' through bulk segregration of impurities during a unidirectional solidification process (figure 1.11).

In principle, each kind of crystal growth process, e.g. CZ, floating-zone (FZ), directional solidification or sheet growth, results in material purification, the amount of which differs for the different growth processes. Appropriate silicon qualities should be associated with respective appropriate growth processes to produce SOG-Si material (figure 1.12).

Various processes in different stages of development are being explored to produce partially refined silicon material for SOG-Si synthesis from a variety of inexpensive industrially manufactured starting materials such as MG-Si, hexafluorosilicic acid which is an inexpensive by-product of the phosphate fertilizer industry and quartz sand (SiO_2).

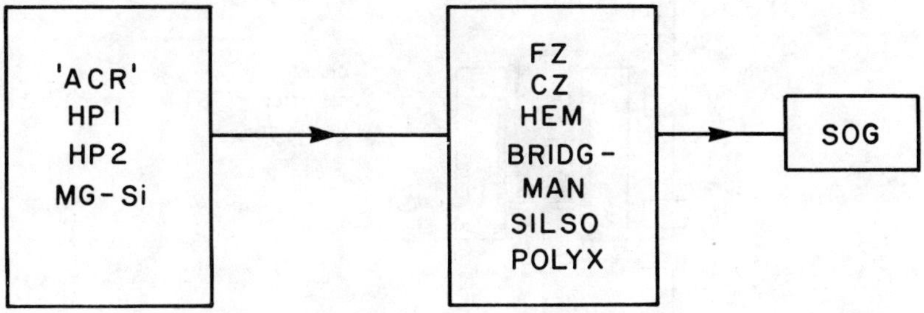

Figure 1.11 Solar-grade synthesis from partially refined silicon by directional solidification. For meanings of abbreviations, see text.

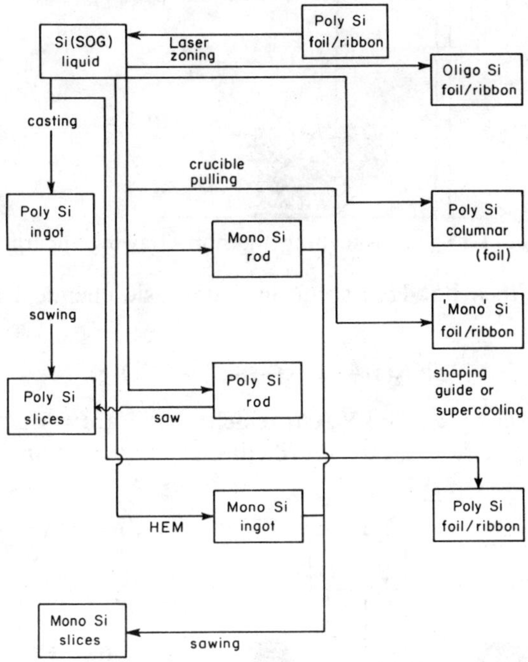

Figure 1.12 Different large-scale process versions to form solar-cell-based material (foils or slices).

1.2.2.2 SiO_2 as the primary silicon source

The advanced carbothermic reduction process (Siemens; Dow Corning)
An alternative process for producing SOG-Si is the advanced carbothermic reduction (ACR) process. It requires the production of high-purity silica and carbon as starting raw materials. The high-purity silica is obtained from inexpensive quartz sand through a glass-melting–fibre-leaching process (figure 1.13) while the high-purity carbon is obtained by leaching carbon

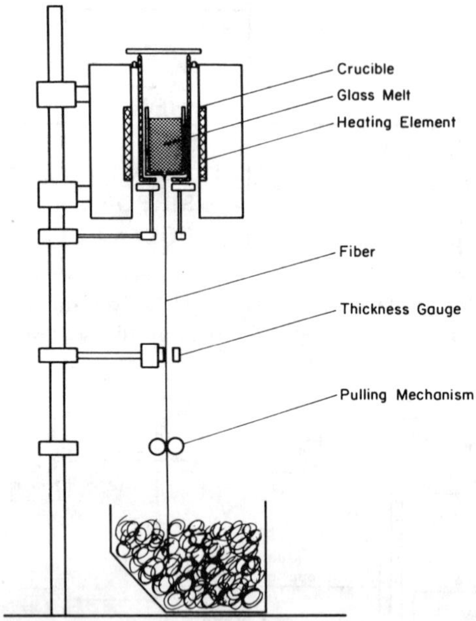

Figure 1.13 Glass-melting and fibre-drawing apparatus.

black pellets with a hot HCl solution. The basic chemical reaction of the process is

$$SiO_2(s) + 2C(s) \rightarrow Si(l) + 2CO(g).$$

Figure 1.14 shows a 550 kVA three-electrode arc furnace used for the carbothermic reduction of the purified silica and carbon in a pilot-scale

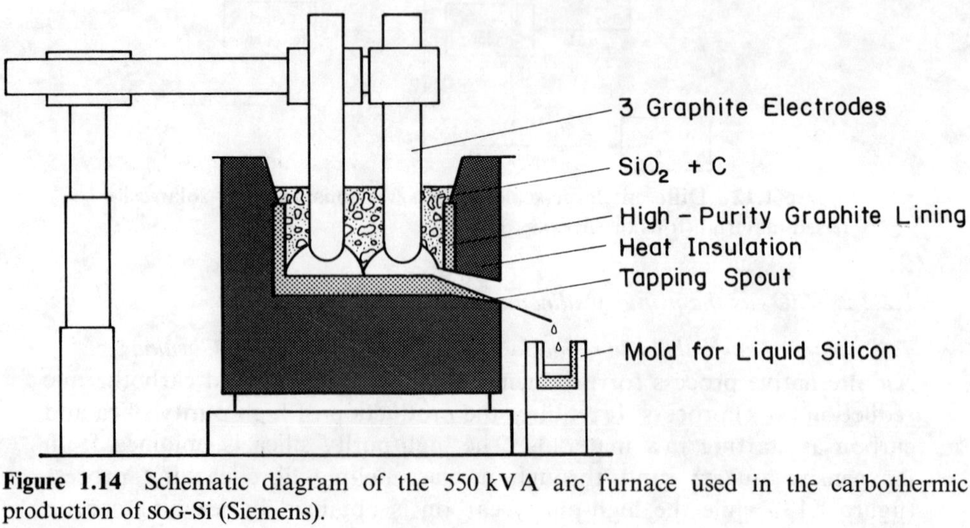

Figure 1.14 Schematic diagram of the 550 kVA arc furnace used in the carbothermic production of SOG-Si (Siemens).

operation. With an inside diameter of 1.2 m it has a capacity for processing 1.2 tons of silica and carbon. Silicon is produced at an approximate rate of 15 kg h^{-1}, collected in graphite mould and then transferred into a CZ crystal puller for further purification. A cost of about US$10 kg^{-1} for the SOG-Si is estimated for an industrial production level of 1000 t Si year^{-1} or more. The best results for solar cells produced from single-crystal ingots obtained from one CZ pull were 13.4% (air mass 1 (AM1)) for 2 cm × 2 cm test cells and 11.1% for cells 100 mm in diameter. A similar process using the carbothermic reduction of silica to produce SOG-Si is used by the Dow Corning Corporation and is presented in figure 1.15.

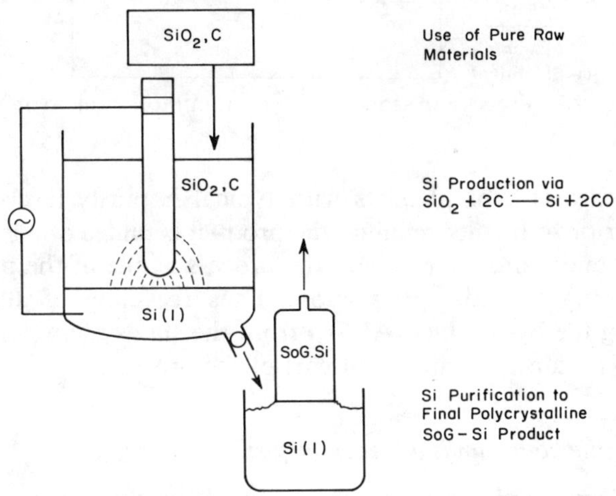

Figure 1.15 Schematic drawing of the Dow Corning process.

The aluminothermic reduction of quartz sand (Heliotronic GmbH; Union Carbide)

The production of silicon by aluminothermic reduction of SiO_2 does not utilize chunks of quartz rock as required in the conventional electric arc process but is based on particulate matter such as powder or sand, the latter being available in large quantities and high purity.

The reduction of quartz with aluminium is a pronounced exothermic reaction taking place in a temperature range 1100–1200 °C:

$$3SiO_2(l) + 4Al(l) \rightarrow 2Al_2O_3(l) + 3Si(s).$$

A suitable technical process on this basis (figure 1.16) requires liquid phases of both oxidic and metallic nature acting as solvents for the reaction products. The two liquid and immiscible systems are represented by an aluminium–silicon alloy and an Al_2S_3–Al_2O_3–SiO_2 slag system. Silicon is

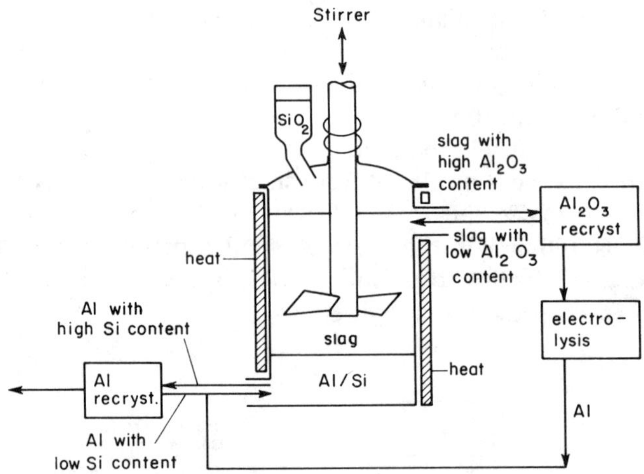

Figure 1.16 Processing sequence of the aluminothermic reduction.

obtained in the form of platelets with typical impurity levels lower than 10 at.ppm. Prior to further refining, the product is milled and leached.

The semi-continuous character of the process is one of the preconditions for guaranteeing low-cost production. Thus recycling of aluminium by recrystallizing the by-product (Al_2O_3) from the slag system and reducing it by electrolysis is a rather important part of the process.

1.2.2.3 Volatile compounds as secondary silicon sources

A different class of reduction processes stems from the conversion of silicon halides (halosilanes) into high-purity silicon. The following section describes a composite of unique processes which aims at producing a low-cost high-purity silicon for solar cell manufacturing. Some of these processes are already at the 'pilot plant' stage. To some extent, the future potential of cost-effective silicon PV cells is directly related to the success of one of these emerging alternative silicon technologies.

Zinc reduction of $SiCl_4$: Battelle Colombus Laboratories
In reviving the so-called duPont process of the early 1950s, a fluidized-bed reactor (FBR) was adapted to harvest silicon:

$$2Zn + SiCl_4 \xrightarrow{FBR} Si + 2ZnCl_2.$$

Furthermore a non-aqueous electrolysis step for recovery of zinc was incorporated:

$$ZnCl_2 \xrightarrow{\text{electrolysis}} Zn + Cl_2.$$

In the proposed 50 mt year^{-1} plant, chlorine gas (Cl_2) will be reacted with soda (NaOH) to form (NaOCl) a saleable product:

$$Cl_2 + 2NaOH \rightarrow NaOCl + NaCl + H_2O.$$

This process is a continuous multi-step closed-loop process in which the raw feed is HG-Si and ($SiCl_4$) the secondary source.

Silane process (Union Carbide Corporation)
The Union Carbide process under development is a multi-step continuous closed-loop process in which the raw material is MG-Si.

The primary-process steps are as follows.

(i) The preparation of chlorosilanes, predominantly ($HSiCl_3$) by hydrogenation of a mixture of $SiCl_4$, which is a recycled by-product, and MG-Si:

$$Si + 3SiCl_4 + 2H_2 \underset{\text{catalyst}}{\overset{\text{Cu}}{\rightleftharpoons}} 4HSiCl_3.$$

(ii) The two-step disproportionation of $HSiCl_3$ into SiH_4 with formation of $SiCl_4$ as the by-product:

$$2HSiCl_3 \xrightleftharpoons{\text{amine catalyst}} H_2SiCl_2 + SiCl_4$$

$$2H_2SiCl_2 \xrightleftharpoons{\text{amine catalyst}} SiH_4 + SiCl_4.$$

(iii) The conversion of silane (SiH_4) into silicon. Two deposition reactions have been investigated, i.e. the free-space reactor and the FBR, the fluidized-bed approach being the most attractive for a continuous process. The silane process developed by Union Carbide is already in the pilot (100 mt year^{-1}) and production plant (1200 mt year^{-1}) stages. The flow diagram of the silane process is shown in figure 1.17.

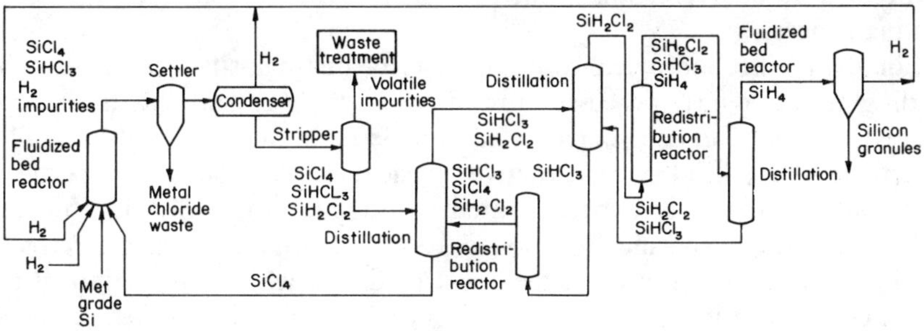

Figure 1.17 Union Carbide silane process flow diagram for the production of SOG-Si.

Fluosilicic acid (SiH_2F_6) reduction (SRI International)
The SRI process under development is the reduction of SiF_4 with sodium. SiF_4 is obtained at a low cost from H_2SiF_6 which is a by-product of the fertilizer industry.

1.3 SILICON SOLAR CELLS FEEDSTOCK MATERIAL

An important effort has been devoted to the definition and the preparation of SOG-Si for replacing the expensive EG-Si. However, solar-grade as well as electronic-grade materials have to be shaped prior to being processed for PV cells production (figure 1.12).

Two kinds of method to produce silicon solar cell feedstock material are investigated although their development stages are quite different.

(i) Bulk growth, leading to ingots which have to be cut into wafers with optimized slicing techniques.
(ii) Ribbon pulling for which all the limitations bound to slicing are definitely overcome. The silicon consumption in ribbon-pulling techniques is between two and eight times smaller than in sliced wafers depending upon the different processes.

There is of course an economic advantage to using ribbons as against ingots (to minimize waste material) but the ribbon-pulling techniques are much less mature and most of them still have to be proved viable before being industrialized. Moreover, bulk growth techniques are already in industrial production, starting with a semiconductor-grade as well as an electronic-grade feedstock material. Bulk growth techniques can then be divided into two approaches.

(i) Ingot-pulling technology.
(ii) Ingot-casting technology.

In casting technology, the material is shaped by using a unidirectional crystallization process in a fixed crucible. For EG-Si casting the process involves a seed of crystalline silicon which is not required in the SOG-Si casting process.

Finally, two technological routes have to be distinguished. In the first case leading technological processes for making single-crystal ingots are considered. The second case corresponds to the elaboration of polycrystalline structures of any kind since attractive physical and economical properties of such material have been corroborated worldwide. The a-Si thin-film technology, although specifically applied today for the consumer electronics market, has yet to acquire the technical readiness which will present it as a major competitor in the PV power market and this will be reviewed at the end of this chapter.

1.3.1 Single-crystal silicon fabrication techniques

Crystal-growing processes produce single-crystal silicon in the form of ingots, sheets or ribbons. The CZ and FZ techniques grow ingots from molten silicon starting from single-crystal seeds, which must then be sawn into wafers. New fabrication processes such as dendritic web growth and edge-defined film-fed growth (EFG) attempt to reduce cost by growing ribbons that can be directly formed into usable wafers:

- single-crystal silicon
  - ingot
    - CZ and advanced CZ methods
    - FZ technique
    - heat exchanger method (HEM)
  - sheet
    - dendritic web growth
  - ribbon
    - EFG.

1.3.1.1 Ingot-pulling technology

Czochralski and advanced Czochralski methods

The CZ method is still the major commercial method of manufacturing ingots of single-crystal silicon. A seed crystal of silicon is slowly withdrawn under rotation from a bed of molten silicon contained in a crucible (figure 1.18(*a*)). If the rate of rotation and pulling (normally less than 10 cm h^{-1}) is carefully controlled, the pulled rod will solidify upon cooling into one large crystal ingot of typically 7.5 cm diameter and 1 m in length. Normally, a p-type crystal is manufactured directly by adding boron to the melt. A schematic diagram of a conventional CZ furnace is shown in figure 1.19.

The CZ method is effective in segregating impurities as the impurities in the molten silicon tend to remain in the melt rather than be drawn into the crystal. Boron does not become segregated as much as other impurities and becomes incorporated into the silicon crystalline structure, resulting in a p-type crystal.

Significant cost reductions have been attained from improvements in the

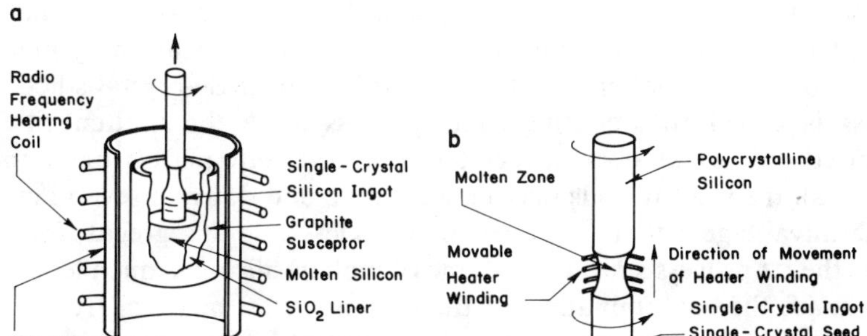

Figure 1.18 Converting polycrystalline silicon to single-crystal silicon: (*a*) CZ growth apparatus; (*b*) FZ processing.

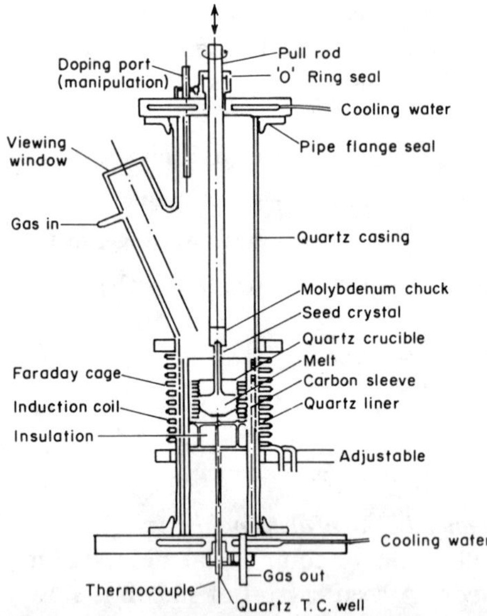

Figure 1.19 Diagram of a conventional CZ furnace.

CZ process which include a rechargeable feature allowing continuous operation. Crystals with diameters as large as 15 cm and of length 1.5 m have been grown using the continuous process. The rechargeable process also allows the crucible to be re-used several times, thus reducing the cost of crucible replacement. Subsequent sawing of the round or rectangular ingots with diamond blades or slurry result in standard finished wafers 0.2–0.35 mm thick. Kerf and etching losses are still greater than 50%.

Floating-zone technique
An ingot of polycrystalline silicon and a single-crystal silicon seed are needed to start the process. The polycrystalline ingot is placed on top of the seed and the interface between the two is melted by movable heating coils (figure 1.18(*b*)). As the heating coils are moved up the stationary ingot, the material below the molten interface solidifies into single-crystal silicon. The process is called the 'floating-zone' process since the molten silicon is supported only by its own surface tension. To ensure perfect uniformity of the crystal, the heating coils may be moved up and down in several cycles.

A disadvantage is that the FZ technique is less able to segregate impurities than is the CZ process since the volume of molten silicon is much less. On the other hand, the contamination of the silicon crystal material from contact with crucible material is avoided as the FZ method is crucible free. In the CZ method, contamination of the silicon by oxygen can occur during the crystal-pulling process.

1.3.1.2 Ingot-casting technology

Heat exchanger method

The HEM is a low-cost directional solidification technique introduced by Crystal Systems. It has been used for commercial production of sapphire 32 cm in diameter and of mass 50 kg. This method has been adapted for the growth of silicon crystals and is actually a unique technique which gives a single-crystal silicon ingot of square cross section.

In the HEM process, molten silicon is allowed to freeze from a seed crystal to form a nearly perfect cubic single-crystal ingot under highly controlled conditions. The seed crystal is placed at the bottom of the crucible (made of silica with a graded structure) which is seated on a high-temperature heat exchanger. The seed is prevented from melting by forcing gaseous helium through the heat exchanger. The HEM is therefore the only directional solidification technique in which there is independent control of the temperature gradients in the solid and liquid with no moving parts. The ingots are cast in a silicon crucible which breaks apart upon cooling. The process which involves a helium gas heat exchanger has resulted in 15.5% efficient rectangular solar cells fabricated from the ingot. The Crystal Systems process produces silicon blocks of 34 cm $\times$ 34 cm $\times$ 17 cm which are more than 95% single crystal at a 1.9 $kg\,h^{-1}$ growth rate with a yield of higher than 95%.

1.3.1.3 Ribbon growth: edge-defined film-fed growth

The EFG method developed by Mobil Solar employs graphite shaping dies to shape the crystal growth to a desired form as well as to separate the growth interface from the main melt surface (figure 1.20). The best shaping dies for silicon are made from dense high-purity graphite. As molten silicon rises by capillary action in the die, it is drawn through a narrow slot, producing a single-crystal ribbon. Mobil has achieved 40 $cm^2\,min^{-1}$ growth of ribbons 10 cm wide. Cells with more than 12% efficiency have been produced from simultaneous multi-ribbon growth of three ribbons at 10 cm width or five ribbons at 5 cm width and 0.15 mm ribbon thickness.

The problem of edge control is avoided in a new growth technique by the EFG process whereby a hollow polygonal section shaped tube called a nonagon of 150 mm diameter with nine flat sides, each of 50 mm width and 0.3 mm thickness, is grown. A capillary shaping die achieves shape control over all crystal dimensions. Large-area growth rates are possible as a consequence of the absence of discontinuities in the meniscus. Very wide sheets of silicon can also be grown without the need for accurate control of temperature over long linear distances. After growth, the nonagon is cut into flat segments for solar cell fabrication by the use of high-intensity CO_2 lasers or high-speed diamond saws.

Inherent advantages of the EFG process include efficient utilization of

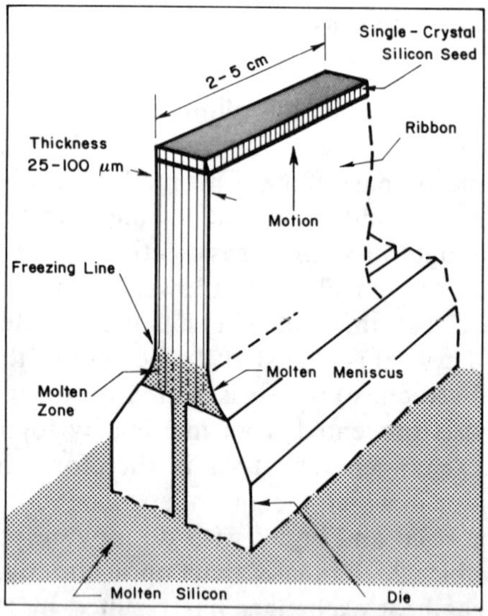

Figure 1.20 EFG: a seed crystal of single-crystal silicon contacts the surface of the melt as it gets drawn through the die by capillary action. As the crystal is slowly raised (at about 15 cm min^{-1}), the silicon cools and freezes into a single-crystal ribbon.

silicon material with no loss involved in slicing or machining operations and efficient utilization of consumable materials such as crucibles, shaping dies and insulation elements. Disadvantages include problems with impurity segregation which requires that the crystal ribbon be drawn from a highly pure silicon melt and contamination by carbon impurities of the shaping dies.

1.3.1.4 Sheet growth: dendritic web growth

A silicon ribbon forms between a pair of dendrites which provide stability to the edges for width control (figure 1.21). The dendrites of silicon propagate into a supercooled silicon melt during growth to shape the ribbon. This process requires that the isotherm at the point of solidification be precisely shaped so that supercooling under the dendrites takes place without supercooling the solid–liquid interface where ribbon growth occurs (figure 1.22). The method avoids the problem of impurity contamination since no die or crucible is used and has yielded best cell efficiencies of 16.9%. The dendrites cannot be incorporated into the final cell structure and must be cut off. Disadvantages include necessity for precise temperature control and slow growth rates. The dendritic web growth by Westinghouse Corporation has achieved growth rates of 13 cm^2 min^{-1} for short ribbon lengths. The ribbon

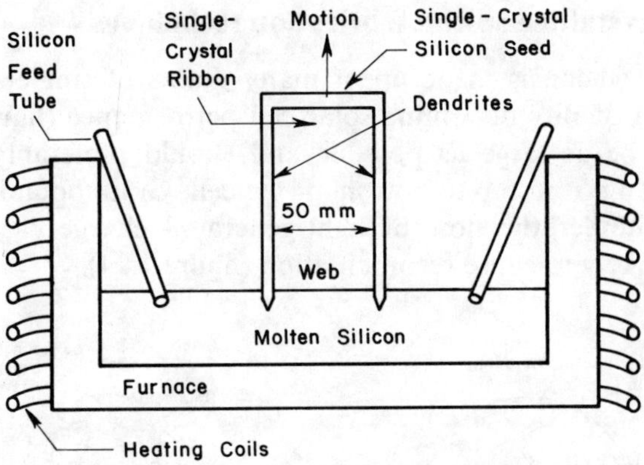

Figure 1.21 Dendritic web growth: a web of single-crystal silicon forms as dendritic seed crystals are withdrawn from the melt.

surfaces are nearly perfect crystallographic (111) facets which require no mechanical treatment prior to device fabrication. The growth apparatus consists of a furnace chamber, a molybdenum susceptor which holds the fused quartz crucible, a molybdenum lid assembly with a slot through which the ribbon is withdrawn, and a reel which serves as a pull mechanism as well as material storage.

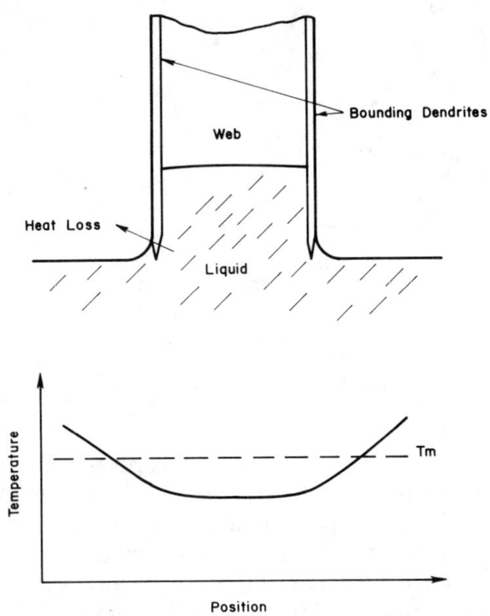

Figure 1.22 Schematic illustration of a growing web compared with the temperature profile at melt surface.

1.3.2 Polycrystalline silicon fabrication techniques

Polycrystalline silicon is made up of many grains of single-crystal silicon whose size and quality determine solar cell performance (figure 1.23). The grains should be as large as possible and should preferably be oriented vertically running from top to bottom of the cell. Grain boundaries are then less likely to hinder the flow of light-generated charge carriers and are therefore less likely to cause recombination (figure 1.24).

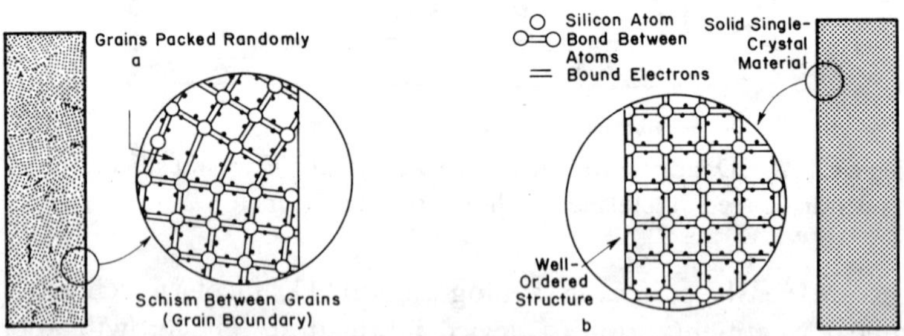

Figure 1.23 (*a*) Polycrystalline silicon is made up of randomly packed grains, each of which is (*b*) single-crystal silicon. The disorder of polycrystalline silicon is responsible for its performance problems.

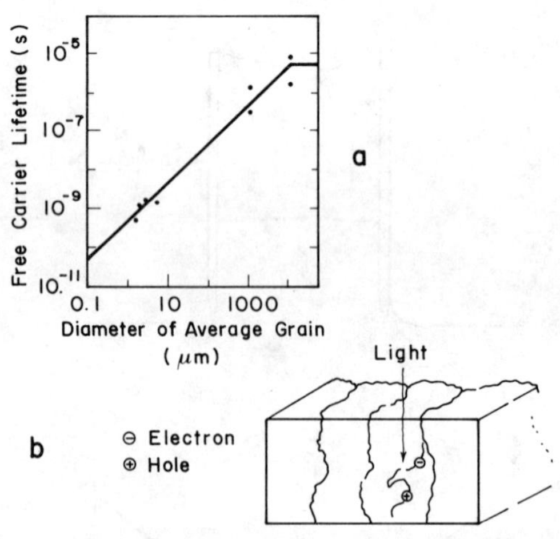

Figure 1.24 (*a*) The lifetime of free charge carriers (which indicates the time that they are likely to be free before recombining) is strongly dependent on grain size in polycrystalline silicon. Since it is important to keep charge carrier combination to a minimum, large-grained polycrystalline material is necessary for good cell performance. (*b*) Wide vertical grains spanning a cell from top to bottom aid cell performance because light-generated carriers tend to flow vertically without encountering grain boundaries.

In a process called directional solidification used to form columnar grains (about a centimetre in size), molten silicon is cooled slowly along one direction so that, as the silicon cools, the solidifying crystal structures are orientated relatively uniformly along one direction. This process is analogous to the HEM method but is a seedless process.

The grain size is the major determinant of polycrystalline silicon cell characteristics and efficiencies. To achieve efficiencies above 10%, grain sizes greater than 0.5 mm are needed and grain boundaries must be orientated perpendicular to the film.

Almost all methods used to make single-crystal silicon described in the previous section can also be used to manufacture polycrystalline silicon under less controlled conditions. Two additional methods unique to polycrystalline silicon exist: edge-supported pulling (ESP) and the sheet growth methods. Also, cell fabrication processes after crystal growth (junction formation, contacts and antireflection coating) are similar to those for single-crystal cells. The various growth methods are given below:

- polycrystalline silicon
  - ingot
    - casting method
      - heat exchanger
      - Silso
      - Semix
      - Polyx
  - ribbon
    - EFG
    - dendritic web growth
    - ribbon-to-ribbon growth
    - ESP
    - ribbon-against-drop (RAD) process
  - sheet
    - silicon-on-ceramic method
    - silicon-on-silicon (SOS) method
    - silicon on inexpensive substrates
    - supported web method
    - ramp-assisted foil-casting technique (RAFT)
    - horizontal supported web (HSW).

1.3.2.1 Ingot-casting technology

During the past 10 years it has been demonstrated that melt-grown polycrystalline silicon could be used for the fabrication of crystalline silicon wafers as a base material (figure 1.25). Thus low-cost technologies which produce, via a controlled directional solidification, multi-crystalline silicon ingots provide a viable alternative to the conventional CZ technique. The

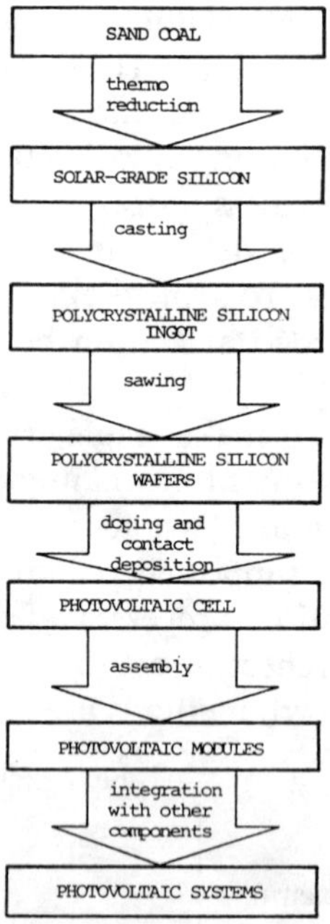

Figure 1.25 Polycrystalline silicon solar cell route.

relatively low crystallization rate inherent to all ingot technologies can be utilized for an *in situ* purification step, easing off the quality criteria for the starting material that is SOG-Si. On the other hand, the casting process requires additional cutting and wafering steps which imply extra costs and loss of silicon. A few technologies are described below with emphasis on crucible re-usability.

Heat exchanger method (Crystal Systems)
The method developed by Crystal Systems is capable of producing polycrystalline silicon solar cells of up to 15% efficiency using a seedless process.

Semix process (Semix)
The ingot-casting process by Semix produces a granular or polycrystalline ingot with large grain size and high intragrain order solidified from a vat of molten silicon with or without a single-crystal seed at the bottom of the vat. This method has been found to be effective in separating impurities and has produced solar cells of up to 15% efficiency similar to the HEM method.

Silso process (Wacker)
Polycrystalline ingots are produced through directional solidification in preheated moulds where the silicon has been melted previously in a quartz crucible. In contrast with the procedures deriving from the Bridgman–

Stockbarger technique, the Wacker ingot-casting process uses two different containers for melting and crystallization. It is a recharging process, with possible re-usability of the mould after casting of the molten silicon. The use of two separate containers for melting and crystallization allows for the decoupling of the two processing cycles. In the upper part, melting and casting take place whereas in the lower part the pre-conditioning of the mould and the crystallization are carried out. The starting material is melted using radio-frequency (RF) heating; the melt crucible can be charged repeatedly by recharging equipment. After casting of the molten silicon, crystallization is carried out in a controlled manner so that a crystalline shell is generated which caulks the mould from the silicon melt. This chilled layer of polycrystalline silicon is produced first in the mould to prevent reaction of the melt with the graphite walls. Crystallization of the remaining silicon melt is then finished in this crystalline container. The method produces typically polycrystalline ingots of size 43 cm × 43 cm and an efficiency of about 50 W kg^{-1}.

The ingot-casting equipment contained in a single water-cooled recipient carries out all the processing steps involved in the melting and crystallization cycles. A schematic diagram of casting equipment for the production of Silso ingots is given in figure 1.26 where the various operations in the Wacker ingot-casting process take place in a reduced inert gas atmosphere. The linear

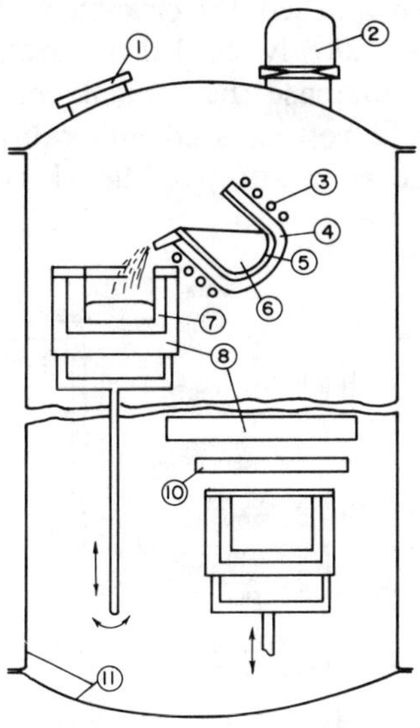

Figure 1.26 Schematic diagram of casting equipment for the semi-continuous production of Silso® ingots: 1, view port; 2, recharging chamber; 3, RF coil; 4, graphite crucible: 5, quartz crucible; 6, silicon melt; 7, mould; 8, thermal insulation; 10, heating element; 11, recipient.

growth rate of the polycrystalline silicon ingot should not exceed a critical threshold of 1 mm min^{-1} since the crystal structure begins to collapse at higher growth rates.

This efficient ingot-casting technology is completed with highly efficient cutting and wafering techniques which produce low-cost silicon wafers at minimum material loss. Cutting and slicing of the finished ingots are accomplished by band-sawing and multi-blade slurry sawing. The band-saw technique contours and portions the ingots and is able to reduce kerf losses to about 2 mm from the 8 mm incurred in outer-diameter (OD) sawing. By using these techniques a 100 kg ingot can be cut into 16 columns of 10 cm × 10 cm cross section with 82% columns output.

Polyx process (Photowatt)

The furnace allows us to obtain multi-crystalline silicon ingots with a basaltic structure by means of an unidirectional crystallization method in a fixed crucible (figure 1.27). The seedless crystallization is operated in a crucible stationary with respect to the working coils (direct induction in the crucible walls). Regulation allows us to stabilize the silicon bath temperature, and silicon solidification is obtained when opening the thermal insulation at the bottom of the crucible. In comparison with the straight Bridgman process in which the crucible moves down the thermal gradient, this modified process leads to a steadier solidification rate and the shape of the solid–liquid interface is more regular—nearly flat over a substantial extent of its surface. When the crystallization is finished, the crucible is removed and can be used again. The ingot is then slowly cooled naturally following a thermal programme. In order to increase the capacity of industrial furnaces, the section of the ingot should be increased preferably to increase the height. Therefore an appropriate adaptation of the RF heating process has been

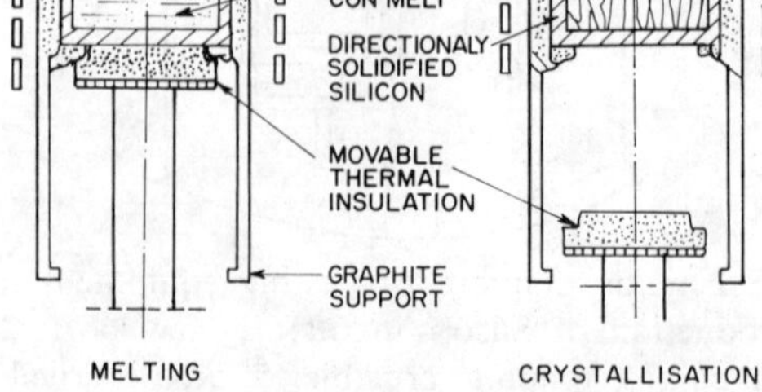

Figure 1.27 Polyx crystallization method.

developed. The RF coil is divided in two separate RF coils: a vertical one for the crucible walls and a horizontal one for the crucible top wall. Therefore, better heating control can be achieved by adjusting separately the induced heat of the two coils. The graph in figure 1.28 shows an example of a crystallization cycle used to obtain 30–60 kg ingots. The first phase of this cycle is the fusion of the polycrystalline silicon and takes approximately 2–3 h; the next phase is stabilization which is carried out for $\frac{1}{2}$ h. Then the thermal insulation of the bottom is slightly opened in order to obtain heat dissipation and to start the crystallization (6–8 h). The ingot is unmoulded from the crucible at the end of the crystallization process and is allowed to cool through controlled cooling.

Such an industrial casting furnace can produce multi-crystalline ingots of mass up to 60 kg with 10% conversion efficiency for up to 90% of the ingot.

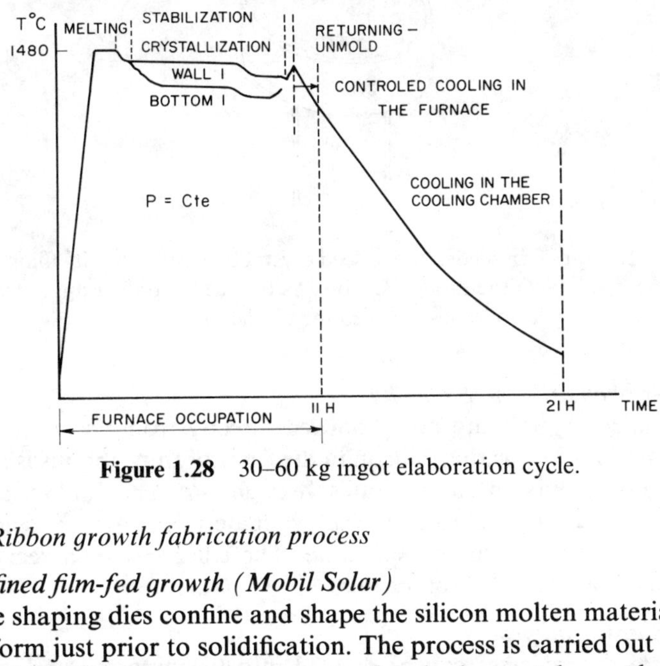

Figure 1.28 30–60 kg ingot elaboration cycle.

1.3.2.2 Ribbon growth fabrication process

Edge-defined film-fed growth (Mobil Solar)

Graphite shaping dies confine and shape the silicon molten material into the desired form just prior to solidification. The process is carried out under less controlled conditions than are necessary for single-crystal growth.

Dendritic web growth (Westinghouse Corporation)

Edge-stabilizing dendrites of silicon are used to grow ribbon-shaped crystals from the melt. Specially designed radiation shields are used to shape the thermal fields around the crystal liquid interface to achieve the differential growth conditions required for the dendrites and the ribbon. Continuous growth of the polycrystalline silicon ribbon is essentially possible through melt replenishment.

Edge-supported pulling

ESP differs from the EFG method in that, here, a foreign material becomes an integral part of the growing ribbon to stabilize the edges (figure 1.29). Graphite filaments are immersed in the silicon melt and withdrawn at a controlled rate, causing a web or ribbon to form between the filaments. The rigid graphite filaments control the ribbon width and edge geometry in a manner analogous to the silicon dendrites in the dendritic web process.

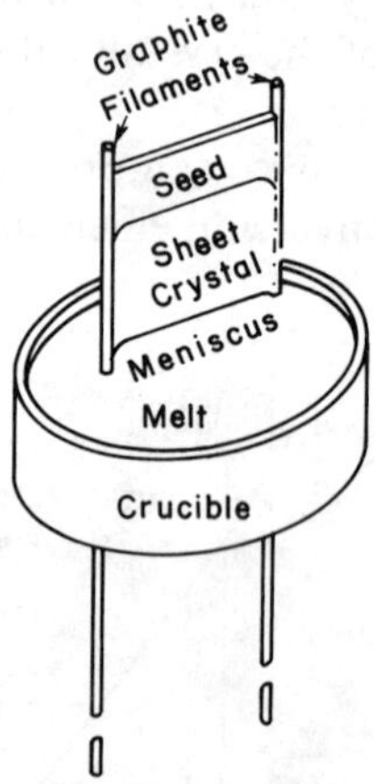

Figure 1.29 ESP: in seeded ESP, a seed crystal between graphite filaments originally contacts the molten silicon. As the seed is pulled up, a nearly single-crystal web solidifies between the filaments.

Ribbon-to-ribbon growth (Solavolt)

Ribbon-shaped crystals are not produced directly from the melt as in the previous methods but, rather, silicon in the form of thin ribbons is deposited on graphite or molybdenum substrates through CVD. The difference between coefficients of thermal expansion of the two materials allows the silicon to be physically separated from the substrate. The silicon is then recrystallized using scanned laser or electron beams.

Ribbon-against-drop process

Ribbon growth processes can be divided into unsupported and supported growth techniques, the RAD process being of the latter kind. The RAD process is a vertical growth process whereby thin polycrystalline films crystallize on both sides of a carbon ribbon which enters the melt via a slot in the bottom of the silica crucible. A schematic diagram of the process designed for growth of ribbons 5 cm wide is shown in figure 1.30. The crucible is heated by RF induction in a graphite susceptor. A pyrocarbon coating 2–5 μm thick grown in an open furnace by pyrolysis at 2100 °C gives the ribbon shaper its wetting properties and its chemical resistance to molten silicon.

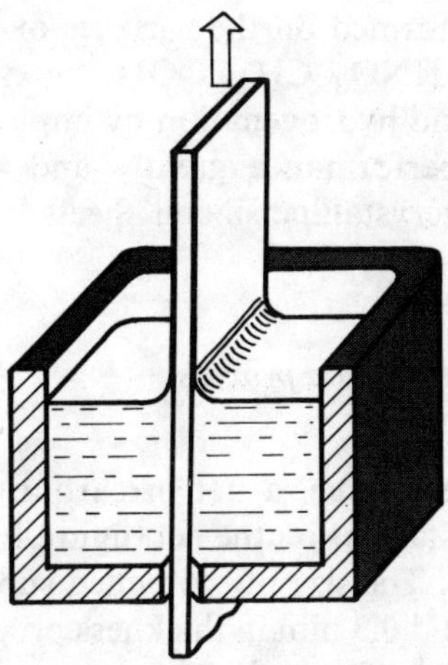

Figure 1.30 Growth of polycrystalline silicon ribbon by the RAD process.

Self-supported sheets of polycrystalline silicon are obtained by burning the carbon ribbon in oxygen at high temperatures. The series of operations involved in the preparation of unsupported silicon sheets is indicated in figure 1.31. The carbon is burned off in a conventional resistance furnace at 1020 °C by exposing the as-grown RAD ribbons to an oxygen-bearing gas for

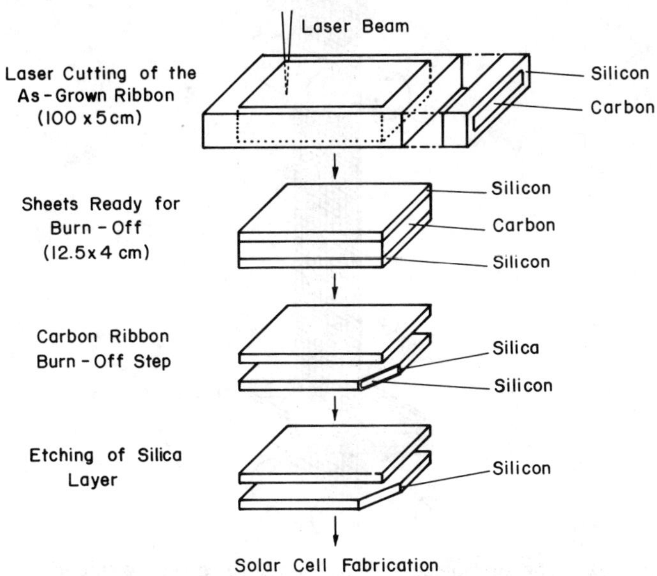

Figure 1.31 Concept of the preparation of unsupported sheets from as-grown RAD ribbons.

1 h. The thin silica film formed on the surfaces of the sheets is then etched away using an $HF–HNO_3–CH_3COOH$ mixture. Subsequent low-temperature annealing and hydrogenation by implantation have been found to improve solar cell performance greatly and 4 cm × 4 cm solar cells fabricated from the polycrystalline silicon sheets have shown efficiencies of 10%.

1.3.2.3 Sheet growth fabrication process

Supported-web method

In the supported-web technique, a net of carbon fibres is pulled through molten silicon which crystallizes on the net (figure 1.32), thus forming a sheet of polycrystalline silicon. The size of the pulled supported-web sheet can be typically 6 cm in width and 0.3 mm in thickness produced at a low-speed pull rate of about 3 cm min^{-1}. Through the conversion of carbon fibres into silicon carbide (molten silicon reacts with the carbon), potential cracks in the silicon sheets from thermal stresses caused by different thermal expansion rates of carbon and silicon are avoided. Thermal stresses due to non-linear thermal profiles during cooling of the ribbon can also be reduced by raising the pulling speed for short time intervals, thus effectively interrupting the silicon coating process on the net. This results in rectangular sheets of polycrystalline silicon.

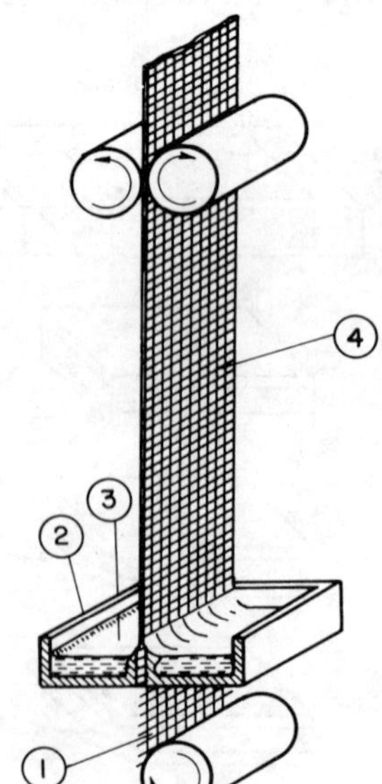

Figure 1.32 The supported-web technique for growing silicon ribbons: 1, carbon-fibre net; 2, crucible; 3, silicon melt; 4, crystallized silicon ribbon.

Ramp-assisted foil-casting technique
The RAFT involves the direct production of self-supporting silicon wafers. The slicing step necessary in ingot-casting techniques is therefore avoided and feedstock silicon consumption is also reduced. Figure 1.33 shows the basic principle involved in the RAFT and a schematic diagram of a continuously operating set-up is given in figure 1.34. The method makes use of a

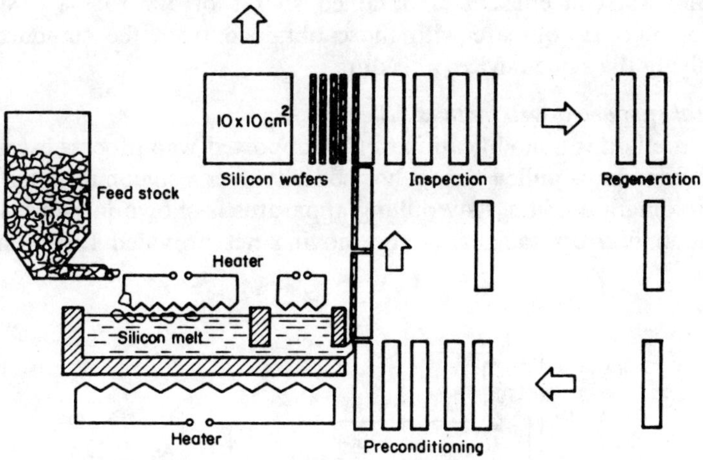

Figure 1.33 Schematic diagram of the RAFT: basic principle.

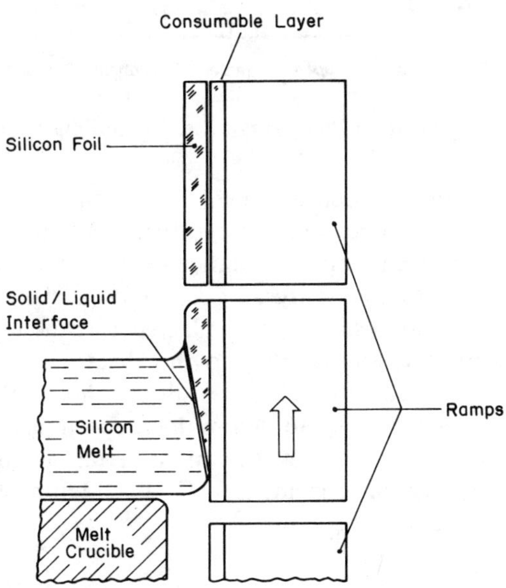

Figure 1.34 Survey of complete RAFT processing.

temporary shaping guide called a ramp on which a silicon melt film is crystallized. The ramp can be re-used several times before an expendable coating has to be regenerated. 10 cm × 10 cm wafers have been grown on ramps of the same surface area. Efficient removal of the heat of crystallization occurs through heat conduction rather than through heat radiation as is normally the case in ribbon technologies. The solid–liquid interface is inclined to allow for high casting rates at moderate temperature gradients. The best cell efficiencies obtained so far of 5.8–10.3% (AM1.5 for 5 cm × 5 cm cell) compare with those obtained from the standard Silso process using the same device procedure.

Horizontal supported-web method (Siemens)
The HSW method is a modification of the supported-web process except here a carbon fibre net is pulled almost horizontally over a silicon melt surface. A schematic diagram of the HSW pulling apparatus is shown in figure 1.35. A solid silicon film crystallizes on the moving net provided that a suitable

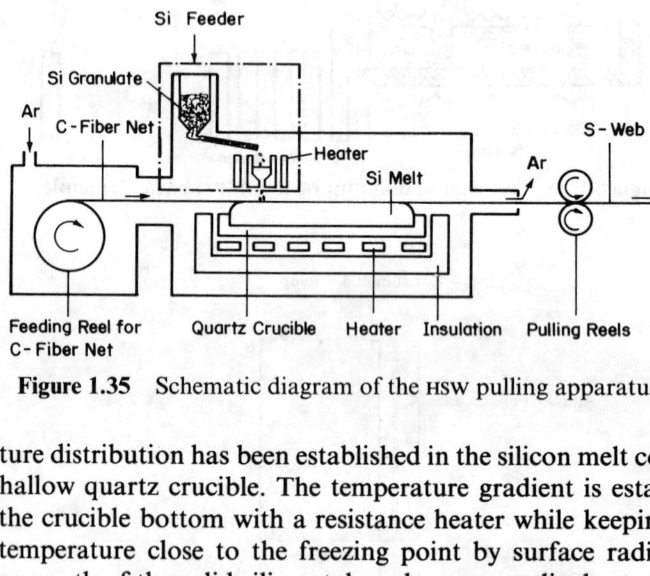

Figure 1.35 Schematic diagram of the HSW pulling apparatus.

temperature distribution has been established in the silicon melt contained in a long shallow quartz crucible. The temperature gradient is established by heating the crucible bottom with a resistance heater while keeping the melt surface temperature close to the freezing point by surface radiation. The thickness growth of the solid silicon takes place perpendicular to the pulling direction and continues as long as the moving sheet is in contact with the melt. With a pulling speed of 1 m min^{-1}, silicon sheets of up to 6 cm in width and with grain sizes up to 4 mm have been grown. Solar cells (2 cm × 2 cm) made from the HSW silicon material by the conventional diffusion process of phosphorus at 850 °C for 30 min show maximum efficiencies of 9.4% under AM1.5.

Silicon-on-ceramic method
Dip coating. Dip coating is one of the silicon-on-ceramic methods in which a polycrystalline silicon film solidifies on a coated ceramic ribbon as it is

withdrawn from the silicon melt (figure 1.36). Polycrystalline silicon ribbons about 100 μm thick have been pulled at a rate of 5–6 cm min^{-1}. Cells with large columnar grains 4–5 mm in diameter have been formed with cell efficiencies of 10.5% for a film 100 μm thick.

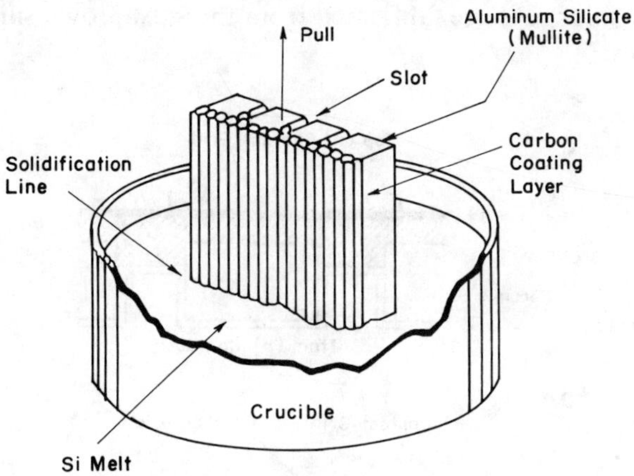

Figure 1.36 Dip coating is a simple SOC process in which a coated ceramic form is slowly withdrawn from a silicon melt while polycrystalline silicon solidifies on it.

Silicon coating by inverted meniscus. In the process of silicon coating by inverted meniscus (SCIM), ceramic sheets are coated with a polycrystalline layer of silicon as they are drawn out through the SCIM furnace. The ceramic sheets are drawn over a trough into which a controlled amount of molten silicon has been released by a plunger lowered into a crucible containing the liquid silicon (figure 1.37). The silicon forms a raised meniscus in the trough

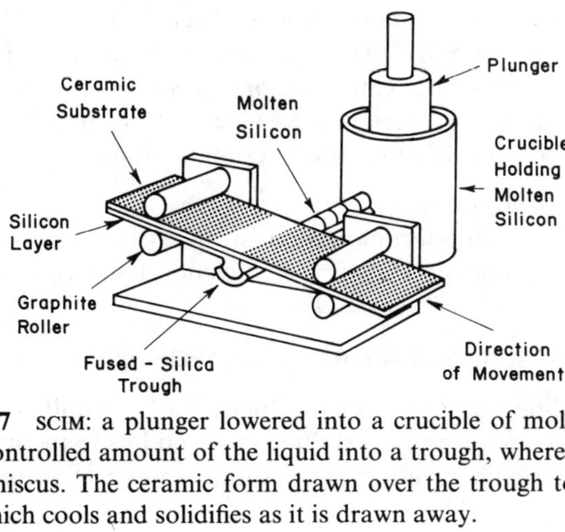

Figure 1.37 SCIM: a plunger lowered into a crucible of molten silicon drives a controlled amount of the liquid into a trough, where it forms a raised meniscus. The ceramic form drawn over the trough touches the silicon, which cools and solidifies as it is drawn away.

which touches and coats the ceramic sheets as they are pulled past at a rate of 4 cm min^{-1}. The method results in less exposure to impurity contamination and ribbon silicon with thicknesses of 100–300 μm have been formed. The SCIM method developed by Honeywell has achieved growth rates of 60 cm^2 min^{-1} and cell efficiencies of 10.5%. Figure 1.38 shows how polycrystalline silicon solar cells are fabricated from the SCIM-grown silicon ribbon.

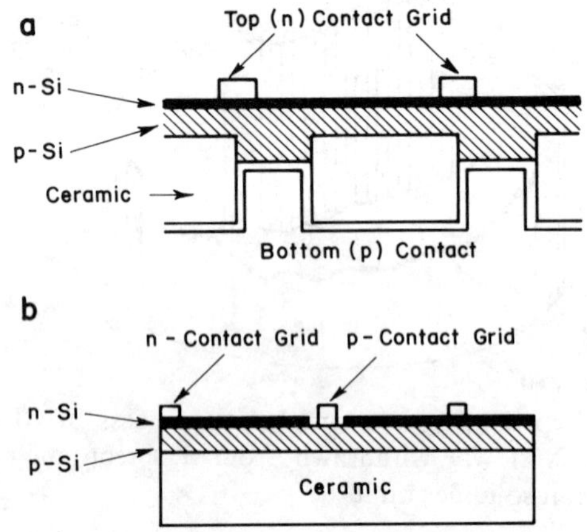

Figure 1.38 (*a*) Contacts can be applied on both sides of an SOC cell by pre-slotting the ceramic base (which does not adversely affect the SOC process) or (*b*) contacts can be attached at the front by cutting through the cell's top layer. The series resistance resulting from the awkwardness of contacting remains a problem in SOC cells.

Silicon-on-silicon method

In the SOS method, a polycrystalline silicon film is deposited on impure silicon substrates through CVD by the reduction of $SiHCl_3$ in H_2 at elevated temperatures. The substrates can be CZ or cast silicon and 10% efficient cells over 100 cm^2 in area have been grown. Cheap MG-Si may also be a suitable substrate for deposition of purer polycrystalline films.

Silicon-on-inexpensive substrates

Silicon vapour deposition techniques on cheap, easily handled materials such as steel or glass are applied with a best efficiency achieved of 23.6%.

1.3.3 Silicon-cutting and silicon-slicing techniques

An integrated line for low-cost production of crystalline silicon ingots whether single-crystal or polycrystalline material has to incorporate efficient cutting and wafering techniques. One disadvantage of using ingots rather than thin or thick films in solar cell fabrication processes is that 50% of the silicon is wasted by the wafering operation.

The technical state of the art in silicon slicing is represented by five different methods. Some of them are used in practice while the others are still in the laboratory testing stage.

(i) Internal-diameter (ID) sawing.
(ii) Conventional multi-blade slurry sawing (MBS).
(iii) Multi-wire slurry sawing (MWS).
(iv) High efficiency multi-blade slurry sawing (HESS).
(v) Fixed-abrasive (multi-wire) sawing technique (FAST).

Table 1.2 presents a comparative and qualitative evaluation on operational conditions of alternative silicon-slicing methods.

Table 1.2 Silicon-slicing methods.

Slicing method	Mode of application	Production rate (wafers h^{-1})	Yield (%)	Surface quality	Technical reliability
ID sawing	Conventional	Average	High	Average	High
ID sawing	Advanced	High	Very high	Good	Very high
MBS	Conventional	Low	Average	Average	Good
MBS	Advanced	Average	High	Good	Fair
MWS	Conventional	Low	Average	Good	Fair
FAST	Testing stage	Average	Average	Good	Low

Two trends in material removal processes are emerging from these methods.

(*a*) Conventional sawing as in ID or FAST methods.
(*b*) Slurry sawing as in MBS, MWS or HESS methods.

1.3.3.1 Sawing process

In the case of the ID sawing technique, silicon stock is removed by synthetic diamond grains which are firmly bonded in a nickel base applied on the internal tool rim by electroplating. Thus the cutting action is in essence a plunge grinding operation, characterized by high cutting speeds v_s of 15–20 m s^{-1} and a rather large contact zone between tool and workpiece. ID sawing results in specific removal rates Z' per tool of 50–80 $cm^2\,mm^{-1}$ which is up to 1000 times higher than in wire sawing with diamond-clad tools

(FAST). Also with regard to the so-called wear ratio G (i.e. work material removed divided by abrasive layer lost by wear), ID sawing with $G \approx 500\,000\ cm^2\,cm^{-2}$ is superior to the FAST process with $G \approx 7500\ cm^2\,cm^{-2}$. In spite of high specific removal rates, ID sawing as well as other slicing processes require a moderate application of coolant when cutting silicon.

As mentioned above, a similar removal process also using electroplated diamond abrasives bonded to the surface of a wire-type tool is effective in the FAST. Figures 1.39 and 1.40 show the general principle of this method, the main advantage of which is the relative low mass of the moving tool frame and the relatively high cutting speed v_s of 100–150 m min^{-1}. However, in spite of the pre-tensioning of the wires, the effective tool load is rather limited owing to the elastic deformation of the wires perpendicular to the work surface. The resulting specific normal forces do hardly exceed 5 g cm^{-1} of contact length while in the case of ID sawing the respective values are 50–100 times higher. Thus, the necessary grooving action with a brittle type of fracture to separate small particles from the work surface cannot take place in fixed-abrasive wire sawing with suitable efficiency.

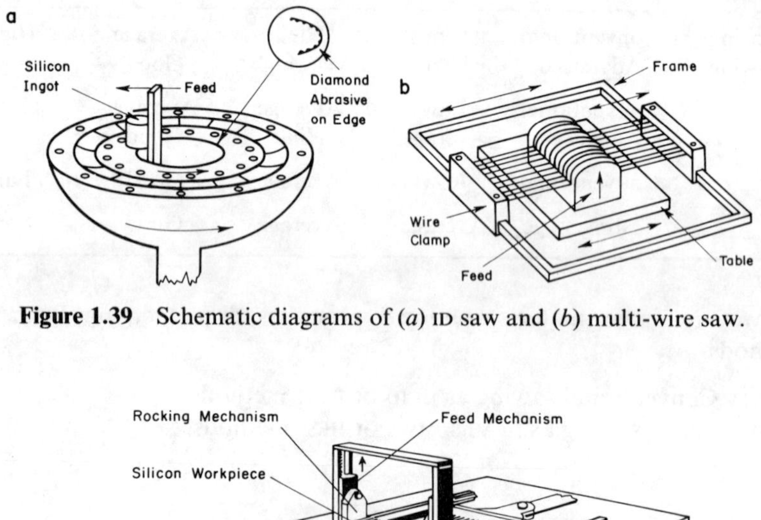

Figure 1.39 Schematic diagrams of (*a*) ID saw and (*b*) multi-wire saw.

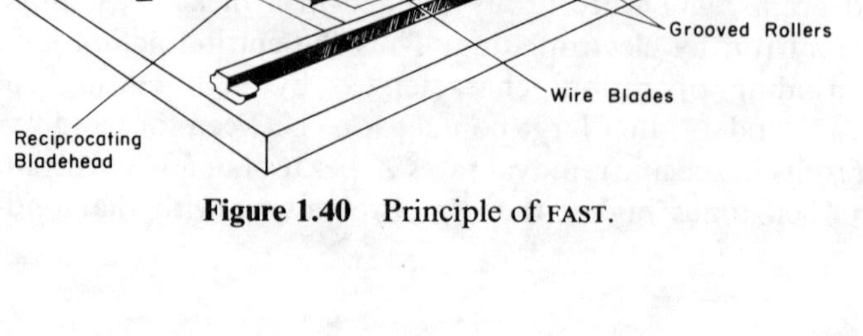

Figure 1.40 Principle of FAST.

1.3.3.2 *Slurry sawing process*

A very different removal process is effective in the so-called slurry sawing process and has not very much in common with a real sawing process which is currently based on cutting like a lathe tool. In fact the slurry process represents a specially modified lapping process, characterized by loose abrasive particles which are conveyed into the contact zone between tool and workpiece by means of a lapping suspension. The lapping grains are forced to roll through the contact zone, affecting the work surface with an extremely high impact frequency, which can reach a few hundred thousand impacts (Hertzian fractures of the very brittle silicon) per second and per square millimetre of the work surface. HESS is a rather attractive sawing technology. Pertinent machines with increased cutting speeds and blade loads have been built and provide a promising alternative to conventional ID sawing.

Technology of high-efficiency slurry sawing

Basically there are two ways to increase the production rate of multi-blade slurry: to increase the number of cutting blades or to increase the cutting rate.

It appears practically that blade packs with more than 300 blades are impractical as tensioning forces become enormous and problems with tool handling and tool accuracy intensify. Therefore the most promising way to boost productivity and to lower production cost at the same time is to increase the infeed rate of conventionally sized blade packs by means of higher cutting speeds, to reduce the stroke and blade length and accordingly to increase blade loads.

1.3.3.3 *Band-saw technique*

For the ingot-casting technique, a tailored technique for contouring and portioning the 100 kg ingots with about 43 cm × 43 cm cross section is essential. A band-saw technique is applied for contouring and portioning the ingot in pre-shaped columns of 10 cm × 10 cm cross section; the columns are then wafered using the HESS technique.

Contouring and portioning of large ingots is favourably carried out with a band-saw technique which allows the machining of larger dimensions than does OD sawing.

The cutting of a 100 kg ingot (43 cm × 43 cm cross section) with a circular saw requires a saw blade with a diameter of 1240 mm at least. Owing to the necessary stabilization the core of the saw blade should have a thickness of about 6 mm; the kerf loss is about 8 mm. In the case of a band-saw the cutting width is only about 2 mm. As a result of this lower cutting width the absolute cutting force in band-sawing is smaller than in OD sawing. Lower cutting forces induce less strain in the brittle material silicon, i.e. the quality of the cut and the yield are improved. The tool used in band-sawing is a toothed endless steel belt with the diamond cutting segments soldered onto

Table 1.3 Operational criteria of different silicon-slicing techniques.

Slicing method	Surface speed ($m\ min^{-1}$)	Cutting force (N)		Surface cut (m^2)		In-feed rate ($cm\ min^{-1}$)	Cutting rate ($cm^2\ min^{-1}$)		Wafers per time (h^{-1})	Wafers per length (cm^{-1})	Yield (%)	Surface damage (μm)
		Per tool	Total	Per tool	Complete pack		Per tool	Complete pack				
ID (1 blade)												
Average	1000	28	—	30	—	5	50	—	30	14	90	10
Maximum	1200	56	—	40	—	8	80	—	48	18	95	10
MBS (300 blades)												
Average	30	0.8	240	0.01	3.0	0.008	0.08	24	15	16	85	10
Maximum	50	1.5	450	0.01	3.0	0.015	0.15	45	27	21	90	8
MBS or HESS (300 blades)												
Average	80	1.5	450	0.01	3.0	0.120	1.20	360	216	16	90	10
Maximum	120	2.0	600	0.01	3.0	0.180	1.50	450	270	21	95	8
MWS (150 wires)												
Average	75	1.0	150	—	1.2	0.007	—	10	6	22	95	5
Maximum	85	1.0	150	—	1.2	0.020	—	30	18	25	98	3
FAST (250 wires)												
Average	100	0.4	100	0.03	7.5	0.012	0.12	30	18	22	90	5
Maximum	140	0.4	100	0.05	12.5	0.016	0.16	40	24	25	96	3

the carrier body. The carrier influences dominantly with its form and material quality the lifetime of the tool and with it the overall output. The saw band is made from low-carbon steel with a strength of about 1400 $N\,mm^{-2}$. It has the dimensions 10 500 mm × 80 mm × 1 mm. The concentration and grain size of the abrasive diamond and the hardness of the bronze bonding define the cutting rate and quality of the cut surface. In general the abrasive grains are synthetic diamonds of medium grain size and concentration. Cutting rates depend on the feeding rate and contact zone between tool and workpiece. On contouring and portioning 100 kg ingots with 43 cm × 43 cm cross section and a height of about 25 cm, cutting rates of more than 120 $cm^2\,min^{-1}$ have been obtained (for a saw blade it is around 70 $cm^2\,min^{-1}$). The cut surfaces are flat; geometric deviations are less than 0.1 mm. The lifetimes of saw bands are comparable with those of OD saw blades for silicon processing.

1.3.3.4 Trends and perspectives

The total production cost for solar silicon wafers is composed of different cost elements, related to material, tool, machine, labour and minor expenses. The dominant cost factor is the material. Then, of the remaining add-on costs, machine and labour have the strongest influence while the tool costs are of secondary importance. Therefore, minimal production costs are likely to be achieved with slicing techniques which provide low value for wafer thickness and kerf losses, and which at the same time produce a maximum number of wafers per unit of time. A comparative survey of operational criteria of different silicon-slicing techniques is presented in tables 1.3 and 1.4.

Many attempts have been made to increase cutting rates and yields mainly by modification of existing slicing processes. From the technological and

Table 1.4 Comparison of performances of slicing techniques.

	Surface Speed ($m\,min^{-1}$)	Silicon feed rate ($cm\,min^{-1}$)	Productivity ($wafers\,h^{-1}$)	Damage (μm)	Kerf loss (μm)
ID saw	3000	5	30	10	350
MBS (300 blades)	30	0.08	12	10	250
MWS (150 wires)	100	0.02	20	5	220
Present MWS (300 wires ×2)	300	0.03	150	5	230

economic comparison of alternative silicon-slicing methods, two trends can be extracted which most probably will determine the future state of application of slicing techniques.

(i) Regarding future mass production of solar silicon wafers, ID sawing with diamond-plated saw blades has an effective advantage against all other processes based on its well introduced and accepted machine and tooling technology. With increasing blade diameters (up to 27 in and eventually more) the output will be increased by cutting larger ingots or by cutting more than one ingot at the same time. In process control systems, as already available with the most advanced ID systems, increasing yield and cutting quality considerably reduce the labour cost as well.

(ii) HESS is the second slicing technique which has a definite chance to play a major role in the mass production of solar cells. Of all slicing techniques it provides the highest production rate and the lowest add-on costs and is expected to become an attractive alternative to ID sawing.

1.3.4 Surface polishing and cleaning

A surface layer about 30 μm thick is damaged after ingot sawing in conventional cell technology. Generally this layer is removed isotropically and rapidly by acid etching, e.g. typically a rate of 1 μm min^{-1} at 20 °C is reached with the following mixture: 1HF (48%) + $1CH_3CO_2H$ (99.8%) + $5HNO_3$ (65%).

The surface can also be polished with alumina powder of increasingly smaller grain size and then with mechanochemical polishing such as 10 min contact on a rotating belt with a mixture of equal parts of 25.9 g of NH_4F + 100 cm^3 of H_2O and 9 g of $(NO_3)_2Cu$, $3H_2O$ + 100 cm^3 of H_2O.

Basic etches, i.e. KOH–NaOH–H_2O mixtures, are also used in industry because they cause less pollution than acid etches do when used in large quantities. Another advantage of basic etches is that somewhat texturized surfaces (with smaller reflectivity) are obtained.

All these polishing etches are quite different from those used to reveal grain boundaries and dislocations (1 volume of (50 g of CrO_3 + 100 ml of H_2O) + 1 volume of HF) or to remove silicon oxide ($\frac{1}{2}$HF + $\frac{1}{2}$ distilled water).

Finally, after polishing, it is necessary to clean the surface from all impurities, e.g. by one of the following procedures: simple cleaning (hot trichlorethylene–hot acetone + hot methanol + rinsing in distilled water + 30 s in $\frac{1}{2}$HF + $\frac{1}{2}$ distilled water + rinsing + drying); 'RCA cleaning' (10 min at 8 °C in each of $1H_2O_2$ (30%) + $4SO_4H_2$ (96%), $1H_2O_2$ (30%) + $1NH_4OH$ (30%) + $5H_2O$, $1H_2O_2$ (30%) + 1HCl (37%) + $5H_2O$ and then $\frac{1}{2}$HF + $\frac{1}{2}$ distilled water).

1.3.5 p–n junction formation

A p–n junction can be formed through either a high-temperature diffusion (HTD) process or an ion implantation process (the CJP). There exist two different approaches for the diffusion process, i.e. diffusion from the vapour phase, or diffusion out of a solid phase. One disadvantage of diffusion from the vapour phase is that diffusion occurs on both sides of the wafer such that the back parasitic junction has to be removed. Diffusion out of a solid phase consists of the deposition of a dopant layer at ambient temperature followed by the diffusion process itself at a high temperature (800–900 °C). The deposition can be performed using, for example, the following techniques.

(i) CVD.
(ii) Spin-on process.
(iii) Spray-on process.
(iv) Screen printing (phosporous-doped paste).

Parasitic junctions at the edges of the wafer are removed by chemical etching, by plasma etching or by laser scribing and breaking.

The CJP is still in the development stage; the doping species are introduced by ion or molecular implantation or are evaporated or spray deposited onto the surface of the semiconductor and driven in by pulsed annealing (laser or electron beam).

Using this process, only a very thin layer is molten just underneath the surface and the bulk of the material is kept at a low temperature, preventing

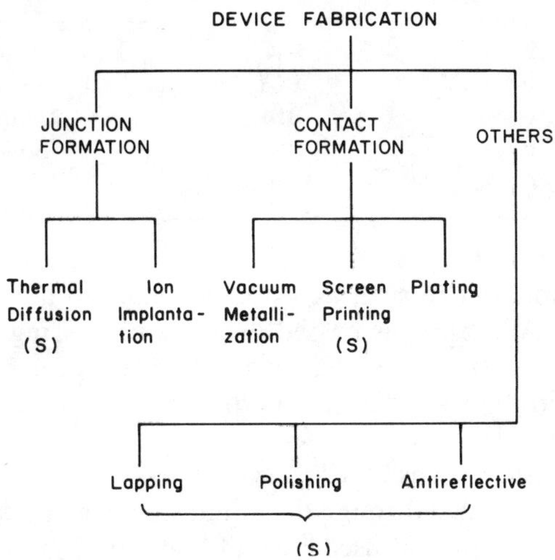

Figure 1.41 Review of technical options for the solar cell fabrication process: S, standardized.

degradation of the minority-carrier lifetime; such a lifetime degradation can be very important during HTD in the case of a lower-grade polycrystalline silicon.

The main other advantages of CJP over HTD are as follows.

(i) Lower energy consumption.
(ii) Process amenable to full automation.
(iii) Lower overall cost.

A review of the technical options for the solar cell device fabrication process is presented in the flow chart in figure 1.41.

1.3.5.1 High-temperature diffusion

Standard solar cell technologies introduce a base layer dopant during crystal growth. The p- or n-type silicon must then be formed upon the base layer depending on the impurity type of the initial wafer (table 1.5). To form p-type silicon, boron is generally used as the impurity from sources such as B_2O_3, BCl_3 and BBr_3. Sources of phosphorus, the n-type dopant, include P_2O_5 and $POCl_3$. The process of delivery of the impurity source to the wafer are dependent on whether it is a gas, liquid or solid.

Table 1.5 Impurity types.

Silicon type	Impurity source	Phase at room temperature
p	B_2O_3	Solid
p	BCl_3	Gas
p	BBr_3	Liquid
n	P_2O_5	Solid
n	$POCl_3$	Liquid

A solid diffusion source such as P_2O_5 is heated in a chamber upstream from the wafers. A carrier gas passed over the vaporizing P_2O_5 carries the impurities to the wafer surface. Various apparatuses for phosphorus diffusion are shown in figures 1.42(*a*), 1.42(*b*) and 1.42(*c*) for solid, liquid and gaseous sources respectively.

For a liquid source such as $POCl_3$, a carrier gas (a mixture of N_2 and O_2 in the ratio of 3:1) is bubbled through the liquid source at about 950 °C. After a diffusion time of 10 min, a diffused layer of about 0.25 μm will form. Typical apparatuses for boron diffusion are given in figure 1.43.

The formation of a p–n junction follows Fick's laws of diffusion. According to the laws of diffusion, if a base layer of doped silicon is coated with

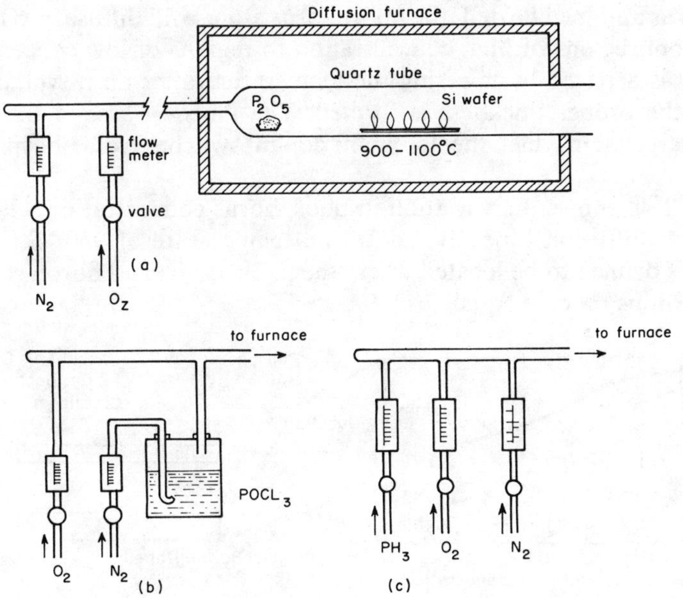

Figure 1.42 Apparatuses for phosphorus diffusion.

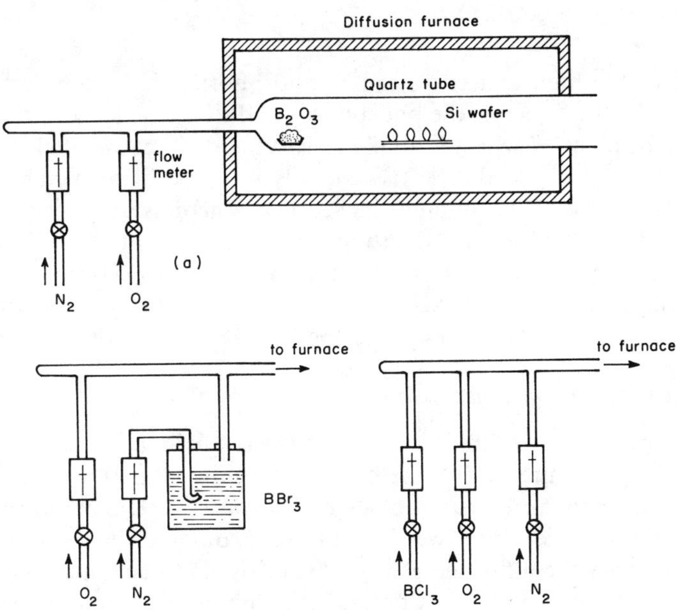

Figure 1.43 Apparatuses for boron diffusion.

phosphorus and then heated, the phosphorus atoms will diffuse into the solid silicon from regions of high concentration to regions of low concentration. Diffusion is arrested by lowering the temperature after an n-type collector layer of the proper thickness is formed. The junction area is the narrow transition region in which the dominant dopant switches from phosphorus to boron.

Figure 1.44 shows the variation in phosphorus concentration with depth for various diffusion times at a diffusion temperature of 950 °C. The p–n junction is defined to be located where the phosphorus and boron respective concentrations become equal.

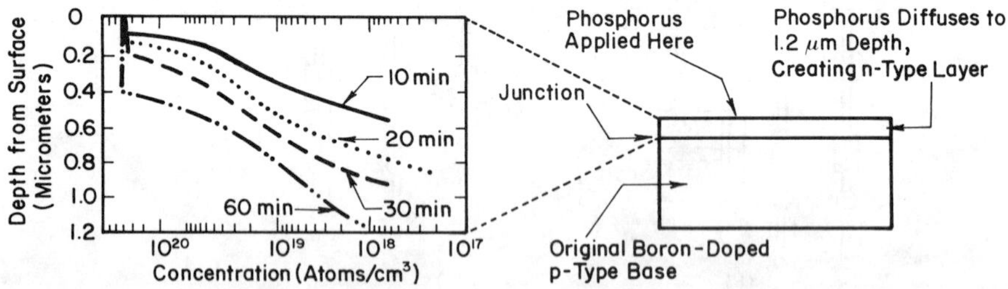

Figure 1.44 Four diffusion profiles for 950 °C diffusion of phosphorus into silicon. The timing (10, 20, 30 or 60 min) can be regulated to produce the right amount and depth of dopant. The base layer is 2×10^{16} B atoms cm^{-3}. The model cell (on the right) shows that the junction region is located where the top-layer doping and the base-layer doping are equal.

Owing to the high concentration of phosphorus at the cell surface even after the diffusion process, the phosphorus and silicon react to form an alloy called the 'dead layer' which is highly resistive and is a cause of electron–hole recombination. The 'dead layer' is generally removed by acid etching.

Since over 50% of the sunlight incident on silicon is absorbed within the first 3 μm of the surface and a depth of 300 μm is needed to absorb the rest, a solar cell should be fabricated so that it has a shallow top collector layer (less than 3 μm thick) and a thick base layer with a high mobility of charge carriers in case light-generated carriers are formed far away from the junction.

1.3.5.2 Cold-junction processing

The ion implantation technique as used in CJP is still in the development stage and provides an alternative method of forming a p–n junction to the conventional, slow and energy-intensive HTD process. During diffusion, the atomic dopants penetrate towards the bulk through an increase in thermal vibrations under the influence of a gradient driving force (energy involved is in the electronvolt range). In this case the dopants are ionized (except for recoil implantation) and accelerated typically between 5 and 100 keV; they penetrate the target through their kinetic energy. When impinging on the solid, the projectiles lose their energy by electronic and nuclear collisions. Nuclear collisions are responsible for the damage which is created in the single-crystal or polycrystalline structure of the base material, i.e. silicon. In order to anneal this damage and to put the dopants in substitutional sites, an annealing procedure is necessary. The silicon wafer is exposed to a beam of high-energy ions of the desired dopant by means of a glow discharge device; for example, if the doped base layer is p type, it is exposed to a phosphorus ion beam. The depth of penetration is adjustable depending on the strength of the beam. The silicon surface must then be annealed with a high-energy laser or electron beam in order to remove the damage caused by the implantation on the crystal lattice and to put the dopant atoms in substitutional sites. The resultant p–n junction does not differ much from a conventional diffusion-formed junction with depths generally about 0.25 μm. Concentration of dopants varies from 10^{16} atoms cm^{-3} at the junction to 10^{21} atoms cm^{-3} at the surface. Advantages over the thermal diffusion process include less energy usage; it is basically a cold process presenting a greater reproducibility on a large scale because of its greater simplicity and greater ease of control of junction depth and dopant concentration as well as dopant profile. In addition, a thin layer of highly doped back-surface field can be formed on the cell at the same time as the front-surface implantation.

Basically there exist four different procedures to perform dopant implantation.

(i) Conventional ion implantation: dopant ions are produced by glow discharge in a source starting from a gas or vapour compound containing the dopant atoms (figure 1.45). These positive ions are then

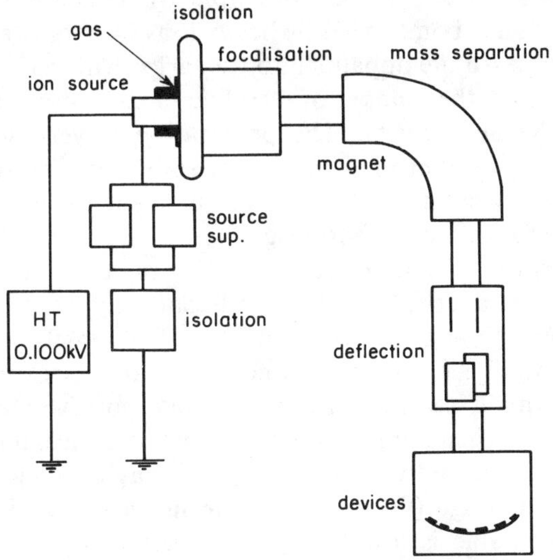

Figure 1.45 Classical ion implantation.

extracted, handled in the correct optics and accelerated before being analysed either by a magnet or a speed filter in order to obtain an isotropically pure ion beam. The typical implant dose is around (1–5) $\times 10^{15}$ cm^{-2}.

(ii) Ion incrustation: this involves the direct use of glow discharge beams issued from a gas including dopant atoms, in which both atomic and molecular ions are implanted simultaneously without any mass selection (figure 1.46). Ion sources of nearly any size can be realized using multi-hole extraction structures as for ion-milling systems.

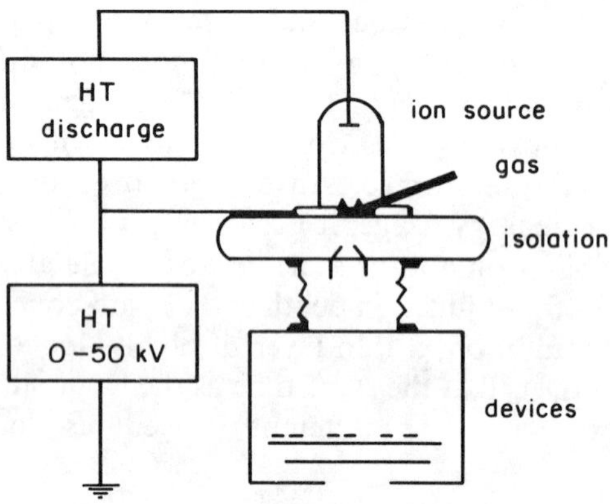

Figure 1.46 Glow discharge implantation.

(iii) Recoil implantation: in this procedure, dopants are first deposited at the silicon surface, at thicknesses typically 50–200 Å, by any technique. Then the film is bombarded by heavy ions (generally rare gases). These heavy ions push the dopants into the target with a yield which is larger than unity, i.e. the number of recoiling dopant atoms is larger than the incident beam intensity. The penetration is very shallow (typically 1000 Å) before annealing; however, severe annealing problems exist and the small gain in yield is not sufficient to make this procedure attractive for cell manufacturing.

(iv) Ionized cluster beam deposition: the material to be deposited is vaporized in a crucible and is adiabatically expanded when going through an exit nozzle into about $N \approx 1000$ loosely coupled atoms. By bombarding this cluster with a beam of either electrons or ions in the kiloelectronvolt range, it attains a single positive charge and can be accelerated. When impinging on the silicon surface, the ionized clusters are broken into individual atoms, each having a mean energy in the electronvolt range (figure 1.47). The mean energy in the systems at present available is low (1–1000 eV); therefore the dopants are just deposited at the surface, a situation intermediate between ion plating and implantation (figure 1.48).

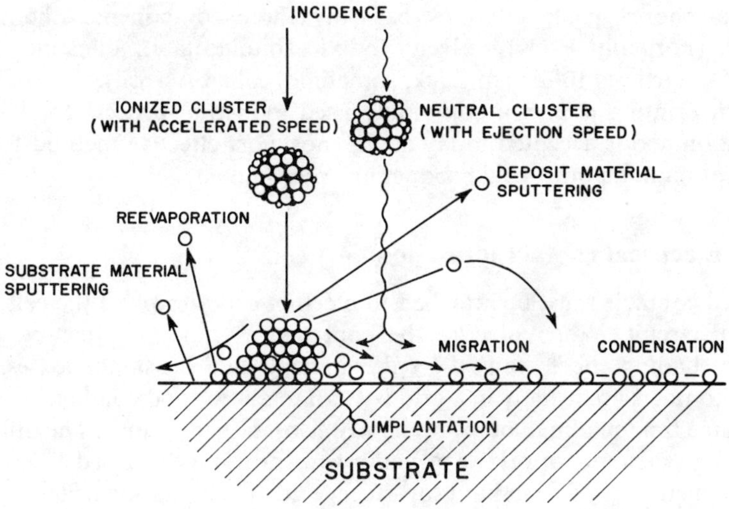

Figure 1.47 Break-up of ionized clusters in atoms.

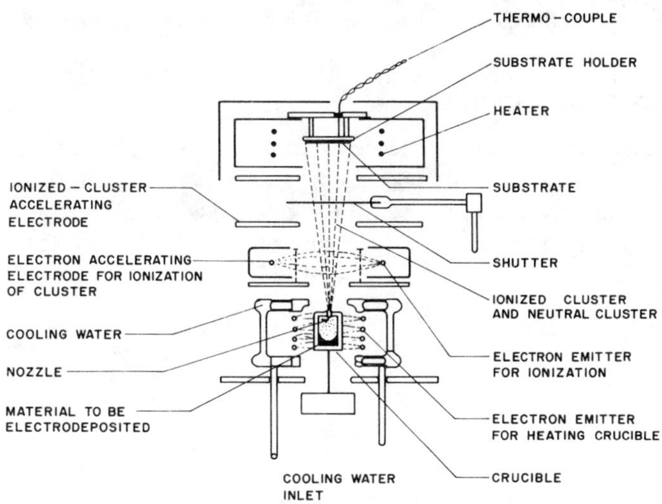

Figure 1.48 Cluster beam deposition.

However, ion implantation is a costly procedure which has to be completed with rapid thermal annealing by laser or strip heater. Strong cost reductions are expected if large production rates can be obtained.

1.3.5.3 Screen-printing junction process

A phosphorus paste is spread on the wafer surface (p-type substrate) using a screen-printing machine. Then diffusion of phosphorus impurities takes place into a conveyor furnace at around 925 °C in an Azote atmosphere (N_2).

The phosphorus paste contains basically three components: the active material (normally P_2O_5); solvent (butylcarbitolacetate); thickener (very pure SiO_2 particles 10–20 μm thick, sometimes called Aerosil).

Screen printing is also a commonly used industrial process for contact elaboration and is accepted today as the most cost-effective method for cell metallization in an industrial production environment.

1.3.6 Electrical contact formation

Electrical contacts must be attached to a cell in order to place the cell in an electrical circuit. When placing the contacts, a trade-off must be made between shading effects caused by the contacts and resistance losses. Top surface contacts are usually in a grid pattern to allow good conduction while at the same time shading not more than 10% of the cell surface. The different techniques (vacuum vaporization and screen printing described below) are being applied today for making top-surface grids. The back-surface contact which can be simply a metal sheet is much less problematic. In certain newer cell designs, light enters from both surfaces of the cell, thus requiring grid contacts and anti-reflection coating on both front and back surfaces.

1.3.6.1 Vacuum vaporization

The metal to be deposited on the cell surface is heated in a vacuum to a temperature high enough to cause melting and evaporation. The metallic vapours are then deposited on a cell through a mask, forming a grid pattern (figure 1.49). Titanium, palladium and silver are typically used for the front

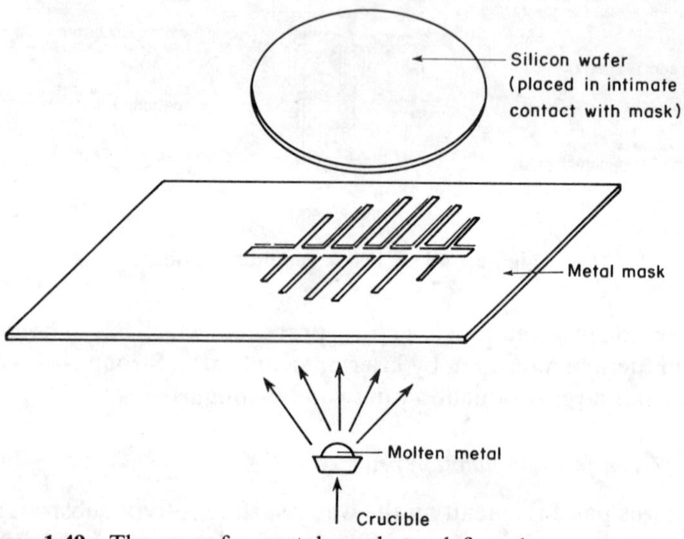

Figure 1.49 The use of a metal mask to define the top-metal grid pattern.

contact with a nickel mask for grid definition. On the back side, aluminium and silver are used successively to obtain Ohmic contacts. Evaporation still leads to the most reliable contacts, but it is far too expensive and has been abandoned by all manufacturers.

1.3.6.2 Screen printing

Two alternative methods for contact formation are electroplating and screen printing. Both techniques should lead to similar costs, provided that one can eliminate noble metals. Screen printing is actually the technique most acceptable worldwide at the production stage.

The metal is painted over the entire front surface of the cell and subsequently etched away from unwanted parts using a photographic technique called photolithography (figure 1.50).

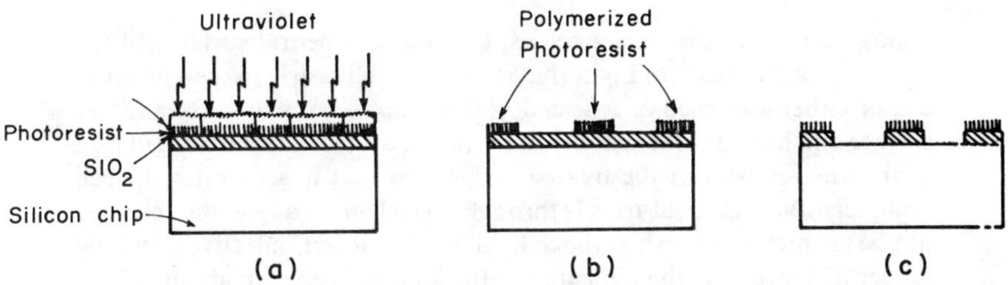

Figure 1.50 Photo-etching technique: (*a*) masking and exposure to ultraviolet radiation; (*b*) the photoresist after (*a*); (*c*) after etching.

The metal contact is usually made up of three separate layers of metal: a thin layer of titanium as a bottom layer for good adherence; a top layer of silver for low resistance and solderability; and a middle layer of palladium to prevent reaction between the titanium and silver layers in the presence of moisture. The contacts must then be sintered at 500–600 °C for good adherence and low contact resistance after the metals have been deposited. An aluminium paste is generally used for the back-side metallization instead of silver and will produce a back-surface field as aluminium is an acceptor. It is also advantageous to deposit the front metallization after the anti-reflection coating.

1.3.7 Anti-reflection coating

Untreated silicon reflects more than 30% of incident light. A single layer of an anti-reflection coating such as silicon oxide is able to reduce cell surface reflection to 10% of incident light while a double layer reduces it to below 3%. Materials such as silicon dioxide (SiO_2), titanium dioxide (TiO_2) and tantalum pentoxide (Ta_2O_5) are also used as anti-reflection coatings. The vacuum process described earlier for contact formation is also used in depositing the anti-reflection coating. Sputtering and chemical spraying can also be used. Figure 1.51 shows how texturing of a cell surface in addition to an anti-reflection coating can reduce reflectivity and improve cell efficiency.

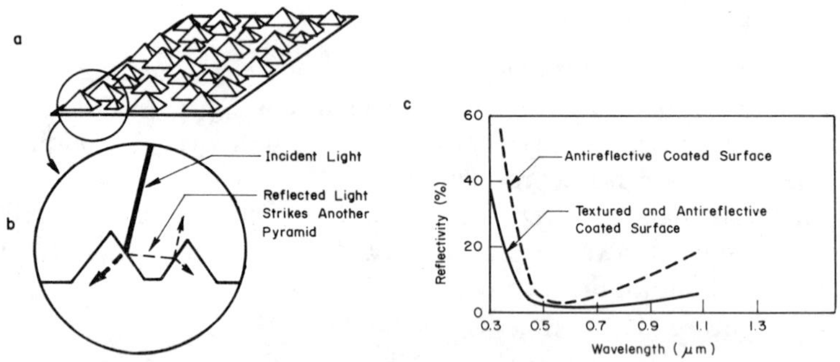

Figure 1.51 (*a*) Texturing a cell exposes the tetrahedral surfaces of the silicon crystal lattice. (*b*) Light that strikes the cell nearly perpendicularly can be either absorbed or reflected; if it is reflected, it will strike another surface and have another chance to be absorbed. Note also that light that is absorbed is bent at the treated surface, so that it penetrates the cell obliquely. Because light travels through the cell at an angle, the cell need not be as thick to absorb as much light. (*c*) The lower reflectivity and the longer distance that the light travels through the cell give about a 15% improvement in cell efficiency compared with smooth uncoated cells.

1.3.8 Cell encapsulation

Cell encapsulation is required to protect metallic contacts and interconnections from corrosive elements in the atmosphere as well as to provide electrical isolation from voltages generated by the solar panel. Encapsulation provides mechanical and chemical protection to the cell and its durability determines the operating life of the module.

Before encapsulation a structural layer at either the back or the front of the module must be attached to provide rigidity. In one module design, cells are attached directly to the structural back and then encapsulated in a flexible pottant (figure 1.52(*a*)). The most commonly used materials for the structural back are anodized aluminium, porcelained steel, epoxy board or window glass. Silicone is widely used for the adhesive and pottant layers.

In another module design, the cells are enclosed and are supported by layers of laminate applied first onto a structural back (figure 1.52(*b*)). The laminate may be either polyvinyl butyral (PVB) or ethylene vinyl acetate (EVA). A final layer at the top of the cell, possibly of low-iron glass, imparts self-cleaning properties and in some designs provide protection from moisture. For moisture-resistant cells, a soft silicone encapsulant may be used instead with a harder silicone surface layer.

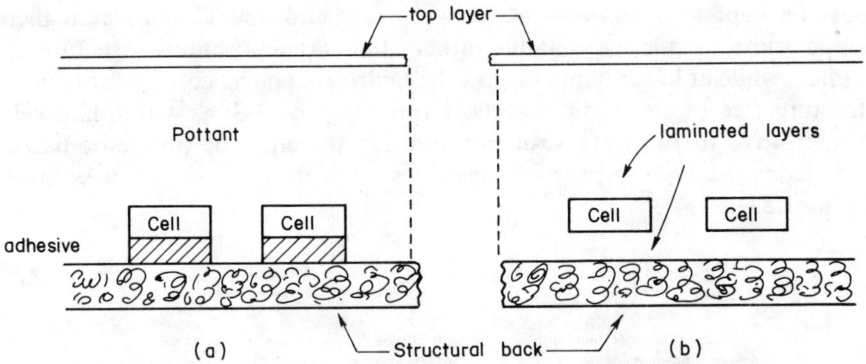

Figure 1.52 Schematic diagrams of the encapsulation of solar cells.

1.3.9 Amorphous silicon thin-film technology

Efficient a-Si solar cells consist of a thin film of the order of 1 μm thick whereas a typical classical cell thickness is 100–200 μm. This is because the absorption coefficient in this disordered material structure is one order of magnitude larger than in crystalline silicon. Moreover the growth temperature for an amorphous material thin film is much lower (200–400 °C) than for the crystalline technological process.

Specific features make a-Si cells very attractive as far as economics is concerned.

(i) The saving of silicon material and consequently a high material yield.
(ii) A low-temperature technological process and consequently a reduction in the energy pay-back period.
(iii) Deposition on inexpensive substrates such as glass, metal or plastic.
(iv) The possibility of producing industrially cells about a square metre in size—a key feature in reducing production costs.

However, a-Si still suffers from a certain instability after exposure to sunlight for some time, i.e. the Staebler–Wronski effect, which together with its low efficiency limit its actual use to the power range of the consumer electronics field such as pocket-sized calculators, wrist watches and battery rechargers.

1.3.9.1 Amorphous silicon fabrication techniques

The a-Si atomic structure compared with single-crystal and polycrystalline silicon does not have a preferential crystallographic orientation. Fabrication of a-Si needs a high degree of control as slight alterations in conditions cause a large change in cell structure. In most of the processes, a gaseous form of a silicon–hydrogen alloy is decomposed by means of a glow discharge system and deposited as a thin film on a substrate. Phosphorus or boron gaseous compounds are included in the a-S_1:H for doping the a-Si film. The substrate must be kept at a temperature between 200 and 350 °C since at higher temperatures a microcrystalline rather than an amorphous structure is formed, while at lower temperatures the hydrogen and silicon do not form a structure free of electronic defects. Fabrication of a-Si as a thin-film cell is insensitive to substrate structure and can therefore be fully automated and deposited on inexpensive substrates such as glass or stainless steel (figure 1.53):

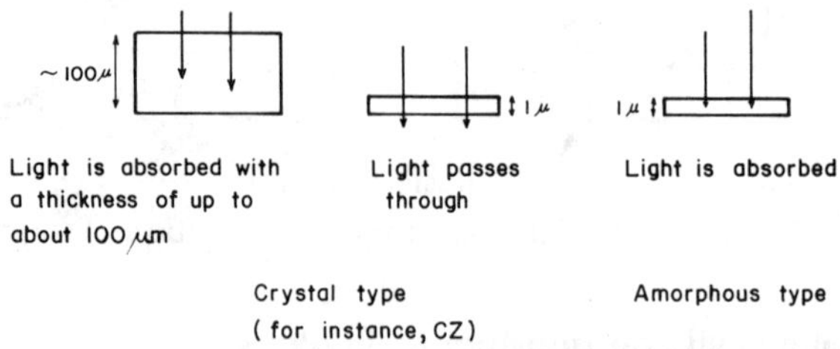

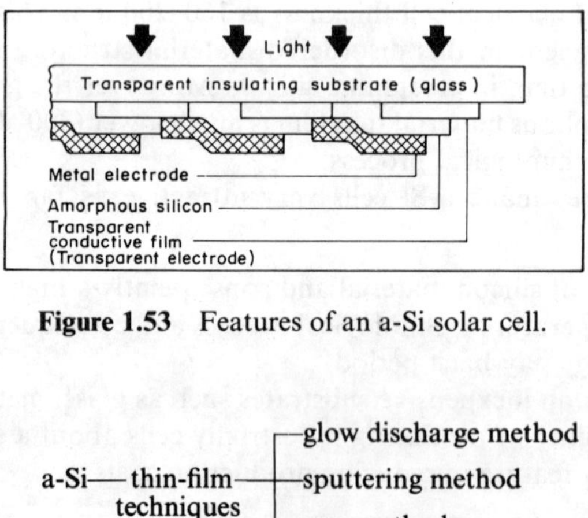

Figure 1.53 Features of an a-Si solar cell.

a-Si — thin-film techniques:
- glow discharge method
- sputtering method
- CVD method.

In most a-Si fabrication processes, silane (S_1H_4) is decomposed in an RF glow discharge reactor near a heated substrate (200–300 °C) (see figure 1.56). The exposure time, discharge power and flow rate all control the film thickness. To produce p-type or n-type silicon, dopants can be introduced to the SiH_4 vapour, e.g. diborane (B_2H_6) for p-type and phosphine (PH_3) for n-type silicon. To produce a-Si:H, SiH_4 vapour is used, while incorporating fluorine into the vapour produces fluorinated a-Si:H (a-Si:H:F). The structural disorder of a-Si gives it a much higher absorption coefficient than crystalline silicon which means that the thickness of a-Si need only be about 1 μm to absorb most of the incident light. The effective band gap is also wider: 1.55–1.80 eV as opposed to 1.1 eV for crystalline silicon.

Major disadvantages lie in the short carrier lifetimes, lower efficiencies and

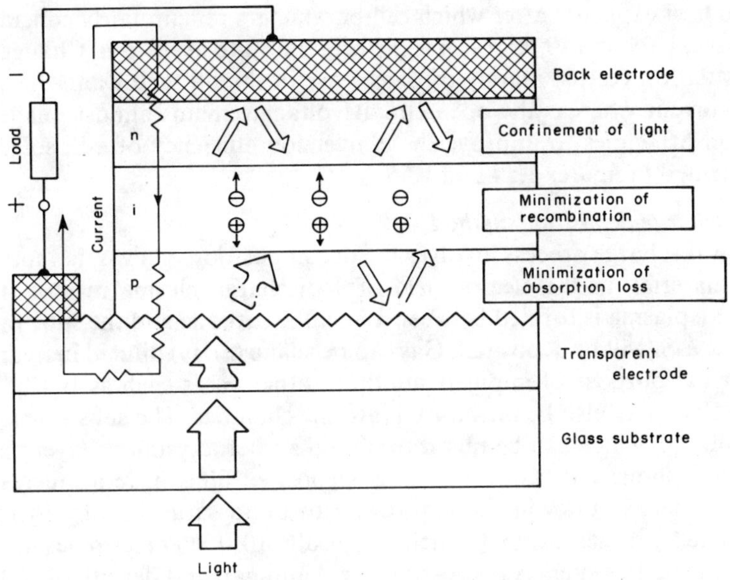

Figure 1.54 Principle of improving the conversion efficiency of an a-Si solar cell.

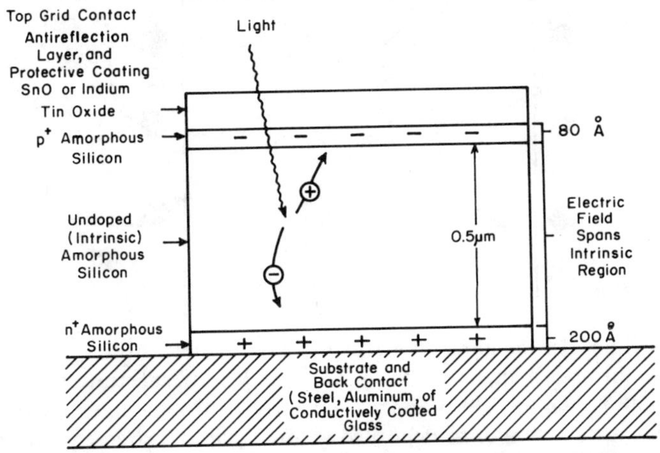

Figure 1.55 A p–i–n device: most light passes through the top layer and generates electron–hole pairs in the intrinsic region. An induced field spans the intrinsic region, sweeping the charge carriers to opposite ends of the cell.

degradation of cell efficiency under prolonged exposure to sunlight. The Staebler–Wronski effect which causes the optically induced degradation of a-Si solar cells is based upon the recombination effect of carriers generated by light in the material. The main decrease in efficiency seems to occur in the first 100 h of exposure after which cell parameters remain fairly constant. A short anneal above 150 °C restores the original state of the cell. Changes due to degradation consist mainly of a decrease in the fill factor and the short-circuit current whereas the open-circuit voltage remains almost unaffected. Common principles to improve the conversion efficiency of a-Si solar cells are presented in figures 1.54 and 1.55.

Glow discharge deposition method

The glow discharge process involves the decomposition of a gas, i.e. source of silicon material, in the electric field of a discharge plasma reactor (figure 1.56): a RF plasma is formed between a metallic cathode and the substrate on which the a-Si will be deposited. Gas can be silane (SiH_4) diluted in hydrogen and, for the purpose of impurity addition, other gases such as B_2H_6, PH_3, FH_4 or CH_4 may also be introduced into the chamber. The substrate is kept at 250–300 °C which can be raised to obtain a microcrystalline layer instead of an amorphous one. Glow-discharge-deposited films have a much lower density of defect states in the gap owing to bond saturation by hydrogen atoms. The hydrogen content which is typically 10–15% determines many of the film properties such as microstructure, band gap and density of states.

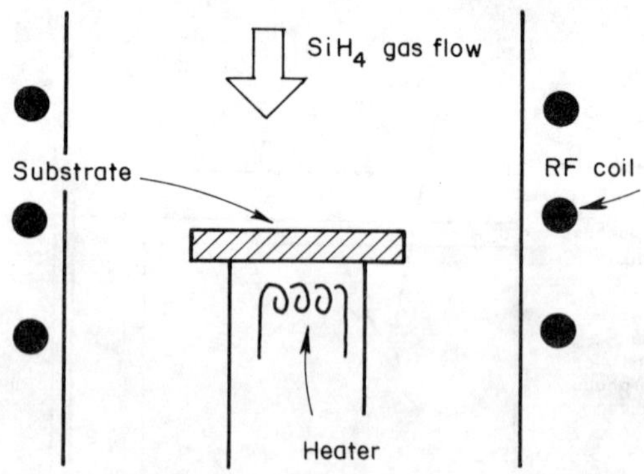

Figure 1.56 In the glow discharge process, silane gas (SiH_4) is passed slowly through a quartz tube in which is held a substrate heated to 200–300 °C. An RF field is applied inductively through an external coil, which causes silicon to be deposited gradually as an extremely hard and stable layer. The thickness of the film can be controlled by varying the exposure and the flow rate. Doping elements are added to the silane gas before starting the discharge.

Relatively low deposition temperatures of 220–300 °C allow for the deposition of a-Si films on different kinds of inexpensive substrate ranging from stainless steel sheets to polyimide films to already formed units such as roof shingles.

For a high a-Si cell efficiency, the intrinsic layer which is the active part of the solar cell device must have a low defect density and low impurity contamination in order to provide good carrier lifetime. Deposition of high-quality intrinsic material by glow discharge of silane is a slow process occurring at a rate of about 0.1 nm s^{-1} which will prove to be a limitation in continuously running automated production lines. Efforts made to increase the deposition rate include the use of disilane (Si_2H_6) rather than monosilane in the glow discharge deposition of intrinsic layers, which approximately doubles the deposition rate, although the quality of films obtained so far using higher silanes is still inferior to that from SiH_4.

Multi-chamber a-Si deposition systems are currently perceived as the best approach to the fabrication of high-quality a-Si solar cells as process gas intermixing and dopant contamination in the intrinsic layer may result from repetitive depositions in single-chamber systems.

A continuous mode of fabrication of flexible a-Si:H solar cells on polyimide substrate by RF glow discharge deposition has been developed by the 3M Company. A schematic diagram of the deposition system is shown in figure 1.57. The best cells produced have a fill factor of 0.601, an open-circuit voltage of 0.919 V and an efficiency of 6.2%. The flexible substrate consists of a polyimide film 2×10^{-3} in thick which has been metallized by sputtering 200 Å of TiN onto 1000 Å of aluminium. The devices consist of p^+–i–n^+ layers 200, 5000 and 100 Å thick on the Al/TiN bottom contact with a quarter-wavelength indium tin oxide layer as a top contact. The multi-chamber RF glow discharge deposition system consists of four reactors within one large chamber: the first reaction chamber for hydrogen plasma etching and bake-out of the substrate before deposition, followed by three chambers for deposition of p^+–i–n^+ layers. Substrate temperatures during deposition are 200–250 °C.

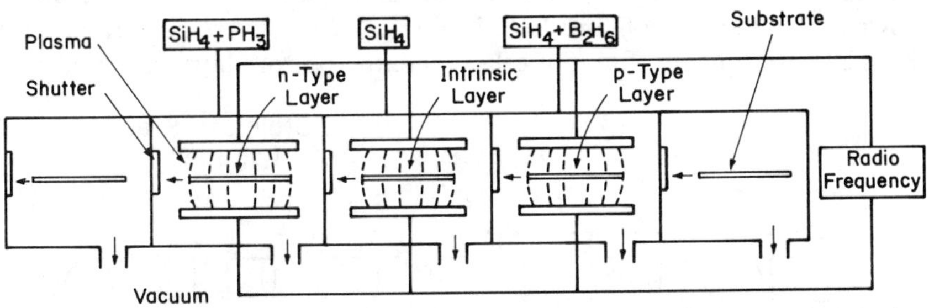

Figure 1.57 Commercial p–i–n cells fabricated by RF glow discharge are deposited in layers as the cells are moved through consecutive isolated chambers.

Chemical vapour deposition, plasma chemical vapour deposition and photochemical vapour deposition

Alternative methods of fabricating a-Si solar cells such as plasma CVD and CVD (figure 1.58) for higher silanes have been successful in raising deposition

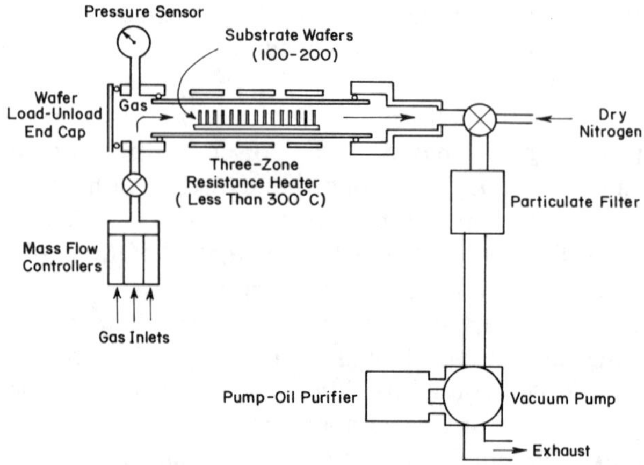

Figure 1.58 CVD: appropriate gases are introduced into a reactor tube. The gases react and are deposited on numerous substrate wafers.

rates to values higher than the 1–5 Å s^{-1} rate of the conventional glow discharge method; however, they have not as yet resulted in films of the same quality.

A continuous mass production process of fabricating a-Si tandem cells by the plasma CVD method has been jointly developed by Energy Conversion Devices (ECD) Solar and Sharp. The roll-to-roll plasma CVD machine schematically shown in figure 1.59 continuously processes a roll of stainless steel substrate which advances from the pay-off chamber through the gas gate into the reaction chamber. In the fabrication of the tandem thin-film PV cell, six amorphous silicon layers are successively deposited onto a roll of stainless steel substrate 400 mm wide and 500 m long. A typical tandem cell is shown schematically in figure 1.60. Higher energy conversion efficiencies are obtained with tandem thin-film cells where the bottom layer has a narrower

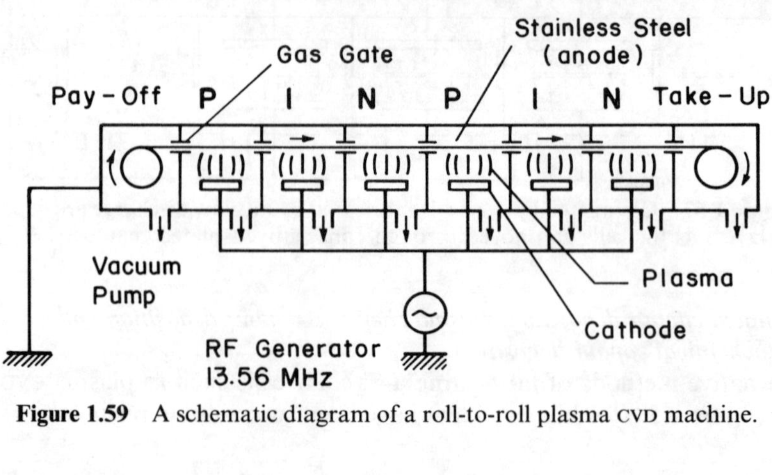

Figure 1.59 A schematic diagram of a roll-to-roll plasma CVD machine.

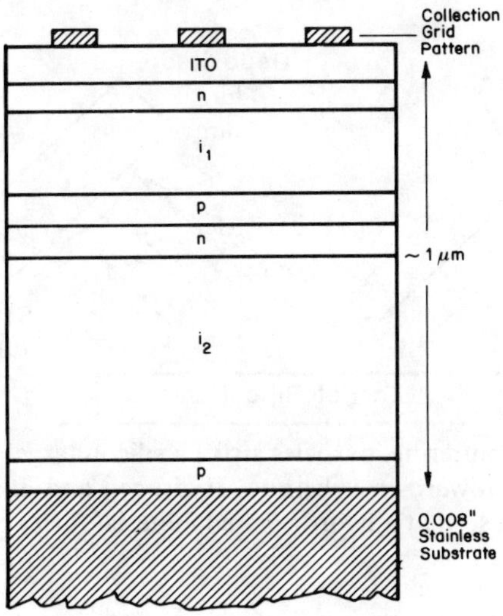

Figure 1.60 Schematic cross section of 1.8 eV/1.8 eV Sharp–ECD Solar tandem PV cell produced by the roll-to-roll plasma CVD method.

band gap because a wider range of incoming solar spectrum is utilized. The roll-to-roll process has been able to achieve high rate depositions of up to 18 $Å\,s^{-1}$. When the intrinsic layer is deposited at a rate higher than 10 $Å\,s^{-1}$, conversion efficiencies of 8.5% and 8.11% have been obtained for the same band-gap a-Si:H tandem cell and the dual-band-gap tandem cell using a-Si,Ge:H and a-Si:H respectively (AM1; 100 $mW\,cm^{-2}$). The roll-to-roll process involves control of process gas isolation between adjacent reaction chambers. Sharp–ECD Solar claim that through the introduction of proper materials and techniques the process has been able to minimize the Staebler–Wronski effect to the extent that cell degradation will be kept below 15% over a period of 20 years.

A new method called photo-CVD avoids bombardment of the films by high-energy particles through the use of ultraviolet light from a low-pressure mercury lamp to decompose the silanes. Admitting mercury vapour into the deposition chamber greatly enhances the deposition rate. The mercury-sensitized photo-CVD method has allowed preparation of a-Si:H layers with properties very similar to those of glow-discharge-deposited films.

Sputtering

Sputtering deposition is the process whereby the coating material or a constituent of the coating material is vaporized from a solid target by high-energy particle bombardment and is deposited as a thin film on a suitable substrate (figure 1.61). In the magnetron case, ions produced in a plasma discharge that is magnetically confined to a region adjacent to the target are

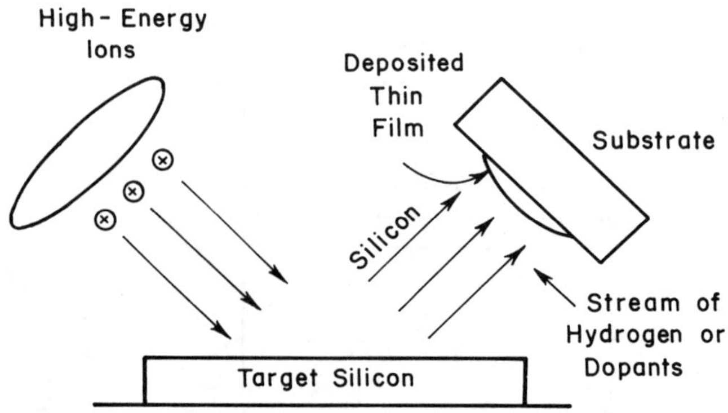

Figure 1.61 Sputtering: particles strike a silicon target, sending a beam of silicon ions towards a substrate. Hydrogen and dopant gases react with the silicon so that a hydrogenated material forms on the substrate.

used as the bombarding particles. The composition of the film condensed from the sputtered flux tends to be the same as that of the target which is an important advantage of the sputtering process.

In reactive sputtering, at least one of the coating constituents is introduced in the gas phase. Reactive sputtering of a-Si:H using silicon targets sputtered in an argon–hydrogen atmosphere has several advantages which include the following.

(i) The ability to control the hydrogen content by adjustment of the hydrogen gas partial pressure.
(ii) The ease of application to alloys such as a-Si,Ge:H.
(iii) The avoidance of using SiH_4 as a working gas.

The explosive nature of silane will necessitate expensive special handling procedures in large production facilities involving glow discharge decomposition of SiH_4.

1.3.10 Options for acquisition of solar photovoltaic technology by Third-World countries

Solar PV technology is undergoing continuous technological development mostly in the area of solar cell fabrication owing to efforts directed at reducing solar cell cost per peak watt and gaining greater economic competitiveness as a decentralized energy source. The present section is aimed at opening a debate on the conditions of acquisition of PV technology by Third-World countries stressing the importance of relevant criteria as capital and human resources availability, industry infrastructures and educational environment.

PV manufacture involves basically the following six components:

(i) Silicon material: electronic grade or solar grade.
(ii) Other raw materials.
(iii) Solar PV cells.
(iv) PV modules.
(v) Balance-of-system (BOS) components and subsystems.
(vi) Design and integration of effective PV energy systems.

The various aspects involved in the acquisition and effective utilization of PV technology by Third-World countries are presented below.

1.3.10.1 Components of photovoltaic technology

Silicon material

The most energy- and cost-intensive stage of the PV manufacturing process is the preparation of the EG-Si in wafer or sheet form. The manufacture of EG-Si feedstock through reduction of quartzite involves highly expensive and complex technology while single-crystal silicon ingot and wafer production, although regarded as high technology, is a relatively simple and adaptable technology.

Other raw materials

(i) Various chemicals such as acids, alkalis, organic solvents and diffusants.
(ii) Conductive pastes made of silver and aluminium.
(iii) High-transmission tempered 'white' glass sheets.
(iv) Various laminating materials such as EVA and Tedlar.
(v) Various mechanical and electrical components such as junction boxes and interconnections.

Except for the laminating materials which are produced only by a few companies in the world, many of the Third-World countries have the necessary infrastructure and expertise to produce the above raw materials.

Solar cells

The manufacture of high-quality solar cells on a commercial scale in the Third-World countries requires acquisition of proper production equipment suitable for conditions prevailing in these countries. Since solar PV technology is not yet fully mature, production equipment which meets all the stringent requirements of large-scale manufacture are generally not available off the shelf. Most solar cell manufacturers build their own production equipment according to their particular design requirements and manufacturing techniques which may differ from company to company. Solar-cell- and module-manufacturing equipment that are supplied by some companies are more or less still in a prototype stage and are not very suitable for use under developing-country conditions.

Photovoltaic modules

PV manufacturing technology based on single-crystal or multi-grain silicon solar cells being offered by PV system manufacturers mostly include the stages of production starting from module manufacture to integration of PV systems. The production equipment available for module manufacture is better designed and more reliable for developing-country conditions.

Balance of system components and subsystems

For most efficient utilization of PV-generated power, BOS components and subsystems should be specially designed for PV power systems. Two features of PV systems, in particular, demand specially designed BOS components to be integrated with the systems.

(i) The high cost of PV power requires that efficient BOS components be used. The availability and abundance of conventional fossil fuels have not encouraged optimization of the operational efficiency of electrical and electronic components.

(ii) PV cells generate DC power. Most electrical appliances available off the shelf use AC power since conventional fossil fuels, nuclear or hydro-power plants generate AC electricity.

With regard to the development and manufacture of BOS components suitable for PV systems, the approach taken by developing countries differs from that of the industrialized nations. The interest of the industrialized nations is in improving overall operational efficiencies rather than on the ruggedness of the products. In contrast with this, the lack of infrastructure in the remote areas of Third-World countries necessitate minimal maintenance of BOS components. In addition, the lower purchasing power of the population might dictate a compromise between achieving maximum operational efficiencies and cost-effectiveness of the power system. BOS components for use under Third-World conditions should therefore be rugged, maintenance free and of low cost. A high-efficiency product manufactured in a developed country and designed to be used under prevailing conditions there may fail to perform with rated efficiency under less reliable Third-World country conditions and may require frequent maintenance and service. BOS components and subsystems are therefore best designed and manufactured in the country of application with due regard to the particular local needs and conditions.

Integration of photovoltaic systems

Optimal PV system design must be carried out by local expertise in the country of application since local environmental conditions as well as specific load requirements must be considered for efficient utilization of PV-generated power. The location-specific and load-specific nature of PV power system design therefore make it crucial that design and integration of PV energy systems be carried out by those familiar with local energy needs in all spheres of activity: social, economic and educational.

The overall efficiency and cost-effectiveness of stand-alone PV systems can be increased with an integrated approach to system design and electrical energy utilization. Such an approach can only be attempted by the user country itself since it requires familiarity and understanding of specific local needs, life-styles, economic, social and educational activities as well as local environmental conditions.

1.3.10.2 Technological options for acquisition of photovoltaic technology

Components of PV manufacturing technology starting from the production of pure silicon material to the finished PV modules and balance of systems are shown in figure 1.62. The level of technical sophistication of the various stages of production is also indicated in figure 1.63. Third-World countries can enter into the PV manufacturing industry at any of the stages of production depending on the available industrial infrastructure and technical expertise.

The technology involved in the PV manufacturing processes becomes progressively more complex as it goes up the path from the final stage, the design and integration of PV energy systems, to the first stage, the production of pure silicon feedstock for solar cell fabrication from quartzite. A particular country therefore has the option to enter into the PV industry along any of the stages starting from the lowest technology, BOS design and

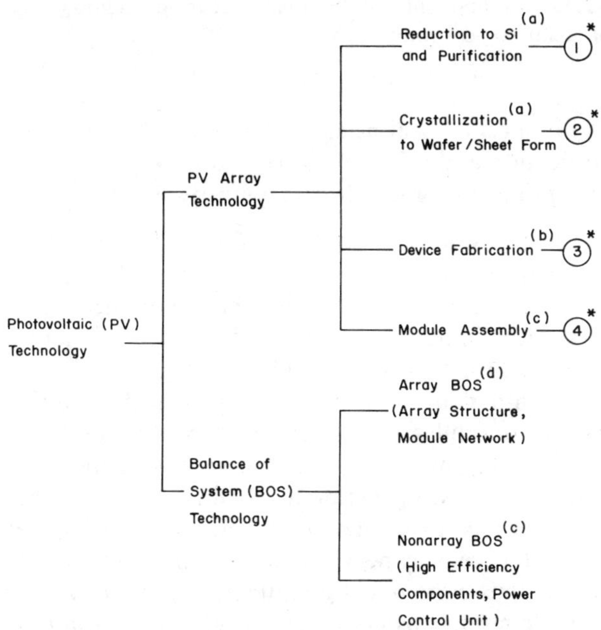

Figure 1.62 Components of PV manufacturing technology.

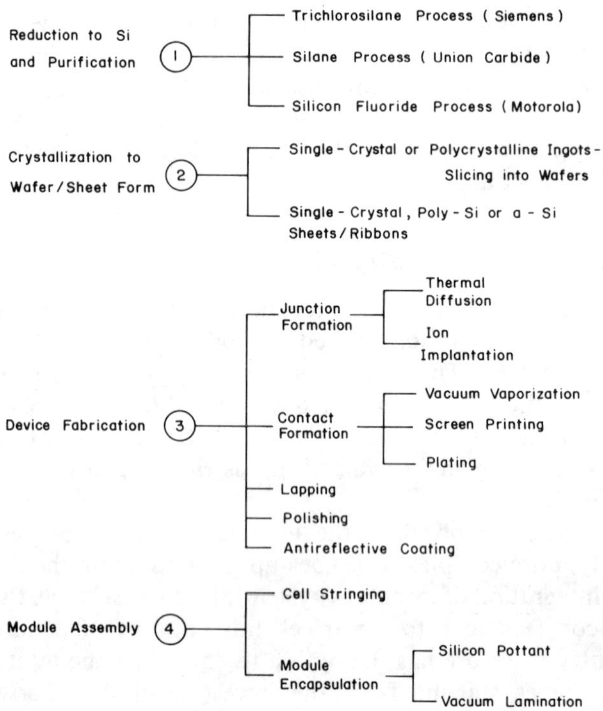

Figure 1.63 Components of PV manufacturing technology and the levels of technological complexity.

development for specific local needs and conditions to the high technology of silicon purification and solar cell fabrication.

The following options for developing countries exist.

(i) The complete PV system can be purchased from PV system manufacturers in developed countries with only the design and integration of the systems performed by local expertise. This strategy is suitable for countries with limited economic and technological resources and a small potential for PV systems applications.

(ii) The BOS components can be developed and manufactured indigeneously. PV modules purchased from other countries can then be designed into a complete system together with the locally made BOS components. This will generate local employment opportunities without a need for high investment in PV among manufacturers.

(iii) Finished solar cells can be imported and assembled locally into modules. A BOS-component-manufacturing industry will then exist alongside the module assembly industry. This option has the potential of reducing PV system cost without the need for very high investment in an indigeneous PV industry and may be suitable for countries with a sufficient local market for photovoltaics.

(iv) Imported single-crystal or polycrystalline wafers or sheets can be fabricated locally into solar cells and assembled into modules. A higher level of investment, technical expertise and infrastructure is necessary for this option. The danger of technological obsolescence of imported production equipment exists since PV technology is still in the developmental stage.

(v) Purified silicon feedstock material can be purchased and single-crystal or polycrystalline silicon ingots or sheets grown from it. The local PV-system-manufacturing industry therefore includes all the stages from ingot or sheet crystal growth to module assembly to PV system design and installation. Although a very high capital investment is needed for this, the potential exists for supplying EG-Si to the semiconductor industry as well as to the local PV industry. This approach is suitable if a large potential demand for PV energy technologies exists locally.

(vi) The last option requiring the highest level of capital investment offers the greatest security in terms of uncertainties in supply of either purified silicon feedstock material or silicon wafers or sheets from PV-system-manufacturing countries. Greater economic viability of the PV industry exists here because of the possibility of supplying EG-Si both to the larger semiconductor industry and to the PV industry. Export potential to other developing countries also exists for any of the products from the various stages of PV manufacture.

BIBLIOGRAPHY

Aulich H A 1986 *Proc. 2nd Int. Photovoltaic Science and Engineering Conf., Beijing, August 1986* pp 214–7

Aulich H A *et al* 1984 *Proc. 1st Int. Photovoltaic Science and Engineering Conf., Kobe, November 1984* pp 777–80

Böer K W 1987 *Solar Wind Technol.* **4** 21–35

Bolton J R 1983 *Solar Energy* **31** 483–502

Delahoy A E 1987 *Solar Cells* **21** 153–66

Ermer W 1985 *Proc. 6th Commission of the European Communities Conf. on Photovoltaic Solar Energy, 15–19 April 1985* (Dordrecht: Reidel)

Ermer W *et al* 1984 *Proc. 1st Int. Photovoltaic Science and Engineering Conf., Kobe, November 1984* pp 781–5

Fuhs W 1987 *Solar Wind Technol.* **4** 7–15

Grabmaier J G 1984 *Proc. 1st Int. Photovoltaic Science and Engineering Conf., Kobe, November 1984* pp 537–42

Grabmaier J G *et al* 1986 *Proc. 2nd Int. Photovoltaic Science and Engineering Conf., Beijing, August 1986* pp 174–7

Helmreich D 1986 *Proc. 2nd Int. Photovoltaic Science and Engineering Conf., Beijing, August 1986* pp 199–202

Hirobe T *et al* 1986 *Proc. 2nd Int. Photovoltaic Science and Engineering Conf., Beijing, August 1986* pp 214–7

Hoyler G 1984 *Proc. 1st Int. Photovoltaic Science and Engineering Conf., Kobe, November 1984* pp 791–5

Ingle W M 19— *Recent Advances in Solar Silicon Purification Technology* (Phoenix, AZ: Semiconductor Products Group, Motorola Inc.)

Kuwano Y 1984 *Proc. 1st Int. Photovoltaic Science and Engineering Conf., Kobe, November 1984* pp 571–6

Marshall S L 1967 *Microelectronic Technology* (Boston, MA: Boston Technical Publishers)

Mufti D 19— *Silicon Technology Revolution—A Challenge for the Developing Countries to Accept or Reject?* (Islamabad: National Institute of Silicon Technology)

Panyakeow S 1984 *Proc. 3rd Asian School on Solar Energy Harnessing* (Bangkok: Asian Institute of Technology—KMITL)

Pernisz U C *et al* 1987 *Solar Cells* **21** 195–203

Ravi K V 1984 *Proc. 1st Int. Photovoltaic Science and Engineering Conf., Kobe, November 1984* pp 43–9

Siffert P 1982 *Proc. 4th Commission of the European Communities Conf. on Photovoltaic Solar Energy, Stresa, 10–14 May 1982* (Dordrecht: Reidel)

Tabata K *et al* 1984 *Proc 1st Int. Photovoltaic Science and Engineering Conf., Kobe, November 1984* pp 809–12

Thorton J A 1987 *Solar Cells* **21** 41–54

Vanier P E and Su F C 1987 *Solar Cells* **21** 167–75

Vernstrom G D 1987 *Solar Cells* **21** 141–6

Werner P G 1982 *Proc. 4th Commission of the European Communities Conf. on Photovoltaic Solar Energy, Stresa, 10–14 May 1982* (Dordrecht: Reidel)

October 1981 *Westinghouse R & D Center; Monitor 8*, vol II, issue IV

1980 *Sunworld* **4**

1988 *Solar Cell Technology, Proc. Workshop on Planning of Network Projects in Materials Science and Solar Energy, Arusha, 21–26 November 1988* (Meudon: Laboratoire de Physique des Solides, CNRS)

December 1985 *Summary of Research and Development of Amorphous Solar Cells—Sunshine Project* (Solar Technology Development, New Energy Development Organization)

1984 *Basic Photovoltaic Principles and Methods* (Golden, CO: Solar Energy Research Institute)

Part II

The Photovoltaic Generator

2 The Photovoltaic Generator

2.1 INTRODUCTION

Energy converters to produce electricity can be generally classified into current sources and voltage sources.

When connected to a load, current sources exhibit relatively little output current variation whereas voltage sources will exhibit a relatively low output voltage variation. In their idealized form, current sources are capable of supplying infinite current, power and energy, all at a constant voltage value. Voltage sources are capable of supplying infinite voltage, power and energy, all at a constant current value. To represent physical sources more closely, voltage sources are always used in connection with a series resistance (voltage drop due to internal resistance) and current sources with a shunt resistance (current drop) as illustrated in figures 2.1(*a*) and 2.1(*b*).

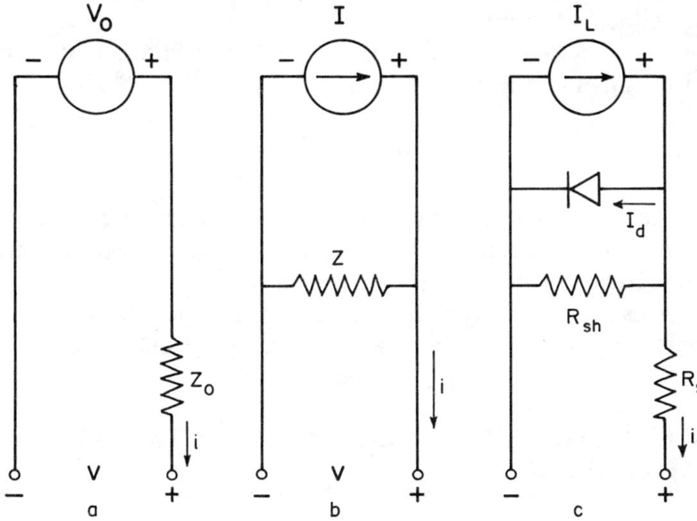

Figure 2.1 Electrical equivalent diagram of a solar cell: (*a*) voltage source; (*b*) current source; (*c*) PV generator.

A photovoltaic (PV) cell is simply a diode of large-area forward bias with a photovoltage (figure 2.1(*c*)). The photovoltage is created from the dissociation of electron–hole pairs created by incident photons within the built-in field of the junction (diode).

The operating equation of current–voltage (I–V) characteristics (where $I=f(V)$) of a solar cell can be easily derived (figure 2.2):

$$I = I_{ph} - I_D$$
$$= I_{ph} - I_0\left[\exp\left(\frac{q(V+R_s I)}{Ak_B T}\right) - 1\right] - \frac{V+R_s I}{R_{sh}}$$

where I_{ph} (A) is the photocurrent, I_D (A) the diode current, I_0 (A) the saturation current, A the ideality factor, q (C) the electronic charge, k_B (J K^{-1}) Boltzmann's gas constant, T (K) the junction temperature, R_s (Ω) the series resistance and R_{sh} (Ω) the shunt resistance.

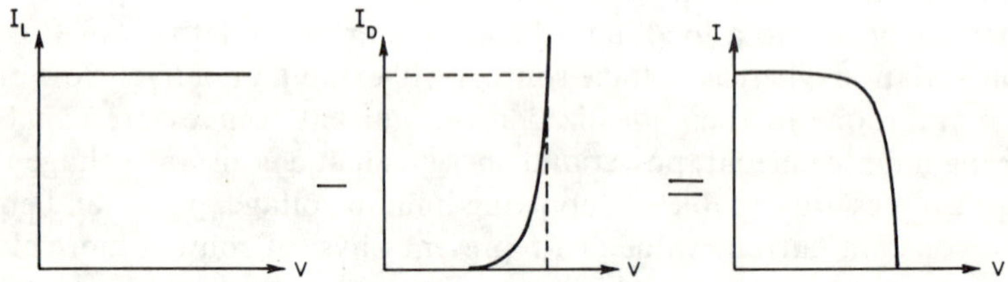

Figure 2.2 Solar cell I–V characteristics.

Under darkness the PV cell is not an active device. It functions as a p–n junction, i.e. as a diode. Externally, it is seen as an energy receiver; it produces neither a current nor a voltage.

If it were connected to an external supply, theory shows that voltage and current are related by the diode equation

$$I_D = I_0\left[\exp\left(\frac{q(V+R_s I)}{Ak_B T}\right) - 1\right].$$

In the case of a PV generator composed of N cells in series and M cells in parallel, the I–V characteristics of the whole generator can be derived by scaling the characteristics of one cell with a factor N in voltage and M in current. This approach is correct in the case of identical cells only. Association of cells in modules and arrays will be considered in chapter 3.

2.2 PHOTOVOLTAIC GENERATOR ELECTRICAL CHARACTERISTICS

The I–V characteristics describe the solar cell electrical terminal characteristics under the influence of meteorological conditions available *in situ* (irradiance level and ambient temperature).

A solar cell $I=f(V)$ graph illustrated in figure 2.3 passes through three significant points: the short-circuit current I_{sc} at M, the open-circuit voltage V_{oc} at S and the maximum power at A.

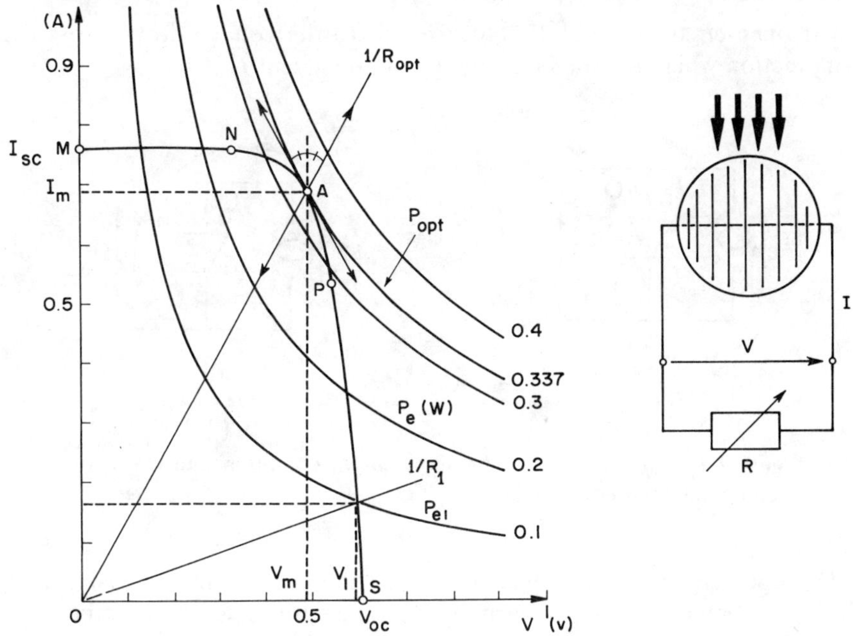

Figure 2.3 Solar cell I–V characteristics with constant-power curves and load lines.

2.2.1 Short-circuit current

For $V=0$,

$$I_{sc}=I_{ph}-I_0\left[\exp\left(\frac{qR_sI_{sc}}{Ak_BT}\right)-1\right]-\frac{R_sI_{sc}}{R_{sh}}.$$

At a normal level of irradiance (no concentration) the series resistance effect is negligible in the short-circuit condition (point M in figure 2.3) and the short-circuit current can be considered equivalent to the photocurrent I_{ph}, i.e. proportional to the irradiance E_e (W m^{-2}) (figure 2.4(*a*)):

$$I_{sc}=I_{ph}=KE_e$$

where K is a constant.

2.2.2 Open-circuit voltage

In the open-circuit condition (point S in figure 2.3) as illustrated in figure 2.4(*b*), the open-circuit voltage can be expressed analytically using the following expression:

$$V_{oc} = \frac{Ak_B T}{q} \ln\left(\frac{I_{ph} + I_0}{I_0}\right).$$

The open-circuit voltage V_{oc} corresponds to the voltage drop across the p–n junction when it is traversed by the photocurrent I_{ph}.

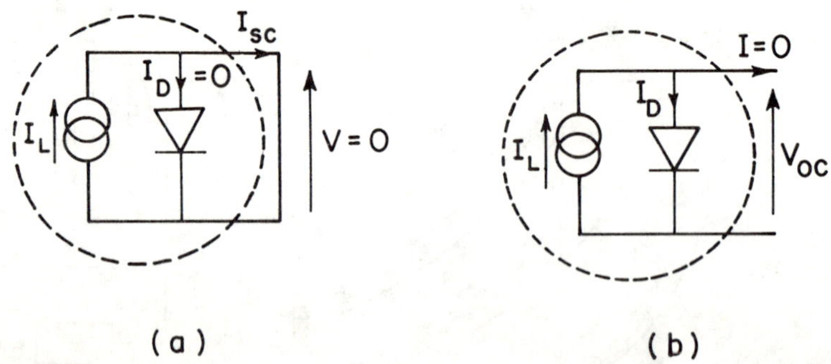

Figure 2.4 Solar cell (*a*) in the short-circuit condition and (*b*) in the open-circuit condition.

The open-circuit voltage increases logarithmically with increasing irradiance level and decreases linearly with an increase in junction temperature (figures 2.5 and 2.6).

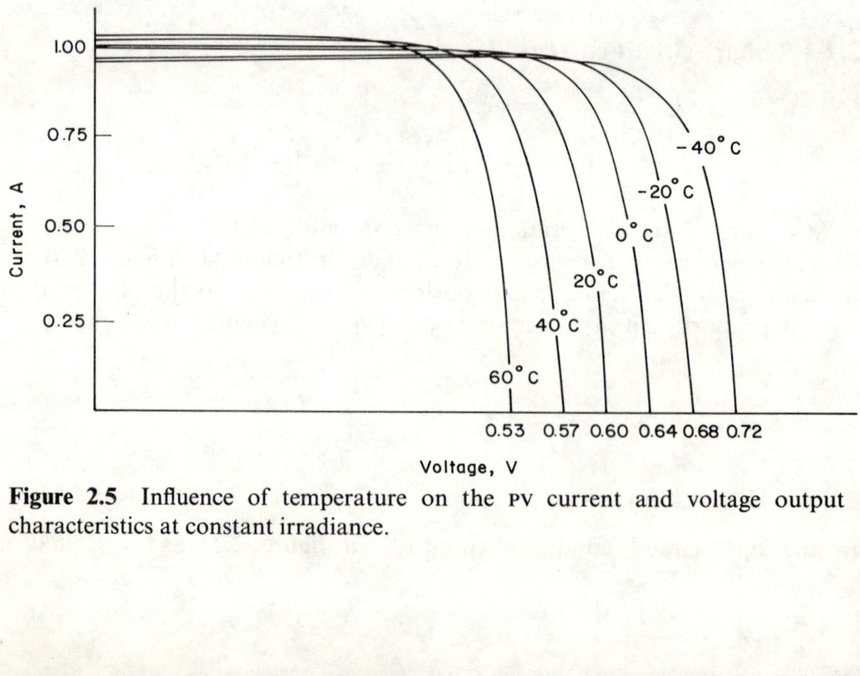

Figure 2.5 Influence of temperature on the PV current and voltage output characteristics at constant irradiance.

This effect of temperature is due to the exponential increase in the saturation current I_0 with an increase in temperature. The saturation current is a current of minority carriers created by thermal excitation and accelerated within the built-in field of the p–n junction.

2.2.3 Maximum-power operation

Let us consider the I–V characteristics of a solar cell for a given irradiance level E_e and a fixed cell temperature T. If the cell's terminals are connected to a variable resistance R, the operating point will be determined by the intersection of the I–V characteristics of the solar cell and the load. For a resistive load, the load characteristics is a straight line with a slope $1/R$ (where $I/V=1/R$).

(i) The power delivered to the load depends on the value of the resistance R only.
(ii) The maximum power is obtained for $R=R_{opt}$, at point A (figure 2.3) where the I–V characteristics of the solar cell is tangent to the slope of the hyperbola of constant power,

$$R_{opt}=\frac{V_{max}}{I_{max}}.$$

For constant irradiance level and temperature, the solar cell imposes neither the operating voltage nor the current while operating within the region NP of the curve in figure 2.3.

(*a*) If the load R is small, the cell will operate in the region MN of the curve. This is a region where the cell behaves as a constant-current source, almost equal to the short-circuit current.
(*b*) If the load R is large, the cell will operate in the region PS of the curve. This is a region where the cell behaves as a constant-voltage source, almost equal to the open-circuit voltage.

2.3 LOADING CONDITIONS

As shown in the previous section concerning the case of a resistive load, the operating point is determined by the intersection of the source (solar cell) I–V characteristics and the load (resistive) I–V characteristics. Electrochemical storage such as a battery is a constant-voltage load while a positive-displacement pump at a given head (with inertia wheel) is a constant-current load. Typical characteristics of devices used with a PV generator are presented superimposed on the I–V characteristics of the PV generator in figure 2.6.

The significance of proper matching between a PV generator and a resistive load can be appreciated by considering the maximum-power operating curve of the PV module (figure 2.6). When connected directly to the PV generator,

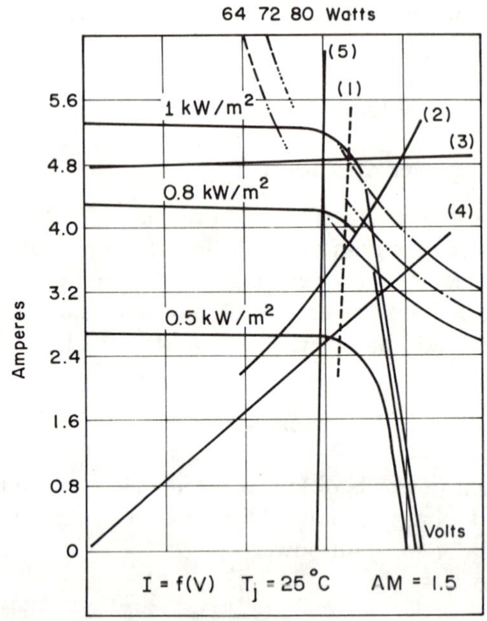

Figure 2.6 Typical load characteristics: curve (1), maximum power line; curve (2), centrifugal pump (at constant head); curve (3), positive-displacement pump at constant head; curve (4), resistor; curve (5), battery.

the combination of the centrifugal pump and a DC permanent-magnet motor as a load system has an operating curve close to the PV generator locus of maximum power at various irradiance levels. Unlike the operating points of the resistor and positive-displacement pump, the DC motor–centrifugal pump combination adapts well to the operating characteristics of a PV generator. Storage batteries are also well adapted to a PV generator when it is designed to have its terminal voltage floating around the range of voltages where the maximum power from the PV generator operates. In common PV systems such as a lighting kit, the three components—array, battery and load—are all connected in parallel, so that the voltage V_S is common to all of them. According to Kirchhoff's current law (figure 2.7),

$$I_P = I_B + I_L$$

where I_P is the PV generator current, I_B the battery current and I_L the load current.

According to the sign of $I_P - I_L$ determined by the relative position of the PV module I–V characteristics with respect to the load characteristics and dependent on the solar irradiance available, the battery current I_B will be positive (charging condition) (figure 2.8) or negative (discharging condition) (figure 2.9).

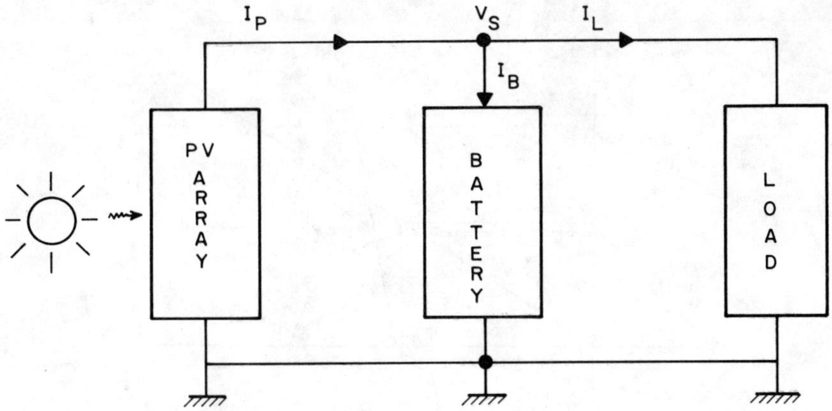

$$I_P + I_B + I_L = 0$$

$I_B > 0$ Battery Charging

$I_B < 0$ Battery Discharging

Figure 2.7 Schematic diagram of a basic PV system.

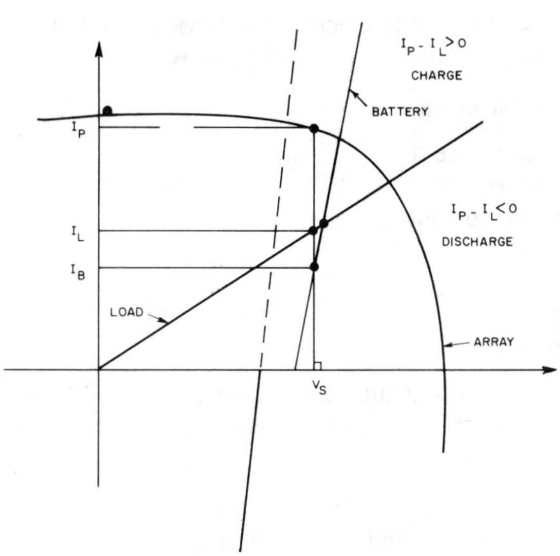

Figure 2.8 *I–V* characteristics curves of a PV module and load during battery charging ($I_B > 0$).

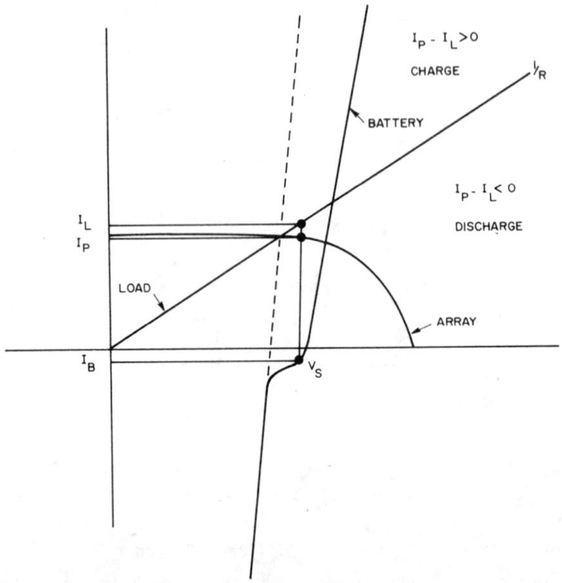

Figure 2.9 *I–V* characteristics curves of a PV module and load during battery discharging ($I_B > 0$).

2.4 PARAMETRIC ANALYSIS OF THE PHOTOVOLTAIC GENERATOR CURRENT–VOLTAGE CHARACTERISTICS

Five parameters govern the electrical behaviour of a solar cell and subsequently of a PV generator. They are as follows.

(*a*) Internal parameters.

(i) Series resistance R_s.
(ii) Shunt resistance R_{sh}.
(iii) Saturation current I_0.

(*b*) External parameters.

(iv) Irradiance E_e.
(v) Temperature T.

Under normal levels of irradiance (no concentration) these parameters can be considered as independent except two, the saturation current I_0 and cell temperature.

(*a*) The shunt resistance is due to the leakage current at the junction and depends on the method of junction construction (recombination of carriers within the bulk of the material). Its value is generally high.

(*b*) The series resistance depends mainly on the resistivity of the grid contact and the surface layer.

(*c*) Graphically, the series resistance and the shunt resistance will modify slightly the shape of the *I–V* characteristics curve (figures 2.10 and 2.11).

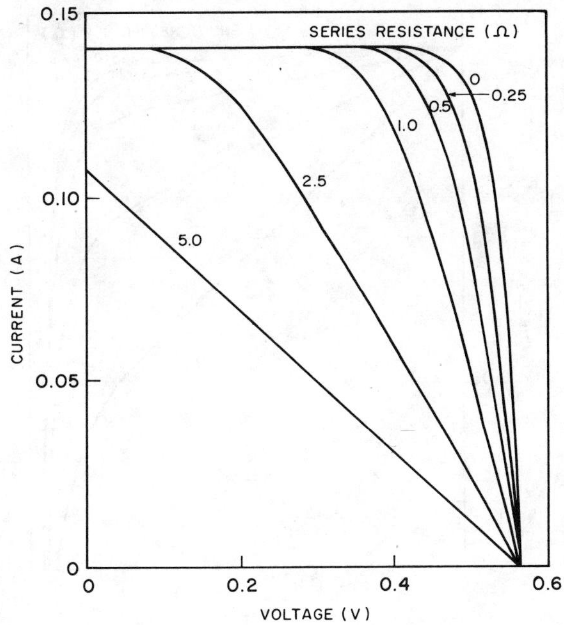

Figure 2.10 Effect of series resistance on the *I–V* curve shape.

(*d*) The series resistance causes a reduction in the slope of the *I–V* characteristics in the region where the cell behaves as a voltage generator (region PS in figure 2.3).

The series resistance of an n^+–p solar cell is the sum of the following components (figure 2.12).

(i) Resistance R_1 of the back Ohmic contact with the p-type silicon.
(ii) The p-type base material resistance R_2.
(iii) The resistance R_3 due to the sheet resistance of the n^+ surface layer.
(iv) The contact resistance R_4 between the top Ohmic grid contact and the n^+-type silicon.
(v) The resistance R_5 along the metallic finger and the resistance R_6 along the grid collector.

Analysing these different contributions shows that the two most important are R_1 and R_3.

The shunt resistance causes an increase in the slope of the module's characteristics in the region where the cell behaves as a current generator (region MN in figure 2.3).

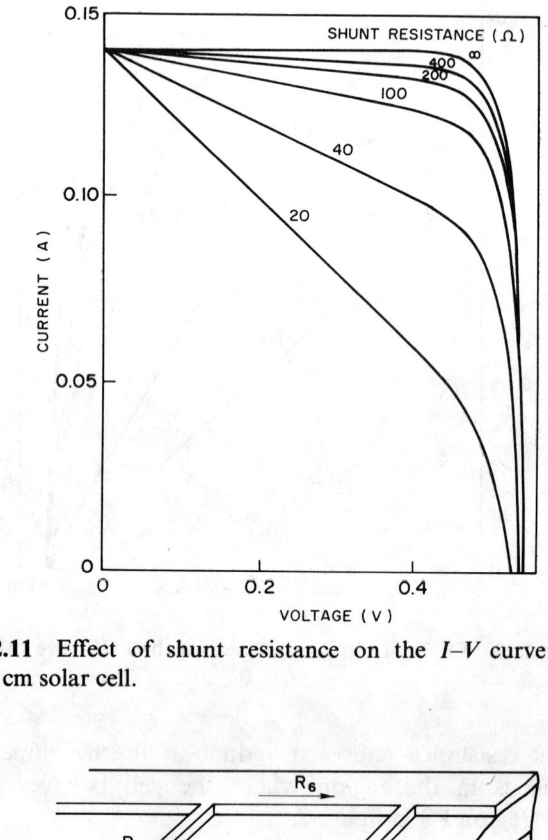

Figure 2.11 Effect of shunt resistance on the I–V curve shape of 2 cm × 2 cm solar cell.

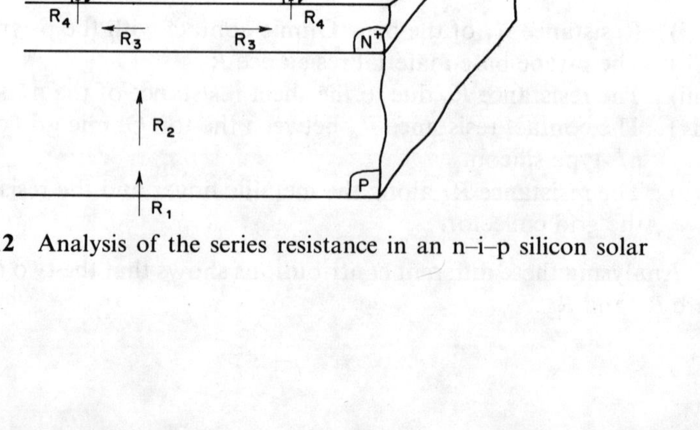

Figure 2.12 Analysis of the series resistance in an n–i–p silicon solar cell.

2.4.1 Influence of solar irradiance intensity

It has already been seen that the photocurrent I_{ph} is proportional to solar irradiance. The current $I_D(V_J)$ of the p–n junction is not affected on the other hand. This results in the characteristics $I=f(V)=I_{ph}-I_D(V_J)$, traced for different values of solar irradiance (figure 2.13).

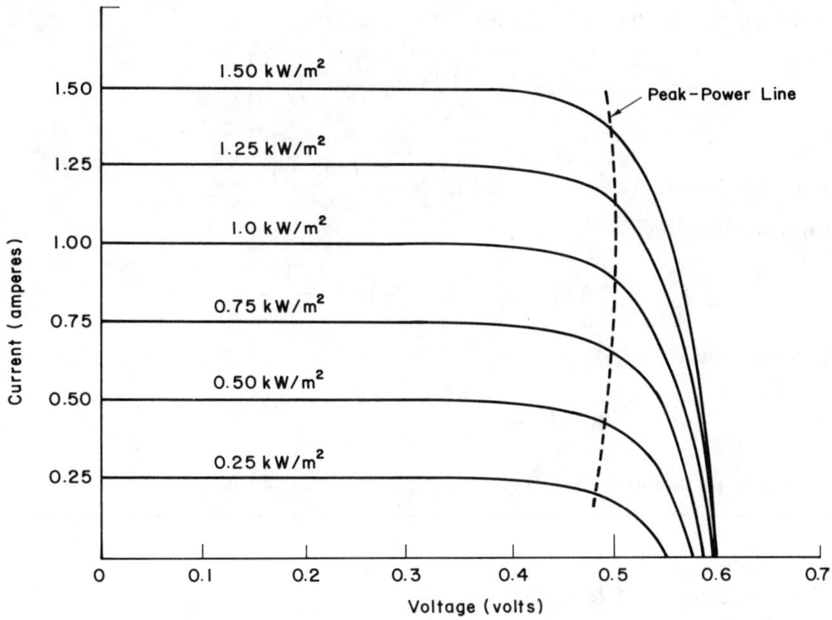

Figure 2.13 Influence of irradiance on the PV current and voltage output characteristics at constant cell temperature.

(*a*) At a low level of irradiance, the short-circuit current I_{sc} is proportional to the solar irradiance (neglecting the series resistance R_s). It is of the order of 30 mA cm^{-2} for an irradiance of 1 kW m^{-2} for a single-crystal silicon cell (at $T=25$ °C).

(*b*) The open-circuit voltage increases slightly with increasing irradiance. It does not depend on the surface area, being a function of the material only. It is of the order of 590 mV for 1 kW m^{-2} of irradiance for a single-crystal silicon cell (at $T=25$ °C).

(*c*) The optimal power of the module is proportional to the irradiance.

We can note from the figure that the points of peak power fall on a vertical line (of constant voltage for all irradiances). Table 2.1 summarizes the irradiance influences at a constant temperature level.

Table 2.1 A summary of irradiance influences.

Mean irradiance level

$$I_{sc}=I_{ph}=KE_e$$

$$V_{oc}=\frac{Ak_BT}{q}\ln\left(\frac{KE_e}{I_0}\right)$$

High irradiance level (X is the concentration factor)

$$I_{sc}=KE-I_0\left[\exp\left(\frac{qR_sI_{sc}}{Ak_BT}\right)-1\right]$$

$$V_{oc}=\frac{Ak_BT}{q}\left[\ln\left(\frac{KE_e}{I_0}\right)+\ln X\right]$$

Low irradiance level

$$KE_e-I_0\left[\exp\left(\frac{qV_{oc}}{Ak_BT}\right)-1\right]-\frac{V_{oc}}{R_{sh}}=0$$

Maximum efficiency

$$\eta(X,E_e)=\frac{\mathrm{FF}(X,E_e)V_{oc}(X,E_e)I_{sc}(X,E_e)}{E_e}$$

(at the mean irradiance level, there is a logarithmic variation in η)

2.4.2 Influence of temperature

Temperature has a non-negligible effect on the solar cell I–V characteristics.

The photocurrent will increase slightly with increasing cell temperature T (0.1% °C^{-1} or 0.03 mA °C^{-1} cm^{-2}). The slight increase in the photocurrent arises from a decrease in the band-gap energy E_g of the material as the temperature increases according to (table 2.2)

$$E_g(T)=E_g(0)-\frac{aT^2}{T+b}. \tag{2.1}$$

Table 2.2 Band-gap energies of silicon and GaAs, together with the parameters a and b in equation (2.1).

Material	$E_g(0)$ (eV)	a (10^{-4} eV K^{-1})	b (K)
Silicon	1.16	7	1100
GaAs	1.52	5.8	300

The open-circuit voltage would decrease linearly with increasing cell temperature owing to the exponential increase in saturation current. The saturation current is a current of minority carriers created by thermal excitation. Its variation with temperature can be expressed as

$$I_0 = A_0 T^3 \exp\left(-\frac{E_g}{k_B T}\right).$$

The following are affected by an increase in cell temperature.

(i) The open-circuit voltage V_{oc} would decrease by 2 mV °C^{-1} between 20 and 100 °C.

(ii) The maximum available power will decrease by 0.35% °C^{-1} approximately and therefore there is a proportional decrease in maximum efficiency (fig 2.14).

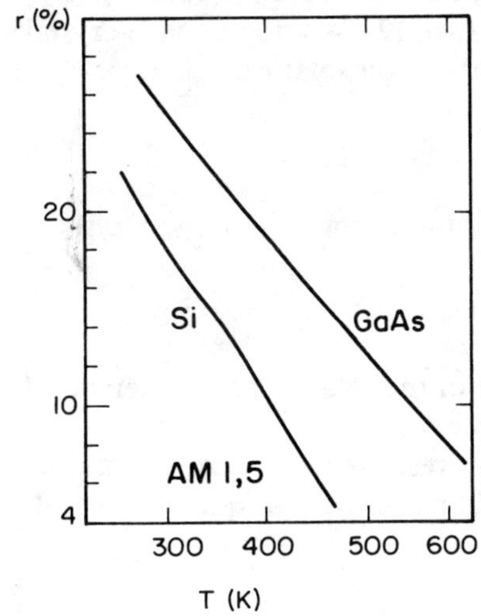

Figure 2.14 Typical influence of temperature on solar cell efficiency.

The maximum efficiency can be expressed as follows:

$$\eta(XE_e, T) = \eta(E_e, T_0)[1 - \beta_0(T - T_0)]\left(1 + \frac{k_B T}{q}\frac{\ln X}{V_{oc}(E_e, T_0)}\right).$$

These effects upon the solar cell I–V characteristics should be taken into account during the sizing stage of a PV generator.

2.5 DEFINITIONS

The particular properties of solar cell modules require a certain number of definitions.

2.5.1 Peak power

The optimal power delivered by the cell for an irradiance of 1 kW m^{-2} and for a junction temperature of 25 °C.

2.5.2 Conversion efficiency

The conversion efficiency is the ratio of the optimal electric power P_{opt} delivered by the PV module from the solar irradiance E_e received at a given cell temperature T, i.e.

$$\eta = \frac{P_{opt}}{AE_e}$$

where the optimal electric power P_{opt} is in watts, the irradiance E_e in watts per square metre and the cell area A in square metres.

Typical values for η are 12–14% for a single-crystal silicon solar cell and 9% for a polycrystalline silicon solar cell.

2.5.3 Fill factor

The fill factor FF is the ratio of the peak power to the product $I_{sc} V_{oc}$:

$$\mathrm{FF} = \frac{I_{max} V_{max}}{I_{sc} V_{oc}}.$$

It determines the shape of the solar cell characteristics $I = f(V)$, and its value is higher than 0.7 for good cells.

The series and shunt resistances account for a decrease in the fill factor. The fill factor is a useful parameter for quality control tests.

BIBLIOGRAPHY

Backus C E (ed) 1976 *Solar Cells* (New York: IEEE)

Lasnier F and Follea D *Lecture Notes* (Bangkok: Asian Institute of Technology)

Rauschenbach H S 1980 *Solar Cell Array Design Handbook* (New York: Van Nostrand–Reinhold)

Roger J *Lecture Notes* (Lyon: University of Lyon)

Rousset P 1983 *Lecture, SDRL–Chulalongkorn University, Thailand, 22 June 1983* (Bangkok: Asian Institute of Technology)

3 Network of Solar Cells, Modules and Arrays

3.1 INTRODUCTION

Important problems occur when solar cells, photovoltaic (PV) modules or arrays are connected in a network, mainly owing to the associated electric and thermal imbalances generated by mismatch of characteristics of individual units. A high-peak-power system necessitates the inclusion of protective devices into the network to cover the higher risk of thermal and electrical hazards. When a high peak power (of the order of several megawatts) is achieved, the optimization of the number of protective components and their location over the network of modules is necessary in order to reduce the total system cost as well as the energy losses inherent in the association of modules into an array network.

Association of solar cells will be considered as well as association of modules. The methods used for the association of solar cells can be extended to the association of large arrays.

Two cases will be considered: the ideal case of identical cells and the practical case of non-identical cells.

Indeed, scattering of the electrical characteristics of solar cells around an average value always occurs. This is due to the imperfect repeatability of the technical processes of solar cell fabrication and/or abnormal conditions of utilization of the solar cells; parasitic occultation or partial or complete shadowing can create electrical imbalances.

Such hazards if not prevented can lead to peculiar working conditions of one or several modules in the network which can behave like a power receptor and consume part of the electric power generated by other modules in the array network.

3.2 SOLAR CELLS IN SERIES

Let us consider the case of two cells in series (figure 3.1(a)) which are connected to the variable load R. In this arrangement, the current I is the

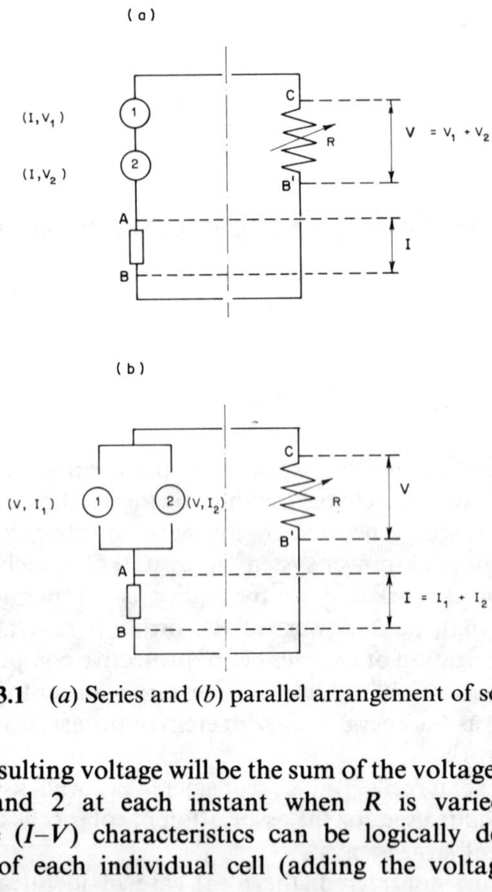

Figure 3.1 (*a*) Series and (*b*) parallel arrangement of solar cells.

same and the resulting voltage will be the sum of the voltages V_1 and V_2 of the solar cells 1 and 2 at each instant when R is varied. The resulting current–voltage (I–V) characteristics can be logically deduced from the characteristics of each individual cell (adding the voltages for the same current).

3.2.1 Identical solar cells in series

In an ideal case when two identical solar cells (cells with identical performance characteristics) are connected in series, the open-circuit voltage is equal to twice the voltage of one individual cell:

$$I > 0 \qquad V = V_1 + V_2$$

$$I = 0 \qquad V_{oc} = 2V_{oc1} = 2V_{oc2}.$$

Under short-circuit conditions (load $R = 0$), the total current $I_{sc} = I_{sc1} = I_{sc2}$ and the respective voltages through the terminals of the solar cells are zero:

$$V = V_1 = V_2 = 0.$$

The cells are working as generators and the generated power is positive.

Figure 3.2 presents the ideal resulting characteristics of two identical cells in series. The subscript G will indicate a PV generator consisting of two or more solar cells in series or in parallel.

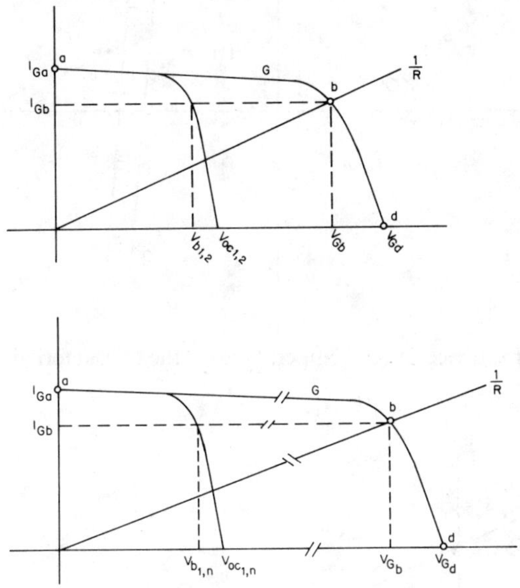

Figure 3.2 *I–V* characteristics curves of identical solar cells in series.

Point b represents the maximum-power point at which

$$I_{max} = I_{max1} = I_{max2}$$

$$V_{max} = V_{max1} + V_{max2} = 2V_{max1} = 2V_{max2}.$$

Figure 3.2 also shows the association of n identical cells in series. (The same reasoning can be applied to the association of identical modules and identical arrays.)

3.2.2 Non-identical solar cells in series

Figure 3.3 shows the effect of variation in irradiance and temperature on the *I–V* characteristics of a solar cell.

Shadowing of a cell in a module causes a lower power output owing to reduced energy input to the cell and increased internal losses in the non-illuminated cell portion. The shadowed cell in a series arrangement will reduce the short-circuit current of the whole module.

Figure 3.4 shows the characteristics of a generator made up of two non-identical solar cells connected in series. Curves 1 and 2 represent the

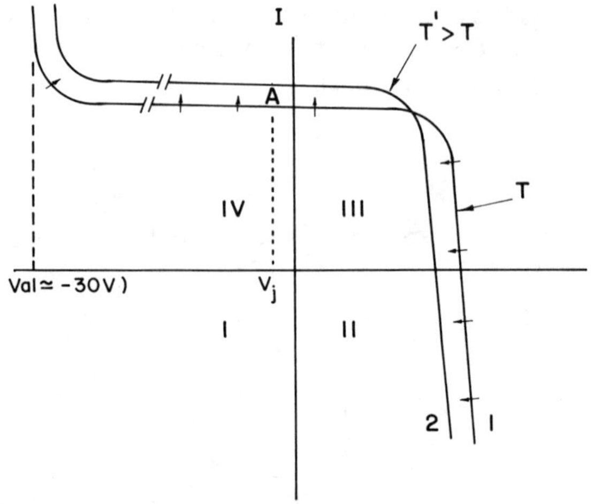

Figure 3.3 Influence of cell temperature on the characteristics of a solar cell.

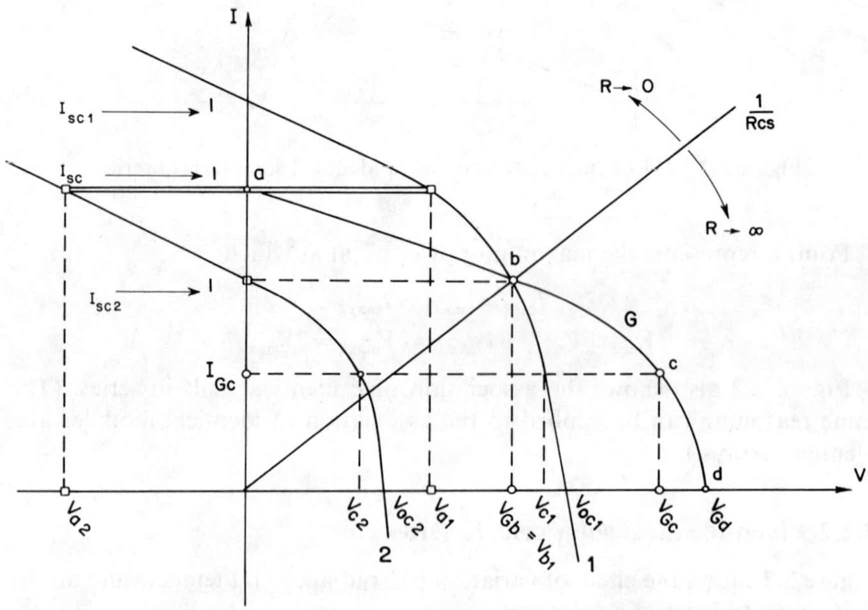

Figure 3.4 Two non-identical cells connected in series.

characteristics of an efficient and a less efficient solar cell respectively. The same current flows in the two solar cells but the total voltage is the sum of the two voltages. If the impedance R of the load is lower than R_{cs}, solar cell 2 becomes a receiver with a reverse voltage.

If we consider the case of two non-identical solar cells connected in a series arrangement and the resulting *I–V* characteristic in figure 3.4, the operating behaviour of the series arrangement can be described as follows.

(i) At point d, the series string is in an open-circuit condition. The total current *I* is equal to zero and the resulting V_{Gd} is the sum of the open-circuit voltages V_{oc1} and V_{oc2}.

(ii) At any point c, between points d and b, the two cells operate respectively as electric generators, i.e. the electric power generated by each cell is positive: $V_{Gd} = V_{c1} + V_{c2}$.

(iii) The point b shows the limit corresponding to the short-circuit current I_{sc2} of the less efficient cell and the resulting voltage V_{Gd} is consequently equal to V_{b1} ($V_{b2} = 0$, short-circuit condition of cell 2). This point is achieved when the impedance of the load reaches a critical value R_{cs} such that cell 2 is neither a generator nor a receptor. If the impedance of the load is less than R_{cs}, cell 2 will operate as a receptor; the electric power corresponding to this cell is negative.

(iv) Point a corresponds to the short-circuit operating point of the series string, $V_{Ga} = 0$, $I_{Ga} = I_{sc}$:

$$V_{Ga} = V_{a1} + V_{a2} = 0 \rightarrow V_{a1} = -V_{a2}.$$

As cells 1 and 2 are non-identical, the voltage at the terminals of solar cell 1 is the same but with reverse polarity as the applied voltage to the terminals of cell 2, such that the resulting voltage equals zero. Cell 2 will operate as a power receptor under a voltage V_{a2}, the maximum reverse voltage which is equal to the minimum operating voltage of cell 1.

The argument is the same for a larger network of cells or modules connected in series with one cell or module less efficient than the rest. The voltage of the entire system can be applied under certain conditions (near short-circuit) to the less efficient cell or module which is going to become

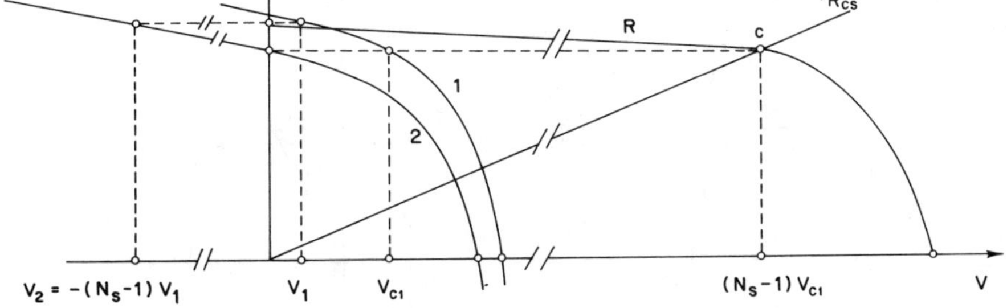

Figure 3.5 $N_s - 1$ identical cells in series.

heated, leading to a thermal imbalance (hot spot) as well as modification of its electric characteristics. This happens when the critical value $1/R_{cs}$ of the load admittance has been exceeded. This phenomenon is illustrated in figure 3.5. Here

$$V_2 = -(N_s - 1)V_1.$$

3.3 SOLAR CELLS IN PARALLEL

3.3.1 Identical solar cells in parallel

Consider the case (figure 3.6) with two identical cells mounted in parallel and connected to a variable load; the resulting voltage V_{Gb} is the same for each cell and the resulting current I_{Gb} will be the sum of the respective currents I_1 and I_2 of the cells:

$$I_{Gb} = I_{b1} + I_{b2}$$
$$V_{Gb} = V_{b1} = V_{b2}.$$

If the two cells are identical, their I–V characteristics will be the same and each of them will generate the same current, the same short-circuit current and the same open-circuit voltage:

$$I = 2I_1 = 2I_2$$
$$V = V_1 = V_2$$
$$I_{sc1} = I_{sc2}$$
$$V_{oc1} = V_{oc2}.$$

The resulting characteristics of the configuration can be easily determined; the voltage is the same and, whatever the value R of the load is, the total current I is equal to twice the current generated by one cell (figure 3.6).

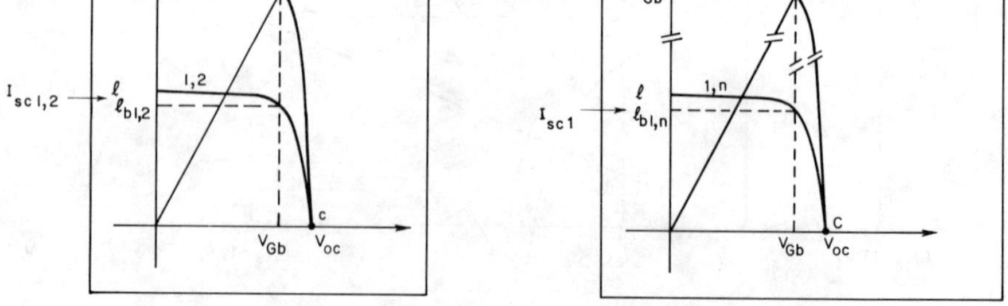

Figure 3.6 Identical cells in parallel.

3.3.2 Non-identical solar cells in parallel

Figure 3.7 gives the detail of the operating characteristics for two non-identical cells connected in parallel.

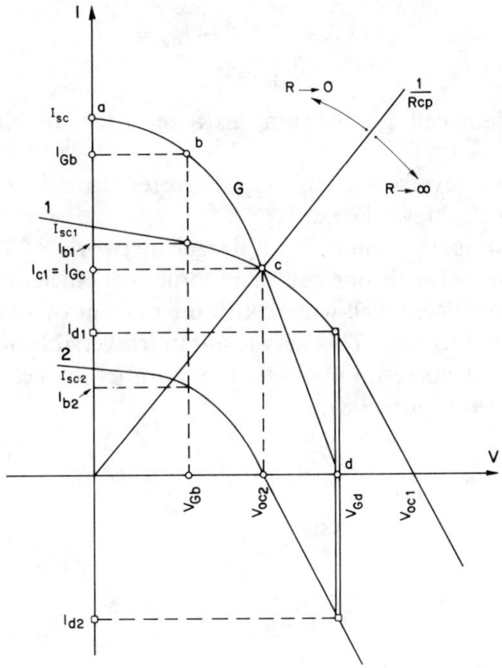

Figure 3.7 Two non-identical cells in parallel.

(i) At point a, the output terminal is short-circuited:

$$I_{sc} = I_{sc1} + I_{sc2}$$
$$V = 0.$$

(ii) For any point b located between points a and c, the resulting voltage is V_{Gb} and the current is

$$I_{Gb} = I_{b1} + I_{b2}.$$

Both cells operate as electrical generators. The delivered electrical power is positive.

(iii) Point c is the critical point corresponding to the open-circuit voltage of the less efficient cell 2:

$$V_{Gc} = V_{oc2}$$
$$I_{Gc} = I_{c1}.$$

At this point, the value of the load resistance is R_{cp}. At this value, cell 2 is neither a generator nor a receptor. For a value R of the load greater than R_{cp}, this cell 2 will operate as a receptor.

(iv) Point d corresponds to the parallel configuration in open circuit; the total current is equal to zero:

$$I_{Gd} = I_{d1} + I_{d2} = 0 \rightarrow I_{d1} = -I_{d2}$$
$$V_{Gd} = V_{d1}.$$

The less efficient cell 2 is working as a receptor. Its electrical power is negative.

The maximum reverse current I_{d2} delivered to cell 2 is equal to the minimum current achievable, i.e. I_{d1}.

The argument is the same for a larger network of cells or modules connected in parallel with one cell or module less efficient than the rest. In this case, the less efficient cell will absorb the current of all the other cells or modules, causing it to heat. This may result in irreversible deterioration if the critical impedance corresponding to the number of cells or modules in parallel is exceeded (figure 3.8).

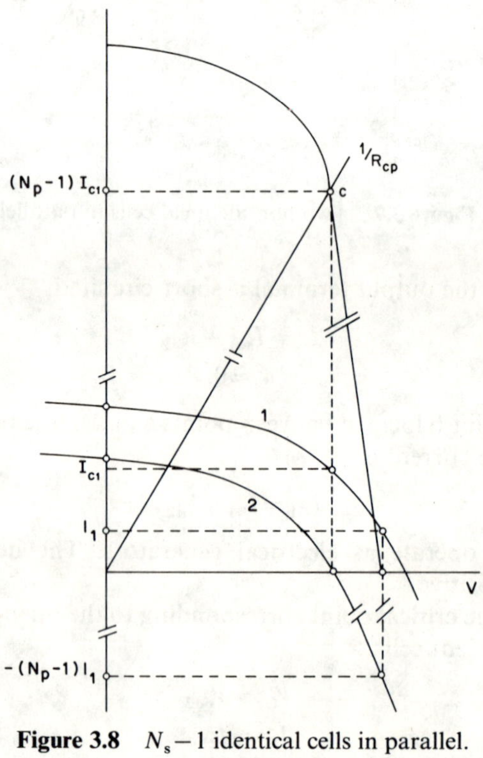

Figure 3.8 $N_s - 1$ identical cells in parallel.

It will be seen in the following that, in order to prevent this problem, the manufacturer introduces diodes in series for each branch in parallel.

For the same reason, a diode is introduced in parallel across cell or module to prevent a series effect.

However, these diodes contribute to energy losses, which can be non-negligible in the case of a large network. For that reason, it is necessary to study the critical points within the network in order to minimize the number of diodes.

3.4 THE PHOTOVOLTAIC ARRAY AND ITS PROTECTION

The PV array is defined as a mechanical integrated assembly of PV modules together with support structure (but exclusive of foundation, tracking, thermal control and other components), as required to form a DC power supply unit.

We have seen that, in a series arrangement of solar cells (solar cells inside a module are generally connected in series), power is absorbed in the shadowed or less efficient cell. To prevent this problem, one can put a diode D_p parallel to this solar cell C_i as shown in figure 3.9. So, this solar cell C_i, if less efficient or shadowed, will produce no energy but will not be a supplementary load in series with R and will not reduce the load current. If the solar cell is identical with the others and if it is not shadowed, no current will flow through the diode D_p and no loss of power will occur. The choice of the diode is such that the value of the diode threshold voltage must be as small as possible, and its current rating should be able to carry the generated current within the series string.

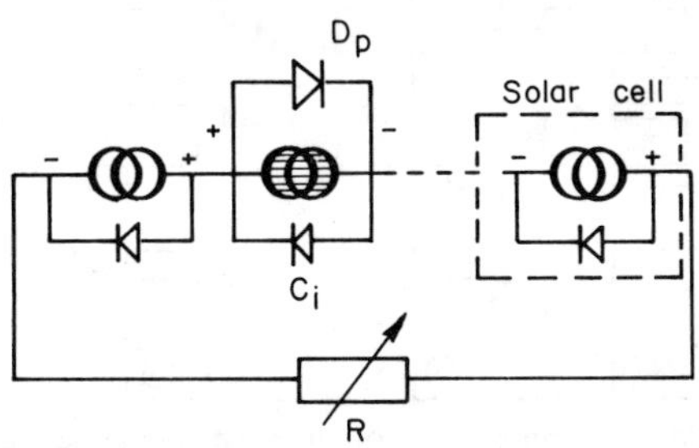

Figure 3.9 Protection of a module made of solar cells in series.

Practically, one does not put a diode in parallel with each solar cell but the protection should be carried out on the module level.

The protection of an association of modules in series is achieved by connecting bypass diodes in parallel with modules (figure 3.10) and the protection of an association of parallel modules is achieved by connecting

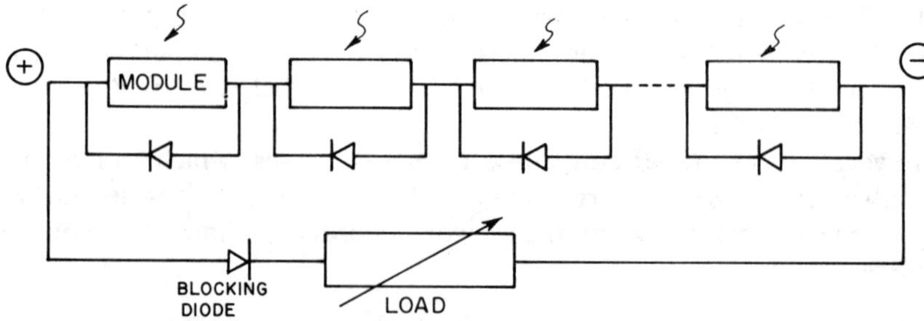

Figure 3.10 Array composed of a series arrangement of modules.

blocking diodes in series (figure 3.11). Figure 3.12 shows the protection of a PV array composed of a series–parallel association of modules. If the load is a storage battery, the first requirement is to prevent the discharge of the elements into the PV array at night by using a blocking diode.

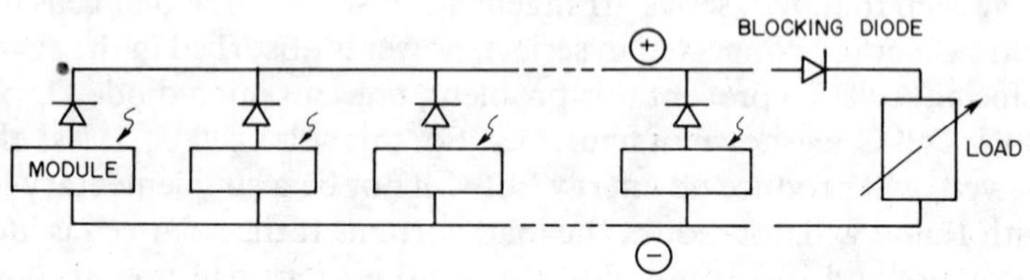

Figure 3.11 Array composed of a parallel arrangement of modules.

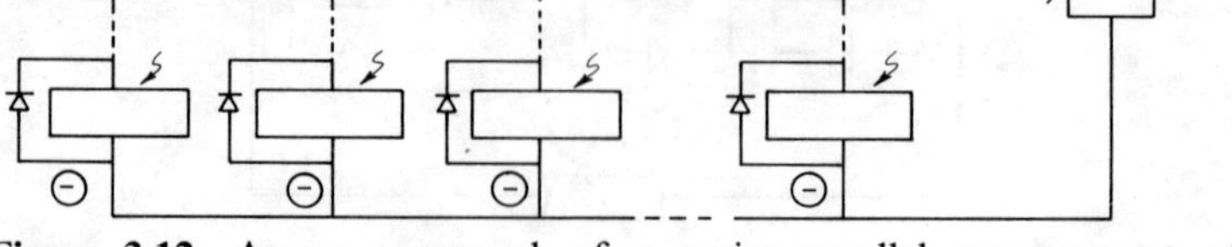

Figure 3.12 Array composed of a series–parallel arrangement of modules.

3.5 INFLUENCE OF TEMPERATURE

Figure 3.3 shows the temperature effect on the *I–V* characteristics of a solar cell. It affects mainly the voltage in a very sensible manner, modifying the energy balance of the cell (figure 3.13).

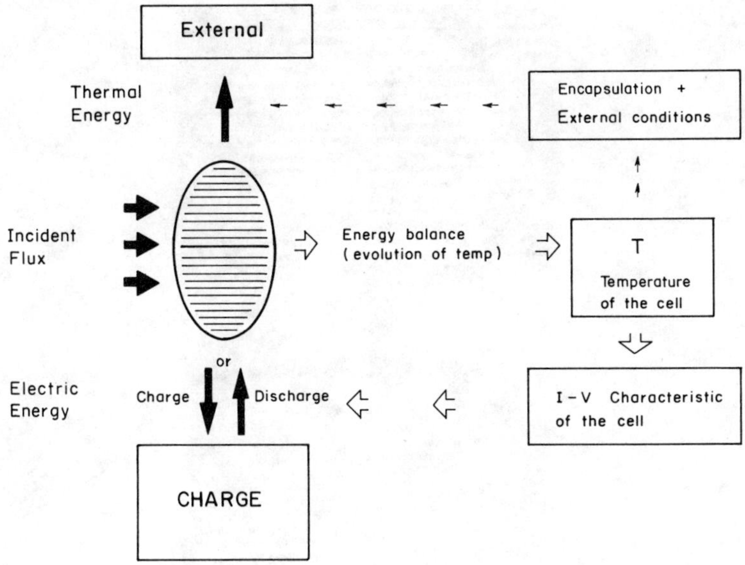

Figure 3.13 Energy balance of a solar cell.

This energy balance is related to the encapsulating material, its thermal dissipation and absorption properties, the generated or absorbed electrical power, the level of irradiance, etc; a kind of closed loop occurs (temperature–dissipated power–temperature) which can be autoregulated as otherwise it leads to thermal overheating which may destroy the cell and more often the contact (figure 3.14).

Figure 3.15 shows a simulation result of the effect of the inertia of the solar cell to reach the temperature of equilibrium at different wind speeds between 0 and 5 $\mathrm{m\,s^{-1}}$ for a load of 25 Ω and under an irradiance of 1 $\mathrm{kW\,m^{-2}}$.

Figures 3.16 and 3.17 show another simulation result of the evolution of the limiting cell temperature as a function of the level of irradiance for a wind speed of 1 $\mathrm{m\,s^{-1}}$ corresponding to the sole effect of natural convection.

Figures 3.18 and 3.19 are simulation results for a module of 34 cells in series; 33 are assumed to be identical, and one is partially shadowed:

$$\text{solar irradiance} = 1\ \mathrm{kW\,m^{-2}}$$
$$\text{temperature} = 293\ \mathrm{K}$$
$$\text{wind speed} = 1\ \mathrm{m\,s^{-1}}.$$

Figures 3.18 and 3.19 show the evolution of the operating point of each cell (V_{c1} non-occulted and V_{c2} occulted) for the elevation of the working point as a function of the elevation of temperature for a 25 Ω load and the short-circuit condition respectively. η is the non-occultation ratio; for $\eta = 0.9$, the cell operates as a receptor at $t = 0$ and becomes a generator before reaching thermal equilibrium.

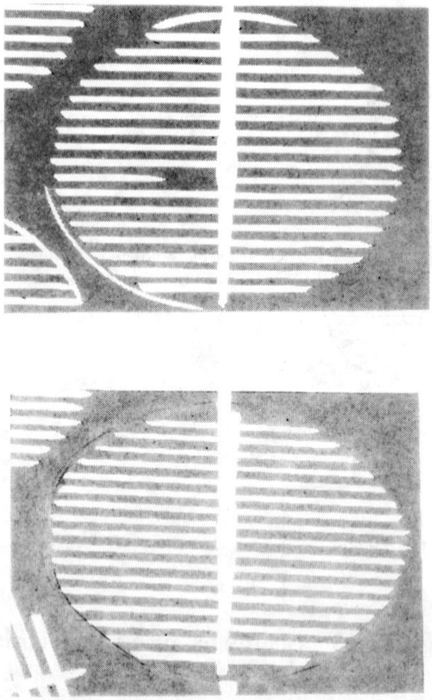

Figure 3.14 Cell contact damage due to heating.

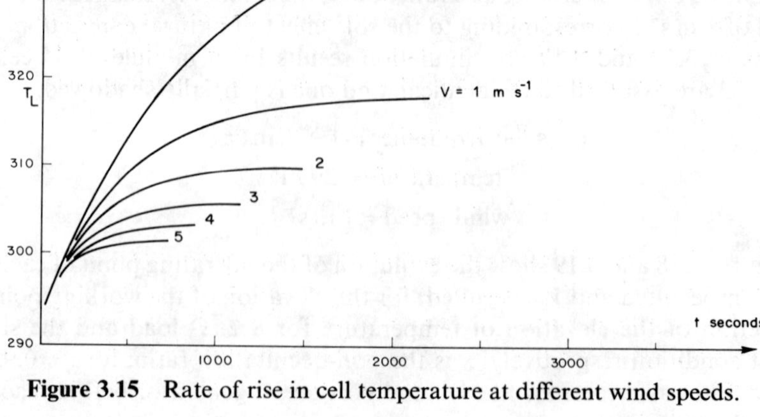

Figure 3.15 Rate of rise in cell temperature at different wind speeds.

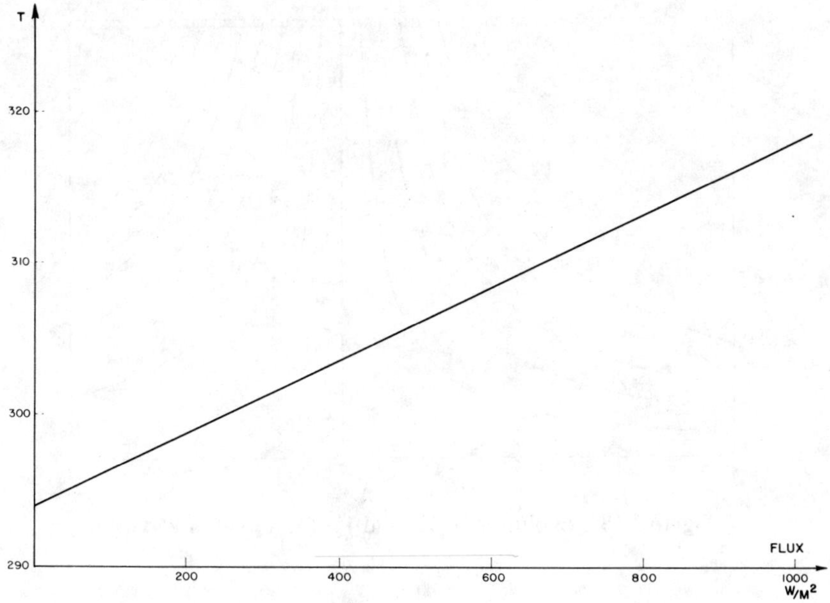

Figure 3.16 Limiting cell temperature as a function of irradiance at a wind speed of 1 m s^{-1}.

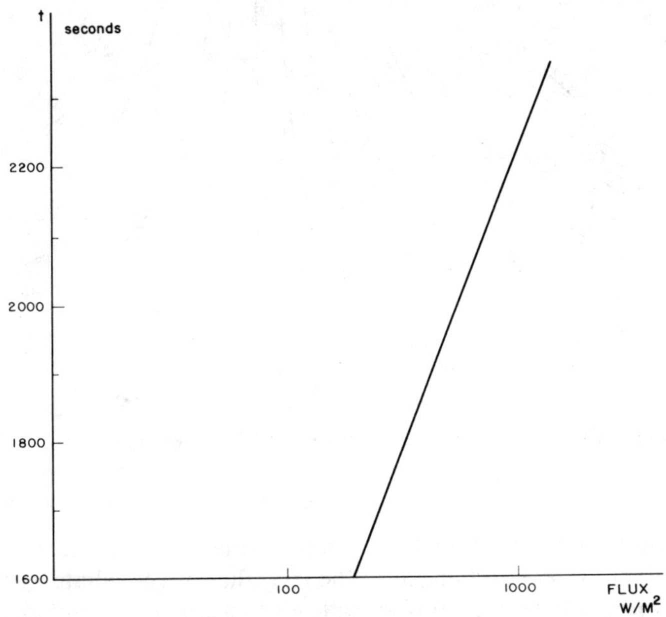

Figure 3.17 Time to reach the limiting cell temperature corresponding to an irradiance level at a wind speed of 1 m s^{-1}.

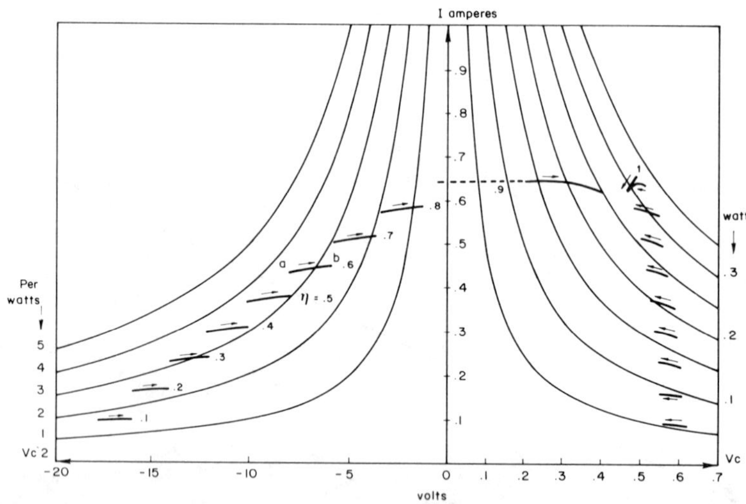

Figure 3.18 Evolution of V_{c1} and V_{c2} for a load of 25 Ω.

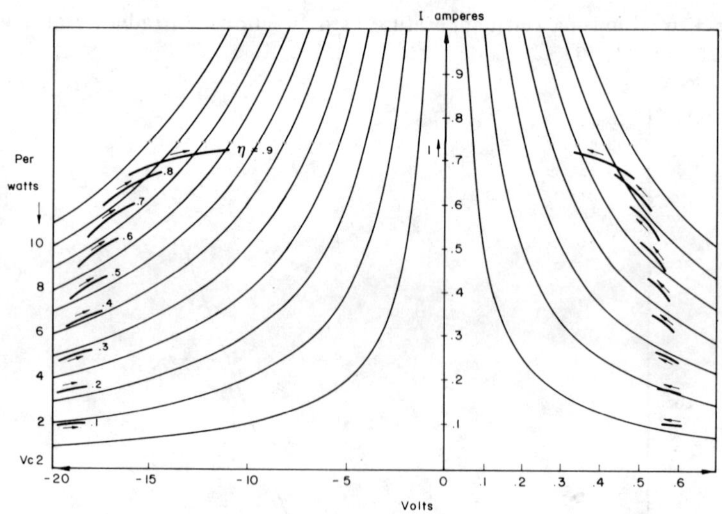

Figure 3.19 Evolution of V_{c1} and V_{c2} for the short-circuit condition.

This result shows generally that, when thermal equilibrium is reached, the energy balance is less prejudicial to the less efficient cell whatever the degree of occultation. The temperature effect leads to a cure of the imbalance due to the mismatching effect, by its secondary action on the characteristics. This is for series arrangements.

BIBLIOGRAPHY

Favolle D P 1979 *Thesis* Université Claude Bernard Lyon I, France

Part III

Photovoltaic Systems Engineering

4 Storage Batteries for Photovoltaic Power Systems

4.1 BATTERIES SUITABLE FOR PHOTOVOLTAIC POWER SYSTEMS

(i) Nickel–cadmium.
(ii) Lead–acid.
(iii) Nickel–iron.
(iv) Sodium–sulphur.

4.2 LEAD–ACID BATTERIES

Lead–acid batteries are most commonly used for residential-scale photovoltaic (PV) systems. At present, only the lead–acid battery and in some cases the nickel–cadmium battery have acceptable performance characteristics and life-cycle costs for PV power system applications.

4.2.1 Battery types

4.2.1.1 Starting, lighting and ignition batteries

(i) Starting, lighting and ignition (SLI) batteries are used primarily in automobiles for engine starting.
(ii) They are designed to deliver a very large current for a short time in order to operate the starter.
(iii) Paper-thin plates allow a great number to fit into a small space. The resulting large surface area allows currents of 200 A or more to be drawn for a few seconds at a time without damage to the battery.
(iv) Normal automotive batteries are designed to take 20 or so deep-discharge cycles before battery life expires.
(v) Normally lead–antimony grids are used. Many types have reduced grid antimony content to minimize self-discharge rate and small

quantities of silver, copper, tin and arsenic added for grid strength and resistance to anodic corrosion.

(vi) There is a good cycle life at shallow depth of discharge (DOD) and poor cycle life at deep DOD.

(vii) Sealed automative batteries are also available using lead–calcium grids. A catalyst may be placed at the top of the cell to react with gas released during charging and to convert it back to water (see figure 4.30).

(viii) Battery life under normal use is about 4–5 years.

A typical 12 V car battery has an energy capacity of approximately 0.78 kW h, delivers 300 A for about 20 s per start (or 0.02 kW h) corresponding to a DOD of 2.56% of rated capacity (0.78 kW h).

4.2.1.2 *Traction, motive power batteries*

(i) These are used in electric lift trucks, mining cars, heavy-duty material handlers, golf carts and electric vehicles.

(ii) Deep-discharge batteries are capable of being repeatedly deep-discharged to 50–80% of their capacity.

(iii) They are built with thick lead plates and greater electrolyte capacity to make them more reliable under difficult operating conditions.

(iv) They are designed for long life with typical cycle lives of 1000–2000 cycles.

(v) Normally lead–antimony plates are used.

(vi) Most suffer from self-discharge problems and may lose up to 30% of rated capacity within several days under no-load conditions.

(viii) Those with lead–antimony grids can be expected to provide about 5–15 years of service in the absence of deep-cycle operation while lead–calcium grid batteries yield a slightly longer service life depending on temperature and other operating conditions.

4.2.1.3 *Stationary batteries*

(i) These are used for emergency stand-by power for communications, power systems, computer systems, emergency lighting, etc.

(ii) Commonly pure lead grids are used with low strength although also available with lead–antimony or lead–calcium plates.

(iii) They have a self-discharge rate.

(iv) They are designed for float service.

(v) Poor cycling performance but capable of occasional deep discharges.

4.2.1.4 *Photovoltaic batteries*

(i) These are designed to meet the specific requirements of terrestrial remote stand-alone PV applications and are optimized to provide the low rate (i.e. $C/500$) operation typical of these systems.

(ii) Large electrolyte volumes and/or high specific gravity (1.300 at full charge) may be employed to provide added protection against freezing and low maintenance requirements.

(iii) They are available with either pure lead or lead–calcium alloy grids.

(iv) They are either 'low-rate' PV batteries for remote daily shallow-discharge applications or 'medium-rate' PV batteries for PV power systems with on-site back-up or utility interface requiring frequent deep-cycle operation.

4.2.1.5 *Gelled electrolyte batteries*

(i) These are manufactured in sealed configuration.

(ii) Lead–calcium grids are employed.

(iii) The sulphuric electrolyte solution is immobilized by the use of additives. The battery is therefore spill proof.

(iv) Need for water addition is eliminated.

4.2.2 Grid alloy specifications

4.2.2.1 *Lead–antimony batteries*

Advantages

(i) Antimony adds strength to the lead.

(ii) They are capable of accepting fast charge rates and of delivering fast discharge rates.

(iii) They can be repeatedly deep-discharged to 50–80% of their capacity.

(iv) Lead–antimony grids adhere better to the PbO_2, resulting in less shedding of active material.

Disadvantages

(i) The battery life is short.

(ii) They have a high self-discharge rate (12% mol^{-1}).

(iii) Because of the high gassing rate, the battery begins to bubble (gas) early during recharge.

(iv) They need a high maintenance; the water level must be restored periodically.

(v) There is slumping of the positive plate.

(vi) 'Plating' occurs.

'Plating' (antimony in the positive plates transferred to and deposited on the negative plates) occurs under continuous charge–discharge cycles. Over a period of years, this results in reduced battery capacity and greater amounts of power are required for charging. The self-discharge rate also increases significantly with age.

4.2.2.2 Lead–calcium batteries

Advantages

(i) Calcium adds strength to the lead.
(ii) Unlike antimony, the presence of calcium reduces gassing and loss of water.
(iii) Low maintenance is required.
(iv) They can be 'maintenance free' if enough extra electrolyte has been added to the battery.
(v) They have a low self-discharge rate (1–3% $month^{-1}$).

Disadvantages

(i) There is poor charge acceptance after deep DOD.
(ii) The battery life is shortened if deep-discharged repeatedly. Usually 15–25% DOD is recommended.

4.2.2.3 Pure lead batteries

Advantages

(i) They have an extremely low self-discharge rate (1% $month^{-1}$).
(ii) A long life in float applications is obtained.

Disadvantages

(i) They are weak grids suitable only for stationary applications.
(ii) They cannot be discharged too deeply or recharged too fast.
(iii) There is a high initial cost.

4.3 LEAD–ACID BATTERY CONSTRUCTION AND CHEMISTRY

4.3.1 Battery construction

A cut-away view of a conventional lead–acid battery is shown in figure 4.1.

4.3.1.1 Battery components

(i) Plate grids (positive and negative plates).
(ii) Separators.
(iii) Elements.
(iv) Container.
(v) Vent plugs.
(vi) Cell connectors.

4.3.1.2 Plate Grids

The plate grids are the supporting framework for the active materials of the plates (figure 4.2). They can be made of either pure lead or an alloy of lead

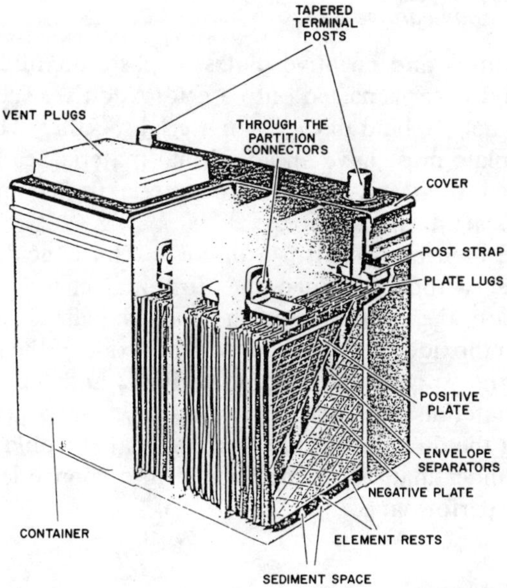

Figure 4.1 Cut-away view of a conventional battery.

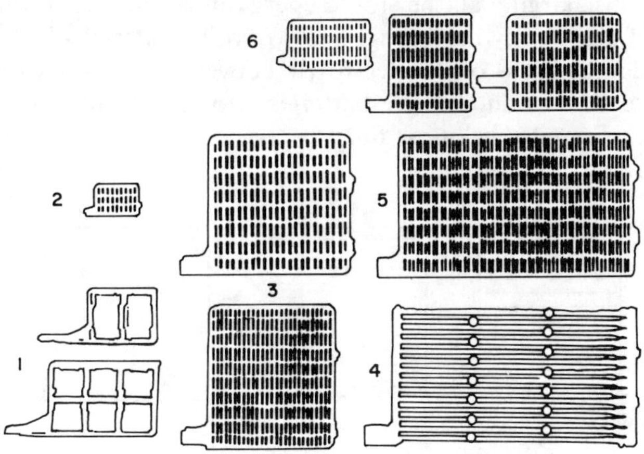

Figure 4.2 Various types of grid: 1, heavy grids for 'low-loss' cells; 2, miniature grid; 3, automotive battery grids; 4, grid for tubular positive plate; 5, grid for negative plate of tubular cells; 6, scooter and motorcycle battery grids.

and a material such as antimony which is added to strengthen and stiffen the soft lead. Batteries with grids containing calcium or strontium for strength have reduced gassing, and, consequently, have less water usage and low self-discharge rates.

4.3.1.3 Positive and negative plates

To form the positive and negative plates, a paste mixture of lead oxide, sulphuric acid and water is pasted onto a grid. Additives such as fibres may also be added to help to bind the active material together. The paste applied to the negative plate must have an additional material, called 'expanders', which are required to prevent the negative material from contracting back to a dense inactive state during service.

The plates are then immersed in a dilute sulphuric acid solution and a 'forming' charge is applied. When the 'forming' charge is applied in a particular direction, the lead oxide on the positive plate is electrochemically converted to lead dioxide (PbO_2) while the lead oxide of the negative plate is converted to a grey sponge lead. Both the PbO_2 and the sponge lead are highly porous materials that allow the electrolyte to penetrate the plates freely. Reversing the direction of the forming charge would cause the plate but not the expander material to be formed into sponge lead, resulting in impaired battery performance.

4.3.1.4 Separators

Separators placed between the positive and negative plates prevent a short circuit from discharging all the stored energy in the plates. A separator is a thin sheet of electrically insulating, finely porous material which allows the passage of charged ions of the electrolyte between the positive and negative plates (figure 4.3). In automotive batteries, separators are commonly made with resin-impregnated cellulose fibre.

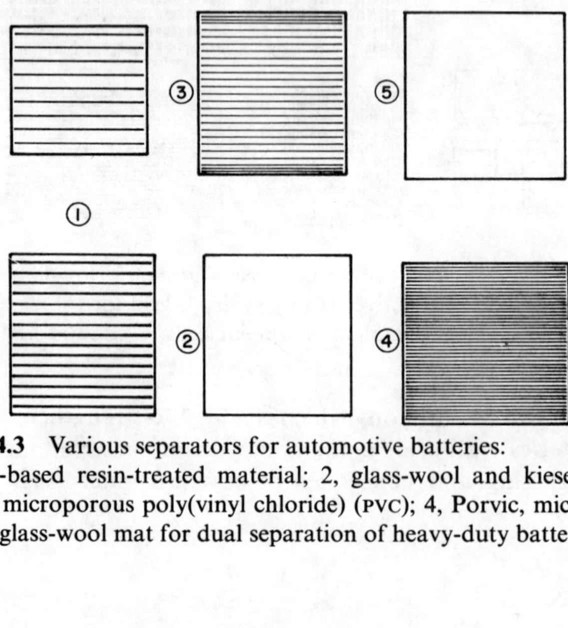

Figure 4.3 Various separators for automotive batteries: 1, fibre-based resin-treated material; 2, glass-wool and kieselguhr; 3, Porvic, microporous poly(vinyl chloride) (PVC); 4, Porvic, microporous PVC; 5, glass-wool mat for dual separation of heavy-duty batteries.

The essential features of a separator are as follows.

(*a*) High porosity ensuring low resistance to passage of current between the plates and free diffusion of acid.
(*b*) Good insulation to prevent metallic conduction between plates of opposite polarity.
(*c*) Inertness to the action of sulphuric acid and electrochemical oxidation.
(*d*) Absence of harmful impurities.

Suitable separator materials are the following.

(*a*) Microporous rubber.
(*b*) Microporous plastic.
(*c*) Paper base (resin treated).
(*d*) Certain kinds of wood with the most suitable being Port Orford Cedar or Douglas Fir.
(*e*) Glass-wool mats.
(*f*) Glass-wool and kieselguhr.

Envelope separators placed around either the positive or the negative plates rather than sheets are also used in some battery designs. They are closed on three sides with only the top open to allow gases generated on the plates to escape to the top of the cell.

4.3.1.5 *Elements*

Each cell of a battery contains an element assembled from placing one group of positive plates and one group of negative plates together with separators in between each positive and negative plate (figure 4.4). An element can contain

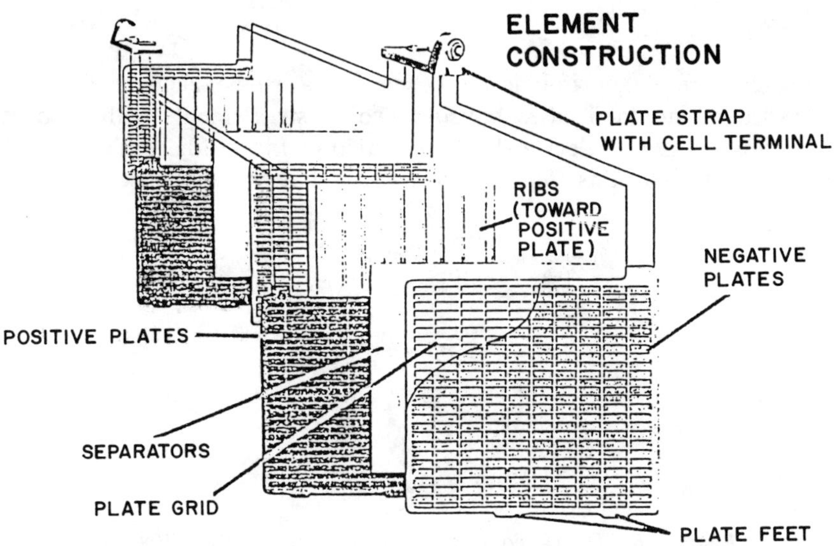

Figure 4.4 Element construction: conventional battery. (From BCI (1982).)

any size or any number of plates although the open-circuit voltage of a cell will remain approximately 2.1 V regardless of the number or size of the plates (figure 4.5). An increased total plate surface area per element due to either a greater quantity of plates or larger plates results in an increased voltage output during discharge at high rates. The cells are connected together in series by welding the post straps of one cell together with the post straps of the adjacent cell (positive group to the negative group of the next cell) and the voltage of the battery equals the sum of the voltages of the individual cells. The battery is not active until the electrolyte mixture of sulphuric acid and water is added.

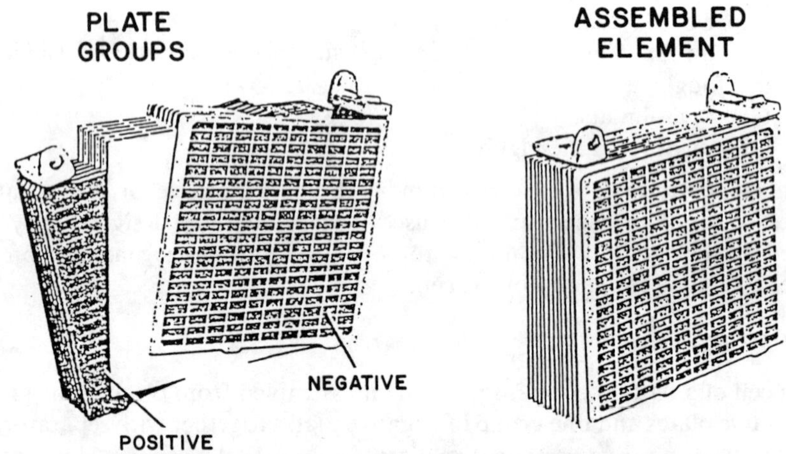

Figure 4.5 Battery element. (From BCI (1982).)

4.3.1.6 Containers

Containers are either individual boxes for a single element or multi-compartment moulded cases housing two to six elements with each cell comprising a separate electrical unit insulated from the adjacent cells by the compartment partitions (figure 4.6). A 12 V battery would consist of six

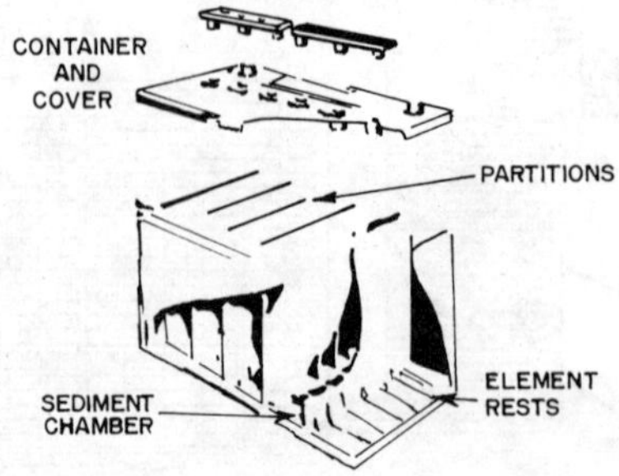

Figure 4.6 Typical container and cover. (From BCI (1982).)

elements housed in a six-compartment container. The plates rest on ribs or grids moulded into the base, thus providing a space for the accumulation of sediment. The sediment largely consisting of lead sulphate acts as an electrical conductor and should not be allowed to come in contact with the plates to avoid a short circuit. Portable battery containers are normally made of hard rubber, polypropylene or other plastic-like materials.

4.3.1.7 *Vent plugs*

Vent plugs fitted in the holes of the battery cover provide access to the battery interior for the purpose of adding electrolyte or water and for taking hydrometer readings. They are baffled to allow gases to escape freely while returning acid spray-back into the cell. Modern vent plugs with 'flame arresters' decrease the danger of accidental ignition of released gases and battery explosion by sparks or flame that enter the cell through the vent opening.

4.3.1.8 *Cell connectors*

Connectors are intercell straps made of antimonial lead or lead-plated copper that are burned or bolted to positive and negative posts of adjoining cells or elements. They are designed to be of low resistance and are capable of carrying currents equal to about $5C$, where C is the nominal ampere–hour capacity.

4.3.2 Battery chemistry

4.3.2.1 *Active chemical materials*

(i) Sponge lead (negative material).
(ii) Lead dioxide (positive material).
(iii) Electrolyte of sulphuric acid and water.

4.3.2.2 *Chemical reactions*

(i) Negative plate:

$$Pb + SO_4^{2-} \underset{\text{charge}}{\overset{\text{discharge}}{\rightleftarrows}} PbSO_4 + 2e^-.$$

(ii) Positive plate:

$$PbO_2 + SO_4^{2-} + 4H^+ + 2e^- \underset{\text{charge}}{\overset{\text{discharge}}{\rightleftarrows}} PbSO_4 + 2H_2O.$$

(iii) Net reaction:

$$PbO_2 + Pb + 2H_2SO_4 \underset{\text{charge}}{\overset{\text{discharge}}{\rightleftarrows}} 2PbSO_4 + 2H_2O.$$

4.3.2.3 *Battery discharge cycle*

When two unlike metals such as the positive and negative plates are immersed in sulphuric acid, a voltage develops between the plates (approximately 2.1 V per cell in a typical lead–acid battery). Electrical energy flows from the cell owing to the chemical reactions (figure 4.7) between the metals and electrolyte as soon as a circuit is connected between the positive and negative terminals, i.e. whenever a load is connected to the battery.

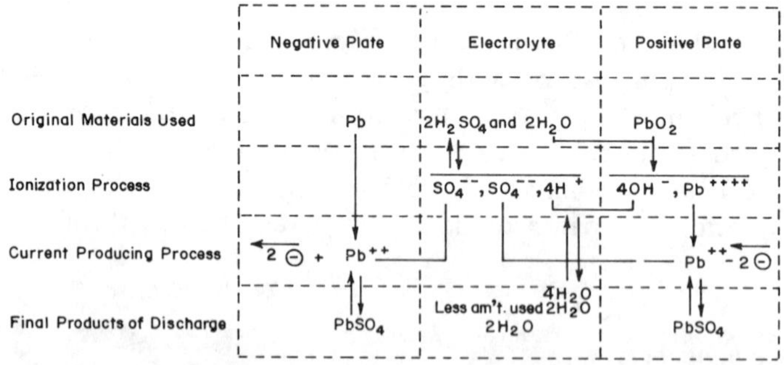

Figure 4.7 Discharge reactions.

During discharge when $PbSO_4$ is being formed at both the positive and negative plates, the sulphuric acid becomes diluted by the water formed at the positive plate, causing a decrease in the specific gravity of the electrolyte. The measurement of the battery specific gravity with a hydrometer is therefore an accurate and convenient method of determining the battery's state of charge (SOC). As the plates become more similar in composition (owing to build-up of $PbSO_4$ on the surfaces) and the acid becomes weaker, the voltage difference between the plates decreases. Eventually, the battery can no longer deliver electricity at a useful voltage and is said to be discharged.

When the battery is subjected to a high discharge rate, it becomes discharged more quickly than at a lower discharge rate. This is due to the fact that, at a high discharge rate, the acid circulation into the pores of the plates and the diffusion of water from the pores are too slow to sustain the discharge. Only the materials at or near the plate surfaces are able to take part in the chemical reactions and consequently only a small percentage of the electrolyte and active plate materials are utilized during the short duration of the discharge. Slow discharge rates, on the other hand, result in the consumption of most of the active material including that near the centre of the plates and most of the acid in the chemical reactions.

4.3.2.4 *Battery charge cycle*

During charge, a current is forced through the cell in a direction opposite to that of discharge causing the discharge reactions to reverse and active materials to be restored to their original state (figure 4.8).

The SO_4^{2-} ions leaving the plates combine with the hydrogen ions and are restored back to H_2SO_4, thus increasing the specific gravity of the electrolyte. Gas is evolved during charge, H_2 at the negative plate and O_2 at the positive plate, owing to the decomposition of water. Irreversible changes occurring during each charge–discharge cycle cause a gradual reduction in available cell capacity.

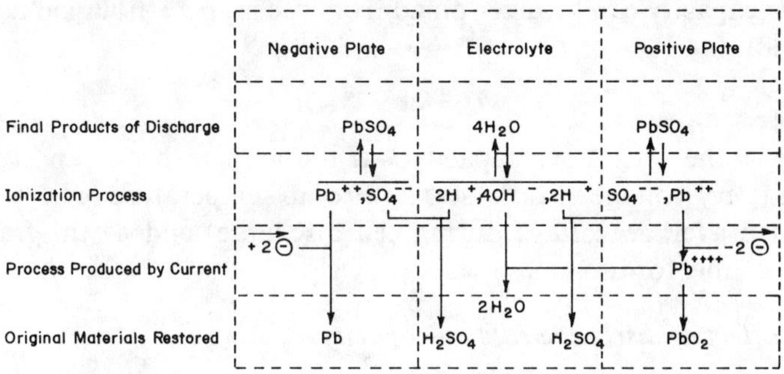

Figure 4.8 Charge reactions. (From Vinal (1955).)

4.4 LEAD–ACID BATTERY CHARACTERISTICS

4.4.1 Storage capacity

4.4.1.1 Ampere–hour capacity

The amphere–hour capacity is the quantity of discharge current available for a specified length of time valid only at a specific temperature and discharge rate. For example, a 12 V battery rated at 100 A h over 20 h can deliver 5 A h^{-1} for 20 h equivalent to 1.2 kW h of power (12 V × 100 A h). At a 5 h discharge rate, the same battery will deliver a maximum of 70 A h equivalent to 0.8 kW h of power. The discharge curve in figure 4.9 illustrates the

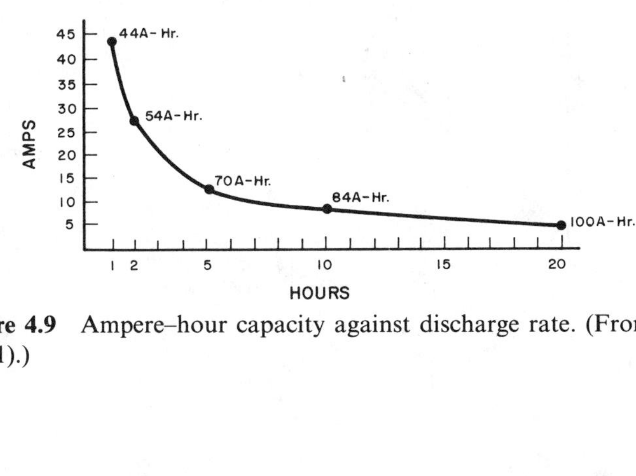

Figure 4.9 Ampere–hour capacity against discharge rate. (From Paul (1981).)

relationship between capacity and discharge rate for a typical 100 A h, 20 h rated, 12 V battery. Battery size, construction, temperature, concentration of electrolyte, plate history as well as temperature and discharge rate all determine battery capacity.

4.4.1.2 Watt–hour capacity

The watt–hour capacity or energy capacity is the time integral of the product of the discharge current and voltage from full charge to cut-off voltage. The watt–hour capacity can be determined from the ampere–hour capacity from the equation

$$\text{WC} = \text{AH} \times V \times \text{BF}$$

where WC is the watt–hour capacity, AH the ampere–hour capacity, V the average battery voltage at 60% SOC and room temperature (or 2 V per cell, the assumed average voltage during cell discharge) and BF the fraction of battery available for use.

4.4.1.3 Effect of discharge rate on capacity

A battery has a larger ampere–hour capacity at longer discharge rates because more time is available for the acid in the electrolyte to penetrate deeper into the battery plates. At high discharge rates, only a small percentage of the electrolyte and active plate materials are utilized because the acid circulation into the pores of the plates and diffusion of water from the plate pores are too slow to sustain the high-rate discharge during its short duration. On the other hand, deeper penetration of the acid into the plates during slow discharges shortens battery life. Available lead–acid battery capacity as a function of discharge rate for various types of battery is shown in figure 4.10.

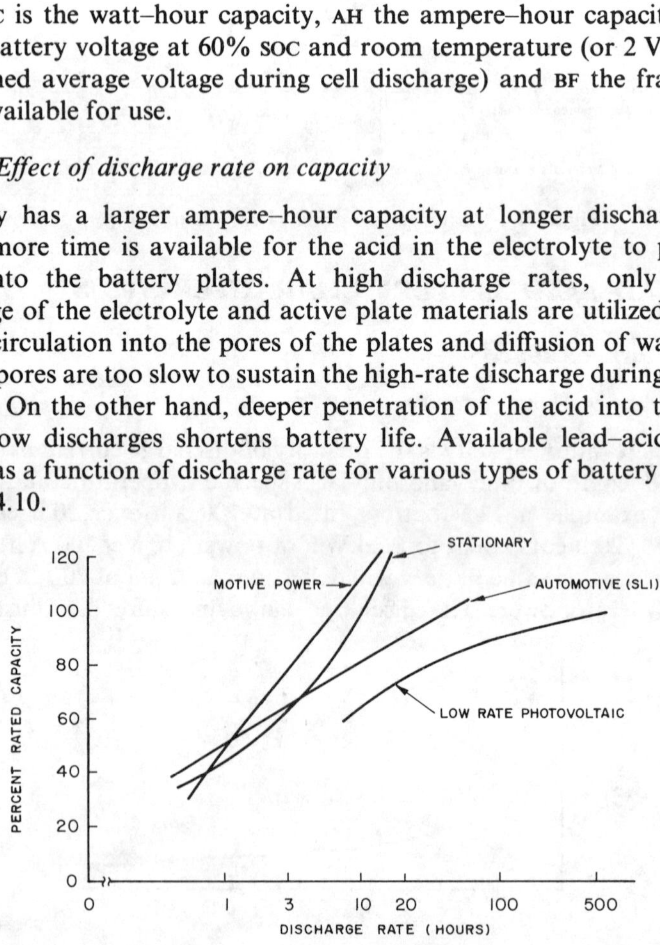

Figure 4.10 Available lead–acid battery capacity as a function of discharge rate.

4.4.1.4 Effect of temperature on capacity

A lead–acid battery's storage capacity decreases about 1% for every 1 °C drop in temperature. Lower temperatures result in less available capacity

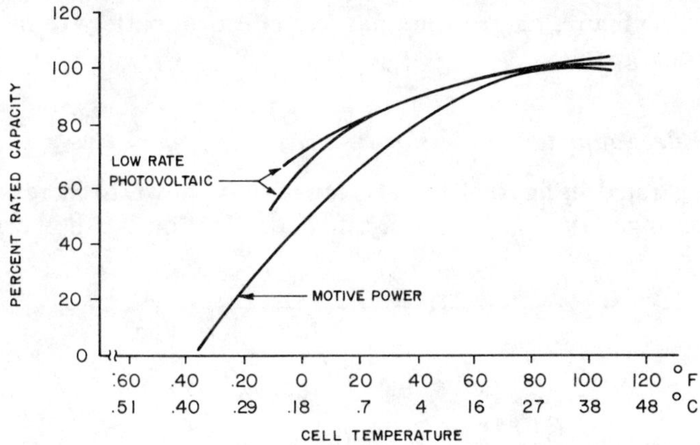

Figure 4.11 Lead–acid capacity as a function of temperature.

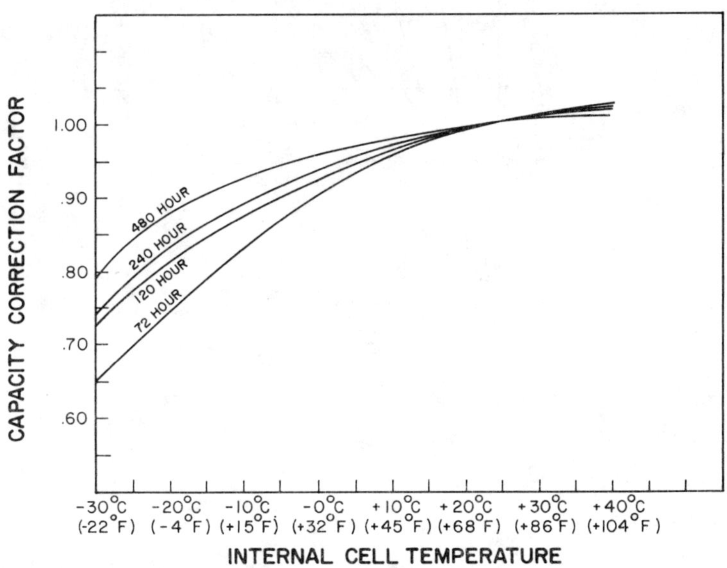

Figure 4.12 To correct for temperature variances (worst-case condition), multiply the battery's rated capacity at 25 °C (77 °F) by the correction factor above to yield the expected battery capacity at a given temperature (internal temperature). To use ambient air temperatures, average the temperature readings over 24 hour period and use the correction factors above.

owing to slower chemical reactions. There is therefore less penetration of ions into the plate material and less of the active material is available for reaction. The increase in the battery capacity with cell temperature is shown in figure 4.11. Capacity correction factors given in figure 4.12 are used to adjust the rated capacity at 25 °C for a different cell temperature. As can be seen from the figure, temperature has less effect on battery capacity at the longer discharge rates.

4.4.1.5 Effect of plate thickness on capacity

This is illustrated in figure 4.13. The capacity is shown to increase linearly with increasing plate thickness at a rate dependent on the discharge rate of the battery.

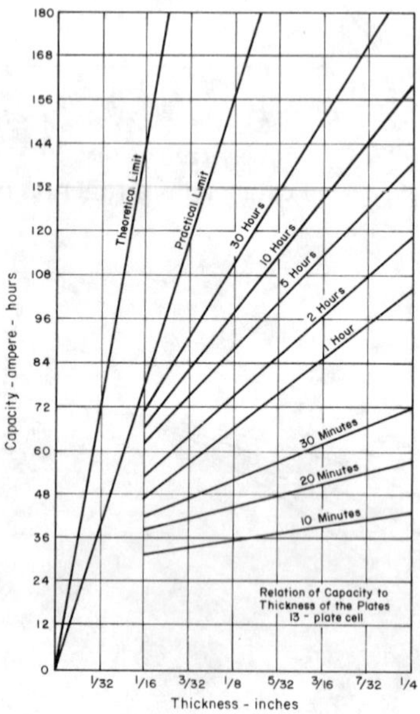

Figure 4.13 Dependence of capacity on the thickness of the plate (13-plate cell).

4.4.2 Cycle life

The cycle life of a battery indicates the number of cycles to a specified DOD, at a specified temperature and at a specified discharge rate until the battery's 'end of life' is reached.

The 'end of life' is defined differently by different manufacturers but is typically when the battery capacity falls below 20% of the rated capacity.

The DOD and battery temperature affect the battery life, as can be seen in figure 4.14. The life of the battery is enhanced by a shallower DOD. In the

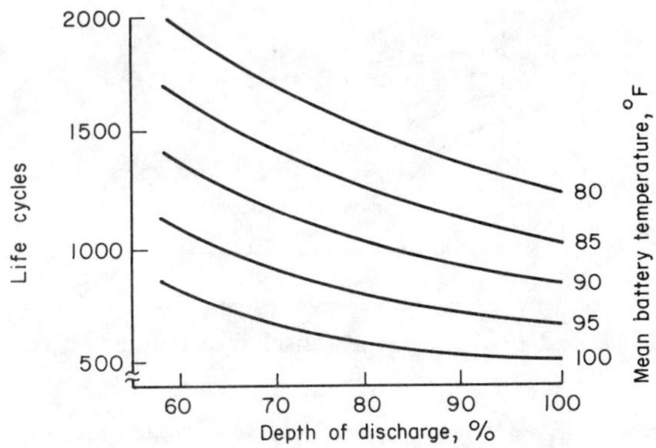

Figure 4.14 The effect of DOD and temperature on lead–acid battery life.

same manner, a deeper DOD shortens the battery's life owing to larger internal stresses resulting from a more complete utilization of active materials. The cycle life is seen to decrease with increasing battery temperature.

Longer discharge rates though increasing available battery capacity shortens battery life because of the deeper penetration of the acid into the plates. The standard for the cycle life of a battery in the industrial battery business is based on the 8 h discharge rate for shallow-cycle batteries and the 6 h discharge rate for deep-cycle batteries.

A shortened battery life is caused by a number of factors.

(i) External corrosion increases the interconnect resistance.
(ii) Grid (internal) corrosion reduces the physical size of current-carrying lead wires (grid) owing to the action of H_2SO_4. This action is faster towards the bottom of the plates because of the phenomenon called 'stratification'.
(iii) Stratification is caused by very low charging rates where the SOC is never high enough to cause gassing and resultant mixing of electrolyte.
(iv) Excessive gassing causes loss of electrolyte and plate damage. Loss of capacity may result because the electrolyte level goes below the top of the plates, thus preventing active materials from reacting there. Violent gassing may also physically damage plates by 'scrubbing' off active materials.

4.4.3 Charge and discharge voltages

4.4.3.1 Discharge curve

As can be seen from figures 4.15–4.18, the cell discharge voltage varies over a certain range with DOD as well as discharging rate.

Discharge voltage curves of actual cells may vary from the curves shown depending on the particular manufacturing techniques and materials used.

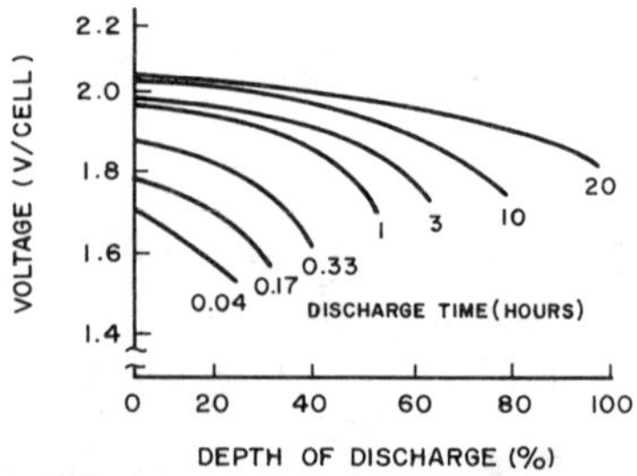

Figure 4.15 Lead–acid SLI battery discharge voltage. (From USDOE (1979).)

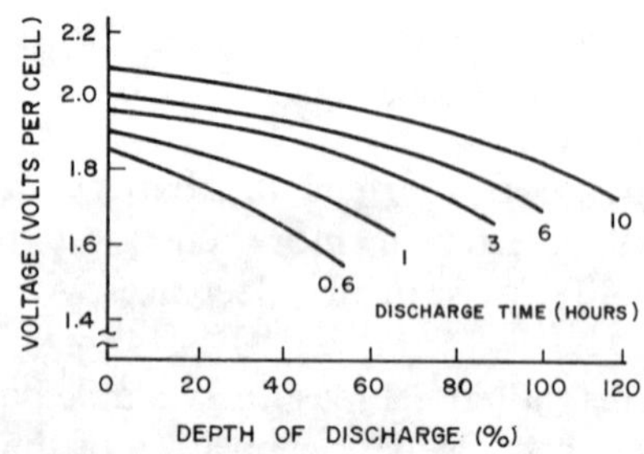

Figure 4.16 Lead–acid motive power battery discharge voltage. (From USDOE (1979).)

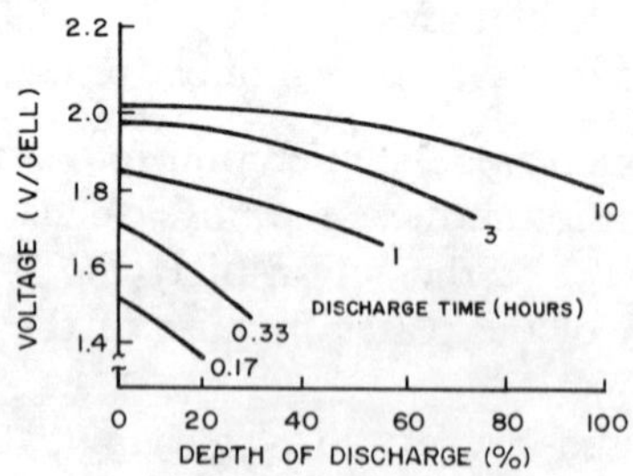

Figure 4.17 Lead–acid stationary battery discharge voltage. (From USDOE (1979).)

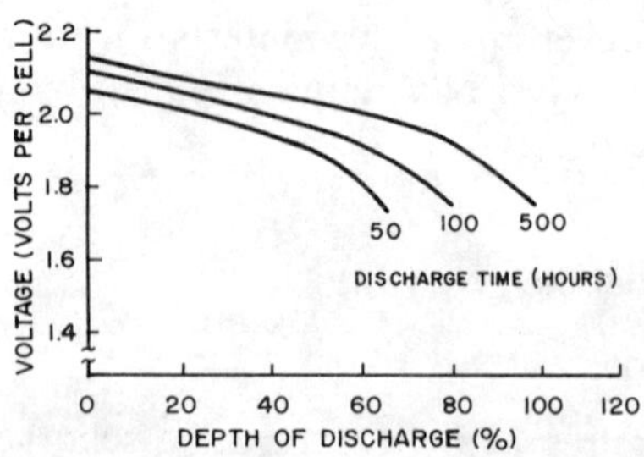

Figure 4.18 Lead–acid low-rate PV battery discharge voltage. (From USDOE (1979).)

To ensure maximum battery life, discharge past the nominal cut-off voltage of approximately 1.8 V per cell should be avoided.

Figures 4.15–4.18 illustrate typical discharge voltage profiles for SLI, motive power, stationary and PV lead–acid cells. The curves present cell terminal voltage as a function of DOD for various discharge rates at a constant operating temperature. For the low discharge rates, voltage falls gradually until a point referred to as the knee of the voltage curve is reached. Further discharge after this results in rapid voltage reduction. A discharge cut-off voltage is usually specified just pass the knee because little additional energy can be withdrawn from the cell beyond this point and further discharge can result in permanent damage. It can be seen that the knee of the curve occurs at higher states of charge for increasing discharge rates.

Discharge voltage is also affected by the operating temperature of the cell. Figure 4.19 presents cell voltage as a function of temperature at various DOD for a constant discharge rate. It can be seen that, at a given DOD, voltage decreases with decreasing temperature with the difference in voltage becoming much greater at deeper DODs.

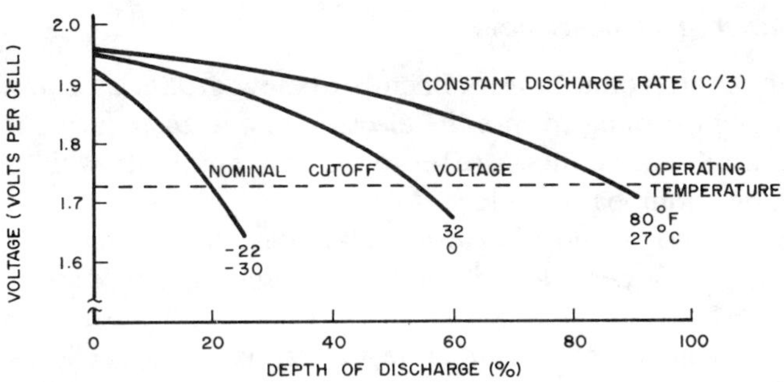

Figure 4.19 Lead–acid battery discharge voltage as a function of temperature.

4.4.3.2 *Charge curve*

The charge voltage curves (figure 4.20) as a function of the battery's SOC for different charging rates show that the charge voltage increases with both increasing SOC and increasing charge rate.

In region A of the curve, efficient charging takes place. The end of charge is being approached in region B and electrolysis begins. When the terminal voltage starts to climb rapidly at near 100% SOC in region C, excessive gassing and loss of electrolyte take place. This overcharging can cause the active material of the plates to loosen and flake off, reducing the capacity and life of the battery.

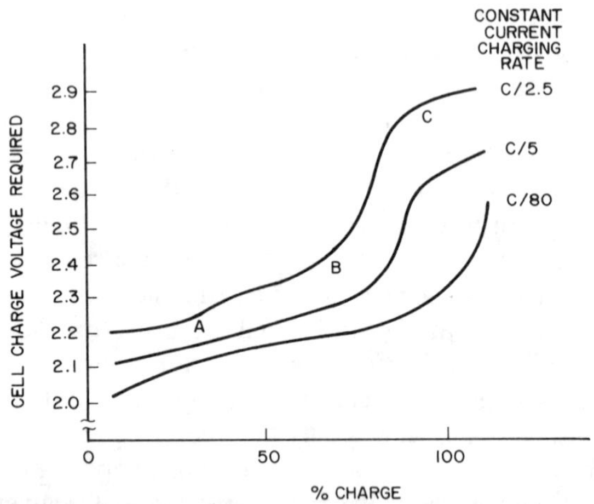

Figure 4.20 Typical charge voltage characteristic. (From USDOE (1950).)

4.4.3.3 Battery voltage window

The end-of-charge voltages recommended below are set at a level that will produce enough gassing to mix the electrolyte and to reduce stratification. The end-of-discharge voltages are set to maximize capacity while at the same time retaining a long battery life.

The recommended end-of-charge voltage is, for a shallow-discharged battery, 2.50–2.54 V per cell and, for a deep-discharged battery, 2.60–2.65 V per cell.

The recommended end-of-discharge voltage is, for a shallow-discharged battery, 1.90 V per cell and, for a deep-discharged battery, 1.70 V per cell.

The performance characteristics of a typical stationery lead–acid battery are shown in figure 4.21.

4.4.4 Specific gravity of the electrolyte

Specific gravity is a unit of measurement for determining the sulphuric acid content of the electrolyte. Specific gravity data presented in table 4.1 are based on an electrolyte temperature of 59 °F (15 °C). Since measured specific gravity values depend on the temperature of the solution, a compensation correction must be made using the following equation:

$$\mathrm{SG} = \mathrm{SG}_{59} + C(59 - T)$$

where SG is specific gravity at the desired temperature, SG_{59} the specific

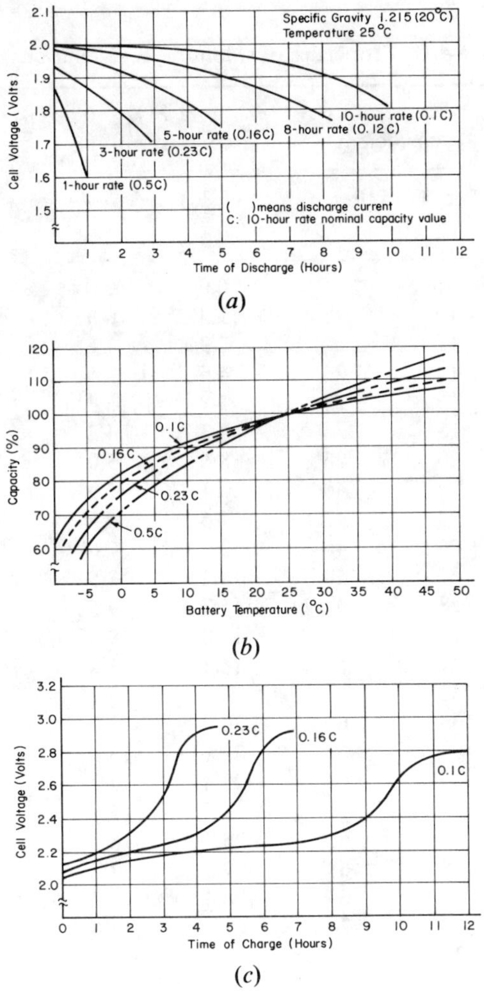

Figure 4.21 Performance characteristics of a typical stationary lead–acid battery: (*a*) discharge characteristics at various rates; (*b*) battery temperature and capacity at various rates; (*c*) charge characteristics at constant current.

gravity at 59 °F (15 °C), C (°F^{-1}) the temperature coefficient and T (°F) the temperature of the electrolyte.

Table 4.2 gives the ranges of specific gravity for four different types of lead–acid battery in both temperate and tropical climates.

Table 4.3 gives typical specific gravity values and battery open-circuit voltages for various SOCs. A fully charged specific gravity of 1.265 corrected to 80 °F (26.7 °C) is assumed.

Table 4.1 Properties of sulphuric acid solutions.

Specific gravity at 59 °F (15 °C)	Temperature coefficient (10^{-5})		H_2SO_4 concentration		Freezing point	
	Per °F	Per °C	(wt. %)	(vol %)	(°F)	(°C)
1.000	—	—	0.0	0.0	32	0
1.010	10	18	1.4	0.8		
1.020	12	22	2.9	1.6		
1.030	14	26	4.4	2.5		
1.040	16	29	5.9	3.3		
1.050	18	33	7.3	4.2	26	−3.3
1.060	20	36	8.6	5.0		
1.070	22	40	10.1	5.9		
1.080	24	43	11.5	6.7		
1.090	26	46	12.9	7.6		
1.100	27	48	14.3	8.5	18	−7.8
1.110	28	51	15.7	9.5		
1.120	29	53	17.0	10.3		
1.130	31	55	18.3	11.2		
1.140	32	58	19.6	12.1		
1.150	33	60	20.9	13.0	5	−15
1.160	34	62	22.1	13.9		
1.170	35	63	23.4	14.9		
1.180	36	65	24.7	15.8		
1.190	37	66	25.9	16.7		
1.200	38	68	27.2	17.7	−17	−27
1.210	38	69	28.4	18.7		
1.220	39	70	29.6	19.6		
1.230	39	71	30.8	20.6		
1.240	40	72	32.0	21.6		
1.250	40	72	33.4	22.6	−61	−52
1.260	40	73	34.4	23.6		
1.270	41	73	35.6	24.6		
1.280	41	74	36.8	25.6		
1.290	41	74	38.0	26.6		
1.300	42	75	39.1	27.6	−95	−71
1.310	42	75	40.3	28.7		
1.320	42	76	41.4	29.7		
1.330	42	76	42.5	30.7		
1.340	42	76	43.6	31.8		
1.350	43	77	44.7	32.8	−56	−49

Table 4.2 Specific gravity range for various types of lead–acid battery.

Battery type	Temperate climate		Tropical climate	
	Fully charged	Discharged	Fully charged	Discharged
SLI	1.260–1.280	1.120	1.210–1.240	1.080
Motive power	1.260–1.280	—	1.210–1.240	—
Stationary	1.200–1.225	—	1.200–1.225	—
Photovoltaic	1.300	1.075–1.130	1.225	1.040

Table 4.3 Typical open-circuit voltage and specific gravity values at various charge levels.

Charge level	Specific gravity	Voltage (12 V battery) (V)	Voltage (6 V battery) (V)
100%	1.265	12.7	6.3
75%	1.225	12.4	6.2
50%	1.190	12.2	6.1
25%	1.155	12.0	6.0
Discharged	1.120	11.9	6.0

When using a hydrometer to measure specific gravity, a temperature correction factor must be applied to any reading made when the electrolyte is not at 80 °F (26.7 °C) since hydrometer floats are calibrated to give a true reading only at one fixed temperature. Regardless of the reference temperature used as a standard a correction factor of 0.004 specific gravity is used for each 10 °F (5.5 °C) change in temperature, e.g. 0.004 added to the indicated hydrometer reading for each 10 °F increment above 80 °F and 0.004 subtracted for each 10 °F decrement below 80 °F.

As an example, for a temperature below 80 °F (26.7 °C), a hydrometer reading of 1.250 and an acid temperature of 20 °F (−6.7 °C), the corrected specific gravity is $1.250 - 0.024 = 1.226$.

Batteries prepared for use in tropical climates have slightly different electrolyte specific gravity values than batteries prepared for use in temperate or cold climates (table 4.4). Since ambient temperature never reaches the freezing point in tropical climates, a milder-strength electrolyte that does not cause the separators and grids to deteriorate as much as the higher-strength electrolyte is used for batteries in tropical climates, thus increasing the service life of the battery.

Table 4.4 Typical specific gravity values at various charge levels for temperate and tropical climates.

State of charge	Specific gravity used in cold and temperate climates	Specifivc gravity used in tropical climates
Fully charged	1.265	1.225
75% charged	1.225	1.185
50% charged	1.190	1.150
25% charged	1.155	1.115
Discharged	1.120	1.080

4.4.5 Freezing point of the electrolyte

Electrolytes with a low specific gravity have high freezing point temperatures with freezing point decreasing as the specific gravity and therefore the battery's SOC increases (figure 4.22). A fully charged battery can be stored at subfreezing temperatures while a battery approaching discharged condition freezes easily.

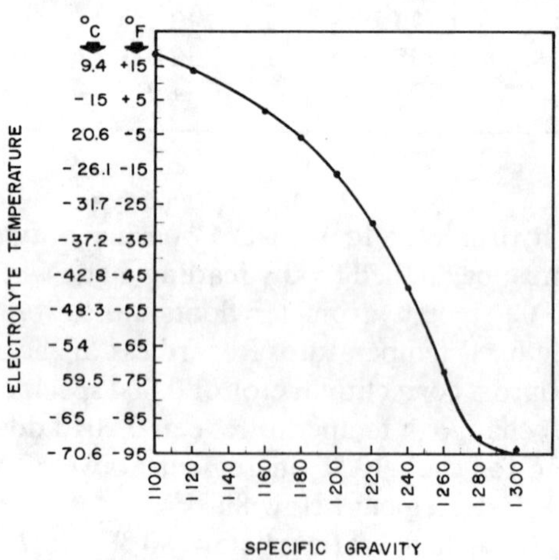

Figure 4.22 Electrolyte freezing point. (From BCI (1972).)

4.4.6 Battery self-discharge

Battery self-discharge is due to the consumption of a small amount of battery capacity by local chemical reactions. A battery will therefore slowly discharge while standing. Figure 4.23 shows how this discharge rate varies with temperature and composition of the plates; the self-discharge rate increases with increasing cell temperature with lead–antimony grids having much higher discharge rates than lead–calcium grids.

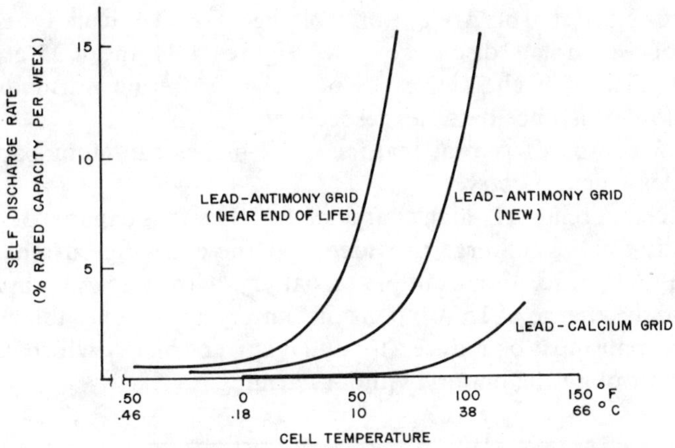

Figure 4.23 Lead–acid battery self-discharge rates. (From BCI (1972).)

4.5 ALKALINE BATTERIES

4.5.1 Nickel–cadmium batteries

4.5.1.1 Electrodes

(*a*) For the positive plate the active material is nickel hydroxide.
(*b*) For the negative plate the active material is a mixture of cadmium and iron.

4.5.1.2 Electrolyte

The electrolyte is a solution of potassium hydroxide.

4.5.1.3 Types

They are manufactured in two basic types.

(*a*) 'Sealed' types have a pressure relief valve built into the cell which prevents explosion if the cell becomes heated or overcharged by releasing the gas into the atmosphere. Some valves reclose after popping open while some stay open, allowing water to evaporate from the battery.
(*b*) Vented types have resealable vents that open and close under small pressure changes.

The electrolyte is a solution of potassium hydroxide in water. It is used only to conduct current between the plates and does not take part chemically in the charge–discharge reactions of the cell active materials.

The voltage output is approximately 1.2 V per cell and remains relatively stable until the battery is almost completely discharged after which the voltage drops off sharply. A cut-off voltage of 1 V is usually used as an indicator of a complete discharge state. Figures 4.24 and 4.25 give typical charge and discharge characteristics of nickel–cadmium normal-resistance and extra-low-resistance batteries respectively.

Table 4.5 compares typical features of a nickel–cadmium battery with those of a lead–acid battery.

It can accept a relatively high charge rate ($C/1$) and is capable of operation under continuous overcharge provided that the charging current does not exceed the $C/15$ rate. For example, a battery with a 15 A h capacity can initially stand a charge of 15 A but, as the battery approaches full charge, the charging current must be reduced to a final current of 1 A which can be run indefinitely through the battery without harming it.

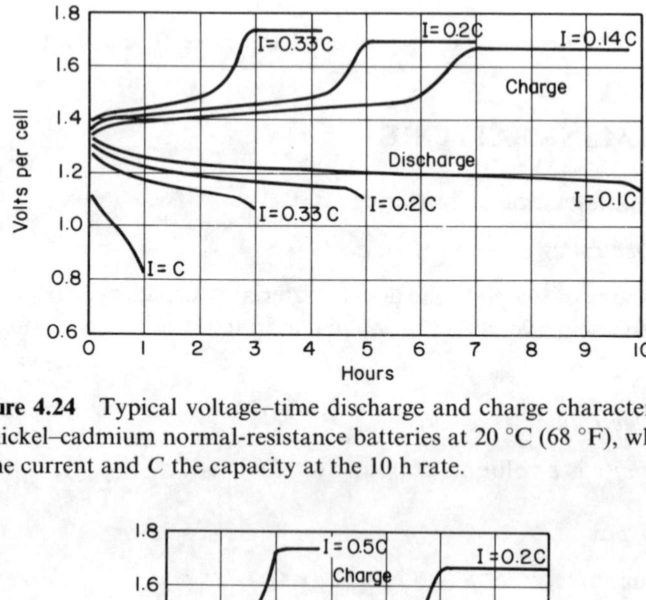

Figure 4.24 Typical voltage–time discharge and charge characteristics of nickel–cadmium normal-resistance batteries at 20 °C (68 °F), where *I* is the current and *C* the capacity at the 10 h rate.

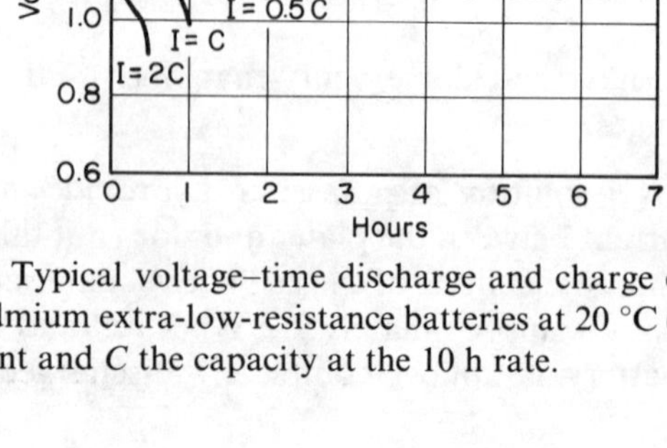

Figure 4.25 Typical voltage–time discharge and charge characteristics of nickel–cadmium extra-low-resistance batteries at 20 °C (68 °F), where *I* is the current and *C* the capacity at the 10 h rate.

Table 4.5 Comparative features of nickel–cadmium and lead–acid storage batteries. The nominal cycle life for lead–acid batteries is the currently available battery value for grid-connected PV systems (with frequent battery cycling). The 1979 cost is included for comparison only; actual prices fluctuate widely.

	Lead–acid	Nickel–cadmium
Type	Medium rate, deep discharge, lead–calcium grid	Medium rate, cycle service, vented pocket plate
Rated capacity at 77 °F, 8 h	100–900 A h per cell	10–400 A h per cell
Nominal discharge cut-off voltage	1.75 V per cell	1.0 V per cell
Nominal voltage	2.45 V per cell, varies with state of charge	1.25 V per cell, fairly constant with state of charge
Available capacity against temperature (% of rated capacity)	70% at 32 °F 20% at −20 °F	90% at 32 °F 65% at −20 °F
Nominal energy efficiency	70–80%	60–70%
Nominal cycle life for 80% discharge cycle	1000–1500	1500–2000
Nominal calendar life without cycling	10–20 years	24 years
Energy density	6–13 W h lb^{-1}	9–10 W h lb^{-1}
Required maintenance	Water replacement; charge equalization; protection against freezing and temperature extremes	Water replacement; occasional full discharge
Charge control	Sensitive to overcharging	Can accept 5–10% overcharge
Hazards	Electric shocks; hydrogen gas evolution; acid leaks or spills	Electric shocks
Cost (1979)	US $165–250 kW h^{-1}	US $300–1000 kW h^{-1}

4.5.1.4 *Chemical reactions*

Reactions in a nickel–cadmium cell are complex but the net result is the transfer of oxygen from the positive to the negative plate during discharge and vice versa during charge. No measurable change in electrolyte occurs as it does not react with either plate. The equation for the complete cell can be written as

$$2Ni(OH)_3 + Cd \underset{\text{discharge}}{\overset{\text{charge}}{\rightleftarrows}} 2Ni(OH)_2 + Cd(OH)_2 .$$

4.5.1.5 *Advantages*

(i) They have a long life and reduced maintenance.
(ii) They are capable of being deep-discharged without incurring any damage to the battery.
(iii) Performance is much less affected by temperature changes than is the performance of the lead–acid battery.
(iv) A voltage regulator is not necessary when a nickel–cadmium battery is used in a solar PV system.
(v) They have excellent charge retention and a high capacity at low temperatures.

4.5.1.6 *Disadvantages*

(i) Cost per ampere hour of capacity is high.
(ii) They display a 'memory' of battery discharge history. For example, a nickel–cadmium battery that has undergone repeated discharges down to 25% DOD will start to behave as though it has been discharged completely when this SOC is reached, i.e. the battery voltage will drop off sharply at 75% SOC as if it has only a quarter of its true capacity. Prolonged periods of overcharge also have this effect.

4.5.2 Nickel–iron batteries

4.5.2.1 *Electrodes*

(*a*) For the positive plate the active material is nickel hydrate.
(*b*) For the negative plate the active material is iron oxide.

4.5.2.2 *Electrolyte*

The electrolyte is a solution of potassium hydroxide.

4.5.2.3 *Chemical reactions*

The reactions consist of the transfer of oxygen from one plate to another, oxygen being taken from the positive plate and added to the negative plate during discharge and the reverse taking place during charge. The electrolyte does not as a whole change in composition or density although changes in electrolyte in the pores of the plates do occur.

The electrolyte is a solution of potassium hydroxide in water to which a small amount of lithium hydroxide has been added. There is no real difference in specific gravity during charge and discharge although a gradual weakening of the electrolyte accompanied by a decrease in capacity of the battery does occur until renewal is required.

The chemical reactions of the alkaline battery are rather complex. The charge and discharge reactions given in figure 4.26 (prepared by the Edison Storage Division of Thomas A Edison) probably do not represent the complete reactions.

The average charge and discharge voltages are as given in table 4.6.

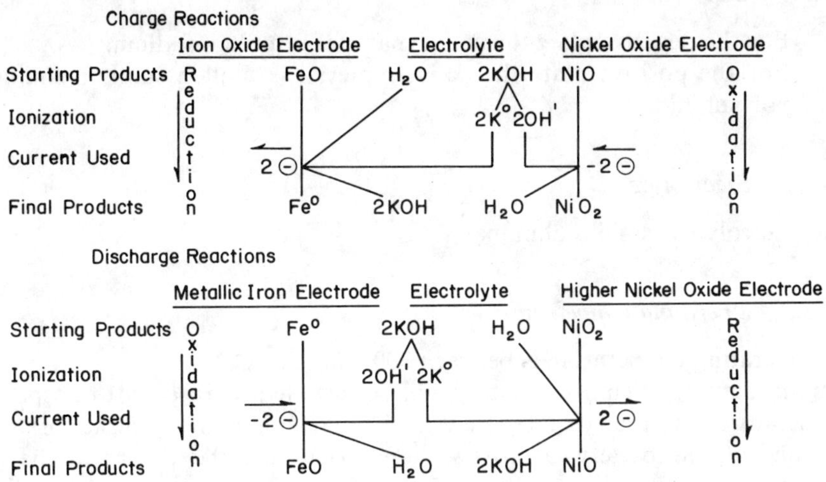

Figure 4.26 Charge and discharge reactions of an Edison cell.

Table 4.6 Discharge and charge voltages per cell.

	Nickel–cadmium	Nickel–iron
Discharge voltage (normal rates, 2–10 h)	1.2 V	1.2 V
Charge voltage (normal rates, 7–10 h)		
Start	1.4 V	1.5 V
Finish	1.7 V	1.84 V

4.5.2.4 Advantages

(i) They have great mechanical strength.

(ii) They are able to stand idle at normal temperatures for long periods in various SOCs with negligible loss of charge and with no deterioration.

(iii) Overcharge at normal charge rates does not affect battery life assuming that the occasional maximum temperature does not exceed 46 °C (116 °F).
(iv) They have a long life (8 years or more) when correctly maintained.

4.5.2.5 Disadvantages

(i) A large plate size is needed for a given storage capacity, resulting in a high weight to ampere–hour capacity ratio.
(ii) They have poor charge retention and low capacity at low temperatures.

4.5.3 Sodium–sulphur batteries

4.5.3.1 Electrodes

(*a*) For the negative plate the active material is molten sodium.
(*b*) For the positive plate the active material is molten sulphur–sodium polysulphide.

4.5.3.2 Electrolyte

The electrolyte is solid β-alumina.

4.5.3.3 Operating temperature

The operating temperature is between 300 and 350 °C.

In an actual system, the battery will be installed in an oven to keep the temperature at the required level. The reaction mechanism of the sodium–sulphur battery, a cross-sectional view and the charge–discharge characteristics of a cell are given in figures 4.27, 4.28 and 4.29, respectively.

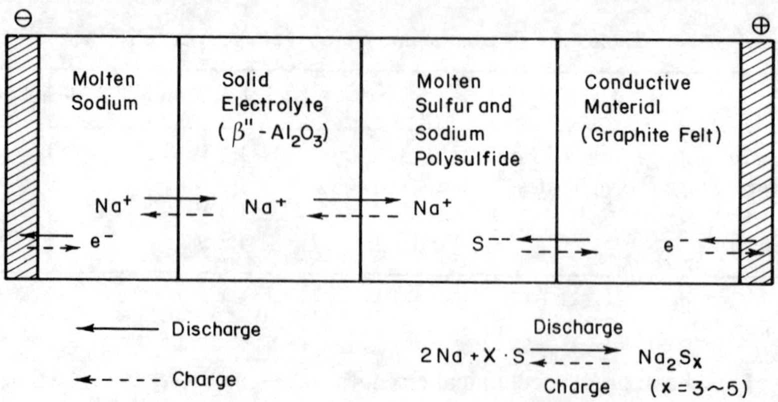

Figure 4.27 Reaction mechanism of sodium–sulphur battery.

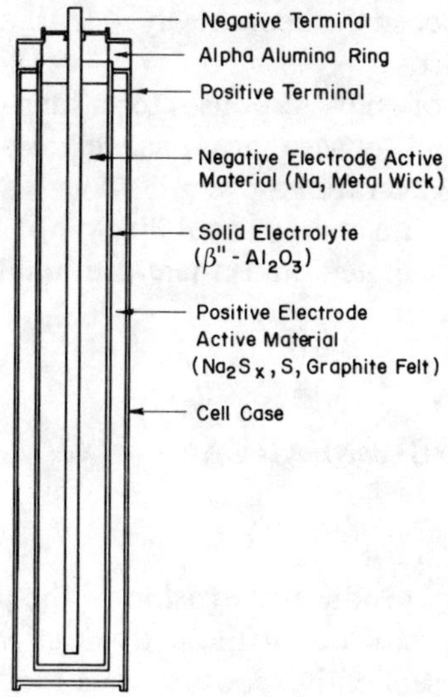

Figure 4.28 Sectional view of sodium–sulphur battery.

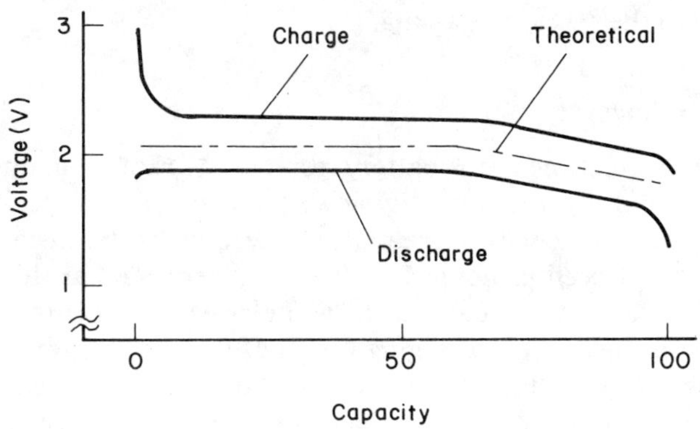

Figure 4.29 Charge–discharge characteristics of sodium–sulphur battery.

4.5.3.4 Advantages

(i) There is no electrode degradation due to shape change.
(ii) They have a long life.
(iii) Maintenance is low.
(iv) No self-discharge occurs.
(v) The battery is completely sealed and therefore no toxic or flammable fumes are emitted.
(vi) The energy efficiency is approximately 85%.

(vii) A high energy density (theoretically 4.3 times that of a lead–acid battery) is obtained.
(viii) They are capable of quick responses to sudden changes from charge to discharge condition (or vice versa) as well as sudden changes in load demand or solar generating power.
(ix) They have a high rate charge capability.
(x) The system is insulated and therefore unaffected by changes in ambient temperature.

4.6 BATTERY PROBLEM AREAS

4.6.1 Overcharging

Overcharging the battery produces corrosion of the positive grids as well as excessive gassing which can loosen the active material in the plates. This loosened material is subsequently deposited as a fine brown sediment at the bottom of the cell between the separators and the plates. Overcharging may also increase the battery temperature to the point of being destructive to the plates and separators. Frequent replacement of lost water also becomes a necessity due to the excessive gassing.

4.6.2 Undercharging

Consistent undercharging of the battery results in a gradual running down of the cells indicated by progressively lower values of the specific gravity readings and by the tendency of the plates to become lighter in colour. When undercharging has been prolonged, sediment is deposited at the bottom of the cells. This sediment is usually a fine white powder consisting principally of lead sulphate. Insufficient charging is one of the most common causes of buckling of the plates. This is due to the strain on the plates because the lead sulphate occupies more space than the original active material does on the plates.

4.6.3 Corroded terminals

The products of corrosion are non-conducting substances which form a layer between the battery terminal and the clamp connector connected to it. This film may not ordinarily interfere with the charging of the battery or with discharging at low discharge currents but will increase resistance to the passage of large currents to the load. The corroded terminals should be cleaned with a dilute solution of ammonia to neutralize the acid and coated with a high-temperature grease or Vaseline. A wire brush can be used to remove dirt, corrosion or rust from battery parts. A terminal cleaning brush is also available to clean tapered posts and the mating surfaces of the cable clamps.

4.6.4 Short circuits

Short circuits may be caused by a breakdown of one or more of the separators, by excess accumulation of sediment at the bottom of the cells or by the formation of tree-like structures of lead from the negative to the positive plates. The latter phenomenon called 'treeing' results from two basic causes.

(i) The presence of certain materials in the grids, such as cadmium, causes the growth of trees at the side or bottom of the element (the tendency to treeing, however, is counteracted by the presence of antimony).
(ii) A process called 'mossing' in which sediment brought to the surface of the electrolyte by gas settles on the top of the plates and bridges over the tops of the separators occurs.

Evidence of short circuits within the battery is as follows.

(i) Continued low readings of specific gravity even though the battery is receiving the normal amount of charge.
(ii) Rapid loss of capacity after a full charge.
(iii) Low battery open-circuit voltage.

4.6.5 Sulphation

Normal sulphate which forms as a natural part of the process of discharge is finely crystalline and easily reduced by the charging current. The term 'sulphation' refers to a condition where large crystals of lead sulphate grow on the plates in place of the tiny crystals normally present. The large crystals tend to increase the internal resistance of the cell; this results in lower discharge and high charge voltages, thereby lowering the voltage efficiency. A heavily sulphated battery is difficult to recharge and may become permanently damaged by plate fracture due to crystal growth.

Sulphation will occur under the following conditions.

(i) When a completely or partially discharged battery remains unused for any length of time.
(ii) When a battery operates at partial SOC for several days without a finishing or equalizing charge.
(iii) When there are temperature variations in the battery. Sulphation is believed to be at least partially caused by the increase in solubility of lead sulphate at higher temperatures in the electrolyte. The small lead sulphate crystals are dissolved during the high-temperature periods and are slowly recrystallized into large crystals when the temperature is

reduced. Electrolyte temperature cycling can be caused by ambient temperature changes or by heat generated within the battery during charge or discharge.

A sulphated battery may be partially returned to its original condition by following a carefully controlled recharging procedure. Sulphation can be avoided by bringing all batteries up to 100% SOC regularly known as an 'equalization charge' and by minimizing the time of exposure to low-state-of-charge conditions.

4.6.6 Electrolysis and water loss

Electrolysis is the conversion of water into hydrogen and oxygen gas. When the battery reaches full charge, at which time the lead sulphate on the plates and the lead sulphate ions in the electrolyte become exhausted, the rise in the plate potential beyond a certain cut-off voltage causes decomposition of water to take place. Reducing the charging rate at this time reduces gassing by reducing the voltage of the battery. Gassing begins when the terminal voltage of the battery reaches about 2.3 V per cell and the quantity of gas formed depends on the portion of current not absorbed by the battery.

Reactions that occur during electrolysis are, at the negative plate,

$$2H^+ + 2e^- \rightarrow H_2(g)$$

and, at the positive plate,

$$SO_4^{2-} + H_2O \rightarrow H_2SO_4 + 2e^- + \tfrac{1}{2}O_2(g).$$

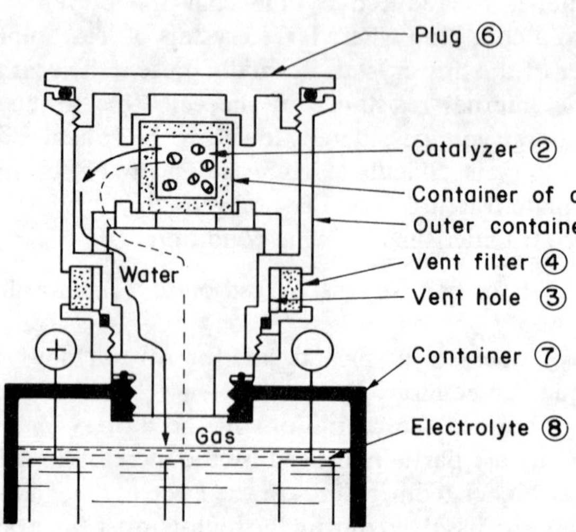

Figure 4.30 Reactor plug–catalyst plug for storage battery.

The net result is

$$H_2O \rightarrow H_2(g) + \tfrac{1}{2}O_2(g).$$

To minimize the water loss, the following approaches can be taken.

(i) Apply a two-step charging process; a fast charge rate followed by a tapering or finishing charge.
(ii) Use a battery protector.
(iii) Limit charge rate to less than the $C/50$ rate.

A reactor plug–catalyst plug is shown in figure 4.30.

4.7 BATTERY MAINTENANCE AND SAFETY PRECAUTIONS

4.7.1 Maintenance

Clean off corrosive slime (escaped sulphuric acid) which has leaked and coated the top of the battery. The extremely corrosive slime must be washed away periodically as it is conductive and can cause leakage currents to discharge in the battery. Wash batteries first with water then with water mixed with a small amount of dissolved sodium bicarbonate and finally with water again.

Add water occasionally to all but sealed batteries. Distilled water is best as tap water contains small amounts of impurities such as calcium and chlorine. A battery should not be overfilled above the level indicator as it will pump acid when it is placed on charge. If there is no level indicator, add water only up to $\frac{1}{2}$ in (13 mm) above the tops of the separators.

Check battery charge condition with a good hydrometer. For sealed batteries, the SOCs can be determined by measuring the battery open-circuit voltage. As the open-circuit voltage is hard to measure (voltage variation from full charge to discharge may be less than 0.2 V), the battery's SOC can also be measured by measuring the battery terminal voltage while the battery is being discharged at the 5–6 h rate (i.e. the rate that completely discharges battery within 5–6 h). Voltage readings taken this way are accurate until 20% SOC (80% DOD) is reached at which point voltage drops off rapidly.

Charge equalization should be carried out periodically (once a month or more depending on the battery) to bring all batteries up to 100% SOC. An equalizing charge is a prolonged charge at the finishing rate or less and can be terminated when the hydrometer and voltage readings have remained constant for several hours (usually 3 h) provided that the charging rate has been held constant during that period.

Overcharge a battery periodically to create a gassing condition in order to prevent stratification from occurring.

4.7.2 Safety precautions

4.7.2.1 Danger of electrolyte spillage

Direct contact with the electrolyte can cause severe burns to the skin and permanent damage to the eyes. When handling, checking, filling, charging or repairing a battery, a rubber apron, gloves, boots and face mask should be worn. Protective clothing and goggles should be worn when mixing acid and water as a violent reaction with considerable release of heat may occur when water is added to high-specific-gravity acid. When preparing an electrolyte of a desired specific gravity, always add acid carefully to water while stirring constantly. Water should never be poured into the acid and use of sulphuric acid solutions of more than 1.400 specific gravity should be avoided.

A fresh-water emergency shower or fountain should be provided within the vicinity of the battery bank. Sealed battery cells should be equipped with pressure release vents. If electrolyte is splashed on clothing or the body, it should be immediately neutralized with a solution of baking soda and water and rinsed off with clean water. If electrolyte is splashed into the eye, flood the eye with clean cool water for about 5 min. If acid is taken internally, large quantities of water or milk followed by milk of magnesia, beaten egg or vegetable oil should be taken.

4.7.2.2 Danger of hydrogen explosion

Hydrogen and oxygen gas liberated during the charge cycle may form an explosive mixture if allowed to accumulate in an unvented room. The battery room or area should be well ventilated and flame-producing or spark-producing devices should be kept out of the battery area or room. A vented cell equipped with a flame arrestor will greatly reduce the likelihood of explosion of the cell caused by accidental gas ignition although even the most modern vent cap designs do not provide complete protection.

4.7.2.3 Damage to battery due to improper handling or storage

A battery should be lifted with hands placed at opposite corners or a battery carrier should be used as excessive pressure on the end walls can cause electrolyte spillage through the vents. Batteries should not be placed on top of each other but should be placed on tiered racks. To prevent danger from freezing of the electrolyte, batteries should be stored in a fully charged condition with the electrolyte up to the level indicator in all cells (wet batteries) or up to about 0.5 in (13 mm) above the tops of the separators. A three-quarters-charged battery is in no danger of freezing. Batteries should be stored in a cool dry place away from hot-air ducts or radiators and shielded from direct sunlight in the summer to minimize the extent of self-discharge. Batteries should not be stored for long periods of time without recharging as the formation of lead sulphate crystals in the plates may cause permanent damage.

4.7.2.4 Facility damage

Spillage of electrolyte on benches, battery terminals, racks, floors and other structures in the room can cause corrosion. Fumes and fine spray of dilute acid given off by the batteries are also very corrosive particularly to metal work. Spilled or leaked electrolyte should be cleaned up with a neutralizing solution such as a mixture of baking soda and water (1 lb of baking soda to 1 gal of water). A corrosion-resistant coating on all structures in danger of coming into contact with electrolyte is recommended.

4.7.2.5 Danger of electrical shock

Danger of electrical shock during inspection, maintenance or servicing of the batteries exists if the terminal voltage of the battery bank is greater than 50 V. Batteries should be installed in groups with a total voltage of not more than 250 V on any one rack with insulation placed between racks. A safety ground disconnect circuit should be provided to allow the battery to 'float' (i.e. the positive and negative terminals of a high-voltage string are disconnected) while any work is being done on the batteries.

4.8 LEAD–ACID BATTERY MODELLING

4.8.1 Battery discharge and charge voltages

Warning The following modelization is only indicative. The equations are derived empirically from manufacturer's data and should not be applied carelessly to any given type of battery operating under transient state conditions as is generally the case for solar systems.

4.8.1.1 Discharge voltage

The current and voltage during discharge can be described in terms of the SOC of the cell (ranging from 0 to 1.0) by the equation

$$V = V_{\mathrm{r}} - \frac{I}{\mathrm{AH}}\left(\frac{0.189}{\mathrm{SOC}} + R_{\mathrm{i}}\right)$$

where SOC is the ratio of the charge at the time of interest to the maximum charge, as measured for the 500 h discharge rate. The other symbols are defined as follows: V_r(V) = 2.094[1.0 − 0.001(T − 25 °C)] is the rest voltage, V (V) is the terminal voltage, I (A) is the current, AH is the ampere–hour rating of the battery for the discharge rate, R_i (Ω) = 0.15[1.0 − 0.02(T − 25 °C)] is the internal resistance of the cell and T (°C) is the ambient temperature. The factor of 0.189 represents the internal resistance due to polarization.

4.8.1.2 *Charge voltage*

During the charging period, the current–voltage relationship is given by

$$V = V_r + \frac{I[0.189/(1.142 - \text{SOC}) + R_i]}{\text{AH}} + (\text{SOC} - 0.9)\ln\left(300\frac{\text{I}}{\text{AH}} + 1.0\right).$$

Note that the last term is included only if the first two terms sum to more than 2.29 V.

4.8.2 Decrease in state of charge

The decrease in the SOC due to nominal self-discharge under the no-load condition (lead–calcium) is given by

$$\text{SOC} = \text{SOC}_0 \exp(-Kt)$$

with

$$K = 300\exp\left(-\frac{4400}{T}\right)$$

where SOC_0 is the initial SOC with T in kelvin, t in hours and K in reciprocal hours. At room temperature, K = 0.0001.

4.8.3 Water loss

Water loss (mm^3 of volume) = 0.336 × (A h of overcharge) + evaporation.

4.8.4 Battery life before corrosion sets in

The battery life is shortened by the following.

(i) Corrosion of the terminals.
(ii) Corrosion of the grid.
(iii) Growth of large lead sulphite crystals.

The life before corrosion sets in is given by

$$\text{life (years)} = (\text{life at } 25\ ^\circ\text{C}) \exp\left[-5070\left(\frac{1}{T} - \frac{1}{298}\right)\right]$$

with the temperature T in kelvin.

4.8.5 Cycle life

The cycle life is affected by the following phenomena.

(i) Shortening by dendrite growth.
(ii) Shortening by sediment at the bottom of the plates.
(iii) Flaking due to vibrations.
(iv) Flaking due to differential expansion.
(v) Dirt.
(vi) Non-uniform plate growth.

The cycle life is given by cycle life $= 9000 \exp[-(1.0 - \mathrm{SOC_{min}})]$.

4.8.6 Cycle life as a function of temperature and depth of discharge

The equation for determining cycle life from temperature and DOD is given below. Usually the manufacturer can supply data on cycle life as shown in figure 4.14. From such a graph or from tabulated performance data, it is possible to obtain a set of three data points of the following form:

$$T_1,\ \mathrm{DOD}_1,\ \mathrm{CL}_1$$
$$T_1,\ \mathrm{DOD}_2,\ \mathrm{CL}_2$$
$$T_2,\ \mathrm{DOD}_3,\ \mathrm{CL}_3$$

where T_n, DOD_n and CL_n represent the battery temperature, DOD and cycle life respectively at day n. These data points can then be utilized to determine values for the parameters a, b and c in an equation of the form

$$\mathrm{CL} = (b - cT)\exp(-a\mathrm{DOD}).$$

Solving for a, b and c yields

$$a = \frac{\ln(\mathrm{CL}_1/\mathrm{CL}_2)}{\mathrm{DOD}_2 - \mathrm{DOD}_1}$$

$$b = \frac{T_2\mathrm{CL}_1\exp(a\mathrm{DOD}_1) - T_1\mathrm{CL}_3\exp(a\mathrm{DOD}_3)}{T_2 - T_3}$$

$$c = \frac{\mathrm{CL}_1\exp(a\mathrm{DOD}_1) - \mathrm{CL}_3\exp(a\mathrm{DOD}_3)}{T_2 - T_1}.$$

To compute the battery life when the DOD varies from cycle to cycle, the following expression is evaluated:

$$L = \left(\sum_{n=1}^{365} \mathrm{CL}(T_n, \mathrm{DOD}_n)^{-1}\right)^{-1}$$

where L is the predicted battery life in years and CL is the battery life in cycles for the battery temperature T_n in degrees Celsius at day n and battery DOD_n at day n.

REFERENCES

BCI 1982 *Battery Service Manual* (Battery Council International)
Paul T D 1981 *How to Design an Independent Power System: Best Energy Systems*
USDOE 1979 *Handbook for Battery Energy Storage for PV Power Systems* (Washington, DC: US Department of Energy)
USDOE July 1980 *Solar PV Applications Seminar: Design, Installation and Operation of Small, Stand-alone PV Power Systems* (Washington, DC: US Department of Energy)
Vinal G W 1955 *Storage Batteries* 4th edn

BIBLIOGRAPHY

Bode H 1977 *Lead–Acid Batteries*
Buresch M 1983 *PV Energy Systems: Design and Installation*
Calloway T M 1986 *Design of an Intermediate-Sized, Autonomous PV–Diesel Power Plant* (Albuquerque, NM: Sandia National Laboratories)
CIRDC 1980 *Compatible Battery for Solar Electric Power Systems Application* (Chloride India Research and Development Centre)
Macomber H L and Ruzek J B 1981 *Photovoltaic Stand-Alone Systems: Preliminary Engineering Design Handbook* (Washington, DC: US Department of Energy)

5 Electronic regulation

5.1 INTRODUCTION

Stand-alone photovoltaic (PV) systems for small or intermediate applications are totally self-sufficient in generating, storing and supplying either DC or AC electricity to local electrical loads.

The management of the energy balance within the system is performed through the regulation process.

PV systems involve a functional circuit composed of different stages.

(i) Conversion.
(ii) Storage.
(iii) Load consumption.
(iv) Regulation.

These systems are subjected upstream to the variable solicitation of the solar energy availability and ambient temperature level and downstream to the load consumption pattern which may vary according to the user's behaviour. Consequently they operate under transient state conditions. Moreover, they operate under technical constraints which may be as follows.

(*a*) The priority given to an auxiliary energy source for some time of the day.
(*b*) The prevention of overcharging or overdischarging conditions of a battery.
(*c*) The starting conditions of a motor, etc.

Regulation is the process which enables the system to provide optimal management of the energy flow within the system.

More generally, at any time t, the regulation process allows the determination of the state of the system which is subsequently used to issue appropriate orders or commands aimed at optimizing the energy transfer function within the system with respect to the constraints.

The regulation of back-up or utility interactive systems can be rather complex and is beyond the scope of the present study.

5.2 STAND-ALONE PHOTOVOLTAIC SYSTEM REGULATION

The energy flow within a simple PV system composed of PV modules, battery storage and load can be described by two main operating modes.

(i) The floating operating condition occurs when the battery voltage fluctuates between maximum and minimum voltage thresholds corresponding respectively to the maximum and minimum state of charge of the battery. In such a case the system is autoregulated during the day and the regulation makes use of a simple protective diode which protects the PV modules from discharge of the battery at night.

(ii) The overcharge and overdischarge conditions occur where the battery voltage reaches some critical values. In that case the regulation which will be later performed by the regulator consists of the following.

(*a*) Collecting information on the state of the system (battery voltage).

(*b*) Knowing the constraints (V_{min} and V_{max}).

This elaborates orders, i.e. to disconnect or reconnect the PV modules or loads to the battery. Details of the process are summarized in table 5.1.

Table 5.1 Summary of the regulation process.

Constraint	Command
(1) $V_B > V_{max}$	Disconnect battery from PV modules
(2) If (1) is true and $V_B < V_{t1}$	Reconnect battery to PV modules ($V_{max} - V_{t1}$ gives hysteresis)
(3) $V_B < V_{min}$	Disconnect battery from load
(4) If (3) is true and $V_B > V_{t2}$	Reconnect battery to load ($V_{min} - V_{t2}$ gives hysteresis)

Different types of regulator can perform this regulation with different degrees of technical sophistication. Elementary types will be described in the next section.

(i) Principle of self-regulation.
(ii) Shunt regulator.
(iii) Series regulator with semiconductor.
(iv) Series regulator by electromechanical cut-off.
(v) Automatic circuit breaking.

5.3 BASIC TYPES OF REGULATOR

5.3.1 Principle of self-regulation

The desired charging profile for a battery is to maintain a moderate charge rate until a relatively high state of charge (SOC) is achieved and then to taper the rate towards a low rate, filling the battery completely without shifting into electrolysis and overheating. To achieve this, an inverse relation between voltage and current is needed under critical conditions of maximum SOC.

The principle of self-regulation is to size the PV generator so that its voltage-sensitive region coincides with the battery's critical region (95–100% SOC) as shown in figure 5.1.

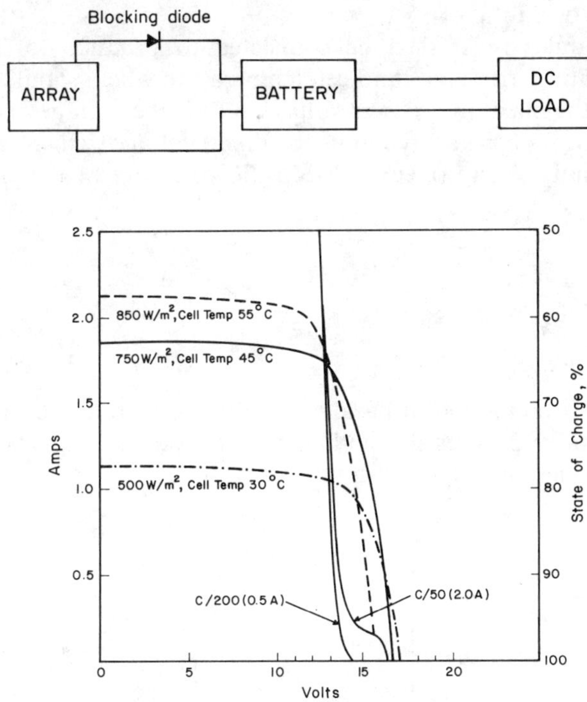

Figure 5.1 Principle of self-regulation.

With an appropriately chosen module, as the battery's SOC approaches 100%, the battery's rising voltage drives the module off the knee of the current–voltage (I–V) curve to a lower current or charge rate, and the desired tapering off of charging rate is automatically maintained. The only requirement is to design the module to match the battery's critical voltage region, taking into account the voltage drop in the blocking diode (about 0.75 V) and in the wiring resistance.

Warning This sizing cannot be standardized easily because it depends on the implementation (wiring) and, above all, is very sensitive to the temperature effect.

A self-regulating system may be the least effective way of extracting energy from the PV array since the battery voltage requirement places specific constraints on the array operating voltage and current, forcing it to operate most of the time away from its maximum-power point.

As an example, consider a 12 V lead–acid battery which has a charge voltage that ranges from 12.8 V at 60% depth of discharge to 14.4 V at full charge.

The voltage operating point of the array that would transfer maximum power from the array is 14.4 V plus the voltage drop across the blocking diode (0.75 V) or a total of 15.15 V.

The PV modules must then be connected to produce 15.15 V at the designed operating irradiance and cell temperature which should preferably coincide with the maximum-power voltage. When the voltage of the battery equals the voltage of the array minus the diode voltage (which occurs when the battery is fully charged), current stops flowing into the battery from the array.

5.3.2 Parallel (shunt) regulation

5.3.2.1 Principle

The regulator is connected in parallel to the PV generator to dissipate any excess energy through a resistor and power components. Figures 5.2 and 5.3 provide circuit design of typical shunt regulators.

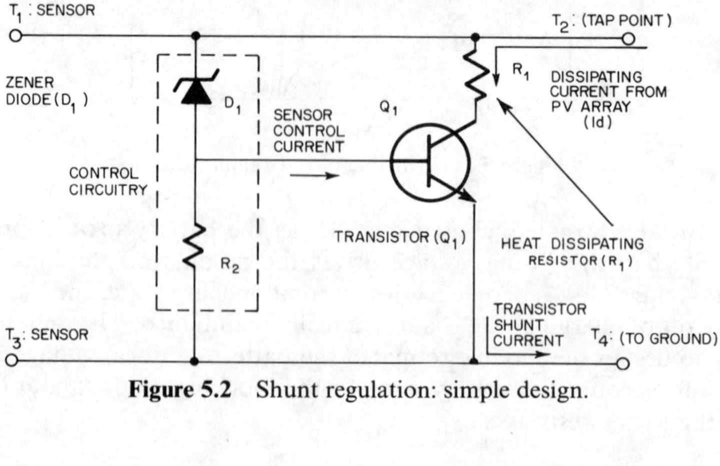

Figure 5.2 Shunt regulation: simple design.

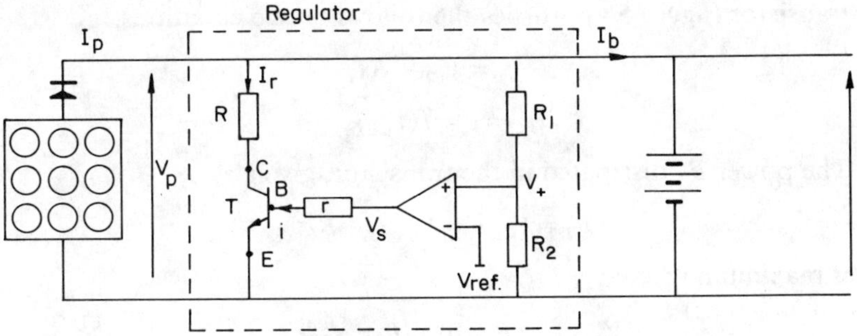

Figure 5.3 Shunt regulator.

5.3.2.2 Study of the operating mode of shunt regulation

The PV array when connected to the battery behaves as a current generator whose current I_P is a function quasi-proportional to solar irradiance. For a given irradiance level, I_P is constant.

Without regulation

The voltage V_B of the battery is less than the limitation voltage V_{lim} corresponding to the maximum state of charge. In this case $V^+ < V_{ref}$, $V_s = 0$, $i = 0$. Therefore the transistor T is off ($I_r = 0$). The shunt regulator does not interfere; as no component is in series with the power circuit of the PV generator, no voltage drop occurs and the current is entirely used for charging the battery ($I_P = I_B$).

Regulation

Suppose that $V_B > V_{lim}$: in this case, $V^+ > V_{ref}$, $V_s > 0$, $i > 0$. The transistor T conducts: $I_r > 0$ which gives $I_P = I_r + I_B$. The occurrence of I_r limits the charging current I_B of the battery and therefore the battery voltage V_B.

Equilibrium

Equilibrium is achieved when the current I_B is just sufficient to maintain the battery at the limitation voltage V_{lim}. In this case, $V_B = V_{lim}$ and $V^+ = V_{ref}$.

Advantages

(i) The regulation is of proportional type.
(ii) There is no voltage drop in the charging unit.
(iii) The power consumption of the regulator is negligible during the no-regulation period.
(iv) A failure in the regulator would not interrupt battery charging.

Disadvantages

(i) The regulator should be able to dissipate the peak power of the PV generator. Indeed, during regulation the operating point of the transistor (figure 5.4) satisfies the following two equations:

$$V_{\text{lim}} = V_{\text{CE}} + RI_{\text{r}}$$

$$I_{\text{C}} = I_{\text{r}} = f(V_{\text{CE}}).$$

The power P_{T} dissipated in the transistor, given by

$$P_{\text{T}} = V_{\text{CE}} I_{\text{r}} = V_{\text{lim}} I_{\text{r}} - RI_{\text{r}}^2$$

is maximum when

$$V_{\text{lim}} - 2RI_{\text{r}} = 0 \rightarrow I_{\text{r}} = \frac{V_{\text{lim}}}{2R}$$

and

$$V_{\text{CE}} = \frac{V_{\text{lim}}}{2}.$$

At these operating points,

$$P_{\text{Tmax}} = \frac{V_{\text{lim}}^2}{4R}.$$

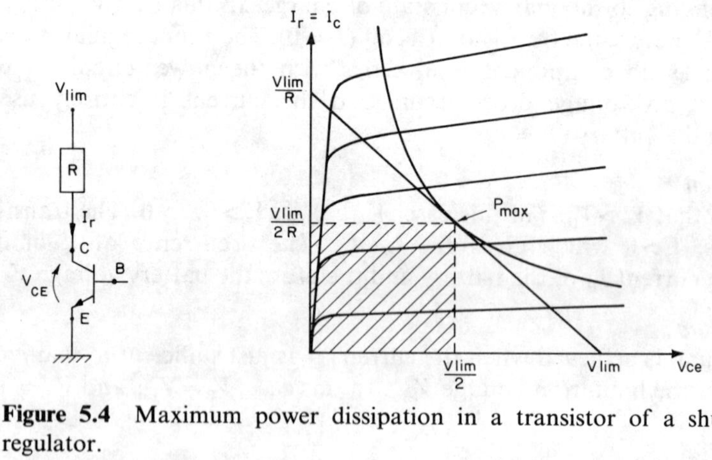

Figure 5.4 Maximum power dissipation in a transistor of a shunt regulator.

(ii) The transistor should be able to dissipate one quarter of the total power.

(iii) The resistance should be able to dissipate the total power when the transistor is in saturation.

Observations

(i) The shunt-type regulator can function without a battery (figure 5.5). Suppose that $I_{\text{r}} = 0$; the PV module has a load of high resistance

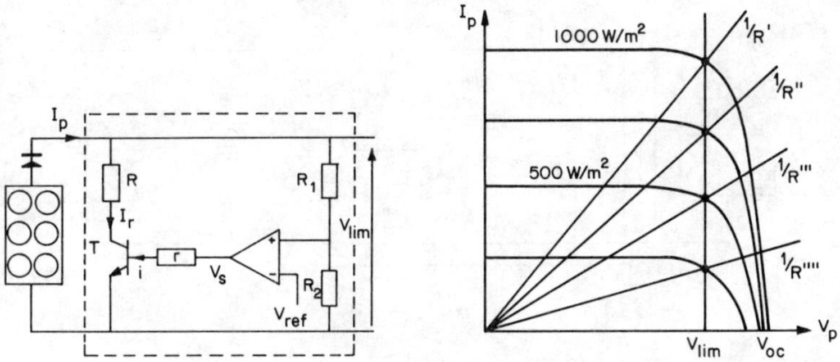

Figure 5.5 Shunt regulator as a variable load.

$R_1 + R_2$; hence $V_P = V_{oc} > V_{lim}$. Therefore, $V^+ > V_{ref}$ and $i > 0$, $I_r > 0$. Equilibrium is achieved when $I_r = I_P$, giving $V_B = V_{lim}$.

(ii) Regardless of the irradiance the output voltage of the PV module is maintained at the limitation voltage.

(iii) The shunt regulator behaves in this case as a variable load whose value varies with irradiance.

These observations should allow verification of the correct operation of a shunt regulator in a solar-powered system.

5.3.3 Series regulation with a semiconductor

5.3.3.1 Principle

The series regulator utilizes a transistor in series with the PV generator. The transistor behaves like a variable resistance whose value is a function of the state of charge of the battery. Figures 5.6 and 5.7 provide a circuit design of typical series regulators.

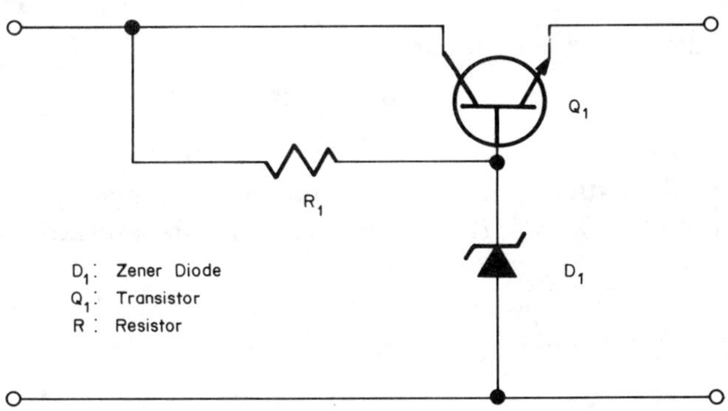

Figure 5.6 Series regulator: basic design.

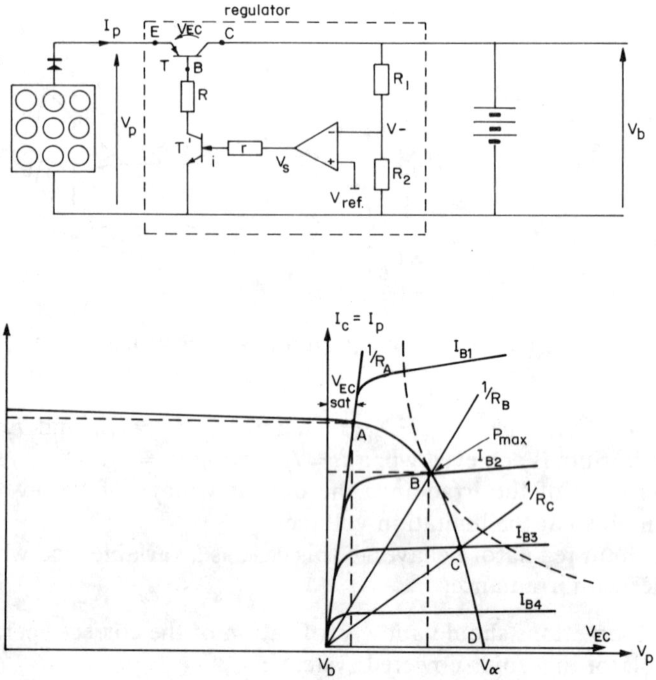

Figure 5.7 Series regulator.

5.3.3.2 Study of the operating mode

The following three equations determine the operating point of the system. Neglecting the voltage drop in the wires,

$$V_P = V_B + V_{EC}.$$

The transistor characteristics are

$$I_P = f(V_{EC}).$$

The PV generator characteristics are

$$I_P = f(V_P).$$

Without regulation
The voltage of the battery is less than the limitation voltage V_{lim}. In this case, $V^- < V_{ref}$; hence $V_s > 0$, $i > 0$, $i_B > 0$. The transistor T is saturated; $V_{EC} = V_{ECsat}$ (a few tenths of a volt).

The operating point is at point A; the charging current of the battery is maximum for a given solar irradiance.

Regulation
Suppose that $V_B > V_{lim}$. In this case, $V^- > V_{ref}$ and $V_s = 0$, $i = 0$, $i_B = 0$. The

transistor T is not conducting, $I_P = 0$ (point D). Because of the absence of a load current, the voltage of the battery decreases and equilibrium is achieved by regulation ($V^- = V_{ref}$ and $V_B = V_{lim}$).

The operating point corresponds to the current which maintains the battery at the limitation voltage.

The transistor behaves as a resistance whose value increases with battery charge ($R_A < R_B < R_C \ldots$ with $R = V_{EC}/I_P$).

Advantages

(i) The regulation is a proportional type.

(ii) The dissipated power at the transistor terminals is low compared with the peak power of the PV module:

$$P_{Tmax} = V_{EC} I_P \text{ at point B.}$$

Disadvantages

(i) During the no-regulation period, the regulator introduces a voltage drop V_{ECsat} in the circuit and a current consumption (base current $I_B = I_{B1} > 0$).

Observation

(i) The power loss can be reduced by using a field-effect transistor of low passing resistance (voltage control).

5.3.4 Series regulation by electromechanical cut-off

5.3.4.1 Principle

The charge regulator acts in series with the solar panel. It stops the charging of the battery by electromechanical cut-off when it reaches its maximum threshold level. It is reset for charging automatically when the lower threshold voltage is reached (figure 5.8).

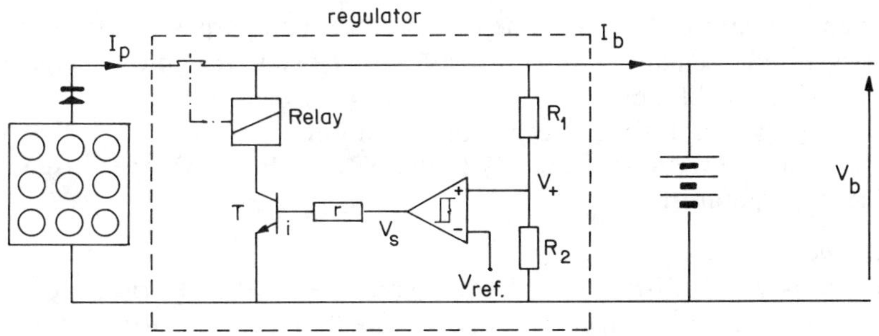

Figure 5.8 Series regulation by electromechanical cut-off.

5.3.4.2 A study of its operation

The electronic control circuit of the relay is a comparator with a double threshold (figure 5.9).

The output of the comparator passes to 'high' when $V^+ > V_{ref}$ and rests in this state as long as V^+ is not lower than $V_{ref} - H$.

The hysteresis H represents the difference between the two threshold voltages of comparison (the threshold being different according to whether V^+ is increasing or decreasing).

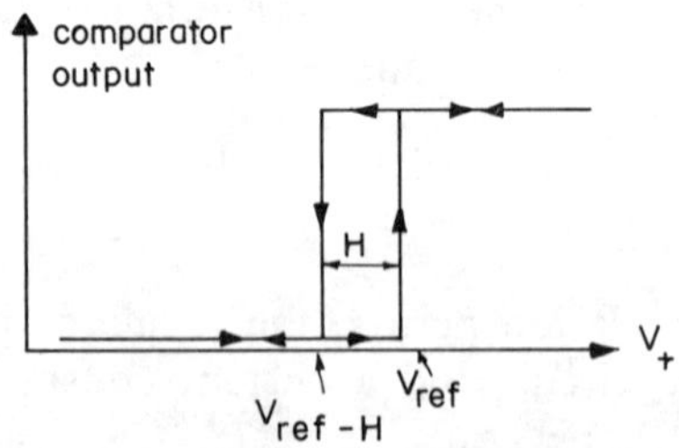

Figure 5.9 Hysteresis loop of comparator.

Without regulation
The battery voltage V_B is lower than the limitation voltage V_{lim}. In this case, $V^+ < V_{ref}$; hence $V_s = 0$, $i = 0$. The transistor T is not conducting, the relay is not powered and the contact is closed. The battery charges with a current I_p which is a function of solar irradiance.

Regulation
Suppose that $V_B > V_{lim}$, $V_s > 0$; the relay changes state and the contact opens which causes the interruption of battery charging. This interruption of charging causes a drop in the voltage of the battery (the effect of the internal resistance). There must be a sufficient voltage drop $V_{lim} - H$ to reset the charging of the battery by commutation of the relay. During regulation, the battery voltage is maintained between two preset thresholds (about 2.1 V and 2.3 V per element for lead–acid batteries).

If the comparator had only one threshold for comparison, a phenomenon of very rapid oscillations of the relay would occur (ringing effect) which would impair its lifetime.

During regulation the frequency of commutation of the relay will depend on the charge current, the capacity of the battery, the discharge current, the ambient temperature, etc.

Advantages

(i) No power is dissipated in the charge regulator (the electrical energy not recovered in open circuit is dissipated in calorific form in the PV module).

(ii) There is the possibility of regulating high powers without any problem (a relay with several contacts could control several branches of modules).

Disadvantages

(i) Regulation is by on and off (less favourable conditions of charging).
(ii) Electromechanical commutation occurs (the lifetime of relays is related to the current and frequency of commutation).

5.3.5 Variants on parallel and series regulation

From the three principles of regulation reviewed, it is possible to conceive 'variants' in order to overcome certain inconveniences (regulation by cutting, regulation by fractional cutting, etc).

Figure 5.10 is a charge controller adapting the concept of series and shunt regulation to ensure longer lifetime of the battery. The relay F controls the series switch disconnecting the PV modules from the battery when its voltage is charged beyond 2.35 V per element. The series switch recloses when the battery voltage decreases until 2 V per element and charging resumes.

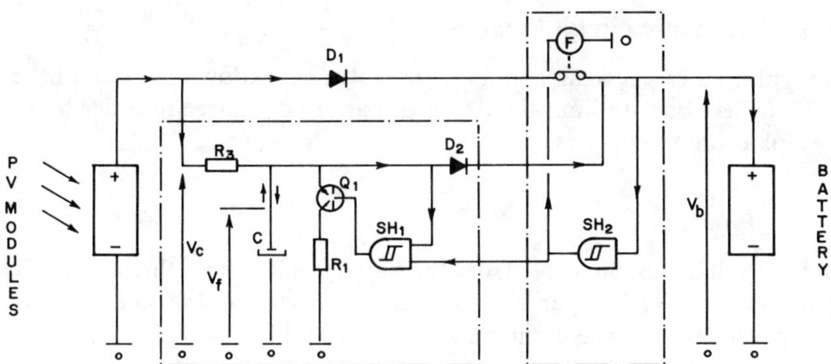

Figure 5.10 Charge controller.

Within the 2 V and 2.35 V region, the regulator is in the floating mode. The transistor Q_1 regulates the amount of shunt current diverted from charging the battery. The battery voltage oscillates at around 2.2 V per element.

Figure 5.11 is a discharge controller. After pressing the button BPM_1, the relay S is energized, closing the series switch and activating the transistor Q_2. As long as the battery voltage is higher than 1.85 V per element, Q_2 conducts the current to energize relay S, thereby maintaining the battery and load connected. As the battery discharges below 1.85 V per element, Q_2 cuts off, no current passes through to energize the relay S and the series switch

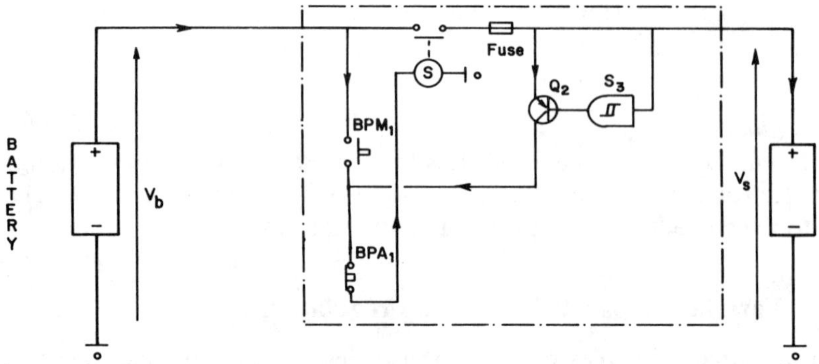

Figure 5.11 Discharge controller.

between battery and load is disconnected. The discharge regulator conducts again when the battery voltage is above 1.85 V and only after button BPM_1 is pressed. BPA_1 is the stop button.

5.3.6 Automatic circuit breaking

In exceptional cases (weak sunlight, generator breakdown or overconsumption) it is necessary to limit the depth of battery discharge in order to avoid the sulphation of its plates.

5.3.6.1 Principle

The load will be disconnected from the battery at a certain threshold voltage (of the order of 1.9 V per element) and is reset automatically when the battery reaches a sufficient charge level (figure 5.12).

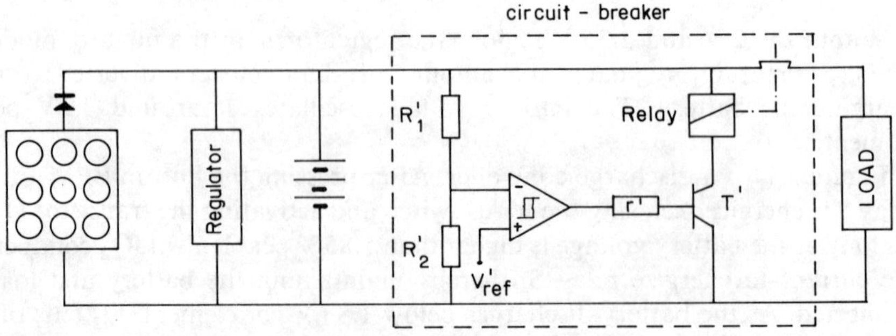

Figure 5.12 Circuit breaker.

5.3.6.2 A study of its operation

The state of charge of a battery is determined by the measurement of its voltage ($V_B < 1.9$ V per element for lead–acid batteries).

The principle of circuit breaking is identical with that of electromechanical cutting.

A double threshold electronic circuit would control a relay whose contact is situated in the discharge circuit.

Observation
Circuit breaking of the utilization device is generally preceded by an alarm whose aim is to warn the user of a temporary breakdown of the system.

5.3.7 Regulator with DC–DC converter

This regulator is a composite of the previously described regulators (figure 5.10). As long as the battery has not reached its maximum state of charge V_{lim}, all the current produced by the PV array is stored within the battery. When the battery voltage reaches V_{lim} (2.35–2.4 V per element) the relay R is powered, the contact opens and the DC–DC converter operates. The DC–DC converter provides the floating voltage of the battery (2.25 V per element).

The charging rate I_s of the battery is small and completely charges the battery.

The current I_P provided by the PV array is also small and there is no excess of energy to dissipate. Such a regulator presents a very good performance although quite expensive and sophisticated. For this economic reason, it is advisable to use it to regulate a large PV array (peak power greater than 1 kW_p).

5.4 CONCLUSION

Electronic regulation is in practice a necessity in most solar PV systems with batteries.

For important installations the charge regulator provides several functions.

(i) Charge limitation.
(ii) Protection from discharging (circuit breaking).
(iii) Monitoring devices (ammeter and voltmeter).
(iv) Protection of the utilization (fuses, etc).

An important point to keep in mind is that in general the voltage of the battery is used as the parameter representative of the state of charge of the battery. This can present some limitation since, in fact, only the electromotive force E (polarization of the electrode) is the state variable as far as the SOC is concerned whereas the terminal voltage is a function of its internal resistance and charging–discharging rate ($V_B = E + RI$).

BIBLIOGRAPHY

Lasnier F 1985 *Lecture Notes ET24* (Bangkok: Asian Institute of Technology)

6 Power Conditioning

6.1 INTRODUCTION

At a given temperature and illumination level, a photovoltaic (PV) generator has current–voltage (I–V) characteristics as in figure 6.1.

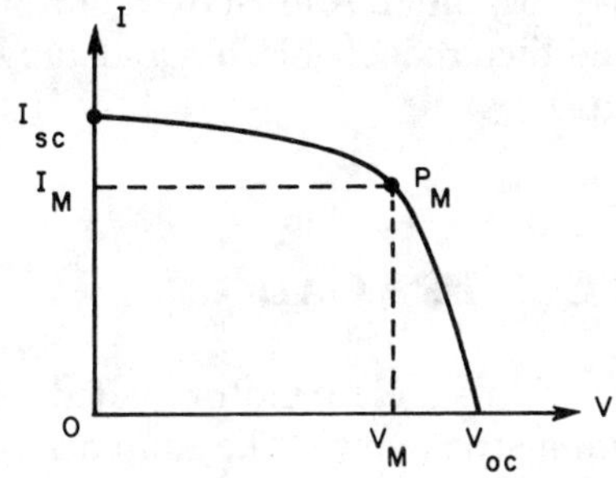

Figure 6.1 I–V characteristics of a PV generator.

Clearly, in order to maximize the transfer of energy, the electrical load must be such that its own I–V characteristics intercepts that of the generator as close to the point of maximum power as possible. There are basically two kinds of load: the voltage receiver and the current receiver, without mentioning the purely resistive load which is the worst case in terms of its matching to the photovoltaic generator characterization.

6.2 VOLTAGE-RECEIVER LOAD

The best example is the battery. The battery voltage V_B is stable and the module's maximum operating voltage V_{max} does not vary much with the operating conditions. The main parameters that may affect both V_{max} and V_B are as follows.

(i) Illumination: variations in the solar irradiance can make V_{max} change by more than 10%.

(ii) Temperature: the temperature effect on V_{max} is about $-2\ \mathrm{mV\ ^\circ C^{-1}}$, about $-0.4\%\ ^\circ\mathrm{C}^{-1}$ for a standard solar cell producing a voltage $V_{max} = 500$ mV. The temperature coefficient of each 2 V battery element is about $-2\ \mathrm{mV\ ^\circ C^{-1}}$, about $-0.1\%\ ^\circ\mathrm{C}^{-1}$. Since the battery is usually placed inside a shade, its temperature is substantially lower than that of the solar cells'. If we take into account a 25 °C difference in temperature, the subsequent difference between V_{max} and V_B is around 10%.

(iii) Charge level of the battery: in the case of a lead–acid battery, the variation in V_B between the end of charge and the end of discharge is about 20%.

(iv) Aging: for both battery and generator, aging affects the efficiencies and creates irreversible degradations. In its lifetime, the difference in voltage between V_B and V_{max} may reach 20%.

On the average, the resulting power loss may be assessed at about 20% of the power capacity. For medium- to high-power systems, the financial loss on the peak-power rating is substantial. A low-cost adaptor having an efficiency of better than 90% would then reduce this loss and allow one to decrease the installed peak-power capacity.

6.3 CURRENT–RECEIVER LOAD

One example is a permanent-magnet DC motor loaded with a constant-torque mechanical load (e.g. volumetric pump). The adaptation is almost impossible since the generator current I_{max} varies proportionally to the illumination level. Again an adaptor would enhance the efficiency of this energy transfer.

6.4 DC–DC CONVERTERS

The adaptors mentioned in the preceding sections are basically DC–DC converters. Since their main purpose is to maintain both generator and load at their respective optimum operating conditions, they are often referred to as maximum-power-point trackers (MPPTS). Their development is now at the industrial stage. It has benefited from developments in solid-state electronics.

The role of DC–DC converters is primarily to convert an input power $P_1 = V_i I_i$ into an output power $P_o = V_o I_o$ with the best possible efficiency $\eta = P_o / P_i$ (figure 6.2).

Without mentioning the galvanic insulation inverters (no common potential between input and output) the efficiency of which is at first penalized by about 10%, the DC–DC converters can be divided into voltage boosters ($V_o > V_i$) and voltage reducers or buck converters ($V_o < V_i$).

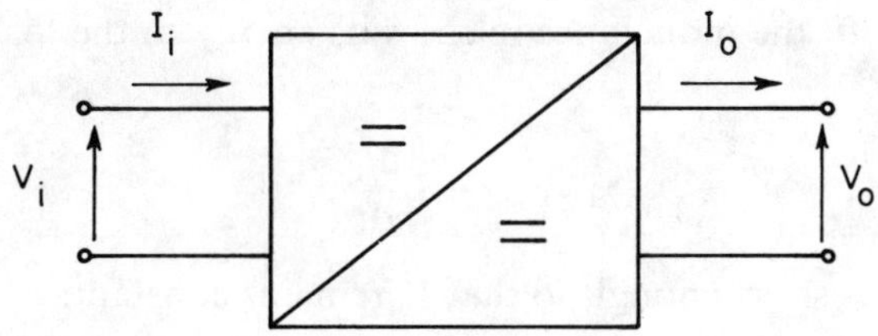

Figure 6.2 Schematic diagram of a DC–DC converter.

6.4.1 Buck converters

Voltage reducers—which are, in fact, current boosters—are generally employed to drive current-receiver loads. Their function is fulfilled by an inductance (figure 6.3).

The switch K oscillates with a period $T = T_{ON} + T_{OFF}$ (figure 6.4).

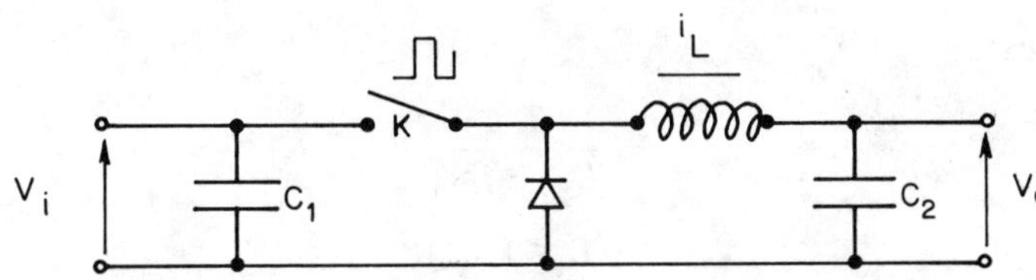

Figure 6.3 Basic circuit of the buck inverter.

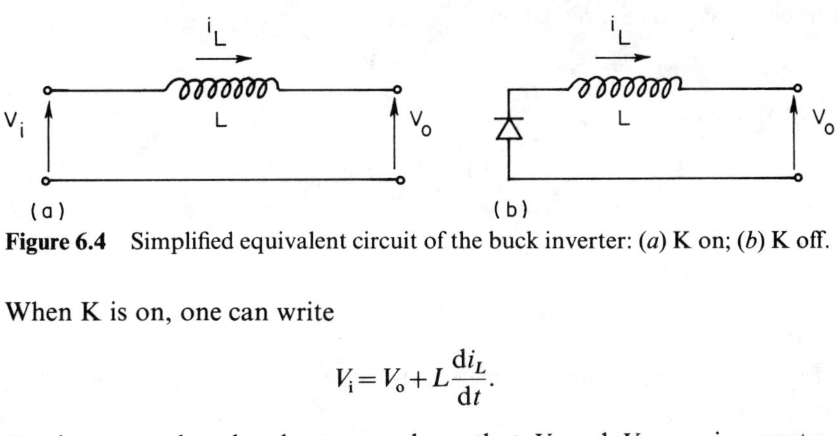

Figure 6.4 Simplified equivalent circuit of the buck inverter: (*a*) K on; (*b*) K off.

When K is on, one can write

$$V_i = V_o + L\frac{di_L}{dt}.$$

T_{ON} is assumed to be short enough so that V_i and V_o remain constant. Therefore,

$$i_L(T_{ON}) - i_L(0) = \frac{V_i - V_o}{L} T_{ON}.$$

During the time T_{ON}, the inductance L has stored (and not dissipated) the energy $\frac{1}{2}Li_L\,(T_{ON})^2$.

When K is off, the inductance releases its energy to the load through the diode:

$$V_o = -L\frac{di_L}{dt}.$$

Again, T_{OFF} is short enough so that V_o remains constant:

$$i_L(T_{ON}+T_{OFF}) - i_L(T_{ON}) = -\frac{V_o}{L}T_{OFF}.$$

Under steady-state conditions,

$$i_L(0) = i_L(T_{ON}+T_{OFF}).$$

Therefore,

$$\frac{V_i - V_o}{L}T_{ON} = \frac{V_o}{L}T_{OFF}$$

which gives

$$V_o = \frac{T_{ON}}{T_{ON}+T_{OFF}}V_i$$

$$V_o < V_i.$$

Since the current i_L cannot be negative, the condition of continuous conduction can be expressed as

$$i_L(0) > 0$$

which yields

$$-\frac{V_o}{L}T_{OFF} > -i_L(T_{ON})$$

or

$$T_{OFF} < \frac{Li_L(T_{ON})}{V_o}.$$

Figure 6.5 displays the difference between the two states of continuous and

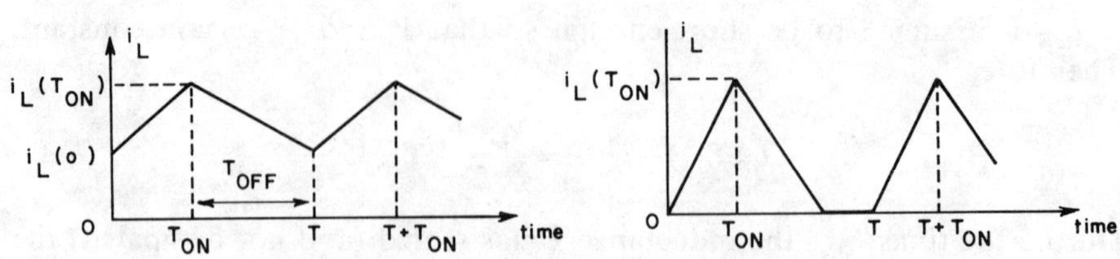

Figure 6.5 Load current in the buck inverter.

intermittent conduction. Obviously the continuous-conduction regime is the smoothest. Then, at a given frequency $1/T=1/(T_{ON}+T_{OFF})$, proper adjustment of T_{ON} allows one to adjust V_i at the maximum-power point voltage V_{max} according to the load current I_L. The average load current is then given by

$$I_L=\frac{1}{T}\int_0^T i_L\,dt=i_L(T_{ON})-\frac{V_o T_{OFF}}{2L}.$$

The role of the two capacitances C_1 and C_2 is to absorb the voltage variations and to assist current boosting and smoothing.

6.4.2 Boost converters

Voltage boosters are generally employed for battery-charging applications. Similar to buck converters, their function is fulfilled by an inductance (figure 6.6).

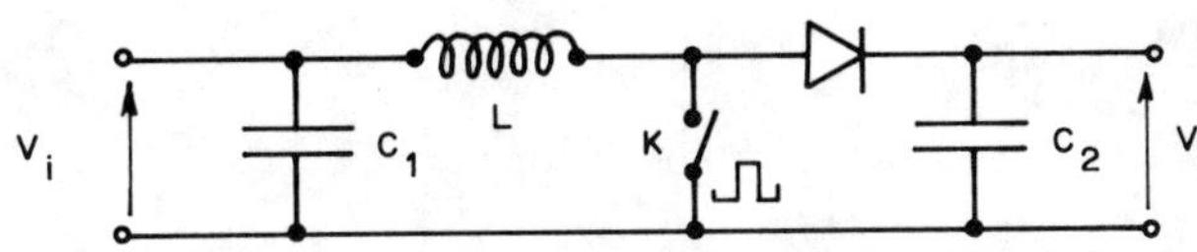

Figure 6.6 Basic circuit of the boost inverter.

The switch K oscillates with a period $T=T_{ON}+T_{OFF}$ (figure 6.7). When K is on, one can write

$$V_i=L\frac{di_L}{dt}.$$

On the assumption that V_i remains constant during T_{ON}, the current variation is

$$i_L(T_{ON})-i_L(0)=\frac{V_i}{L}T_{ON}.$$

During the time T_{ON}, the inductance has stored (and not dissipated) the energy $\frac{1}{2}Li_L(T_{ON})^2$.

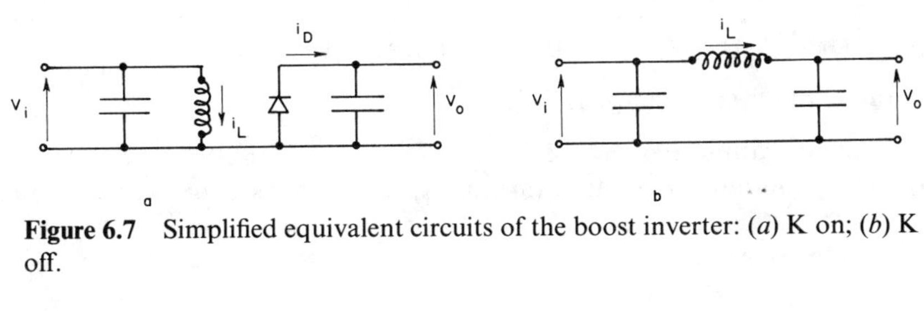

Figure 6.7 Simplified equivalent circuits of the boost inverter: (*a*) K on; (*b*) K off.

When K is off, the inductance releases its energy to the load through the diode:

$$V_i - V_o = L\frac{di_L}{dt}.$$

Again, T_{OFF} is short enough so that both V_i and V_o remain constant:

$$i_L(T_{ON}+T_{OFF}) - i_L(T_{ON}) = \frac{V_i - V_o}{L} T_{OFF}.$$

Applying the steady-state conditions

$$i_L(0) = i_L(T_{ON}+T_{OFF})$$

leads to

$$\frac{V_i}{L} T_{ON} = -\frac{V_i - V_o}{L} T_{OFF}$$

which gives

$$V_o = \frac{T_{ON}+T_{OFF}}{T_{OFF}} V_i$$

$$V_o > V_i.$$

As in the case of the buck converter, the condition of continuous conduction is $i_L(0) > 0$, which yields

$$i_L(T_{ON}) > \frac{V_i}{L} T_{ON}$$

or

$$T_{ON} < \frac{L i_L(T_{ON})}{V_i}.$$

Then, at given frequency $1/T = 1/(T_{ON}+T_{OFF})$, proper adjustment of T_{ON} in the limits of the continuous-conduction regime allows one to adjust V_i at the maximum-power-point voltage V_{max} according to the load voltage V_o.

6.4.3 Design and sizing of DC–DC converters

The three major criteria are as follows.

(i) The operating frequency $1/T$.
(ii) The commutation voltage (at least V_i for the buck converter and V_o for the boost converter).
(iii) The peak current.

Clearly, a higher operating frequency and smaller sizes of inductance and capacitors can be used. However, as concerns the switch, there is a compromise between this frequency and the magnitude of the peak current. Electromechanical relays can switch very high currents but at a low frequency. Among electronic statistic relays, thyristors (widely used for very-large-power applications) are also limited in frequency. The most convenient switching devices seem to be transistors, bipolar transistors or metal–oxide–semiconductor field-effect transistors (MOSFETS), and in the near future multi-channel MOSFETS (figure 6.8).

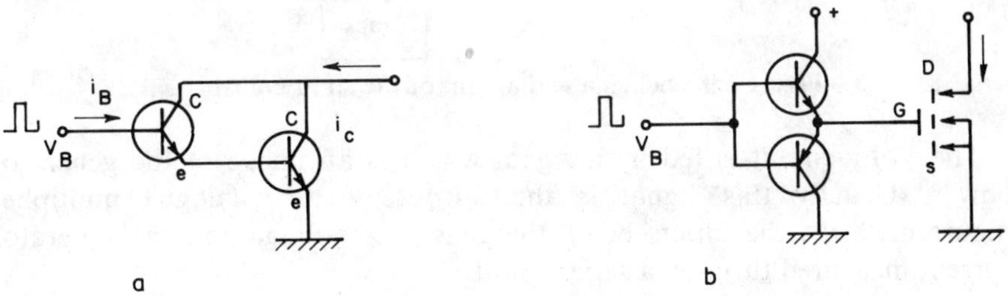

Figure 6.8 Transistor switching devices: (*a*) the Darlington circuit; (*b*) totem-pole-driven MOSFET.

In the Darlington circuit, the current in the base controls the current in the collector–emitter circuit: $I_B = I_C/K$. The gain K can be as high as 500, so that the switch is operated from a very small amount of energy. However, when it is on, the collector–emitter voltage is of the order of 2 V and thus induces a power loss $V_{CE}I_C$. The switching time is also high (about 0.5 μs), which limits the operating frequency.

In the MOSFET circuit, driven by complementary (n–p–n and p–n–p) bipolar transistors, the current in the grid modulates the resistance of the drain-to-source canal. The grid voltage control is very simple and, when the switch is on, this resistance R_{ON} can be a slow as 100 mΩ; the power dissipation is then $R_{ON}I^2$, which requires a current-limiting circuit and the overdimensioning of the characteristics of the transistor. The switching time is very short (around 50 ns), which allows high-frequency operation (up to 50 kHz).

As concerns the diode, it must stand high reverse voltages and be fast enough to operate according to the switching frequency.

The capacitances absorb relatively high peak currents and must have a low series resistance.

6.5 MAXIMUM-POWER POINT TRACKER

Whatever the type of converter used (buck or boost), a special circuit must provide adequate modulation to the switch, in order to maintain the PV generator voltage at its optimum value V_{max}. Since the operating conditions are highly variable (e.g. with illumination and temperature), this circuit has to be a closed-loop control (figure 6.9).

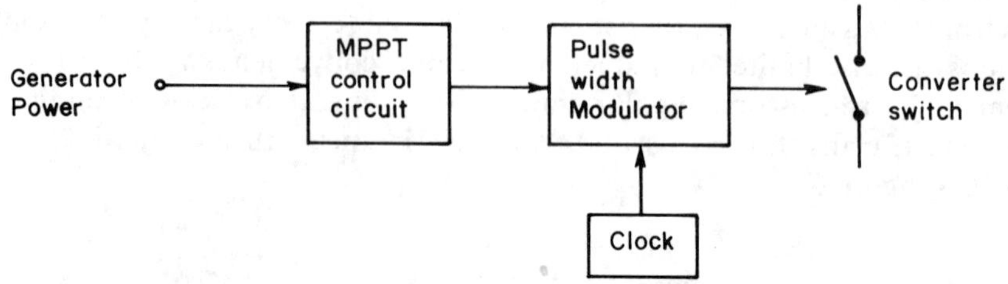

Figure 6.9 Schematic diagram of the MPPT control.

The MPPT circuit is fed by a signal which is an image of the generator power. Usually this signal is the output of an analogue multiplier (figure 6.10(*a*)), the inputs being the generator voltage and the generator current measured through a series shunt.

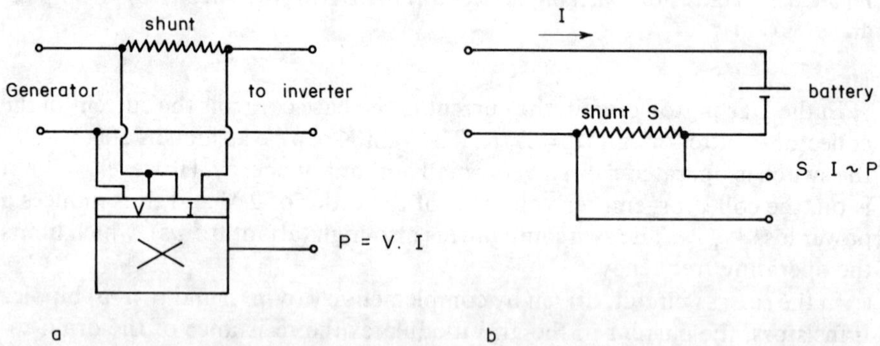

Figure 6.10 Circuits providing signal proportional to input generator power: (*a*) with analogue multiplier; (*b*) with shunt for battery charging.

In the case of a boost converter for battery charging, the output current is proportional to the input power since the battery voltage is seen to be constant at the operating frequency. Therefore the required signal may simply be measured across a series shunt in the load circuit (figure 6.10(*b*)).

There are two types of MPPT circuit. In the first (figure 6.11), alternative modulation is applied to the generator voltage. Synchronous demodulation of the generator power signal allows one to display on which side of the maximum-power voltage the generator is at present operating. The circuit then provides suitable information to the pulse width modulator to act on T_{ON} and hence to correct the voltage.

The second type of MPPT circuit applies the fact that, at the maximum-power point, the steady-state condition is expressed as $dP = 0$. As shown in

University of Southampton
Hartley Library
Self-issue Receipt

Customer ID: ***314855

Title: Photovoltaic engineering handbook
ID: 94-009851
Due: 5/6/2015,23:59

Total items: 1
06/11/2014 15:32

Please keep your receipt as proof of items loaned. You may be asked for this at the exit.

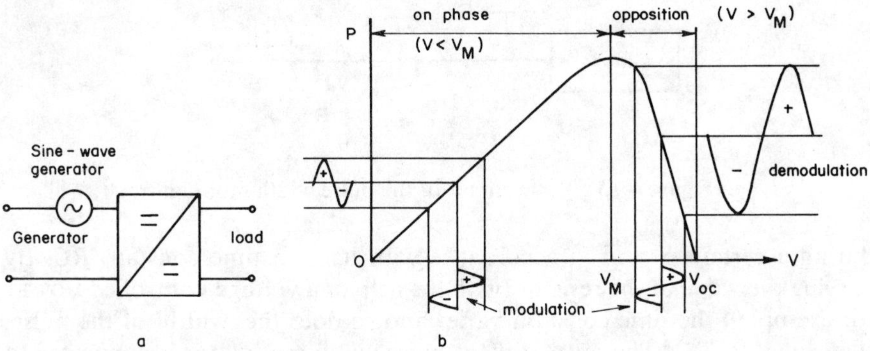

Figure 6.11 Maximum-power point tracking by modulation–demodulation: (*a*) circuit; (*b*) modulation–demodulation principle on the generator's power–voltage characteristic.

figure 6.12, a differentiating amplifier provides a signal proportional to dP/dt.

A ramp generator provides a continuous variation in the generator through the pulse width modulator that acts on T_{ON}. If the subsequent power variation dP/dt is positive, the operating point moves continuously to the maximum-power point. However, if the power variation is negative, the bistable flip–flop output switches to negative and changes the sign of the ramp of the integrating amplifier, i.e. reverses the generator voltage variation. The system keeps oscillating around the maximum-power point, with a frequency compatible with the accuracy of the amplifiers and with the frequency of the pulse width modulator. For example, with a 20 kHz operating frequency of the DC–DC inverter, a 500 Hz frequency of the MPPT loop is quite suitable, leading the values of the time constants of both differentiating and integrating amplifiers.

The pulse width modulator is now a very rapidly developing electronic device; its interesting performances have ensured that it is used in a wide range of applications. Basically, it is made of a clock providing pulses which are memorized by a first-order *RC* filter (figure 6.13). Its output remains high

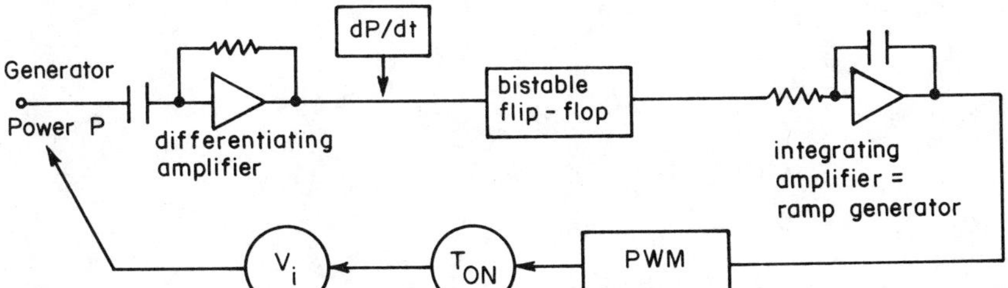

Figure 6.12 Schematic diagram of a power-differentiating MPPT loop.

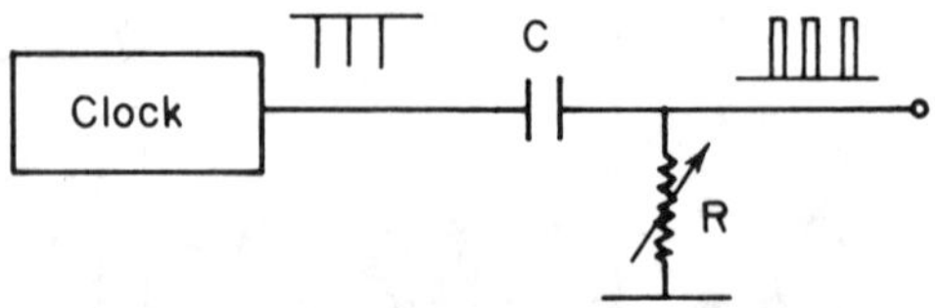

Figure 6.13 Basic circuit of the pulse width modulator.

during a certain period in accordance with its own time constant RC. By varying the value of the resistor (with the help of a voltage-controlled MOSFET for example), the time constant varies and so does the 'width' of the pulse, then equal to T_{ON}. The output of the pulse width modulator is connected to the gate of the switch K.

Maximum-power point tracking circuits (all complementary metal–oxide–semiconductor technology) coupled with buck and boost converters display very high efficiencies. Although it is a recent development, efficiencies over 95% have already been reported; this obviously requires careful design and selection of the components. The cost of such devices is still high. However, if a 500 W DC–DC MPPT inverter allows one to save 20% of the installed peak-power capacity, it is worth it. Finally, the modern electronic technologies applied to photovoltaics look very attractive and safe, but their on-the-spot reliability has yet to be proved.

BIBLIOGRAPHY

Lasnier F and Follea D *Lecture Notes ET24* (Bangkok: Asian Institute of Technology)

7 Adaptation of a Positive-displacement Pump Directly Connected to a Photovoltaic Generator

7.1 INTRODUCTION

The electrical power supplied by a photovoltaic (PV) generator depends on solar irradiance intensity, ambient temperature and load requirement. In order to maximize the transfer of electrical power, the electrical load must be such that its current–voltage (I–V) characteristics curve intercepts the PV generator I–V curve as close to the maximum-power point as possible. Depending on their nature, some loads are more adapted to the PV generator than others; this is the case for the centrifugal pumping system or battery. Some others (such as the surveyed positive-displacement pump, which is typically a current receiver) cannot be operated without an additional matching device called an adaptor of impedance or DC–DC converter. This device has two functions: the tracking of the maximum-power point and the transfer of power to the load. These two functions were realized using a pulse-width-modulated series-switch step-down regulator. An experimental efficiency of 85% was achieved using locally available technology. This device was tested on a locally made positive-displacement pump, providing satisfactory results.

7.2 SYSTEM CHARACTERIZATION

The system comprises a PV array of eight modules (type BPX 47C) in two parallel arrangements of four modules in series. It is capable of supplying a maximum current I_{max} of 3.5 A at an irradiance level of 1 kW m^{-2} and an open-circuit voltage of nearly 70 V. The surveyed pump is a locally made piston pump coupled to a permanent-magnet motor (nominal characteristics 51 V, 1500 rev min^{-1}, $R_i = 1\ \Omega$, $L = 40$ mH and $I_{max} = 10$ A). A flywheel is coupled to the motor shaft to keep the cyclic torque virtually constant. In such a motor–pump arrangement, the torque (and consequently the motor current) is virtually independent of the speed (and consequently the motor voltage) for a given head under nominal operating conditions.

In order to simulate varying discharge heads up to 50 m, a hydropneumatic tank was installed at the outlet of the pump (figure 7.1). The *I–V* characteristics of the motor–pump combination were recorded using an external power supply and are presented in figure 7.2. The load characteristics (figures 7.2 and 7.3) and related efficiencies (table 7.1) show that this motor–pump combination is suitable for a minimum manometric head of 20 m, requiring a motor current close to 3.9 A (table 7.1). The comparison between the PV generator and the load $I=f(V)$ (figure 7.2) characteristics at different heads shows that the load cannot be connected directly to the PV array for total manometric heads higher than 15–20 m. Moreover, the higher the head, the higher will be the efficiency of the pump if the PV array is capable of supplying the required input power.

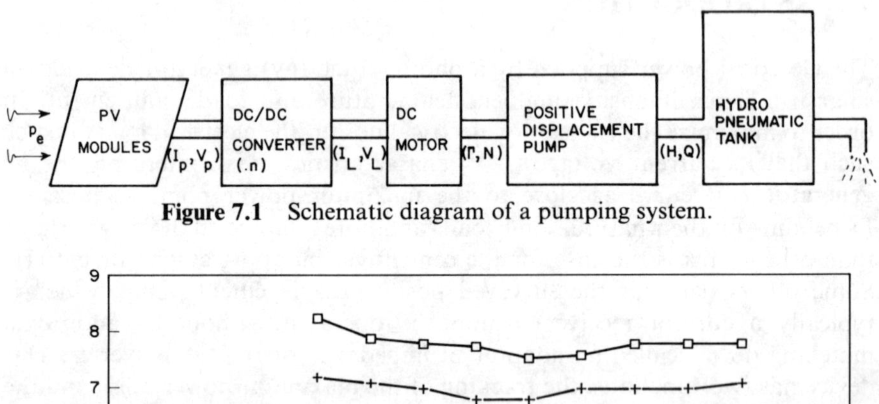

Figure 7.1 Schematic diagram of a pumping system.

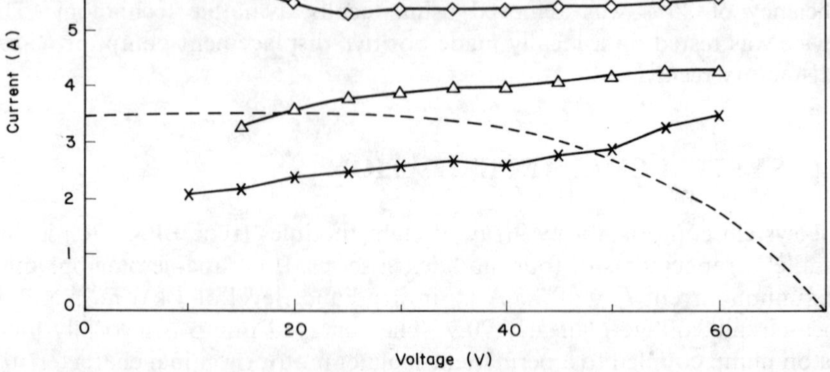

Figure 7.2 *I–V* characteristics for a positive-displacement pump: □, H = 45 m; +, H = 40 m; ◇, H = 30 m; △, H = 20 m; x, H = 10 m; ---, PV modules *I–V* characteristics at 1 kW m^{-2} irradiance.

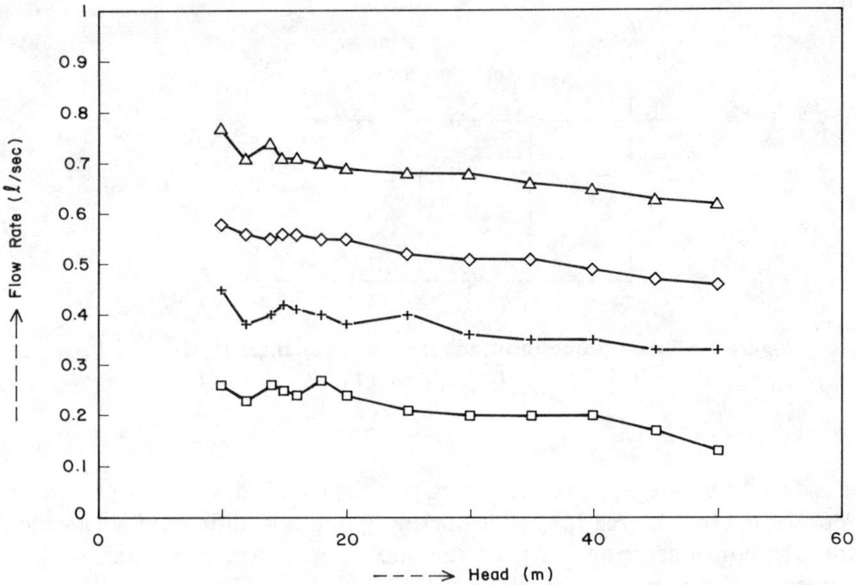

Figure 7.3 $\dot{Q}$–H characteristics for a positive-displacement pump: □, $V = 20$ V; +, $V = 30$ V; ◇, $V = 40$ V; △, $V = 50$ V.

Table 7.1 Motor–pump efficiency for each head at different speeds (voltages).

Head (m)	η (%) for the following voltages				
	20 V	30 V	40 V	50 V	60 V
10	52	56	55	52	47
15	62	65	62	58	57
20	65	64	67	64	63
25	55	73	69	69	70
30	54	65	69	73	71
35	55	65	71	74	79
40	55	65	71	73	72
45	46	62	69	71	73

Given the respective characteristics of the PV array and the load, an adaptor of impedance was designed and tested in order to allow direct connection between the modules and motor–pump system so that the energy transfer between PV array and motor–pump system could be maximized.

The basic concept of this device is to determine the maximum-power point of the PV generator by following its isopower curve until it intersects the load characteristic. As illustrated in figure 7.4, the power consumed by the load is

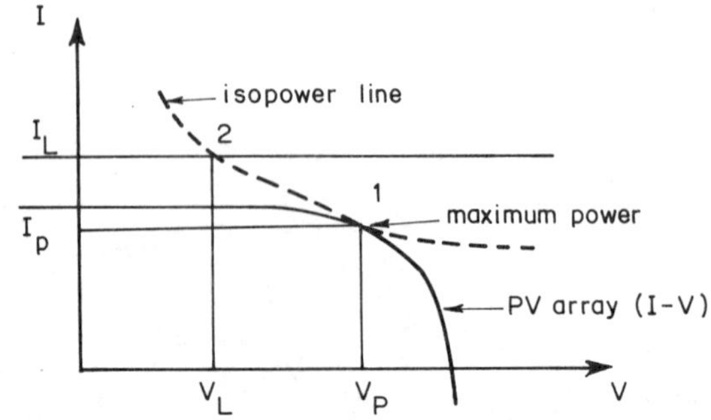

Figure 7.4 Basic concept of adaptor of impedance: $P_m = \tau N$; $P_h = \rho g H \dot{Q}$; $P_L = I_L V_L$; $P_P = I_P V_P$; $P_L = \eta P_P$; $V_L = V_P (T_C/T)$; $I_L = I_P (T/T_C)\eta$.

equal to the area $I_L V_L$. The maximum power produced by the PV generator is represented by the area $I_P V_P$. The operating points at different loads are such that the consumed power equals the maximum power produced by the PV generator (i.e. area $I_L V_L$ equals area $I_P V_P$ in figure 7.4).

For this prototype, the maximum power was tracked assuming that the maximum power occurs at a fixed reference voltage V_{ref} independent of the irradiance level. The lack of accuracy and subsequent decrease in efficiency are considered small in relation to the achieved simplicity and high degree of reliability of the device. Since the nominal voltage of the motor is 51 V, the DC–DC converter is a voltage reducer and current booster using a pulse-width-modulated series-switch step-down regulator.

7.3 PULSE-WIDTH-MODULATED SERIES-SWITCH STEP-DOWN REGULATOR

An efficient way to obtain a lower voltage from a higher source is shown in figure 7.5. Instead of absorbing the difference between the input and required output with a power-dissipating element, a low-impedance transistor switch is made to open and close periodically between input and output. The variation in energy during the two operating modes of the switch is absorbed by a large capacitance C_1.

If the switch S_1 has zero voltage drop in its closed position, the output as shown in figure 7.5 varies periodically between 0 V and the input voltage. The average value of this waveform is $V_L = V_P T_C/T$, where T_C is the switch closed time and T is the switching period. This is the voltage that would be read with a DC voltmeter at the output terminals. The ripple component has a peak-to-peak value of V_p V and would not be observed with a DC voltmeter. By adding the $L_1 C_1$ filter shown in figure 7.5, the ripple component of the

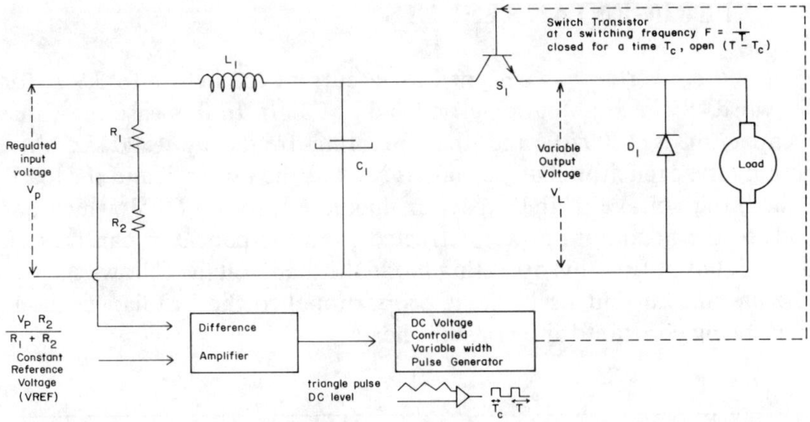

Figure 7.5 Adaptor of impedance: conceptual design (DC–DC converter). $I_L = I_P(T/T_C)$; $V_L = V_P(T_C/T)$; $V_P = V_{ref}$.

input voltage can be reduced to a value acceptable to the motor. In particular, V_p can be chosen to correspond to the maximum power of the PV modules and regulated by the duty cycle of the switch (the ratio T_C/T of the closing time over the total time of 1 cycle).

During the time T_C that the switch is closed, the load current I_L is supplied by both the PV modules and the capacitance C_1. When the switch is opened, no current is supplied to the load. This means that the current taken from the input source is a series of pulses of amplitude I_L lasting for a time T_C out of every T units of time. The average value of the input current is $I_L\,(T_C/T)$.

If the commutation losses in the switch are neglected, input power must equal output power. The circuit thus acts like a step-down transformer—it takes from the input source voltage V_p a current I_P of average magnitude $I_L(T_C/T)$. It transforms this at the output to a lower voltage $V_L = V_p(T_C/T)$ at a stepped-up average output current $I_L = I_P(T/T_C)$. I_L is nearly constant at a given manometric head for the surveyed positive-displacement pump.

The duty cycle T_C/T of the DC–DC converter is regulated such that (I_P, V_p) corresponds to the maximum-power points of the PV array. As a consequence, V_L at nominal operating conditions will vary corresponding to the current I_P of the panels, which is proportional to the irradiance.

The voltage step-down is achieved at a very high efficiency, since the only losses in such a step-down DC–DC converter are those in the switch S_1 when it is closed. Using a transistor switch, the voltage drop within the switch in the closed position can be as low as 1 V. During the time that the switch is open, the full input voltage exists across it. Since no current flows in the switch, there is no power dissipation. During the transition between open-switch and closed-switch times or vice versa, there is a momentary overlap of high voltage and current, resulting in some losses.

7.4 EXPERIMENTAL RESULTS

The DC–DC converter was designed for a current load close to 3.9 A (or in other words for a total manometric head of 20 m). In this case the value of the capacitor is 1200 μF and the operating frequency 1800 Hz. At the beginning the capacitor charges and, as soon as the PV array and the load are matched, the voltage of the capacitor fluctuates around 60 V, which corresponds to the maximum power extracted from the panels. It can be seen in figure 7.6 that during this operating mode the load voltage follows the trends of the module current itself, i.e. is proportional to the irradiance, the load current being constant for a constant head.

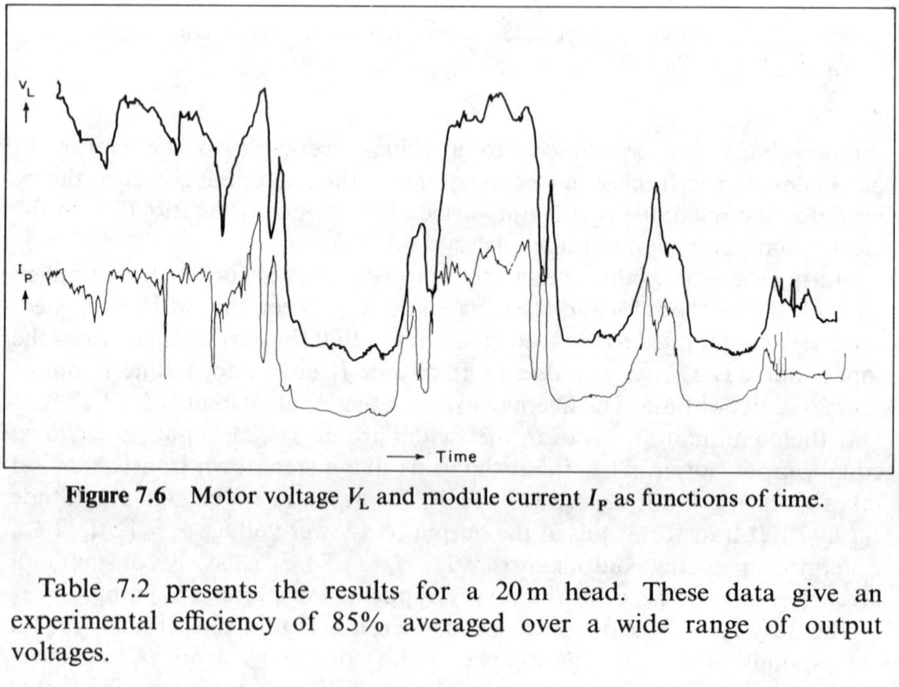

Figure 7.6 Motor voltage V_L and module current I_P as functions of time.

Table 7.2 presents the results for a 20 m head. These data give an experimental efficiency of 85% averaged over a wide range of output voltages.

7.5 CONCLUSION

A DC–DC converter with 85% efficiency has been developed and tested with a locally made positive-displacement pump. The maximum head required for this pump is 20 m, with a corresponding load current of 3.9 A. If the average module efficiency is about 10% and at 20 m the motor–pump efficiency is about 65%, the overall efficiency of the system can be expected to be of the order of 6%.

Table 7.2 Experimental results on the operation of the positive-displacement pump using a DC–DC converter.

Time (hours)	I_P (A)	V_P (V)	I_L (A)	V_L (V)	P_P (W)	P_L (W)	η (%)
9:22	2.40	51.2	3.90	28.0	122.88	109.20	89
9:15	2.55	50.8	3.90	28.0	129.54	109.20	84
9:25	2.68	52.2	3.97	30.4	139.64	120.84	86
9:35	2.79	55.4	3.90	34.0	154.57	132.60	85
9:45	2.85	56.7	3.90	35.0	161.59	136.50	84
9:55	2.85	57.5	3.90	36.2	163.87	141.18	86
10:05	2.95	58.7	3.92	37.5	173.16	147.18	85
10:15	3.00	59.6	3.95	38.5	178.80	152.07	85
10:25	3.02	60.5	3.85	40.5	182.71	155.92	85
10:35	3.17	61.6	3.87	43.5	195.27	168.50	86
10:51	3.24	63.3	3.84	46.2	205.09	177.40	86
11:00	3.02	62.2	3.82	43.4	188.15	166.00	86
11:10	3.15	63.1	3.74	45.8	198.76	171.29	88
11:20	3.15	62.4	3.84	44.0	196.56	168.96	86
11:30	3.10	63.0	3.77	44.8	195.30	169.12	86
11:40	3.12	62.9	3.77	44.8	196.56	169.12	86
11:50	3.12	62.6	3.82	44.0	195.62	168.30	86
12:00	3.19	62.8	3.90	44.5	200.33	173.55	86
12:10	3.15	62.8	3.87	44.2	197.82	171.27	86
12:20	3.12	62.3	3.87	43.4	194.68	167.95	86
12:30	3.12	62.1	3.88	42.5	194.06	164.90	85
12:40	3.15	62.5	3.90	43.4	196.87	169.26	86
12:50	3.12	62.0	3.88	42.5	193.44	164.90	85
13:00	3.11	61.6	3.92	41.8	191.57	164.06	85
13:10	3.08	61.2	3.89	41.4	188.49	161.04	85
13:20	3.04	61.2	3.87	41.2	186.04	159.65	86
13:30	3.00	60.6	3.85	40.0	181.80	154.00	85
13:40	3.01	60.4	3.88	39.5	181.80	153.26	84
13:50	3.00	60.0	3.95	39.0	180.00	154.05	86
14:00	3.05	60.5	3.98	39.5	184.53	157.01	85
14:10	2.94	59.4	3.95	37.5	174.64	148.12	85
14:20	2.83	58.6	3.88	36.4	165.84	141.05	85
14:30	2.75	57.6	3.96	34.0	158.40	134.64	85
14:40	2.62	56.9	3.89	32.0	149.08	124.48	84
14:50	2.56	56.4	3.89	31.0	144.38	120.48	84
15:00	2.47	55.5	3.82	30.0	137.36	114.75	84

The DC–DC converter efficiency depends on the technology used in its design. The system can be improved by using a more sophisticated maximum–power point tracking. Investigations will be pursued in order to assess the performance of the pump and the degree of reliability of the DC–DC converter over a period of at least 1 year.

NOMENCLATURE

g	Acceleration due to gravity ($m\,s^{-2}$)
H	Total manometric head (m)
I_L	Load current (A)
I_P	Module current (A)
N	Speed ($rad\,s^{-1}$)
P_e	Solar irradiance intensity (W)
P_h	Hydraulic power (W)
P_m	Motor power (W)
$\dot{Q}$	Flow rate ($m^3\,s^{-1}$)
T	Pulse period (s)
T_C	Time that switch is closed (s)
V_L	Load voltage (V)
V_P	Module voltage (V)
η	DC–DC converter efficiency (%)
ρ	Mass per volume ($kg\,m^{-3}$)
τ	Torque (N m)

BIBLIOGRAPHY

Matsuo H and Murozono M 1984 *Proc. 1st Int. Photovoltaic Science and Engineering Conf., Kobe, November 1984* Technical Digest
Roger J 1979 *Solar Energy* **23** 193–8
Wagdy R A 1985 *Solar Cells* **14** 27–42

8 Centrifugal Photovoltaic Pumping: Theoretical Study of Direct-coupled Pump

8.1 INTRODUCTION

Despite the present high prices of solar cells, terrestrial applications of small units of a direct-coupled centrifugal pump driven with photovoltaic (PV) electricity have multiplied during the past few years.

In water-pumping systems, the use of batteries and voltage regulators can be eliminated by carefully choosing the components of the system. Direct coupling between the PV modules and the motor–pump arrangement makes the system simple and autonomous. Repairs and maintenance are reduced to a minimum. Moreover, the system starts to pump in the morning and stops in the evening without any external intervention and can operate with a good adaptation to the available solar power. With these features, the direct-coupled system becomes appropriate for remote regions in developing countries.

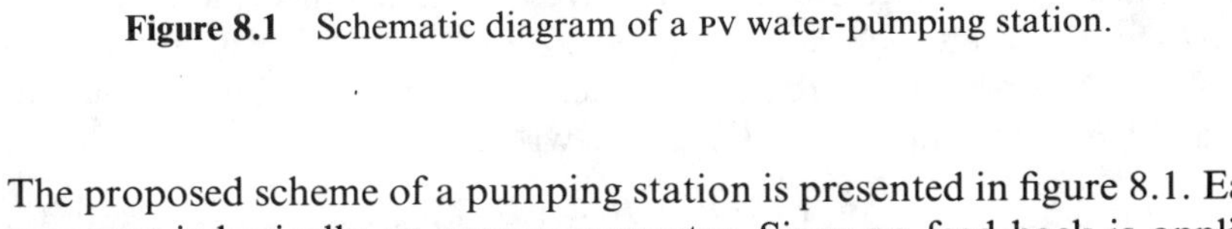

Figure 8.1 Schematic diagram of a PV water-pumping station.

The proposed scheme of a pumping station is presented in figure 8.1. Each component is basically an energy converter. Since no feed-back is applied, the system is called open loop, and static operation can be modelled by writing operating equations for each component.

8.2 MODEL OF THE PHOTOVOLTAIC GENERATOR

The relation between current and voltage may be written for a single solar cell as

$$I = I_{ph} - I_0\left[\exp\left(\frac{q(V+IR_s)}{Ak_BT}\right) - 1\right] \tag{8.1}$$

where I_{ph} (A) is the photocurrent, I_0 (A) is the reverse saturation current, q (C) is the electronic charge, R_s (Ω) is the series resistance, k_B ($J\,K^{-1}$) is Boltzmann's constant, A is the ideality factor, T (K) is the cell temperature, I (A) is the cell current and V (V) is the cell voltage.

In this relation, the shunt resistance between the terminals of the cell is assumed to be very large and is neglected. This is justified in the case of a well constructed small-area solar cell.

The photogenerated current I_{ph} is a function of cell temperature and irradiance:

$$I_{ph} = f(T, R). \tag{8.2}$$

By differentiating the above equation,

$$\Delta I_{ph} = \left(\frac{\partial I_{ph}}{\partial R}\right)_T \Delta R + \left(\frac{\partial I_{ph}}{\partial T}\right)_R \Delta T + \frac{1}{2!}\left(\frac{\partial^2 I_{ph}}{\partial R^2}(\Delta R)^2 + 2\frac{\partial^2 I_{ph}}{\partial T\,\partial R}\Delta T\,\Delta R + \frac{\partial^2 I_{ph}}{\partial T^2}(\Delta T)^2\right) + \dots \tag{8.3a}$$

To represent a solar cell by a more exact model, the higher-order terms should also be considered. However, the error introduced by the higher-order terms are small at high cell temperatures and even smaller at moderate temperatures and irradiances. Hence, neglecting the higher-order terms, we have

$$\Delta I_{ph} = \left(\frac{\partial I_{ph}}{\partial R}\right)_T \Delta R + \left(\frac{\partial I_{ph}}{\partial T}\right)_R \Delta T. \tag{8.3b}$$

The linear dependence of the photocurrent on irradiance and temperature at low to moderate values of irradiance and cell temperature has been established from both theoretical considerations and experimental measurements. Therefore, equation (8.3*b*) can be written as

$$I_{ph} = B_0 + B_1 R + B_2 T \tag{8.4}$$

where B_0, B_1 and B_2 could be easily found from experimental observations or from the manufacturer's specifications.

A PV generator consists of a number of solar cells which are connected in series and parallel to achieve the required voltage and current. The current–voltage (I–V) relation could be approximated by scaling the characteristics of a single solar cell. Suppose that the generator has M cells in parallel and N cells in series; from equation (8.1),

$$\frac{I}{M}=I_{\mathrm{ph}}-I_0\left[\exp\left(\frac{q(V/N+IR_s/M)}{Ak_BT}\right)-1\right]$$

or

$$I=MI_{\mathrm{ph}}-MI_0\left[\exp\left(\frac{q(MV+NIR_s)}{ANMk_BT}\right)-1\right] \tag{8.5}$$

where I and V are now the current and voltage of the PV generator. In this simple model, it is assumed that the cells are identical. The effects due to mismatching of cell characteristics are not considered.

The saturation current I_0 is a function of temperature and may be written in the form

$$I_0(T)=K_{\mathrm{dev}}n_i^2(T) \tag{8.6}$$

where K_{dev} is a device-specific constant, independent of the temperature. The intrinsic carrier concentration n_i^2, which is also a function of temperature, is

$$n_i^2(\mathrm{T})=A_0T^3\exp\left(\frac{E_{g0}}{k_BT}\right) \tag{8.7}$$

where E_{g0} is the energy gap of the device material at 0 K and A_0 is a device-specific constant of the material. If we consider the material to be silicon and assume typical values for the constants, I_0 can be obtained from equations (8.6) and (8.7):

$$I_0(T)=2.119\times10^3T^3\exp\left(\frac{-1.4\times10^4}{T}\right). \tag{8.8}$$

The open-circuit voltage is computed when the output current is zero:

$$V_{\mathrm{oc}}=(MC_1)^{-1}\ln\left(\frac{I_{\mathrm{ph}}+I_0}{I_0}\right) \tag{8.9}$$

where

$$C_1=\frac{q}{AMNk_BT}.$$

The short-circuit current is computed when the output voltage is zero:

$$MI_0\exp(C_1NR_sI_{\mathrm{sc}})+I_{\mathrm{sc}}-M(I_{\mathrm{ph}}+I_0)=0. \tag{8.10}$$

This equation is non-linear. The Newton–Raphson iterative technique is used to solve equation (8.10).

8.3 MAXIMUM-POWER POINT

The power output of the generator is given by

$$P = IV. \tag{8.11}$$

Figure 8.2 shows an I–V curve with three working points a, b and c. As the load resistance varies, the power extracted from the generator varies and, at one value of the resistance, the power reaches a maximum value (point b). At this point the load is 'matched' to the source and the system operates with maximum efficiency.

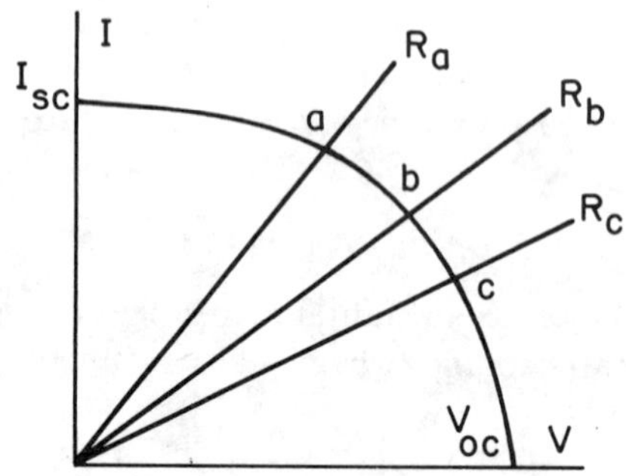

Figure 8.2 I–V characteristics of a solar cell and load resistance.

To obtain the maximum power, the differential dP should be zero. Hence,

$$dP = I\,dV + V\,dI$$

$$0 = I\,dV + V\,dI. \tag{8.12}$$

Equation (8.1) may be written as

$$C_1(MV + NIR_s) = \ln\left(\frac{M(I_0 + I_{ph}) - I}{MI_0}\right). \tag{8.13}$$

Taking the differential of V and I in equation (8.13) and using equation (8.12), we have

$$C_1(MV - NIR_s) = \frac{I}{M(I_0 + I_{ph}) - I}. \tag{8.14}$$

Solving equation (8.14) for V and substituting this expression into equation (8.13) to solve for I, we have

$$\ln\left(\frac{M(I_0 + I_{ph}) - I}{MI_0}\right) - \frac{I}{M(I_0 + I_{ph}) - I} = 2C_1NIR_s$$

or

$$x \ln x = (G_1 x + 1)(G_2 - x) \tag{8.15}$$

where

$$x = \frac{M(I_0 + I_{ph}) - I}{M I_0} \tag{8.16}$$

$$G_1 = 2C_1 N M R_s I_0 \tag{8.17}$$

$$G_2 = \frac{I_{ph}}{I_0}.$$

The non-linear equation (8.15) is solved by the Newton–Raphson method. To find a close approximation to the solution, the intersection of the two curves $Y_1 = x \ln x$ and $Y_2 = (G_1 x + 1)(G_2 - x)$ approximates the solution x. The initial value is set at

$$x_{initial} = G_2.$$

Once x is found,

$$I_{max} = M[(I_0 + I_{ph}) - x I_0] \tag{8.18a}$$

$$V_{max} = N I_{max} R_s M^{-1} + M^{-1} I_{max} C_1^{-1} [M(I_0 + I_{ph}) - I_{max}]^{-1}. \tag{8.18b}$$

The maximum power is then given by

$$P_{max} = V_{max} I_{max}. \tag{8.19}$$

A computer program can be developed to generate the module's I–V characteristics curves at various irradiance levels. The program can also compute the maximum power and the associated current and voltage. Graphics capability is also incorporated to draw the I–V curves and the locus of maximum-power points at various irradiances.

8.4 MODEL OF THE DC MOTOR

The basic equations of a DC motor are

$$E = K\varphi\omega \tag{8.20}$$

$$V_a = E + I_a R_a \tag{8.21}$$

$$\tau = K\varphi I_a \tag{8.22}$$

where K is a constant depending on the construction of the given motor, E (V) is the developed voltage, φ (Wb) is the magnetic flux, ω (rad s^{-1}) is the rotor speed, V_a (V) is the armature (rotor) terminal voltage, I_a (A) is the armature current, R_a (Ω) is the armature series resistance and τ (N m) is the induced torque.

In the case of a permanent-magnet (PM) motor, the flux is constant. The current and voltage are computed as

$$I_a = \frac{\tau}{K\varphi} \tag{8.23a}$$

$$V_a = K\varphi C_0 S + \frac{\tau R_a}{K\varphi} \tag{8.23b}$$

where

$$S = \frac{\omega}{C_0} \quad C_0 = \frac{2\pi}{60}.$$

S is the rotor speed (rev min^{-1}).

For a series motor, on the assumption that the saturation condition is not reached, the flux is directly proportional to the current:

$$\varphi = K_\varphi I_a \tag{8.24}$$

where K_φ is a constant relating the magnetic flux from the series coil to the armature current. Then, from the basic motor equations (8.20)–(8.22),

$$I_a = \left(\frac{\tau}{KK_\varphi}\right)^{1/2} \tag{8.25a}$$

$$V_a = (KK_\varphi C_0 S + R_a)\left(\frac{\tau}{KK_\varphi}\right)^{1/2}. \tag{8.25b}$$

Finally, for a shunt motor, the flux is proportional to the applied voltage:

$$\varphi_1 = K_{\varphi 1} I_f = K_{\varphi 1}\frac{V_a}{R_f}. \tag{8.26}$$

The motor's terminal current is

$$I_{a0} = I_a + I_f. \tag{8.27}$$

If we solve the last two equations with the basic equations for the motor, the current and voltage can be evaluated:

$$I_{a0} = \left(\frac{\tau}{KK_{\varphi 1}K_2(1-K_2)}\right)^{1/2} \tag{8.28a}$$

$$V_a = R_f K_2 I_{a0} \tag{8.28b}$$

where

$$K_2 = \frac{R_a}{R_f + R_a - KK_{\varphi 1}C_0 S}.$$

8.5 MODEL OF THE CENTRIFUGAL PUMP

The mechanical load characteristics can be described by

$$\tau_L = \tau_0 + (\tau_{nL} - \tau_0)\left(\frac{S}{S_0}\right)^x \tag{8.29}$$

where

$x=0$	for a constant torque
$x=1$	for viscous friction
$x=2$	for a centrifugal pump at constant head.

Hence, for a centrifugal pump,

$$\tau_L = \tau_0 + (\tau_{nL} - \tau_0)\left(\frac{S}{S_0}\right)^2.$$

For simplicity, this equation is written as

$$\tau_L = A + BS^2 \tag{8.30}$$

where

$$A = \tau_0$$

$$B = \frac{\tau_{nL} - \tau_0}{S_0^2}.$$

The pump is modelled by the characteristic surface of H, Q and S where H (m) is the head, Q (m^3) is the flow and S (rev min^{-1}) is the speed and these parameters have the relationship

$$H = f(Q, S).$$

Alternatively, the equation may be written as

$$H = A_1Q^2 + 2B_1SQ + C_1S^2 \tag{8.31}$$

where A_1, B_1 and C_1 are constants for a given pump and could be found either from the test data or from the manufacturer's specifications.

Now, the current and voltage can be determined for each value of the speed of rotation at constant head. For the PM motor,

$$I_a = \frac{A + BS^2}{K\varphi} \tag{8.32a}$$

$$V_a = K\varphi C_0 S + R_a \frac{A + BS^2}{K\varphi}. \tag{8.32b}$$

For the series motor,

$$I_a = \left(\frac{A+BS^2}{KK_\varphi}\right)^{1/2} \tag{8.33a}$$

$$V_a = (KK_\varphi C_0 S + R_a)\left(\frac{A+BS^2}{KK_\varphi}\right)^{1/2}. \tag{8.33b}$$

For the shunt motor,

$$I_{a0} = \left(\frac{A+BS^2}{KK_{\varphi 1}K_2(1-K_2)}\right)^{1/2} \tag{8.34a}$$

$$V_a = R_f K_2\left(\frac{A+BS^2}{KK_{\varphi 1}K_2(1-K_2)}\right)^{1/2}. \tag{8.34b}$$

In all three cases the voltage and current can be found independently of the PV module's characteristics. They are only determined by the load and motor characteristics, for each value of speed at constant head. Hence, it is possible to draw a I–V operating curve for the motor–pump system. The operating points are the intersections of the module's I–V curves and the motor operating (working) curve. At the operating point, the power needed to achieve the speed S, at a given head, is extracted from the module at the corresponding irradiance.

By drawing the operating curves of the motor on the PV module's I–V curves, the motor whose operating curve follows closely the locus of the module's maximum-power points at all irradiance levels provides the best adaptation to the module.

8.6 STARTING TORQUE OF THE DC MOTOR

The starting torque developed by a motor should be greater than the frictional torque in order to accelerate the rotational speed of the shaft from zero to the value determined by the irradiance level and load. Therefore the motor which produces a high starting torque at low irradiance levels is always preferred.

The speed–torque relations are obtained for the three types of DC motors from the PV module and motor equations.

For the PM motor,

$$S = \frac{1}{MK\varphi C_1 C_0}\ln\left(\frac{I_{ph}}{I_0} - \frac{\tau}{MI_0K\varphi}\right) - \frac{\tau(MR_a+NR_s)}{MK^2\varphi^2C_0}. \tag{8.35}$$

For the series motor,

$$S = \frac{1}{MKK_\varphi C_1 C_0}\left(\frac{KK_\varphi}{\tau}\right)^{1/2}\ln\left[\frac{I_{ph}}{I_0} - \frac{1}{MI_0}\left(\frac{\tau}{KK_\varphi}\right)^{1/2}\right] - \frac{NR_s+MR_a}{MKK_\varphi C_0}. \tag{8.36}$$

For the shunt motor,

$$\left(\frac{(R_f-KK_{\varphi 1}C_0S)\tau}{KK_{\varphi 1}R_a}\right)^{1/2}\left(\frac{R_fR_a}{R_f-KK_{\varphi 1}C_0S}+\frac{NR_s}{M}\right)$$
$$=\frac{1}{MC_1}\ln\left[\frac{I_{ph}}{I_0}-\frac{1}{MI_0}\left(\frac{(R_f-KK_{\varphi 1}C_0S)\tau}{KK_{\varphi 1}R_a}\right)^{1/2}\right]. \quad (8.37)$$

The derivation of the starting torque that follows assumes that the module reverse saturation current is small compared with the photocurrent and the armature resistance of the shunt motor is small compared with the field resistance. With these assumptions, the starting torque is found when the speed of the motor is set to zero. The following non-linear equations are obtained for the PM, series and shunt motor starting torques.

The starting torque τ_s for a PM motor is

$$MI_0\exp\left(\frac{C_1\tau_s(MR_a+NR_s)}{K\varphi}\right)=MI_{ph}-\frac{\tau_s}{K\varphi}. \quad (8.38)$$

The starting torque τ_s for a series motor is

$$MI_0\exp\left[C_1\left(\frac{\tau_s}{KK_\varphi}\right)^{1/2}(MR_a+NR_s)\right]=MI_{ph}-\left(\frac{\tau_s}{KK_\varphi}\right)^{1/2}. \quad (8.39)$$

The starting torque τ_s for a shunt motor is

$$MI_0\exp\left[C_1\left(\frac{R_f\tau_s}{KK_{\varphi 1}R_a}\right)^{1/2}(MR_a+NR_s)\right]=MI_{ph}-\left(\frac{R_f\tau_s}{KK_{\varphi 1}R_a}\right)^{1/2}. \quad (8.40)$$

The Newton–Raphson method is used to solve these non-linear equations. The initial approximations, computed for the PM, series and shunt motors respectively, are

$$\tau_{s(\text{initial})}=MI_{ph}K\varphi \quad (8.41a)$$

$$\tau_{s(\text{initial})}=KK_\varphi(MI_{ph})^2 \quad (8.41b)$$

$$\tau_{s(\text{initial})}=M^2I_{ph}^2\frac{KK_{\varphi 1}R_a}{R_f}. \quad (8.41c)$$

Starting torques at different irradiance levels are computed by determining I_{ph} for each irradiance level.

8.7 EVALUATION OF THE THEORETICAL OPERATING SPEED

The following two configurations are considered, in order to obtain the theoretical performance of PV pump with a PM motor and centrifugal pump.

(*a*) Direct-coupled system with no battery storage.
(*b*) Direct-coupled system with a maximum-power point tracker (MPPT) as an interface device.

The operating speeds are computed at constant head and at various irradiance levels for the two configurations.

8.7.1 Direct-coupled systems

The PV generator is directly connected to the motor. Operating speeds at various operating points, for given irradiance levels and at constant head, are found by solving the following non-linear equation, which in turn is obtained from the equations for the motor, pump and module:

$$\frac{S^2B(MR_a+NR_s)}{K\varphi}+MK\varphi C_0S+\frac{A(MR_a+NR_s)}{K\varphi}=\frac{1}{C_1}\ln\left(\frac{I_{ph}}{I_0}-\frac{A+BS^2}{K\varphi MI_0}\right)^{1/2}. \tag{8.42}$$

The Newton–Raphson technique is used with the initial approximation

$$S_{initial}=\left(\frac{K\varphi MI_{ph}-A}{B}\right)^{1/2}$$

which is a value close to the solution.

8.7.2 System with maximum-power point tracker

The MPPT considered here uses the module's I–V characteristics as its control variables. It operates such that the power supplied by the PV generator to the motor is always maximum. Hence the operating point is the intersection of the operating curve of the motor at constant head and the curve of constant maximum power at a given irradiance level. The operating point is then found by solving equations (8.19) and (8.23). Figure 8.3 illustrates the different I–V curves. The speed at the maximum power point is computed from

$$\left(K\varphi C_0S+\frac{R_a}{K\varphi}(A+BS^2)\right)\frac{A+BS^2}{K\varphi}=P_m. \tag{8.43}$$

The intersection of the two curves $Y_1=P_mK^2\varphi^2/(A+BS^2)$ and $Y_2=R_aBS^2+K^2\varphi^2C_0S+R_aA$ determines S. The initial value for the Newton–Raphson method is taken as the point of intersection of the line which is tangent to the curve Y_1 at $S=0$ and the curve Y_2. The value is approximately

$$S_{initial}=\left(\frac{P_m}{C_0B}\right)\times 0.333.$$

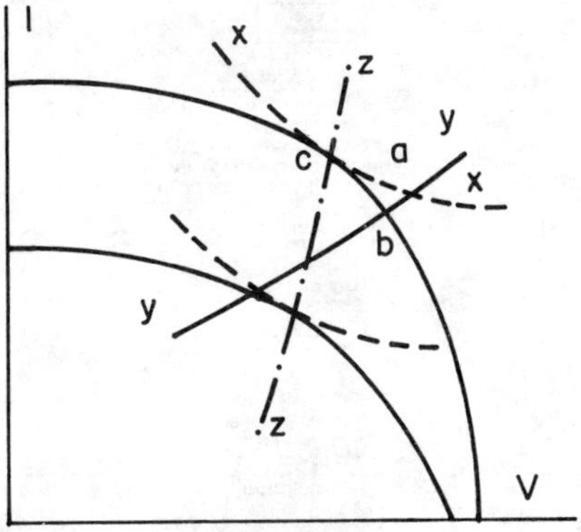

Figure 8.3 Operating points of the direct-coupled and MPPT system: a, operating point with MPPT; b, operating point with direct coupling; c, maximum-power point; xx, constant-maximum-power curve; zz, locus of the maximum-power point; yy, operating curve of motor.

Note that it is considered here that the transfer efficiency of the MPPT device is 100%, which is fairly realistic, since experimental efficiencies of more than 95% have been reported.

8.8 EVALUATION OF THE THEORETICAL EFFICIENCY

Once the operating speed is known, the flow for a given head can be calculated. Then the efficiencies of the system components can be obtained for both configurations. The output and input powers are

$$P_{out} = \rho g H \dot{Q} \tag{8.44}$$

$$P_{in} = RANM \times 10^{-4} \tag{8.45}$$

where ρ (kg m^{-3}) is the density of water, $g = 9.8$ m s^{-2}, $\dot{Q}$ (m^3 s^{-1}) is the flow rate, H (m) is the head, R (W m^{-2}) is the solar radiation intensity, A (cm^2) is the solar cell surface area, and N and M are the number of cells in series and parallel respectively.

The overall efficiency would be

$$\eta_0 = \frac{\rho g H \dot{Q} \times 10^4}{RANM}. \tag{8.46}$$

When the torque is known from the equations for the motor, the efficiencies of the pump and motor can be found. The transmission losses are assumed to be small and neglected. For the pump,

$$\eta_p = \frac{\rho g H \dot{Q}}{\tau C_0 S}. \tag{8.47}$$

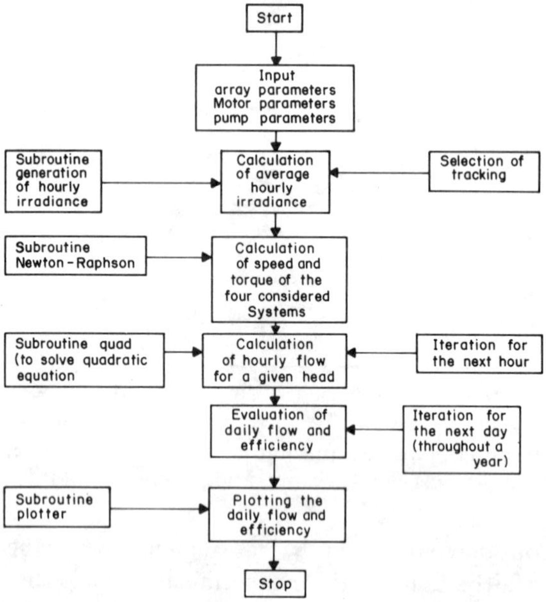

Figure 8.4 Flow chart of simulation program.

For the motor,

$$\eta_m = \frac{C_0 \tau S}{V_a I_a}. \tag{8.48}$$

For the motor–pump subsystem,

$$\eta_{mech} = \eta_p \eta_m. \tag{8.49}$$

With a computer program developed (figure 8.4), the performance curves of both systems are obtained.

8.9 RESULTS AND DISCUSSION

The prediction of daily flow and efficiency have been carried out by computer simulation, using meteorological data for Thailand, for a direct-coupled system and a MPPT system with, firstly, a PV module fixed–tilt at a latitude angle and, secondly, a PV module mounted on a tracking system about the north–south axis.

8.9.1 Theoretical performance of the photovoltaic generator

The parameters of the PV generator needed for the computations were obtained from manufacturer's specifications.

The theoretical variations in the characteristic variables of the PV generator (the open-circuit voltage, the short-circuit current, the maximum-power voltage and current) against irradiance level are shown in table 8.1. The graphical representation of the I–V relationship at various irradiance levels is illustrated later in figure 8.6. It can be seen that, while the short-circuit current varies greatly with the irradiance level, the open-circuit voltage does not. This is because the photocurrent I_{ph}, and hence the short-circuit current, is a linear function of illumination flux. The open-circuit voltage is a logarithmic function of the irradiance level.

Table 8.1 Theoretical performance of the PV generator (at 30 °C).

Irradiance ($\mathrm{W\,m^{-2}}$)	V_{oc} (V)	V_{max} (V)	I_{max} (A)	I_{sc} (A)	P_{max} (W)	Efficiency (%)
1000	37.06	27.96	14.99	16.24	419.21	15.16
900	36.85	28.10	13.53	14.63	380.25	15.27
800	36.62	28.22	12.06	13.02	340.45	15.39
700	36.36	28.32	10.59	11.41	299.84	15.49
600	36.06	28.38	9.11	9.80	258.47	15.58
500	35.70	28.39	7.62	8.20	216.40	15.65
400	35.27	28.33	6.13	6.59	173.73	15.70
300	34.71	28.16	4.64	4.98	130.60	15.74

The relation between current and voltage at maximum-power points is shown in the same figure by an almost vertical line. The voltage at the maximum-power point remains nearly constant at any irradiance level. The evaluation of the maximum-power line is important, in the interest of designing the components of a PV system, so as to maximize power extraction from the modules.

The maximum efficiencies, as can be seen from table 8.1, are almost constant around 15%. This value is slightly higher than the efficiency of the commercial silicon solar cells, which are now up to 14%. This difference is caused by the assumptions made in formulating the model of the PV generator.

8.9.2 Influence of temperature

The effect of temperature on the I–V characteristics is illustrated in figure 8.5 and table 8.2. The open-circuit voltage decreases with increasing temperature, such that

$$\frac{\mathrm{d}V_{oc}}{\mathrm{d}T} = -3.2\ \mathrm{mV\,^{\circ}C^{-1}}.$$

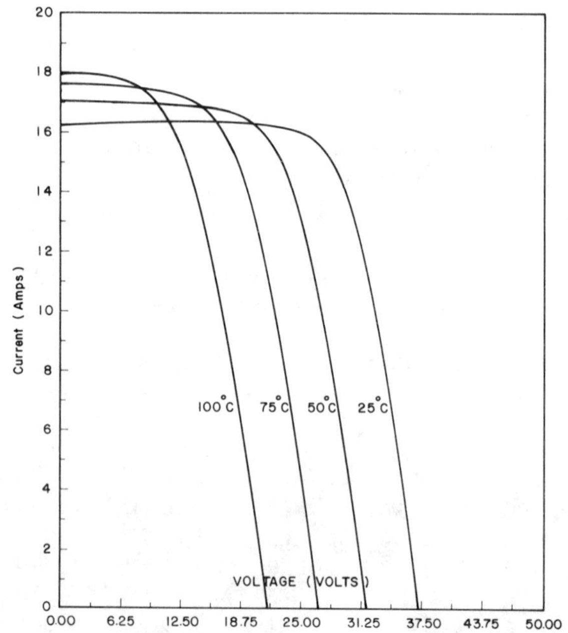

Figure 8.5 *I–V* curves of a PV DC pumping system (at different temperature levels and 1 $kW\,m^{-2}$ irradiance).

The module reverse saturation current, which increases sharply with increasing temperature, causes the open-circuit voltage to drop, whereas the short-circuit current increases slightly with increasing temperature, such that

$$\frac{dI_{sc}}{dT} = +56\ \mu A\ cm^{-2}\ {}^{\circ}C^{-1}.$$

The overall result of these variations is a reduction in the efficiency with an increase in temperature:

$$\frac{1}{\eta}\frac{d\eta}{dT} = 6.7 \times 10^{-3}\ {}^{\circ}C^{-1}.$$

Table 8.2 Influence of temperature.

Temperature (K)	V_{oc} (V)	V_{max} (V)	I_{max} (A)	I_{sc} (A)	P_{max} (W)	Efficiency (%)
300	37.06	27.96	14.99	16.24	419.21	15.16
325	21.99	23.04	15.10	16.81	347.99	12.59
350	26.89	18.33	14.96	17.38	274.28	9.92
375	21.76	13.90	14.40	17.94	200.23	7.24

These values differ slightly from typical values, which are $-2.3\ mV\ °C^{-1}$, $20\ \mu A\ cm^{-2}\ °C^{-1}$ and $5.3 \times 10^{-3}\ °C^{-1}$ respectively. The difference could have originated from the assumptions made in the model.

The influence of the temperature becomes significant in high-temperature applications, such as in a system with a concentrator, in large-scale PV systems and in space applications. However, in small-scale PV systems the effect of temperature can be considered to be small. In this study the cell temperature is assumed to be constant at 30 °C.

8.9.3 DC motor analysis

The components of a direct-coupled system should be chosen such that maximum possible power is extracted from the modules. Moreover, since a PV generator behaves differently from conventional power sources, the performance of the known components should be optimized. Typical values of the motor's parameters, for a rated motor power in the range 300–400 W and a rated motor voltage of around 30 V, are used in the computations.

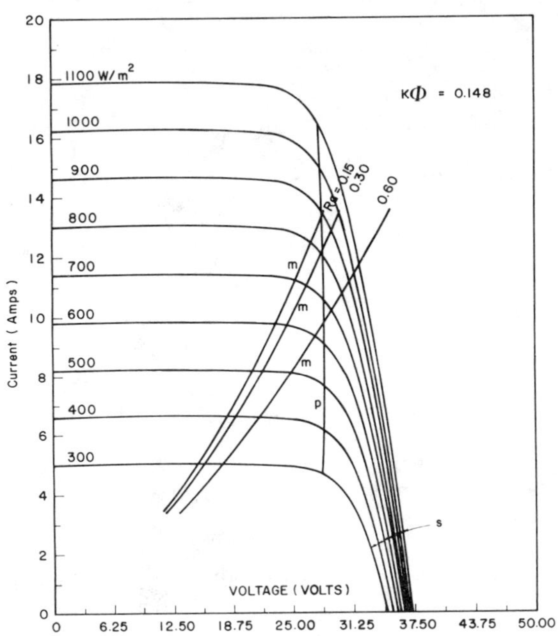

Figure 8.6 *I–V* curves of a PV pumping system (PV module *I–V* curves at different irradiances and DC PM motor *I–V* curves at different series resistance values): p, maximum-power line; s, PV module's *I–V* curve; m, operating curve of motor.

However, the typical values used for the PM motor are closer to the measured parameter values obtained from testing a Guinard motor.

In choosing a suitable motor for the system, AC motors are not taken into consideration because of their lower efficiency at smaller ratings and the need for additional components, such as DC–AC converters and batteries. In this section therefore we consider DC motors of the PM, series and shunt types of motor. The operating curves of these motors for different values of their parameters and at constant head, drawn on the module curves, are shown in figures 8.6–8.10.

The best module–motor characteristic matching would only be obtained if the operating curve of the motor followed the locus of maximum-power points of the module's I–V curves closely. By looking at the operating curves of the PM, series and shunt motors in the above-mentioned figures, it can be seen that the operating curve of the PM motor follows the line of maximum-power points more closely than do the operating curves of the other two motors. This means that the power extracted from the modules with the PM motor is closer to the maximum power generated at more irradiance levels than with the other two types of motor. Moreover, the parameters of the motor also influence the operating curve. Figures 8.6 and 8.7 show that,

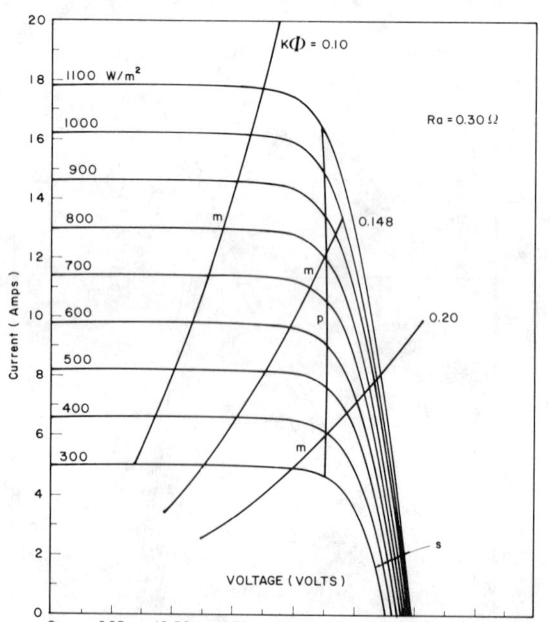

Figure 8.7 I–V curves of a PV pumping system (PV module I–V curves at different irradiances and DC PM motor I–V curves at different magnetic field constants): p, maximum-power line; s, PV module's I–V curve; m, operating curve of motor.

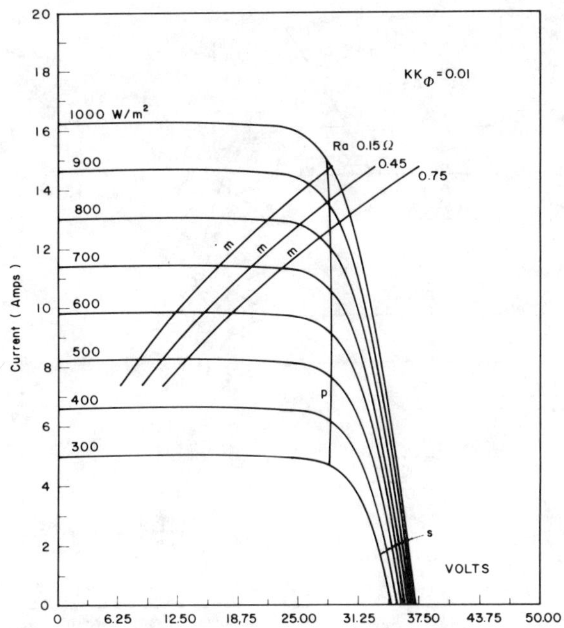

Figure 8.8 *I–V* curves of a PV pumping system (PV module *I–V* curves at different irradiances and DC series motor *I–V* curves at different series resistance values): p, maximum-power line; s, PV module's *I–V* curve; m, operating curve of motor.

although the motor operating curve does not shift very much with changes in armature resistance, it shifts to a larger extent with changes in $K\varphi$.

Hence the parameters R_a and $K\varphi$ for the PM motor should be chosen for better adaptation to the associated PV module. The variation in the starting torque produced by the three motors with the irradiance level is shown in figure 8.11. The figure shows that the starting torques produced by the PM motor at nearly all irradiance levels except higher levels are greater than the starting torques developed by the other two types of motor. This feature, too, enhances the use of PM motors in a direct-coupled system.

Moreover, PM motors are relatively smaller in size, highly efficient and inexpensive. The low cost of this type of motor is due to the cheaper high-flux Alinco magnets, a recent development in ceramic magnetic materials. Series motors also give reasonably good adaptation, if not the best, as a component of a direct-coupled system. The problem with this type of motor is that they can reach very high speeds up to a dangerous limit at low shaft outputs. This is a real possibility if the water level in the well drops below the inlet of the pump and power is still being supplied by the modules; this 'running away', as it is usually called, may damage the motor. Therefore an additional

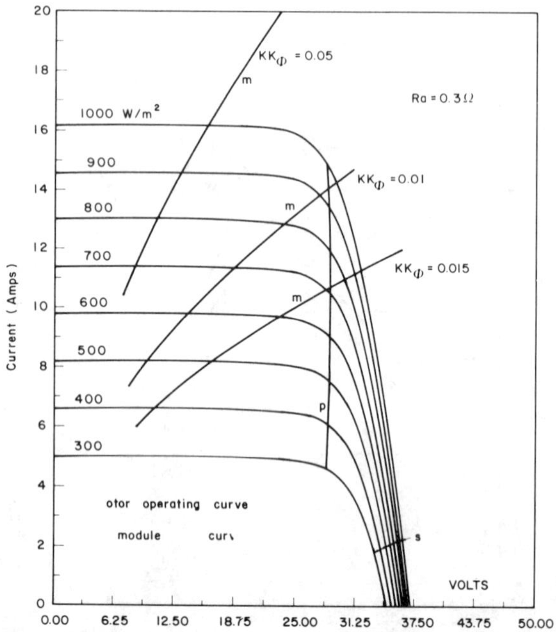

Figure 8.9 *I–V* curves of a PV pumping system (PV module *I–V* curves at different irradiances and DC series motor *I–V* curves at different magnetic field constants).

protective device is required. Hence the series motor is not a reliable component of a PV pumping system which is usually installed in remote areas.

The operating curves of shunt motors do not follow the line of maximum-power points. Moreover, the operating curves are almost flat in the operating ranges of currents and voltages. This indicates that these motors do not respond very much to irradiance even at higher levels. Besides, the starting torque is low (figure 8.11) at nearly all irradiance levels. These observations establish the fact that shunt motors are inappropriate for PV pumping purposes.

The analysis of the interaction between the DC motor and the PV generator provides an interesting design criterion other than the selection of a suitable motor. At the point of intersection of the motor's operating curve and the line of the module's maximum-power points, maximum power is extracted, at a given head, from the PV generator. If the average irradiance level of a region in which the system is to be installed is known, then the parameters of the system components could be chosen such that the operating curve cuts the line of maximum-power points at around this irradiance level. Thereby, a higher system efficiency and in turn higher flow could be obtained.

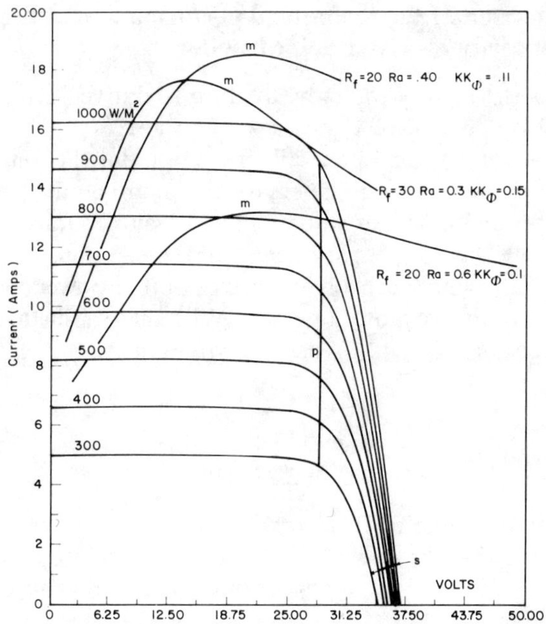

Figure 8.10 *I–V* curves of a PV pumping system (PV module *I–V* curves at different irradiances and DC shunt motor *I–V* curves at different internal parameter values).

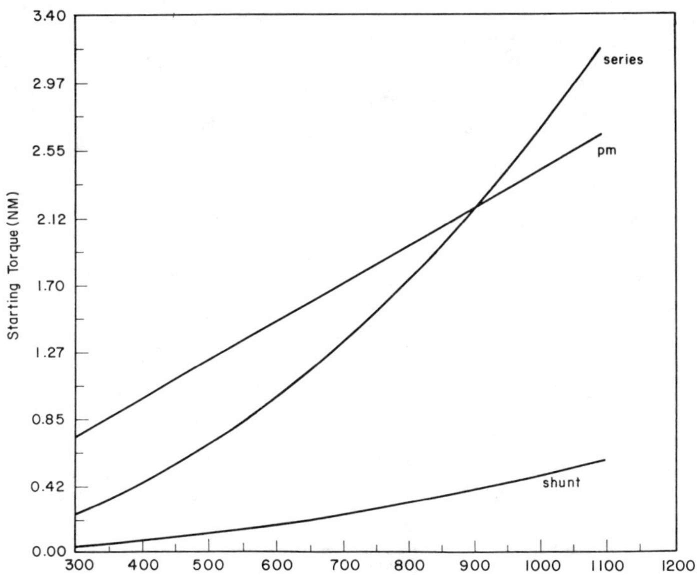

Figure 8.11 Variation in starting torque with irradiance.

8.9.4 Comparison of the theoretical performances of the direct-coupled and maximum-power point tracker

The main idea is to study the individual and relative performances of the direct-coupled system and of a system with a MPPT.

Experimental data were not available when the theoretical study was carried out to estimate the parameters for the pump model (equations (8.29) and (8.31)). Typical parameter values are first assumed for the parameters in equation (8.31). The above parameters are then adjusted to simulate a constant head of 4 m, such that the maximum-power points and the operating curve of the PM motor intersect. With these adjustments, the values of the parameters A_1, B_1 and C_1 used in equation (8.31) are

$$A_1 = 1.85 \times 10^6$$
$$B_1 = 1.870$$
$$C_1 = 2.78 \times 10^{-6}.$$

The parameters estimated from regression analysis of experimental data, collected while testing the direct-coupled system with a Guinard (ALTA-X) pump installed later on, are found to be in the range of values assumed in the theoretical study.

The dependence of the overall efficiency on the irradiance level is illustrated in figure 8.12 for the two considered systems at a constant head of 4 m. Since the system with a MPPT transfers maximum power to the load at all

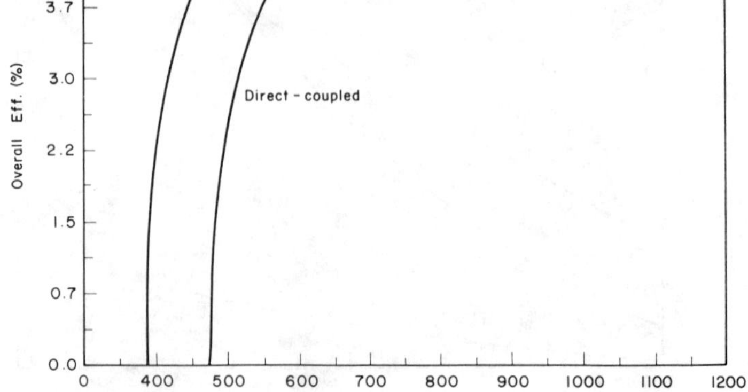

Figure 8.12 Variation in system efficiency with irradiance.

irradiance levels, its efficiency is always higher, except at one point where the efficiencies of both systems coincide. The irradiance at this point, 800 W m^{-2} corresponds to the intersection of the motor's operating curve and the line of maximum-power points of the module of a direct-coupled system. The efficiency of the direct-coupled system reaches a maximum at this irradiance. At higher irradiances the system efficiency starts to drop slightly owing to frictional losses in the pipes which are significantly higher at high flows. This is evident in the flow–speed curve. At lower speeds the increase in flow rate is higher than at higher speeds which correspond to higher irradiance levels (compare figures 8.13 and 8.14).

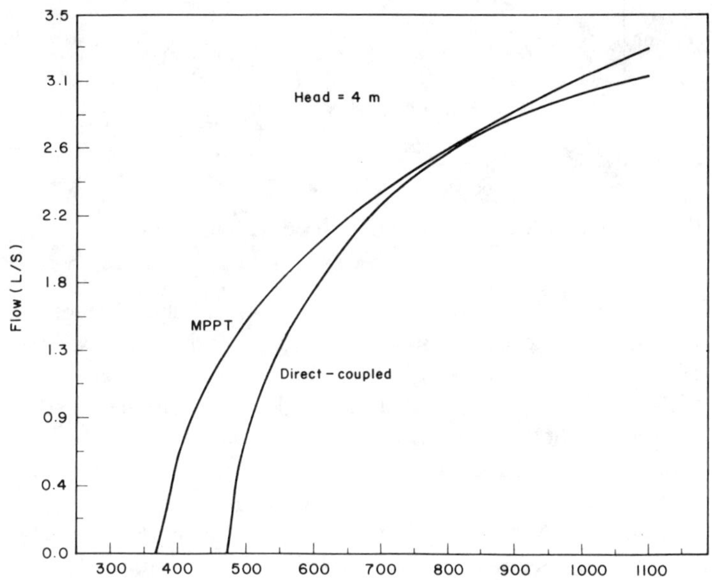

Figure 8.13 Variation in flow with irradiance.

The interesting conclusion one could draw is that with the MPPT the system could pump water at lower irradiances (around 370 W m^{-2}), whereas without the MPPT the pumping starts at a relatively higher irradiance (around 470 W m^{-2}). This remarkable feature also implies that the system with the MPPT pumps water during cloudy periods too. However, the overall system efficiency of the order of 4.5% could only be obtained above 525 W m^{-2} for the system with the MPPT and above 600 W m^{-2} for the direct-coupled system.

The facts discussed above are again evident in figure 8.13, which shows the relation between the flow and irradiance level for both systems at a constant head of 4 m. The flow becomes zero with and without the MPPT at irradiance

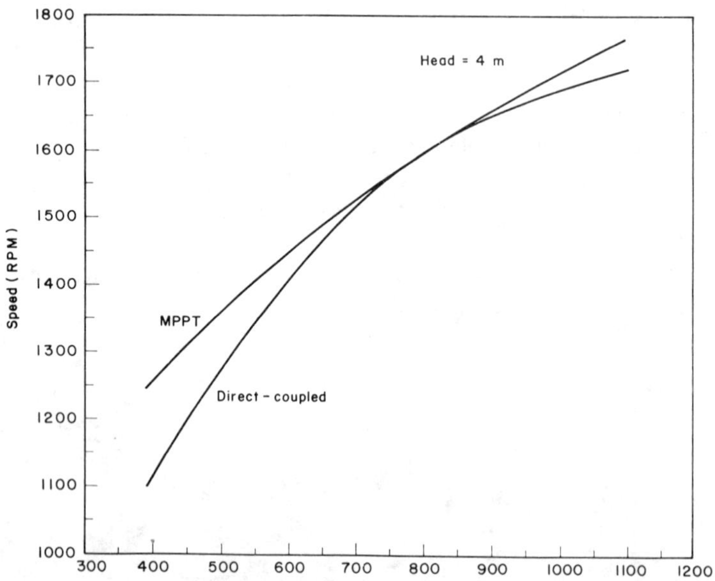

Figure 8.14 Variation in speed with irradiance.

levels of around 370 W m^{-2} and 470 W m^{-2} respectively. Further, the flow is almost the same at around 800 W m^{-2} for both systems. Again, the speed–irradiance curve shows equal speeds of rotation at around an irradiance level of 800 W m^{-2} at a constant head of 4 m. Otherwise, the speed is always greater for the system with the MPPT. When the speed is higher, the flow is higher, and hence the efficiency is higher for the system with the MPPT than for the direct-coupled system at a given irradiance level.

The complete autonomy of both systems can be realized from this theoretical performance study. The motor starts to rotate when the irradiance reaches a certain level. The speed, which does not show any abnormal behaviour with irradiance, increases with increasing irradiance level (figure 8.14). Pumping of water begins at the above-mentioned irradiance levels. Again, pumping stops when the irradiance falls below this same level. The practical implication of this simple self-governing character of the system ensures that the system can be installed in very remote areas where technical know-how of maintenance and operation is very much limited.

Figure 8.15 shows the efficiency of the 'mechanical chain' as a function of light flux at the constant head of 4 m. Here, the two curves are close together at around 800 W m^{-2}. However, the maximum 'chain' efficiency occurs for both systems at around 650 W m^{-2}. This phenomenon explains another remarkable feature in designing the system. Although good efficiencies are essential, the best matching in the system could not be obtained just by choosing the best efficiencies for the motor and pump. Since each compo-

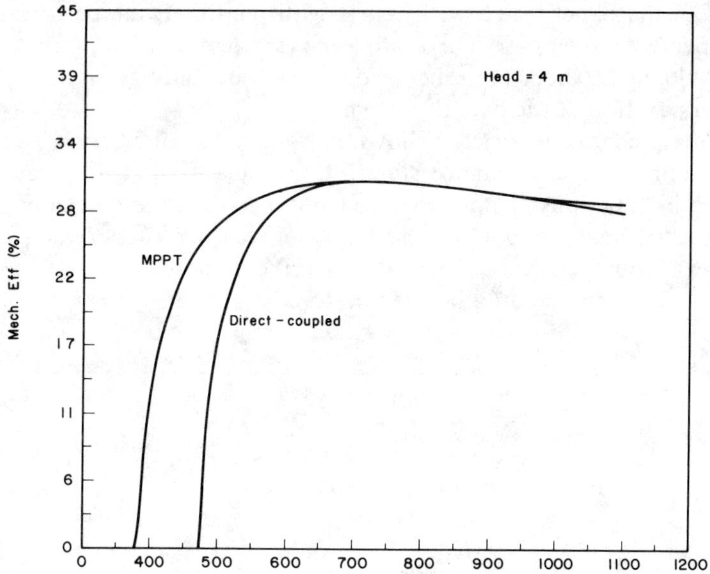

Figure 8.15 Variation in mechanical efficiency with irradiance.

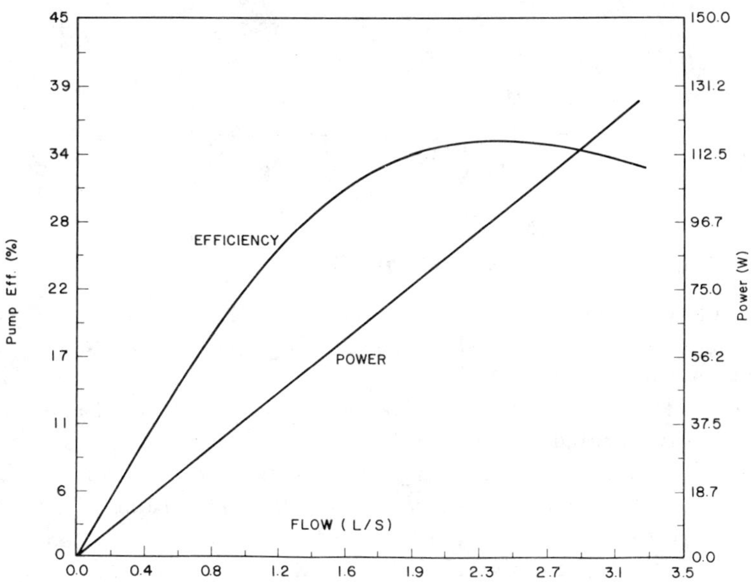

Figure 8.16 Variation in pump efficiency with flow.

nent, from the panel to the well, reacts with the other, the performance of each depends on the other. Therefore, each problem on a new well is specific and should be studied, as is being done in this study to obtain the best efficient operation of the overall system.

The pump characteristics are shown in figures 8.16 and 8.17 with the aim of completing the performance study. These curves with efficiency, power and head as functions of flow are in accordance with the typical characteristics of centrifugal pump. Once the characteristics are known, the appropriate type of pump can be selected from the manufacturer's data sheet. Alternatively, a pump could be ordered to be made.

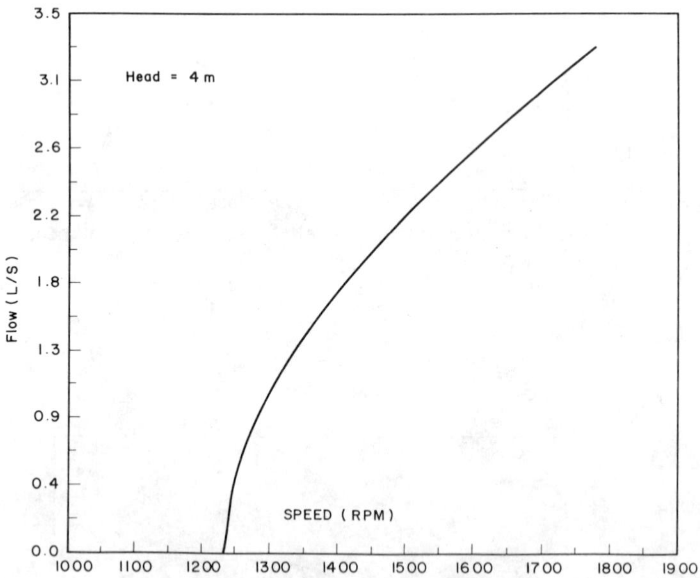

Figure 8.17 Variation in flow with speed.

8.9.5 Comparison of daily flow and efficiency

The amount of water pumped which could be expected throughout a year at the constant head of 4 m is estimated. The hourly total irradiances on an inclined plane are generated and the hourly flows, and hence the daily flows, are computed throughout a year. The daily flows are plotted against the number of days throughout the year for the four systems. Figures 8.18 and 8.19 show the average daily flow variation with and without a sun tracker for the direct-coupled system and the system with the MPPT. Table 8.3 shows the maximum, average (daily) and yearly flow for the four systems. Figures 8.20 and 8.21 illustrate the daily variations in efficiency throughout the year. Table 8.4 compares the increase in yearly flows for the four systems.

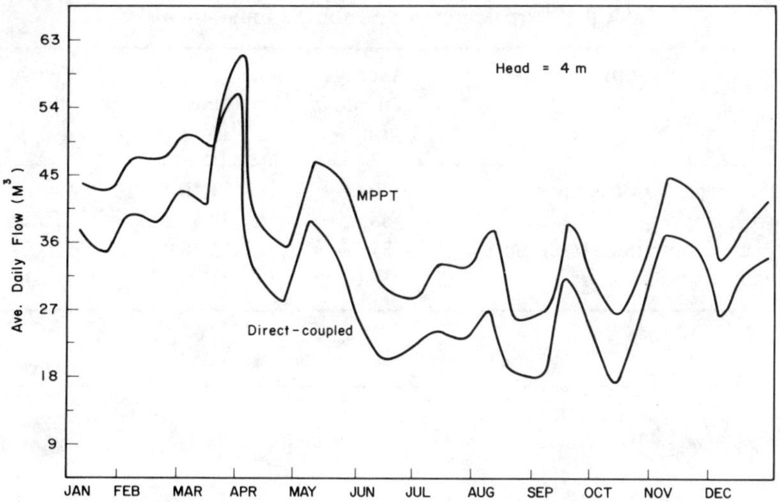

Figure 8.18 Average daily flow variation throughout a year (for direct coupling and MPPT without tracking).

It is obvious, from table 8.4, that the direct-coupled system with the sun tracker is not so attractive, although it allows a 28% increase in flow over the direct-coupled system without the tracker. Instead, the MPPT could easily be used with lower cost and only for the loss of 2.5% in yearly flow. The direct-coupled system with tracking gives an increase of 2.5% over the system with

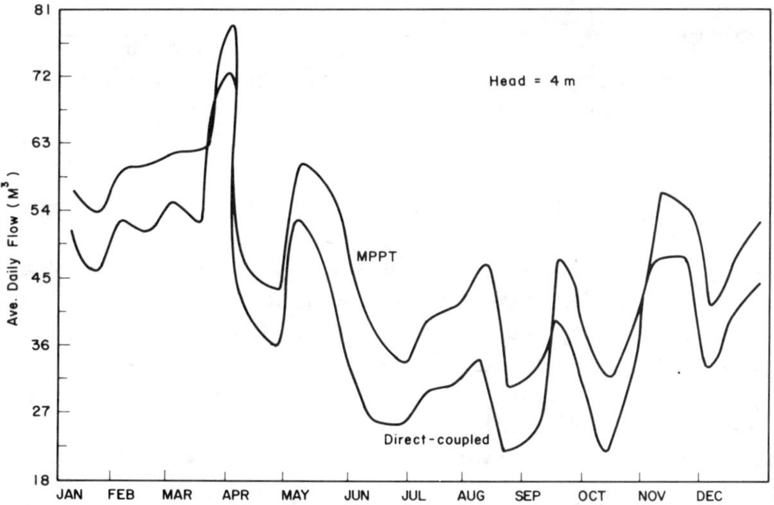

Figure 8.19 Average daily flow variation throughout a year (for direct coupling and MPPT with tracking).

Table 8.3 Prediction of flow for the four systems.

System	Average daily flow (m^3)	Maximum flow (m^3)	Yearly flow (m^3)
Direct coupled with tracking	40.05	71.50	14618
MPPT with tracking	48.55	77.80	17720
Direct coupled without tracking	31.26	55.40	11374
MPPT without tracking	39.04	61.00	14248

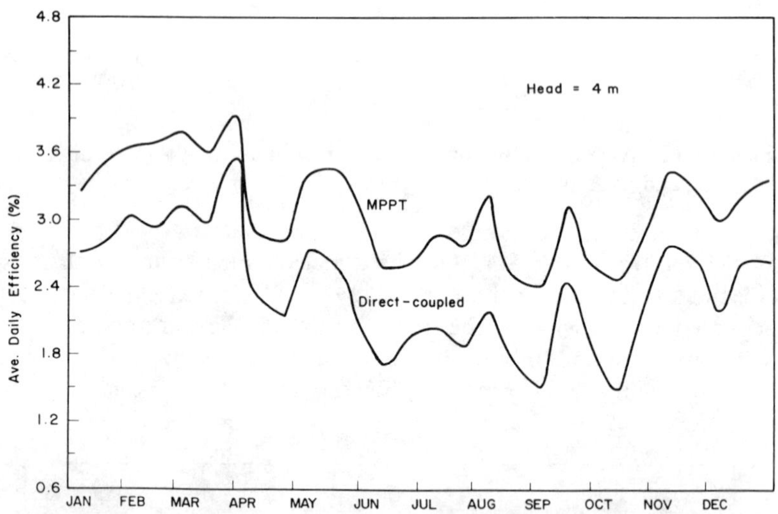

Figure 8.20 Average daily variation in efficiency throughout a year (for direct coupling and MPPT without tracking).

the MPPT without tracking. The result makes the MPPT more attractive to the pumping system.

The selection of a suitable system, however, is a compromise between the total cost, water needs, limits of the well and, finally, the available level of solar radiation in the particular region. The need to optimize the dimensions of the solar array, for a given site, arises because of the relatively high cost of solar cells. This could be computed using the computer program in this study. Once the past meteorological data of the installation site are known, the average amount of water which could be expected, subjected to the well water table and its limits, can be estimated for a given size of PV array. The array size could be varied until the expected amount of pumped water is within the range of average water needs at the given site.

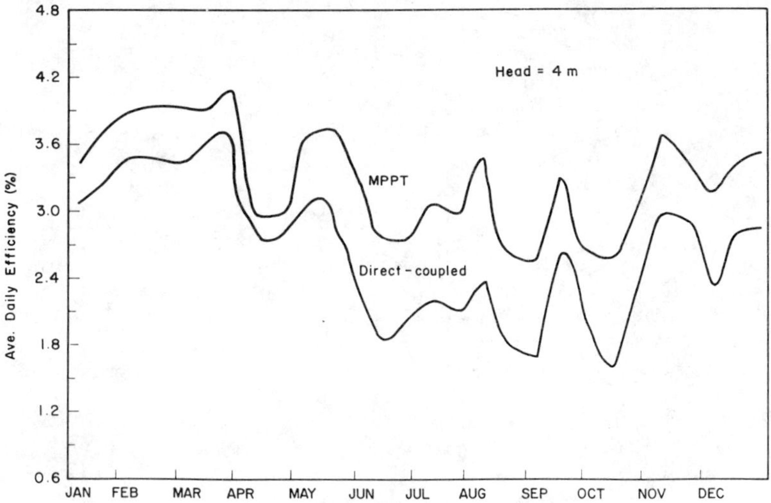

Figure 8.21 Average daily variation in efficiency throughout a year (for direct coupling and MPPT with tracking).

Table 8.4 Comparison of the increase in yearly flow for the four different systems.

	Increase (%) in yearly flow			
	Without tracking		With tracking	
	Direct coupled	MPPT	Direct coupled	MPPT
Direct coupled without tracking	—	25.3	28.5	55.8
MPPT without tracking	−25.3	—	2.5	24.4
Direct coupled with tracking	−28.5	−2.5	—	21.2
MPPT with tracking	−55.8	−24.4	−21.2	—

BIBLIOGRAPHY

Easwarakhanthan T 1981 *Thesis No ET-81-10* Asian Institute of Technology, Bangkok

Part IV

Characterization and Testing Methods

9 International Test Procedures for Photovoltaic Modules

9.1 INTRODUCTION

The main problem in photovoltaic (PV) performance measurement arises from the fact that solar cells have a highly selective spectral response and are therefore very sensitive to the spectral composition of the incident radiation. Outdoors, this varies considerably with location, weather, time of year and time of day. Indoors, it depends on the type of simulator used and it can change as the equipment ages. Unless measurement procedures take account of these variations and other difficulties, such as the marked temperature dependence of solar cells, the results can be grossly erroneous. Performance ratings of the same module by different laboratories have been known to disagree by as much as 20%.

Uncertainties of this magnitude cannot be tolerated at any stage in the development, design and manufacture of PV equipment. They can mislead research and development workers, distort cost-per-watt comparisons between competing manufacturers and result in overdesigned or underdesigned arrays. Discrepancies between manufacturer's and customer's measurements can lead to costly and time-wasting arguments and investigations. Standard measurement procedures are essential if these problems are to be avoided.

The measurement procedures described in this chapter are based on recommendations by Treble (1976) which were embodied in the current European standard specification (Anon 1980) and are now being considered by the International Electrotechnical Commission (IEC), Technical Committee 82, for inclusion in a future *IEC Standard.* Properly applied, they should reduce measurement discrepancies to no more than $\pm 2\%$. These procedures at present apply only to crystalline silicon flat-plate modules.

In a product such as a PV module, it is essential that the buyer is given some assurance that the high capital cost will be offset by a long trouble-free service life. The function of design qualification is to provide that assurance by subjecting representative samples to a series of tests to identify environmental factors and design features which could affect the attainment of a sufficiently long lifetime. Design qualification test schedules for PV modules have been published in the USA (by the Jet Propulsion Laboratory), Europe (by the Commission of European Communities (CEC) Joint Research Centre), Australia, France and Japan. A new international standard is being prepared by IEC Technical Committee 82. For the present purpose, in this chapter we shall briefly describe the current European schedule (Anon 1984), which the IEC Technical Committee is using as its main source material.

Production acceptance tests are carried out by the manufacturer on every module before it leaves the factory, to check performance and workmanship. Quality assurance tests, which may be shortened versions of some of the design qualification tests, are conducted periodically on samples from the production line to ensure that the design standard is maintained. Modifications which may affect performance or reliability should be proved by a repetition of the quality assurance tests or, in the case of major changes, by a repeat qualification. There are at present no international standards for production acceptance and quality assurance. It is left to the manufacturer to specify his own requirements in these matters.

9.2 PERFORMANCE MEASUREMENT

9.2.1 General principles

In current practice, the performance of a module or other PV device is determined by exposing it at a known temperature to stable sunlight (natural or simulated) and tracing its current–voltage (I–V) chracteristic, while at the same time measuring the incident irradiance with a reference solar cell.

The reference cell must have essentially the same relative spectral response as the modules being tested so that it 'sees' in the same way, and it must be carefully pre-calibrated by determining the ratio of its short-circuit current at 25 °C to the irradiance in sunlight having an internationally agreed reference spectral irradiance distribution. Such a cell automatically relates the measured performance to the reference spectral distribution and thus eliminates the effects of spectral variations in the natural sunlight or simulator.

For rating purposes, the measured I–V characteristics is transposed to the following standard test conditions: an irradiance of 1000 $W\,m^{-2}$ (having the reference spectral distribution); a cell temperature of 25 °C.

The power output of the module at the rating voltage (nominal working voltage) under these standard conditions is the rated power, sometimes referred to as the 'peak' power.

9.2.2 Reference spectral irradiance distribution

The current reference spectral irradiance distribution (Anon 1980) is taken from an Energy Research and Development Administration (ERDA)–National Aeronautics and Space Administration (NASA) document (Anon 1977). It is the computed distribution for direct sunlight at air mass 1.5 (AM1.5) under certain weather conditions and corresponds to an irradiance (area under the curve) of 834.6 W m^{-2}. IEC Technical Committee 82 is proposing to change this to the distribution of total (direct + diffuse) sunlight at AM1.5 on a plane surface tilted at 37° to the horizontal, with 0.2 ground reflectance, under certain specified atmospheric conditions (Bird *et al* 1983). The new distribution corresponds to an irradiance of 1000 W m^{-2}, the standard level, and is considered to be more appropriate for rating flat-plate PV devices. Figure 9.1 shows both distributions, the old one normalized to 1000 W m^{-2} for comparison. It can be noticed that the new one is slightly stronger in the blue and weaker in the red. The effect of this will be to reduce ratings by 2 or 3% to a more realistic level.

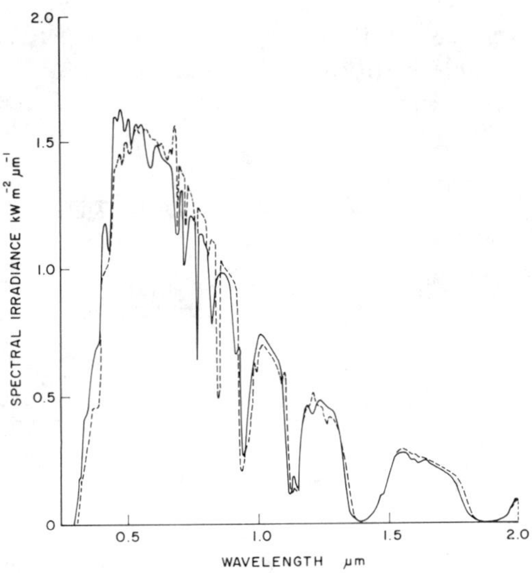

Figure 9.1 Comparison of reference spectra: ——, new AM1.5 total spectrum; ---, old AM1.5 direct spectrum (normalized to 1 kW m^{-2}).

9.2.3 General measurement requirements

The following requirements apply to all measurements.

(*a*) Temperatures must be measured to an accuracy of ±1 °C.
(*b*) Voltage and currents must be measured to an accuracy of ±0.5%, using independent leads from the terminals of the test specimen.
(*c*) Short-circuit currents of PV devices should preferably be measured at

zero voltage, using a variable bias (preferably electronic) to offset the voltage drop across the series resistance. Alternatively, they may be determined by measuring the voltage drop across a precision four-terminal fixed resistor, provided that the measurement is made at a voltage less than 3% of the open-circuit voltage, within the range where there is a linear relationship between current and voltage and, if necessary, the reading is corrected to zero voltage.

(*d*) Voltmeters must have an internal resistance of at least 20 $k\Omega\,V^{-1}$.

(*e*) The calibration of all instruments must be certified to be within the required accuracy at the time of measurement.

9.2.4 Current–voltage characteristics in natural sunlight

Measurements in natural sunlight require stable irradiance conditions, with fluctuations of less than $\pm 1\%$. For rating purposes, the irradiance must be at least 800 $W\,m^{-2}$. The procedure is as follows.

(*a*) Mount the reference cell near to and coplanar with the module on a suitable adjustable platform and connect them to the test instrumentation as shown in figure 9.2.

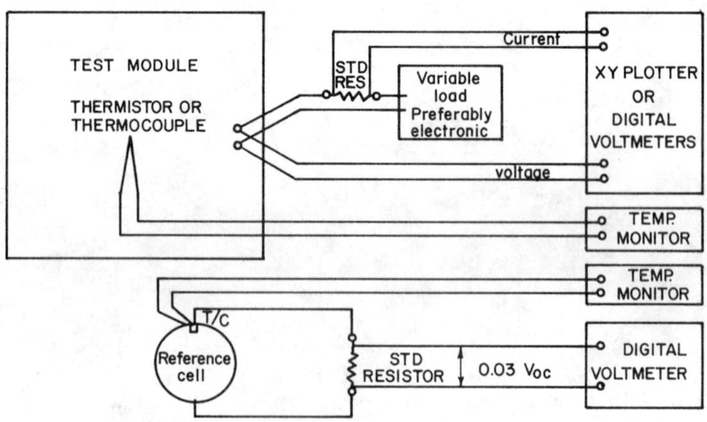

Figure 9.2 Performance test connections.

(*b*) Orient the reference cell and module so that they are normal (within $\pm 10°$) to the direct solar beam.

(*c*) Shade the module from the sun and wind until its temperature is uniformly at the ambient air temperature.

(*d*) Remote the shade and immediately record the *I–V* characteristics and temperature of the module. At the same time, record the short-circuit current and temperature of the reference cell. If the reference cell temperature differs by more than 2 °C from the temperature at which it was calibrated, a suitable adjustment must be made to the calibration value.

(*e*) Transpose the measured characteristic to the desired irradiance and temperature (see § 9.2.7).

9.2.5 Current–voltage characteristics measurement using a steady-state simulator

The simulator must meet the relevant requirements specified in § 9.2.12. The procedure is as follows.

(*a*) Mount the reference cell with its active surface in the test plane and normal (within ±5°) to the centre line of the simulator beam.

(*b*) Set the irradiance at the test plane to the desired level by adjusting it until the short-circuit current of the reference cell is at the corresponding value, according to its calibration (short-circuit current (A) = calibration value ($A\,W^{-1}\,m^2$) × desired irradiance ($W\,m^{-2}$)).

(*c*) Without disturbing the setting, remove the reference cell and mount the module in its place, with its active surface in the test plane and normal to the centre line of the simulator beam. Shade it from the beam and allow it to reach ambient temperature. (Note that, if the beam is sufficiently wide and uniform, the module may be mounted beside the reference cell, in which case the procedure is the same as in natural sunlight.)

(*d*) At the same irradiance setting, remove the shade and immediately record the I–V characteristics and temperature of the module.

(*e*) If the module is not at the desired temperature, transpose the characteristics to this temperature (see § 9.2.7).

9.2.6 Current–voltage characteristics measurement using a pulsed simulator

The simulator must meet the relevant requirements in § 9.2.12. The procedure is the same as for a steady-state simulator, except that the reference cell is mounted beside and coplanar with the module and the shading procedure is not necessary because of the negligible heat input. In some pulse simulators, the pulse is triggered by a separate photocell when the irradiance reaches a level which has been previously set with a reference cell.

9.2.7 Transposition of current–voltage characteristics

Measured I–V characteristics may be transposed to standard test conditions, or other desired temperature and irradiance within ±30% of the levels at which the measurement was carried out, by applying the following equations:

$$I_2 = I_1 + I_{sc}\left(\frac{I_{sr}}{I_{mr}} - 1\right) + \alpha(T_2 - T_1)$$

$$V_2 = V_1 - R_s(I_2 - I_1) - KI_2(T_2 - T_1) + \beta(T_2 - T_1)$$

where I_1, V_1 are coordinates of points on the measured characteristics, I_2, V_2 are coordinates of the corresponding points on the transposed characteristics, I_{sc} is the measured short-circuit current of the module, I_{mr} is the measured short-circuit current of the reference cell, I_{sr} is the short-circuit current of the reference cell at the standard or other desired irradiance, T_1 is the measured temperature of the module, T_2 is the standard or other desired temperature, α and β are the current and voltage temperature coefficients of the module in the standard or other desired irradiance and within the temperature range of interest (β is negative), R_s is the internal series resistance of the module and K is a curve correction factor.

The units of all parameters in the above equations must be consistent. When $T_2 - T_1$ is less than 2 °C, the curve correction factor K may be assumed to be unity.

9.2.8 Determination of temperature coefficients

The temperature coefficient α of current and the temperature coefficient β of voltage vary with irradiance and, to a lesser extent, with temperature. They are best measured in simulated sunlight, using a minimum of two representative solar cells of the same areas and configurations as those in the relevant module. Any mismatch between the cells in a module could adversely affect the accuracy of transpositions of the characteristics of the module.

A pulsed simulator is preferred, as it creates no additional heat and so does not affect the temperature of the test cells.

The procedure is as follows.

(*a*) Attach a suitable temperature sensor to the test cell, so that its temperature can be measured to an accuracy of ± 0.5 °C.

(*b*) Mount the test cell with good thermal contact to a temperature-controlled block and use the attached sensor to provide the control signal.

(*c*) Mount the test cell as near as possible to a suitable reference cell, with their active surfaces both in the test plane. The active surfaces must be normal (within $\pm 5°$) to the centre line of the beam.

(*d*) Set the irradiance at the test plane so that the reference cell produces its calibrated short-circuit current at the desired level. The reference cell temperature must be 25 ± 0.5 °C.

(*e*) With the test cell stabilized at or near the minimum temperature of interest, measure its short-circuit current I_{sc} and open-circuit voltage V_{oc}.

(*f*) Stabilize the test cell at a temperature approximately 10 °C above the previous level and repeat I_{sc} and V_{oc} measurements. Repeat this procedure at approximately 10 °C increments up to the maximum temperature of interest.

(*g*) Repeat steps (*a*)–(*f*) on each of the other test cells.

(*h*) Plot the values of I_{sc} and V_{oc} as a function of temperature and construct a least-squares fit curve through each set of data.

(*i*) From the slopes of the *I*–*V* curves at a point midway between the minimum and maximum temperatures of interest, calculate α_c and β_c, the temperature coefficients for single cells.

(*j*) Calculate the temperature coefficients for the module as follows:

$$\alpha = n_p \alpha_c \qquad \beta = n_s \beta_c$$

where n_p is the number of cells in parallel and n_s is the number of cells in series.

9.2.9 Determination of internal series resistance

R_s may be determined in simulated sunlight by the following procedure.

(*a*) Trace the *I*–*V* characteristics of the module at room temperature and at two different irradiances. (Magnitudes need not be known.) During the two measurements, the cell temperature must not vary by more than 2 °C.

(*b*) Choose a point P on the higher curve at a voltage slightly higher than the voltage for maximum power (figure 9.3). Measure δI, the difference between the current at this point and I_{sc1}.

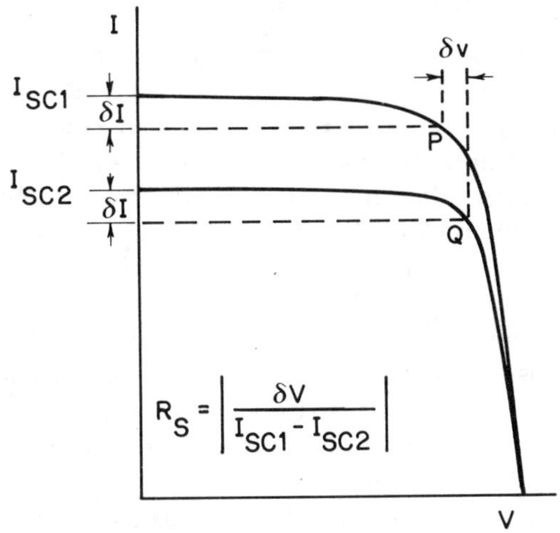

Figure 9.3 Determination of R_s.

(*c*) Determine the point Q on the lower curve at which the current is equal to $I_{sc2}-\delta I$.

(*d*) Measure the voltage displacement δV between points P and Q.

(*e*) Calculate R_{s1} from

$$R_{s1}=\frac{\delta V}{I_{sc1}-I_{sc2}}$$

where I_{sc1} and I_{sc2} are the two short-circuit currents.

(*f*) Repeat steps (*c*)–(*e*), using a characteristic taken at a third irradiance and the same temperature, in combination with each of the first two curves, to determine two more values, R_{s2} and R_{s3}.

(*g*) Take the mean of R_{s1}, and R_{s2} and R_{s3} as the definitive value of R_s.

9.2.10 Determination of curve correction factor

K may be determined in simulated sunlight by the following procedure.

(*a*) Trace the *I*–*V* characteristics of the module at an irradiance within $\pm 30\%$ of the desired level and at three different temperatures (T_3, T_4 and T_5) over an interest range of at least 30 °C. When measuring the characteristics, precautions must be taken (for instance, by enclosing the module in a temperature-controlled chamber with a transparent window) to ensure that the cell temperature is uniform within ± 2 °C of the intended level.

(*b*) Using an assumed value of *K* (say, 1.25×10^{-3} V A^{-1} °C^{-1}, which is typical for crystalline silicon cells) transpose the characteristic measured at temperature T_3 to T_4 by applying the following equations:

$$I_4=I_3+\alpha(T_4-T_3)$$
$$V_4=V_3-KI_4(T_4-T_3)+\beta(T_4-T_3)$$

where I_3, V_3 are coordinates of the T_3 characteristics and I_4, V_4 are the corresponding coordinates of the T_4 characteristics.

(*c*) If the transposed T_4 characteristic does not coincide to the desired accuracy with that obtained by measurement, repeat step (*b*) by inserting different values for *K*. If the transposed T_4 characteristics and the measured characteristics still do not coincide in each case, repeat the transpositions using different values of *K* until a correct fit is obtained.

(*d*) When the proper value of *K* has been determined, transpose the T_3 and T_4 characteristics sequentially to match the characteristics at the temperature T_5. If the transposed and measured characteristics do not coincide in each case, repeat the transpositions using different values of *K* until correct fits are obtained.

(*e*) Use the mean of the three values of *K* thus determined.

9.2.11 Measurement of relative spectral response

The relative spectral response of a solar cell (figure 9.4) is measured by placing the cell on a temperature-controlled mount, irradiating it uniformly from a monochromatic source and measuring the short-circuit current and the irradiance at fixed wavelength intervals over the response range. The currents are then divided by the irradiances or a proportional parameter and plotted as a function of wavelength. Alternatively, the irradiance may be kept constant (for instance, by varying the width of a monochromator exit slit), in which case the relative spectral response is obtained directly from the current readings.

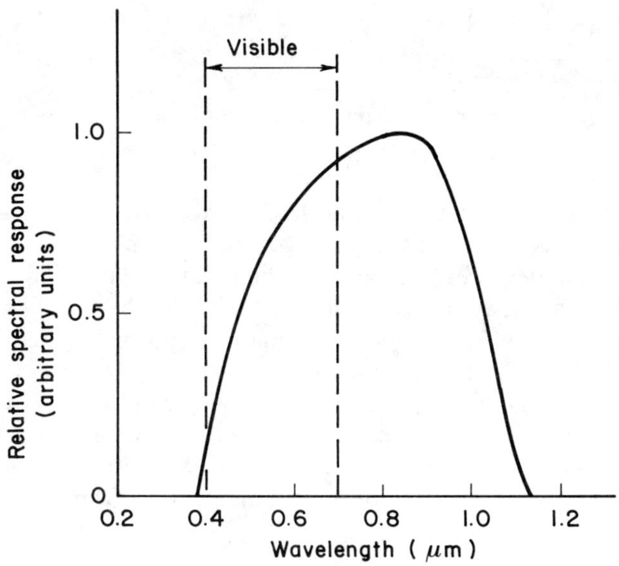

Figure 9.4 Spectral response of a silicon solar cell.

The irradiance monitor may be a vacuum thermocouple, a pyroelectric radiometer or other suitable detector. Another alternative is to use a previously calibrated reference solar cell whose relative spectral response covers the required range.

Figures 9.5 and 9.6 show two examples of test arrangements, the first embodying a quartz prism monochromator and the second a filter wheel as the monochromatic source. In both cases, the radiant source is a 1000 W tungsten–halogen lamp, operated from a stable supply at a colour temperature of 3200 K. The test cell and the irradiance monitor are mounted on opposite sides of a rotatable temperature-controlled block, so that either may be presented to the monochromatic beam in precisely the same place. Alternatively, they may be mounted on a slide with suitable positioning stops.

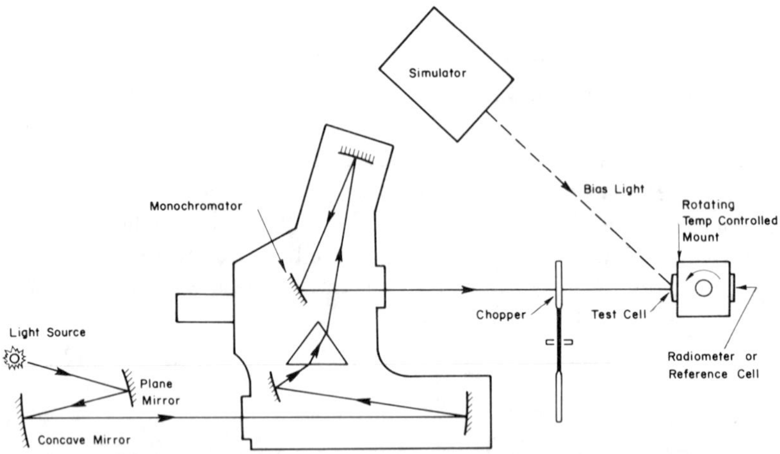

Figure 9.5 Spectral response measurement using a monochromator.

The filter wheel carries narrow-band filters covering the response range of the cell in wavelength steps not exceeding 50 nm. The filters are arranged so that each can be indexed in turn between the radiant source and the test cell or irradiance monitor. It is important that the filters should have negligible (under 0.2%) side bands. The monochromator is normally used with fixed slits and manually set to the same wavelength steps.

With crystalline silicon and other cells where the response has been shown to change linearly with irradiance, the short-circuit current of cell (voltage drop across a standard four-terminal fixed resistor) and the open-circuit voltage of the vacuum thermocouple or radiometer may be measured directly

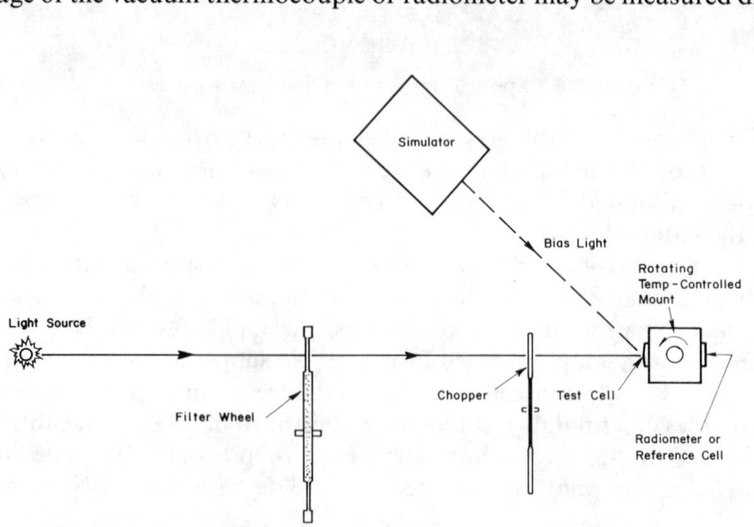

Figure 9.6 Spectral response measurement using a filter wheel.

with a DC digital voltmeter or potentiometer. However, if this is done, the exit beam, test cell and irradiance monitor must be completely enclosed in an anti-reflection light-tight box and meticulous precautions must be taken to avoid thermal and other random electromotive forces which would cause errors. Alternatively, the exit beam may be chopped at a low frequency and the output voltage amplified and rectified. In this case, it is important to ensure that the amplifiers are linear and drift free.

With non-linear devices, such as amorphous silicon solar cells, it is necessary to use a chopped monochromatic beam and to increase the irradiance to the desired operational level with bias light from a suitable steady-state simulator, as indicated by the broken lines in figures 9.5 and 9.6. If there is any doubt about the linearity of the test cell, it is wise to use bias light.

9.2.12 Simulator requirements

The IEC is proposing to group solar simulators for PV testing in three classes.

Class A For accurate measurements in the laboratory and for the calibration of secondary reference cells.
Class B For production acceptance tests.
Class C For bias lighting and other purposes, where lower standards of spectral match, uniformity and stability are acceptable.

Two types are commercially available. The 'steady-state' type, e.g. filtered xenon, dicroic filtered tungsten (ELH) or modified mercury vapour with tungsten electrodes, is suitable for single cells and small modules. The pulsed type, consisting of one or two long arc xenon flash lamps, is better for large modules, as it can irradiate large areas uniformly. Another advantage of the pulsed type is that there is negligible heat input to the module, so that it remains uniformly at the ambient temperature, which can be easily and accurately measured. The pulse-forming network and the data aquisition and data-processing equipment are generally supplied as part of the simulator. With a pulsed simulator, it is important to ensure that the interval between data points is sufficient, bearing in mind the response time of the module and the data aquisition system.

A simulator must, of course, be capable of producing the standard irradiance of 1000 $\mathrm{W\,m^{-2}}$ (as measured by a reference cell) at the test plane and higher or lower irradiances as may be required.

Its spectral irradiance distribution must match the reference spectrum to the extent indicated for the relevant class of simulator in table 9.1.

The irradiance in the test plane, as measured with a suitable detector, must be uniform over the full extent of the nominated test area, to the degree specified for the relevant class of simulator in table 9.1. The detector must be no larger than a single component cell of the module.

Table 9.1 Simulator characteristics. The spectral match is the ratio of percentage specified in table 9.2 for each wavelength interval.

Characteristics	Tolerance		
	Class A	Class B	Class C
Spectral match	0.75–1.25	0.6–1.4	0.4–2.0
Non-uniformity of irradiance	±2%	±5%	±10%
Temporal instability	±2%	±5%	±10%

The non-uniformity is defined as

$$\pm\frac{\text{maximum irradiance} - \text{minimum irradiance}}{\text{maximum irradiance} + \text{minimum irradiance}} \times 100\%.$$

During the time of data acquisition, the irradiance must be stable to the degree specified for the relevant class of simulator in table 9.1.

Table 9.2 Breakdown of reference spectrum.

Wavelength interval (μm)	Percentage of the total irradiance between 0.4 and 1.1 μm (%)
0.4–0.5	18.5
0.5–0.6	20.1
0.6–0.7	18.3
0.7–0.8	14.8
0.8–0.9	12.2
0.9–1.1	16.1

The temporal instability is defined as

$$\pm\frac{\text{maximum irradiance} - \text{minimum irradiance}}{\text{maximum irradiance} + \text{minimum irradiance}} \times 100\%$$

where the maximum and minimum irradiance are those measured with the detector at any particular point on the test plane during the time of data acquisition.

All these characteristics must be rechecked whenever an optical part of the simulator is replaced or there is reason to believe they have been affected by aging.

9.2.13 Computation of spectral mismatch error

The spectral mismatch error is the error introduced in PV performance measurement caused by the interaction of the mismatch between the spectral responses of the reference cell and the module and the mismatch between the reference spectrum and the natural or simulated sunlight used for the test. It may be computed as follows.

Let J_1 (A m^{-2}) be the short-circuit current density of the reference cell in sunlight having the reference spectral irradiance distribution and an irradiance of 1000 W m^2, J_2 (A m^{-2}) be the short-circuit current density of the reference cell, as measured in natural or simulated sunlight, $s_{1\lambda}$ (A W^{-1}) be the absolute spectral response of the reference cell at wavelength λ, $k_1 s_{1\lambda}$ be the relative spectral response of the reference cell at wavelength λ, J_3 (A m^{-2}) be the short-circuit current density of the module in sunlight having the reference spectral irradiance distribution and an irradiance of 1000 W m^2, J_4 (A m^{-2}) be the short-circuit current density of the module, as measured in natural or simulated sunlight, $s_{2\lambda}$ (A W^{-1}) be the absolute spectral response of the module at wavelength λ, $k_2 s_2$ be the relative spectral response of the module at wavelength λ, $E_{s\lambda}$ (W m^{-2} μm^{-1}) be the absolute spectral irradiance at wavelength λ of the reference spectral irradiance distribution, $k_3 E_{s\lambda}$ be the relative spectral irradiance at wavelength λ of the reference spectral irradiance distribution, $E_{t\lambda}$ (W m^{-2} μm^{-1}) absolute spectral irradiance at wavelength λ of the natural of simulated sunlight used for the test and $k_4 E_{t\lambda}$ be the relative spectral irradiance at wavelength λ of the natural or simulated sunlight.

Then,

$$J_1 = \int s_{1\lambda} E_{s\lambda}\, d\lambda$$
$$J_2 = \int s_{1\lambda} E_{t\lambda}\, d\lambda$$
$$J_3 = \int s_{2\lambda} E_{s\lambda}\, d\lambda$$
$$J_4 = \int s_{2\lambda} E_{t\lambda}\, d\lambda$$

Integration of the products of the measured relative spectral responses and the relative spectral irradiances yields the following parameters:

$$A_1 = \int k_1 s_{1\lambda} k_3 E_{s\lambda}\, d\lambda = k_1 k_3 J_1$$
$$A_2 = \int k_1 s_{1\lambda} k_4 E_{t\lambda}\, d\lambda = k_1 k_4 J_2$$
$$A_3 = \int k_2 s_{2\lambda} k_3 E_{s\lambda}\, d\lambda = k_2 k_3 J_3$$
$$A_4 = \int k_2 s_{2\lambda} k_4 E_{t\lambda}\, d\lambda = k_2 k_4 J_4.$$

The error in measuring the short-circuit current density of the module is then given by

$$\text{spectral mismatch error} = \frac{J_4 - J_3}{J_3} \times 100\%$$
$$= \left(\frac{A_1 A_4}{A_2 A_3} \frac{J_2}{J_1} - 1 \right) \times 100\%$$

9.2.14 Reference solar cell

There are two types of reference solar cells.

Primary cells	A reference cell whose calibration is based on a radiometer or standard detector conforming to the current World Radiometric Reference.
Secondary cells	A reference cell calibrated in natural or simulated sunlight against a primary reference cell.

At least two cells of stable performance should be selected for calibration as primary reference cells. Their relative spectral responses much be such that spectral mismatch errors in performance measurement are less than ±1%.

The packaging of reference cells is important. For the testing of fixed modules in natural sunlight where the direct solar beam is not at normal incidence, they should be encapsulated in a module package similar to the modules being tested, surrounded by real or dummy cells of the same size, shape and optical properties, so that they respond in the same way to variations in the geometrical distribution of the incident radiation. A similar requirement applies for measurements in simulators of the type where multiple reflections of light from and to the module under test can affect the irradiance at the test plane. In this case, the module package of the reference cell must be of the same size and shape as the modules under test and the position of the reference cell may be packaged singly or, if not intended for day-to-day use, left unpackaged. Whether the reference cell is packaged or not, the electrical connections must consist of a four-wire system (Kelvin probe) and means must be provided for measuring the junction temperature to an accuracy of ±1 °C.

The calibration of reference cells in frequent use should be cross-checked at intervals of no more than 1 month by comparing their short-circuit currents at the same irradiance. If there is any change in the current ratio beyond ±1%, the cells must be recalibrated. In any case, all reference cells should be recalibrated at least every 12 months.

9.2.15 Calibration methods

The calibration of primary reference cells is the key to good performance measurement. The IEC is currently considering five methods, three outdoor and two indoor. They are as follows.

(i) Direct sunlight method, using filters: the calibration value (ratio of short-circuit current at 25 °C to incident irradiance, with the reference spectral distribution) is computed from the following.

(*a*) The ratio of the short-circuit current of the reference cell to the incident irradiance, when the cell is exposed to direct sunlight

passing in turn through a set of four bandpass filters covering the cell response range.
- (*b*) The irradiance of the reference spectrum in each of the four wavebands.
- (*c*) A factor to correct for the imperfections of the filters.

(ii) Global sunlight method: the calibration value is computed from the following.
- (*a*) The relative spectral response of the reference cell.
- (*b*) Its short-circuit current in global sunlight (on a horizontal plane) of measured irradiance and relative spectral irradiance distribution.
- (*c*) The reference spectrum.

(iii) Total sunlight method: the calibration value is calculated from the short-circuit current of the reference cell in total sunlight of measured irradiance (with the direct beam at or near normal incidence) under specified atmospheric conditions.

(iv) Solar simulator method: the calibration value is computed from the same parameters as in the global sunlight method, except that a class A simulator is used instead of global sunlight.

(v) Differential spectral response method: the calibration value is computed directly from the measured absolute spectral response of the reference cell and the reference spectrum.

The global sunlight method, developed by the Royal Aircraft Establishment, Farnborough, UK, is the only method which has been proved to have a year-to-year repeatability of better than $\pm 1\%$. It involves the on-site measurement of the spectral irradiance distribution of the global sunlight, which is difficult and requires stable irradiance for several minutes, but experience over many years has shown that, provided that the weather is suitable, this measurement can be omitted. The total sunlight method, a development of the global method, takes advantage of this fact and can be expected to have the same good repeatability. Certainly, it is the simplest and cheapest procedure, with the fewest pitfalls, and, unlike direct sunlight and the differential spectral response method, is suitable for all sizes of reference cells and types of package. For these reasons, it is the best candidate for a future international standard.

For calibration by the total sunlight method, the following conditions are necessary.

(i) Clear sunny weather.
(ii) No observable cloud within a 30° half-angle cone surrounding the sun.
(iii) Total irradiance (direct + diffuse) of not less than 900 $\mathrm{W\,m^{-2}}$.
(iv) Air mass between AM1 and AM2.

(v) Radiation sufficiently stable that the variation in reference cell short-circuit current is less than $\pm 0.5\%$ over a period of twice the response time of the pyranometer.
(vi) Turbidity coefficient of 0.265 ± 0.185 at a wavelength of 0.5 μm.
(vii) Precipitable water vapour content of the atmosphere between 0.4 and 2.6 cm.

The last two conditions call for the use of a sun photometer or other suitable device for measuring turbidity and atmospheric water vapour content. If this instrument is not available, it is probably sufficient to ensure, as in the global method, that the ratio of diffuse to total irradiance is less than 0.25.

The procedure simply consists of mounting the reference cell and a pyranometer (previously calibrated to the World Radiometric Reference against a primary absolute cavity radiometer) on a manual tracking platform, with their receiving surfaces coplanar and in close proximity, pointing them at the sun, so that the solar beam is normal (within $\pm 5°$) to the receiving surfaces, and carrying out simultaneous measurements of irradiance and the short-circuit current and temperature of the reference cell under the specified environmental conditions. It may be necessary, in certain circumstances, to cover the foreground with non-reflecting material to reduce ground reflection. If it is not possible to control the temperature of the reference cell to 25 °C, it must be shaded from the sun, except when a measurement is being made.

Measurements are repeated until the ratio of short-circuit current (corrected to 25 °C) to irradiance over five successive sets of readings varies by less than 1%. The calibration is carried out on at least three different days and a mean taken of the acceptable data.

Suitable environmental conditions may be infrequent but it is not necessary to conduct primary reference cell calibrations very often, as any number of secondary reference cells may be generated from one primary cell. Secondary calibrations are carried out by simultaneously measuring the short-circuit currents of primary and secondary reference cells in natural sunlight or a class A simulator. The calibration value of the secondary reference cell is obtained by multiplying the known calibration of the pirimary cell by the ratio of short-circuit current (secondary) to short-circuit current (primary).

9.3 DESIGN QUALIFICATION

EUR Specification No 502 (Anon 1984) lays down the tests which the CEC Joint Research Center, Ispra, have applied for the qualification of modules for projects funded by the CEC or upon special request from industry or other organizations. It is similar to their original specification, *EUR Specification No 501* (Anon 1984) but embodies amendments made in the light of experience and omits certain tests which were found not to be cost effective.

For design qualification, it is necessary to submit eight representative samples of the module. One of these is held untested for control purposes and the others are subjected to the test sequences in figure 9.7. The meanings of the test abbreviations and the test sequences used in figure 9.7 are given in the following brief descriptions.

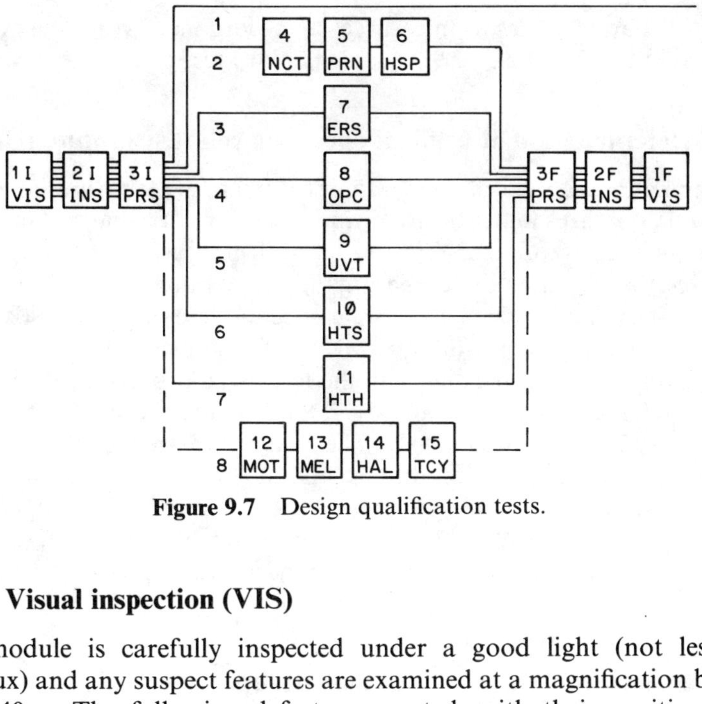

Figure 9.7 Design qualification tests.

9.3.1 Visual inspection (VIS)

Each module is carefully inspected under a good light (not less than 1000 Lux) and any suspect features are examined at a magnification between 8 and 40 ×. The following defects are noted, with their position in the module.

(i) Discolouration.
(ii) Bubbles or voids in the encapsulant.
(iii) Chipped, cracked, broken or displaced cells.
(iv) Broken, missing or displaced interconnects.
(v) Contact corrosion or delamination.
(vi) Defects on front or back surfaces.

9.3.2 Insulation test (INS)

A voltage of 1000 V DC is applied for 1 min between the shorted terminals and the metal frame (or simulated structure in the case of a non-metallic frame). The insulation resistance must be not less than 100 MΩ. Using an insulation tester with a current limitation of 50 A, the voltage between these points is increased at a rate not exceeding 500 $V\,s^{-1}$ to a maximum equal to 1000 V DC plus twice the maximum system voltage and held at that level for 1 min. There must be no evidence of dielectric breakdown or surface tracking.

9.3.3 Electrical performance at standard test conditions (PRS)

The I–V characteristics is measured in accordance with the procedure described in §§9.2.4, 9.2.5 or 9.2.6 and standard test conditions as in §9.2.7.

9.3.4 Determination of nominal operating cell temperature (NCT)

The nominal operating cell temperature (NOCT) is the temperature at which the module stabilizes under the following conditions: irradiance, 800 $W\,m^{-2}$; ambient air temperature, 20 °C; wind speed, 1 $m\,s^{-1}$.

The NOCT is measured by comparing the temperature of the test module at thermal equilibrium with that of a reference module of known NOCT, when positioned side by side in still air in front of a stable (±1%) and uniform (±3%) simulator. The irradiance in the test plane is previously adjusted so that the reference module achieves a final temperature of (ambient air temperature, 20 °C) above its NOCT. The final temperature of the test module is measured under these conditions and its NOCT is then obtained by subtracting (ambient air temperature, 20 °C). The module temperature is best measured by calculation from the open-circuit voltage V_{oc}, the pre-determined V_{oc} at a known temperature and the temperature coefficient of voltage.

9.3.5 Electrical performance at nominal operating cell temperature (PRN)

The I–V characteristics is measured at NOCT and an irradiance of 800 $W\,m^{-2}$. Alternatively, the characteristics under these conditions may be transposed from the characteristics under standard test conditions.

9.3.6 Hot-spot endurance test (HSP)

Hot spots can be caused in the modules of an operational array by partial shadowing, mismatched or cracked cells or interconnect failures. To determine the ability of the module to withstand this effect, it is short-circuited in a uniform irradiance of 1000 $W\,m^{-2}$ and the hottest cell identified using an infrared camera. This cell is then fully shadowed and the percentage of shadowing gradually decreased until the short-circuit current coincides with the pre-determined current of the unshadowed module at maximum power. This is the condition of maximum power dissipation in the shadowed cell. The module is kept in this condition for 1 h. This test can be omitted if not more than 12 cells in series are protected by a shunt diode.

9.3.7 Energy rating at standard day operation (ERS)

Using the *I–V* characteristics of the module at various irradiances and temperatures, a computation is carried out of its total energy output over a 'standard day', with specified changes in irradiance and temperature. At present, there is no international agreement on what constitutes a 'standard day' but IEC Technical Committee 82 is considering the matter.

9.3.8 Pre-conditioning test (OPC)

To test the stability of the module in natural sunlight, it is mounted normal to the sun at noon and left until the irradiation, as measured with an integrating pyranometer, reaches an aggregate of 60 $kW\,h\,m^{-2}$.

9.3.9 Ultraviolet irradiation test (UVT)

To determine whether the module is affected by ultraviolet radiation, it is exposed, at 60 ± 5 °C, to a total ultraviolet fluence of 15 $kW\,h\,m^{-2}$, representing about 40 clear summer days in Mediterranean conditions. The acceleration factor must not exceed 5 ×. The ultraviolet source must match the reference spectral irradiance distribution between 280 and 400 nm with a maximum deviation of $\pm 35\%$ in 80 nm wavebands and its uniformity must be better than $\pm 15\%$. Ozone produced by the ultraviolet must be kept away from the module.

9.3.10 High-temperature storage test (HTS)

To reveal any deficiencies of the encapsulation system, the module is installed in a suitable test chamber and kept at a temperature of 90 ± 5 °C for 20 days.

9.3.11 High-temperature high-humidity storage test (HTH)

To determine the ability of the module to withstand long-term humidity, it is subjected in a suitable test chamber to a temperature of 90 ± 5 °C and a relative humidity of not less than 95% for 20 days.

9.3.12 Mounting twist test (MOT)

To detect faults which might be caused by mounting on a non-ideal supporting structure, one corner of the module is displaced from the plane of the other three by an amount x m, given by

$$x = 0.021(L^2 + W^2)^{1/2}$$

where L (m) is the length and W (m) is the width of the module. This displacement corresponds to a deformation angle of 1.2°.

9.3.13 Mechanical loading test (MEL)

To determine the ability of the module to withstand high winds, snow and ice loads, it is mounted in a test chamber designed so that the air pressure on either side of the module can be varied to produce uniform positive and negative loads. The procedure is as follows.

(*a*) Apply a load of 600 Pa to the front of the module for 10 s. Release the pressure and, after a recovery period of 5 min, record any residual deflection.
(*b*) Apply the load to the front in steps of 300, 600, 900 and 1200 Pa, maintaining it at each step for at least 10 s and recording the deflection. Release the pressure and, after a recovery period of 5 min, record any residual deflection.
(*c*) Repeat (*a*) and (*b*) with the load applied to the back of the module.
(*d*) Increase the load on the front of the module to 2400 Pa (to simulate wind gusts of 130 km h^{-1}) or 5400 Pa (to simulate snow and ice loading) and maintain for at least 10 s. Release the pressure and, after a recovery period of 5 min, record any residual deflection.
(*e*) Repeat (*d*) with the load applied to the back of the module.
(*f*) Apply alternating positive and negative loads of 600 Pa (10 s each) for 10 cycles.

9.3.14 Hail resistance test (HAL)

To verify that the module can withstand hailstones, artificial ice balls, kept in a freezer for at least 8 h before use, are fired at the module window from a suitable pneumatic or spring-actuated gun, so as to reach the module at a velocity within $\pm 5\%$ of the appropriate hailstone terminal velocity. Unless otherwise specified by the module manufacturer, the ice balls are 25 mm in diameter and the corresponding terminal velocity 23 m s^{-1}. Altogether, 10 balls are fired. They are aimed at a corner, an edge just inside the frame, over cell edges near interconnects, over points of minimum spacing between cells, at points near the module fixing locations and at points furthest from the module fixing locations.

9.3.15 Temperature-cycling test (TCY)

To test the ability of the module to withstand temperature-induced stresses, it is installed in a climatic chamber with automatic temperature control and with the surrounding air circulated at a velocity of not less than $2\ \mathrm{m\,s^{-1}}$. There it is subjected to 10 cycles at module temperatures of between 20 ± 2 and 80 ± 2 °C. The rate of change must not exceed $2\ °\mathrm{C\ min^{-1}}$ and the module temperature must remain stable (within ± 2 °C) at each extreme for at least 10 min. (The proposed IEC test is more severe than this, calling typically for 200 cycles between -40 and $+85$ °C.)

9.4 MODULE CATEGORIES

Not all modules will need to withstand all the above qualification tests at the specified severities. It is likely that the IEC will divide modules into three or four categories and will specify appropriate qualification tests for each. For example, one category could be that in which modules are required to withstand the most severe conditions anywhere in the world for periods of 20 or 30 years. At the other extreme would be modules designed for the leisure market, where price is the overriding factor. In between, there could be a category for more benign climates.

9.5 PRODUCTION ACCEPTANCE

As already stated, there are, at present, no international standards for production acceptance tests. However, most manufacturers carry out the following tests on every module.

(i) Visual inspection.
(ii) Insulation test.
(iii) Performance at standard test conditions.

Many manufacturers do not measure the complete *I–V* characteristics but limit their measurements to the short-circuit current, the current at a voltage near the maximum-power point and the open-circuit voltage.

9.6 QUALITY ASSURANCE

A selection of design qualification tests, some shortened to save time, may be used periodically for the purpose of ensuring that the design standard is maintained in long production runs. It should not be necessary to repeat tests NCT, HSP, ERS, OPC, UVT, MEL and HAL, unless major modifications have been made since qualification. Shortened versions of tests HTH and TCY, preceded and followed by VIS, INS and PRS, should suffice in most cases.

REFERENCES

Anon 1977 *ERDA–NASA Document No 1022/77/16* (Washington, DC: Energy Research and Development Administration–National Aeronautics and Space Administration)

—— 1980 *CEC Specification No. 101* Issue 2; *EUR Document 7978 EN* (Luxembourg: Commission of the European Communities)

—— 1984a *EUR Specification No 501* (Luxembourg: Commission of the European Communities)

—— 1984b *EUR Specification No 502* Issue 1 (Luxembourg: Commission of the European Communities)

Bird R E, Hulstrom R L and Lewis L J 1983 *Solar Energy* **30**

Treble F C 1976 *Proc. 2nd ERDA–NASA Terrestrial Photovoltaic Measurement Workshop, Baton Rouge, LA*

—— 1985 *Proc. 4th Asian School on Solar Energy Harnessing, Bangkok, 12–20 December 1985* (Bangkok: Asian Institute of Technology)

10 Asian Institute of Technology Photovoltaic Module Test Bed System

10.1 INTRODUCTION

A complete computerized system for the characterization and performance testing of photovoltaic (PV) modules under natural sunlight and ambient condition was designed and installed at the Asian Institute of Technology (AIT). The 'system' comprises two stages: the first stage is the monitoring stage where module performance is monitored by computer-controlled electronic hardware; the second stage is the data-processing stage. Software was coded for data processing, analysis and characterization of PV modules. The system was designed within the concept of providing a tool for regional PV system manufacturers as well as for academic researchers to monitor the performance of PV modules under natural conditions and to analyse these performance data in order to characterize the module. The system automates daily test procedures and provides the engineer with more time to interpret and analyse results. A numerical technique for PV module characterization using the single-diode lumped-parameter model of a solar cell (Braunstein *et al* 1977) is included.

10.2 OBJECTIVE

The PV module test bed system or simply, the 'system' is developed with the following design objectives.

(*a*) Automation of monitoring and processing of meteorological data and PV module performance data.
(*b*) Accuracy of performance measurements.
(*c*) Adaptability of the system to different users.
(*d*) Coding of numerical PV module characterization techniques into software programs.
(*e*) Acquisition of current–voltage (I–V) characteristic curves of PV

modules under different meteorological conditions to monitor its performance and degradation and to determine the module performance parameters.

10.3 PARTS OF THE SYSTEM

Figure 10.1 shows the functional block diagram of the PV module test bed system. The system is divided into two stages: the monitoring stage and the analysis stage.

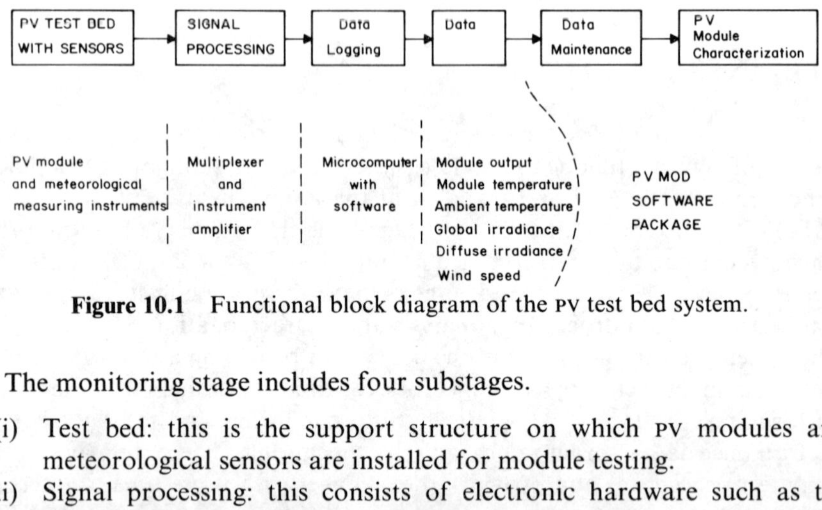

Figure 10.1 Functional block diagram of the PV test bed system.

The monitoring stage includes four substages.

(i) Test bed: this is the support structure on which PV modules and meteorological sensors are installed for module testing.

(ii) Signal processing: this consists of electronic hardware such as the variable load, filters and amplifying circuits to amplify data and filter signal noise.

(iii) Data logging: a personal microcomputer is used to control the mode and frequency of data measurements.

(iv) Data: these include the wind speed, the global and diffuse solar radiation, ambient and module back-surface temperatures, I–V characteristics of the PV module.

The analysis stage includes data processing and PV module characterization software. This stage can be executed on microcomputers away from the data-monitoring installation. The software 'PVMOD' is specifically designed to handle all data file maintenance, data search and data processing and to provide a numerical technique for PV module characterization.

The following subsections will describe in more detail each functional block of figure 10.1. The realization of these functional blocks into actual system components is schematically shown in figure 10.2. The system in

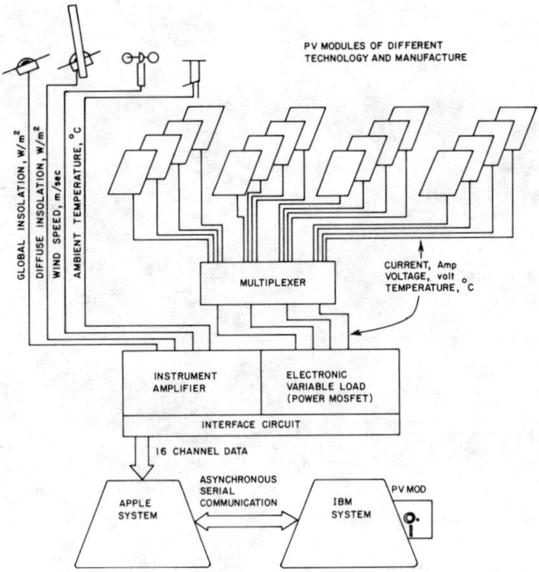

Figure 10.2 PV module test bed instrumentation and data-processing schematic diagram.

figure 10.2 has been installed and is operational at the Energy Technology Division, AIT.

10.3.1 The photovoltaic module test bed

The PV module test bed is a two-axis tracking structure installed with 16 PV module input terminals and 16 temperature surface sensors. A global pyranometer is installed on the same test plane of the module at the uppermost boundary of the test plane. An anemometer is also installed vertically at the uppermost boundary of the test plane. The diffuse pyranometer is installed on a separate stand 2 m due south from the centre of the test bed to avoid shading.

Figure 10.3 shows the PV module test bed installed at the Energy Technology Division, AIT. A weather-proof box is installed at the base to house the relay circuits. Figure 10.4 shows the 16 10 A power relays for the PV modules and the 16 1 A communication relays on the right-hand side for the PV temperature sensors.

16 PV modules can be installed on the test plane of the test bed and they are switched through relays to the multiplexer box (figure 10.5). The relays divide these 16 PV module output signals and 16 PV module temperature signals into four sets. The multiplexer box has four channels (sets) and each channel has four switches. Each switch controls one relay and, thereby, one PV module

Figure 10.3 PV module test bed.

and its temperature sensor. Within one channel, only one of the four PV modules relays can be 'switched on'. Therefore, with each channel having one PV module 'switched on', the system can test four PV modules simultaneously (figure 10.2).

Figure 10.4 Power relay and communication relay.

Figure 10.5 Multiplexer with computer.

10.3.2 Electronic variable load

The main aim of the module test bed is to monitor the I–V characteristics curves of modules under different natural conditions. To extract an I–V characteristics curves which reflects the entire operating range of the module from open circuit to maximum power and then to short circuit, the test module is connected to a load, whose power demand can be modulated by controlling either the voltage or its current supplied. In practice, there are three different ways to achieve this. One way is to use a variable resistor as the load and, by varying the resistance, the amount of current flowing, from zero to maximum, is regulated. A suitably sized capacitor increases its voltage as it draws in current until fully charged when its voltage is constant. A transistor when used as the load also regulates the amount of current passing through it by controlling the amount of base circuit current. The resistor and the transistor both regulate the amount of current allowed to pass through them and the capacitor imposes upon the module its own rising voltage, from zero to maximum, as it is being charged by the module (figures 10.6–10.8).

An electronic variable load using transistors can be easily implemented using metal–oxide–semiconductor field-effect transistor (MOSFET) devices (figure 10.9). The requirement of an extremely low drive current and the associated extremely high power gain of the power MOSFET is a major advantage in PV module testing.

A detailed discussion of the three general types of variable load can be found in the section on instrumentation in chapter 21.

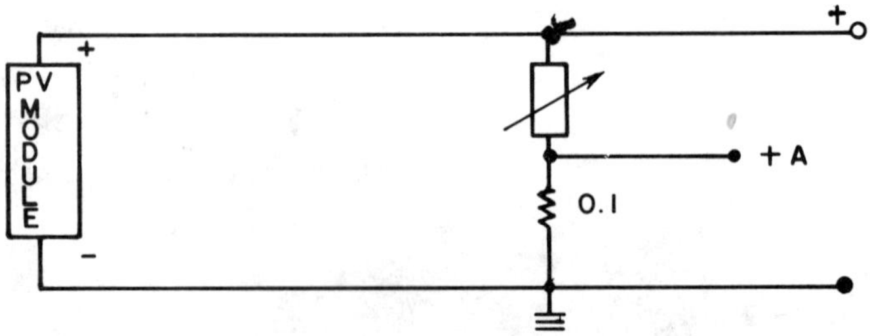

Figure 10.6 Resistor as a load to measure the I–V curve of a PV module.

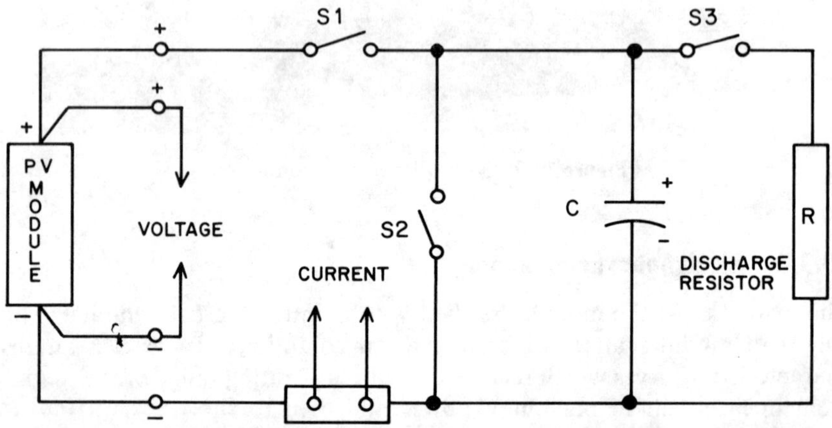

Figure 10.7 Capacitor as a load to measure the I–V curve of a PV module.

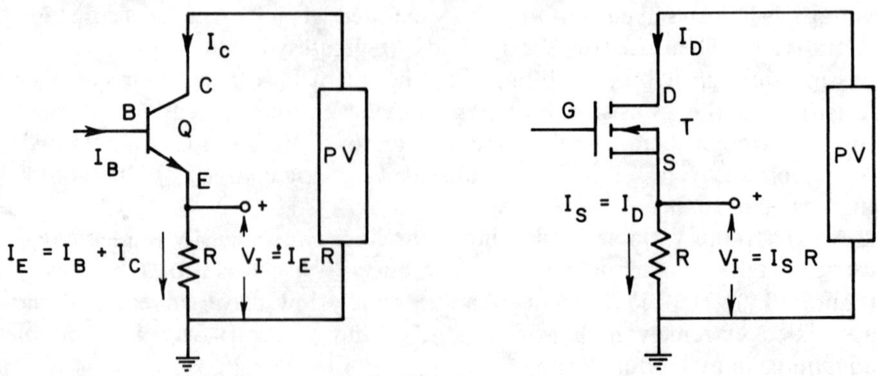

Figure 10.8 Transistor as a load to measure the I–V curve of a PV module.

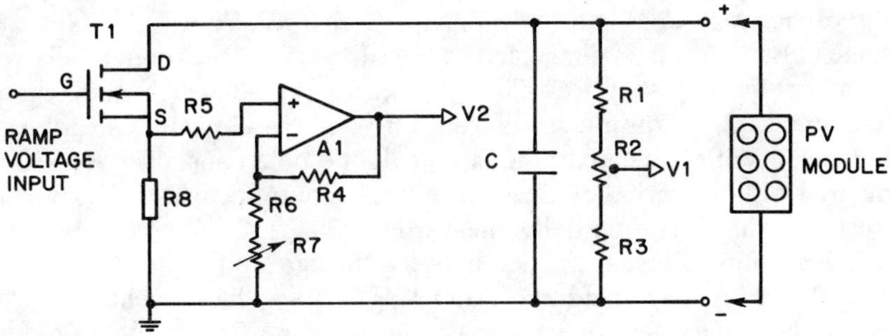

Figure 10.9 MOSFET electronic variable load to measure the *I*–*V* curve of a PV module.

10.3.3 Ramp generator

The MOSFET is a voltage-controlled device. A voltage must be applied between the gate G and the source S terminals to produce a flow of current from the drain D to the source S (figure 10.9). When a voltage increasing at a constant rate is fed to the gate and source terminal, the source current (module current) increases at a rate dependent on the MOSFET characteristic. A small accurate resistance R_8 at the source terminal converts the source current (module current) into a voltage signal and is amplified (by A_1) for read-out (at V_2).

A ramp generator (figure 10.10) generates a voltage rising at a constant rate of what is more popularly called a ramp voltage. The output of the ramp generator is connected to the gate terminal of the MOSFET to drive the device, regulating the module current passing through it. The heart of this generator

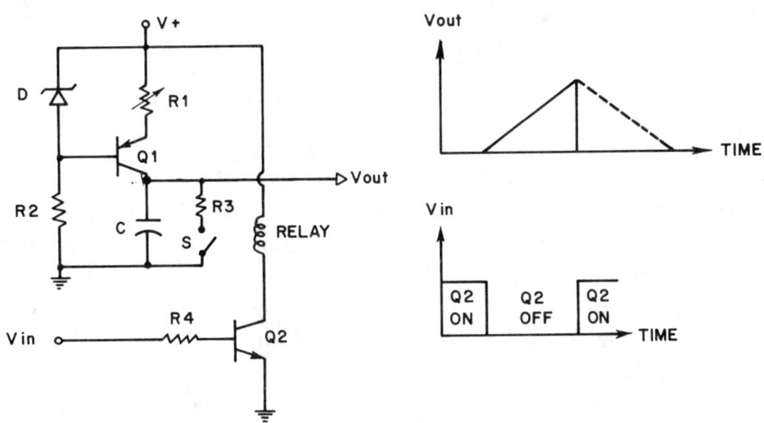

Figure 10.10 Ramp generator schematic circuit and its operation where switch S is a normally closed switch. For a normally open switch, the ramp voltage rises when Q_2 is on.

is the capacitance C. When the capacitor is charged from a constant step-voltage pulse input, the voltage across the capacitor rises exponentially with time at a time constant RC. Since the capacitor voltage cannot change instantaneously, the output starts from zero and rises toward a steady-state value dependent on the charging circuit. If the time constant is very large compared with the period of the pulse input, the output consists of exponential sections which are essentially linear ramp voltage.

When a ramp voltage is applied between the gate and source terminals of figure 10.9, the voltage read at V_2 and V_1 will trace the I–V characteristics curve of the PV module. As the gate voltage increases from zero, the module current increases until the short-circuit current is reached. The transfer characteristics of the MOSFET in figure 10.11 shows that the drain (PV) current starts to rise when the gate-to-source voltage is around 3 V and rises linearly beyond 2 A.

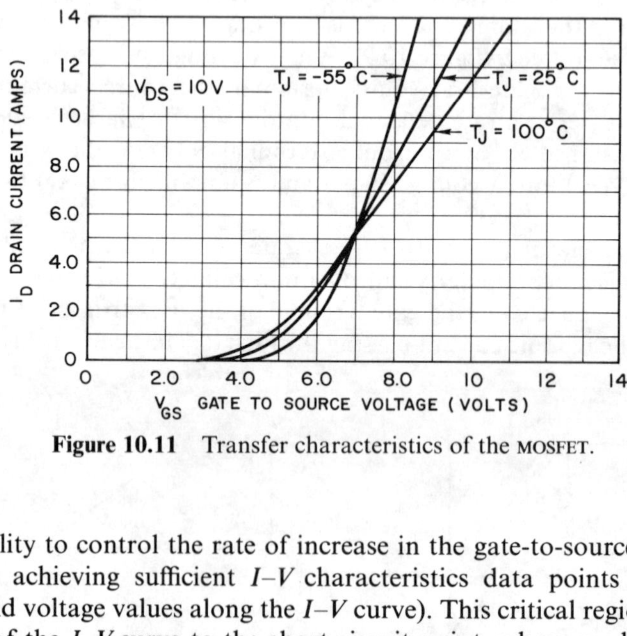

Figure 10.11 Transfer characteristics of the MOSFET.

The ability to control the rate of increase in the gate-to-source voltage is critical in achieving sufficient I–V characteristics data points (a pair of current and voltage values along the I–V curve). This critical region between the knee of the I–V curve to the short-circuit point, where module current variation is small relative to the decrease in module voltage, is generated faster than the rest of the curve. Figure 10.12 is the output characteristics of the MOSFET. Starting from open circuit, the drain current (module current) increases as V_{GS} increases. In the critical region of the I–V curve, the MOSFET sees an apparent constant-current load from the module. This constant-current load line will be a horizontal line cutting through the operating region of the curves to the saturation region in figure 10.12. This is where a sudden drop in V_{DS} (module voltage) occurs upon entering the critical region.

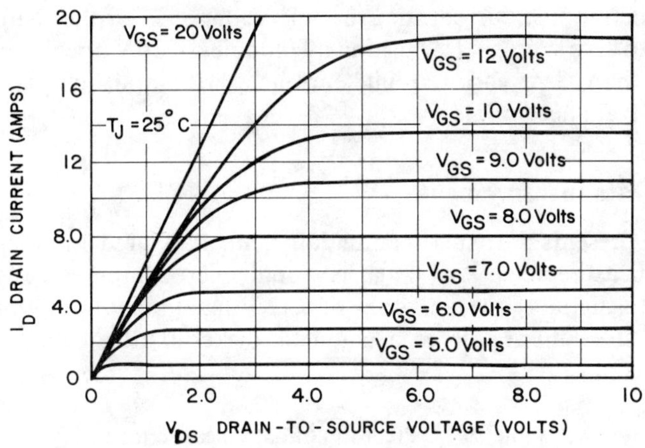

Figure 10.12 Output characteristics of the MOSFET.

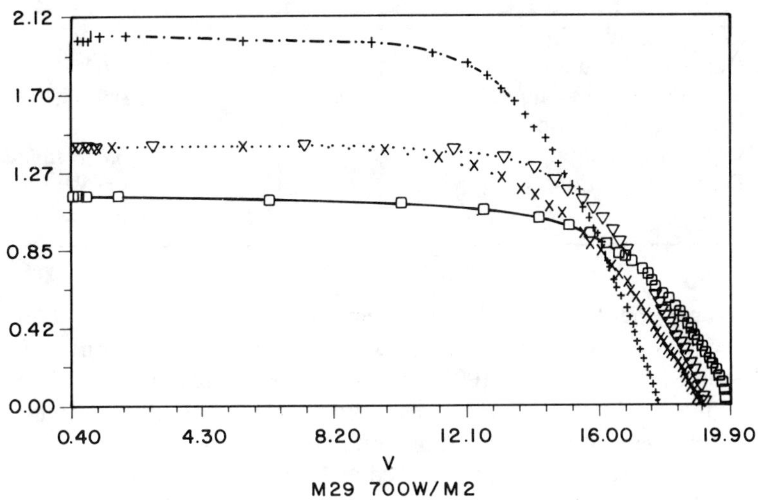

Figure 10.13 Sample of a graphics output *I–V* characteristics curve: □, ——, module 1; ▽, ..., module 2; +, —·—, module 3; ×, · · ·, module 4.

Figure 10.13 is a sample *I–V* characteristics curve. Note that data points are sparsely distributed along the critical region.

The operating region of figure 10.12 also reveals that actual short circuit is not achieved by the MOSFET. At the maximum 20 V gate-to-source voltage, the drain current is approximately a linear function of the drain-to-source voltage with an inverse slope of 0.14 Ω. It was suggested (Treble 1985) that the short-circuit current can be measured at a voltage less than 3% of the open-circuit voltage, within the range where there is a linear relationship between the module current and module voltage. Then the reading can be corrected to zero voltage. For testing PV modules of 20 V open-circuit voltage and less than 3 A short-circuit current, the output characteristics of figure 10.12 are acceptable.

10.3.4 Data measurement

Table 10.1 presents the experimental data being measured and their precision error of measurement. An APPLE personal microcomputer coupled to an analogue-to-digital (A–D) converter of 5.12 V maximum input voltage rating is used to store these data in a magnetic diskette, after being filtered and amplified.

Table 10.1 Precision of data measurement.

Description	Precision
Voltage of PV module	0.1 V
Current of PV module	0.02 A
Temperature of PV module's back surface	0.5 °C
Ambient temperature	0.5 °C
Global irradiation	6.0 W m^{-2}
Diffuse irradiation	6.0 W m^{-2}
Wind speed	0.1 m s^{-1}

The complementary metal–oxide–semiconductor (CMOS) A–D converter chip used has a 16-channel multiplexer and an 8 bit A–D converter. The conversion time is less than 100 μs and the input voltage range is from 0 to 5.12 V. This gives a measurement resolution of 0.02 V. During each cycle, 16 input channels are read within 2 ms, and stored in random-access memory addresses where they are later retrieved and stored in a diskette.

Table 10.2 A–D converter channel allocation.

Channel	Data	Channel	Data
1	Voltage V1 (V)	9	Temperature T1 (°C)
2	Current I1 (A)	10	Temperature T2 (°C)
3	Voltage V2 (V)	11	Temperature T3 (°C)
4	Current I2 (A)	12	Temperature T4 (°C)
5	Voltage V3 (V)	13	Ambient temperature (°C)
6	Current I3 (A)	14	Global irradiance (W m^{-2})
7	Voltage V4 (V)	15	Wind speed (m s^{-1})
8	Current I4 (A)	16	Diffuse irradiance (W m^{-2})

The *I–V* curve generated in figure 10.13 consists of 99 points non-uniformly distributed along the curve. Each point represents one read cycle performed by the A–D converter. Table 10.2 shows the 16 experimental data, with their channel allocations, read during one read cycle. Each input channel is read 100 μs apart. Four PV modules with back-surface temperature sensors and the weather data are monitored serially in one read cycle.

Monitoring, measurement and storage of data are controlled by a BASIC program written for the APPLE computer. A simplified flow chart is presented in figure 10.14. Data monitoring is done in two modes. A fixed time interval

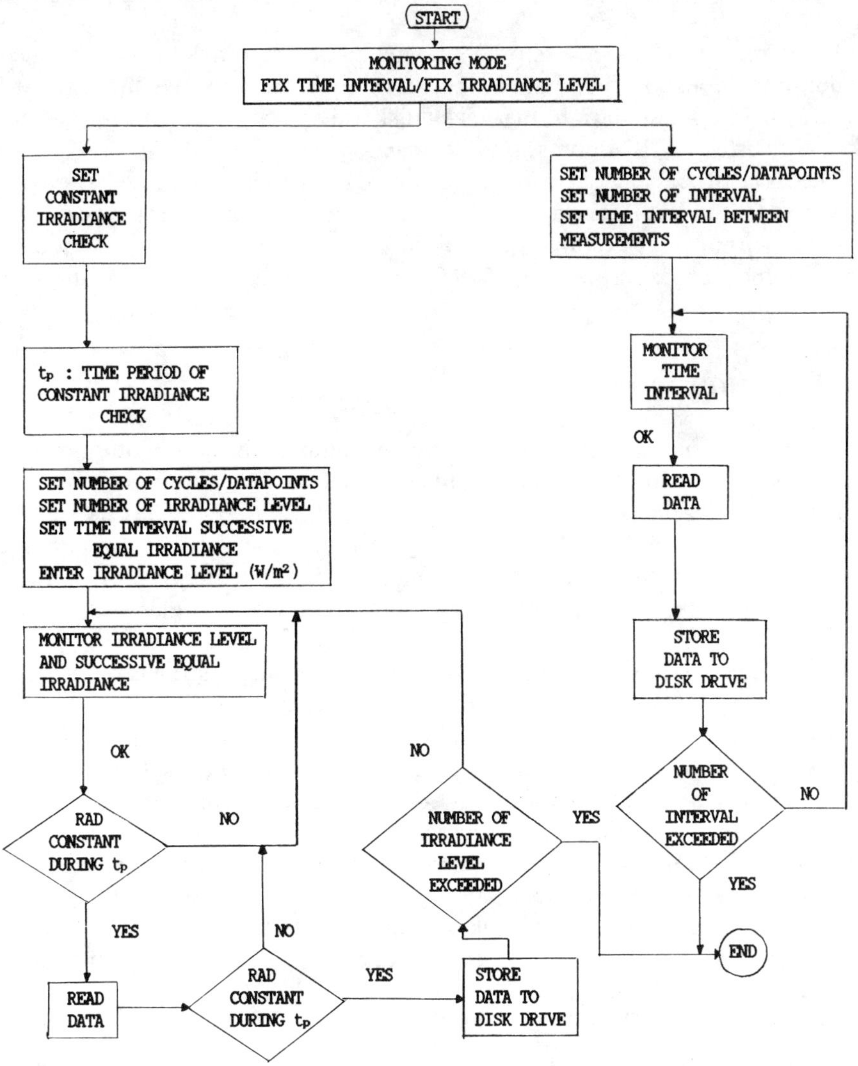

Figure 10.14 Simplified diagram of data-monitoring measurement and storage.

between measurements can be specified from 5 min to several hours. A second mode allows setting of irradiance levels. Up to 10 levels can be set with options for constant irradiance check. Both of these modes requires specification of the number of data points or read cycles.

Global irradiance is checked to ensure constant irradiance level during measurement of an *I–V* characteristics curve. The system monitors, for a specified length of time, the irradiance level before and after an *I–V* curve generation. This ensures constant irradiance during measurement and cancels any error due to the response time delay of thermopile radiometer. Response time delay for a class 1 radiometer is typically not more than 25 s for 99% reading.

Experimental data are stored in magnetic diskettes after specified conditions (constant irradiance checking) prior to storage are satisfied (figure 10.14). These data in the APPLE diskette operating system format are then converted to IBM-compatible format for analysis.

The data-processing stage is implemented in a user-interactive software, PVMOD 3.0. It maintains and processes all the data and provides services such as data formatting, combination of data files, deletion of data, module performance table, graphs of *I–V* curves and PV module characterization.

10.4 PVMOD 3.0

Outdoor performance testing of PV modules requires the compilation of vast amounts of data in order that module characterization can be reliable and statistically acceptable. Module tests at AIT show that regular as well as sudden shifts in sky conditions throughout the day within a month are common occurrences. Clouds can be thick or thin, or scattered or uniform from day to day. The spectral energy distribution of the global radiation varies according to these changing sky conditions even at the same irradiance level. Therefore, more than a single day of performance testing is required to characterize a module in this type of meteorological condition.

Moreover, *I–V* curves are affected by irradiance level as well as by the cell temperature. For a meaningful comparison of *I–V* curves, one of these must be held constant and the other allowed to vary. Since these parameters are not controlled in outdoor performance testing, constructing *I–V* curves at constant cell temperature or at constant irradiance level requires screening each set of measurements every day.

The conceptualization of PVMOD 3.0 started from the consideration of the foregoing needs, together with the objective to automate routine daily processing of vast amounts of data.

PVMOD 3.0 assists the user in the following areas.

(i) It automates daily data processing and provides efficient statistical analysis of data.

(ii) It provides convenience in checking the stability of meteorological parameters such as irradiance, temperature and wind speed during the measurement of an *I–V* characteristic, thereby helping the user to decide on the acceptability of the data.
(iii) It assists the user in grouping *I–V* curves of the same cell temperature with varying irradiance levels or in grouping *I–V* curves of the same irradiance level but varying cell temperatures.
(iv) It provides graphical representation of data easily on the screen or on a printer.
(v) It provides a graphical analysis of the influence of cell temperature and irradiance level on the performance of modules.
(vi) It provides a simple summary of the data or a full print of the module performance report.
(vii) Without PVMOD 3.0, visual checking of data becomes wearisome and time consuming; plotting of graphs using available worksheet software is also laborious and takes up too much time that will be more valuable if spent in the analysis of these data.
(viii) PVMOD 3.0 also provides a theoretical model and a numerical solution technique to characterize PV modules.

PVMOD 3.0 is the third version of computer software coded at AIT in conjunction with PV module performance evaluation and their analysis. The complete package was coded on an installment nature. Functional software subprograms were added as the needs arose throughout its development stage from version 1.0 to version 3.0. Finally, functional keys and screen presentations were standardized in the latest version.

The two basic functions of PVMOD 3.0 are to process data and to provide an analytical tool to study these data. They are implemented in seven subprograms shown on the main menu screen in figure 10.15.

```
PVMOD 3.0

BY TONY GAN ANG
ENERGY TECHNOLOGY DIVISION
ASIAN INSTITUTE OF TECHNOLOGY

MENU

0. EXIT PVMOD
1. FORMAT AND LINK DATA FILES
2. COMBINE FILES
3. PV ARRAY PERFORMANCE
4. PLOT GRAPHS
5. VIEW DATA FILES
6. ANALYSIS
7. SYSTEM CONFIGURATION
```

Figure 10.15 Main menu screen of PVMOD 3.0.

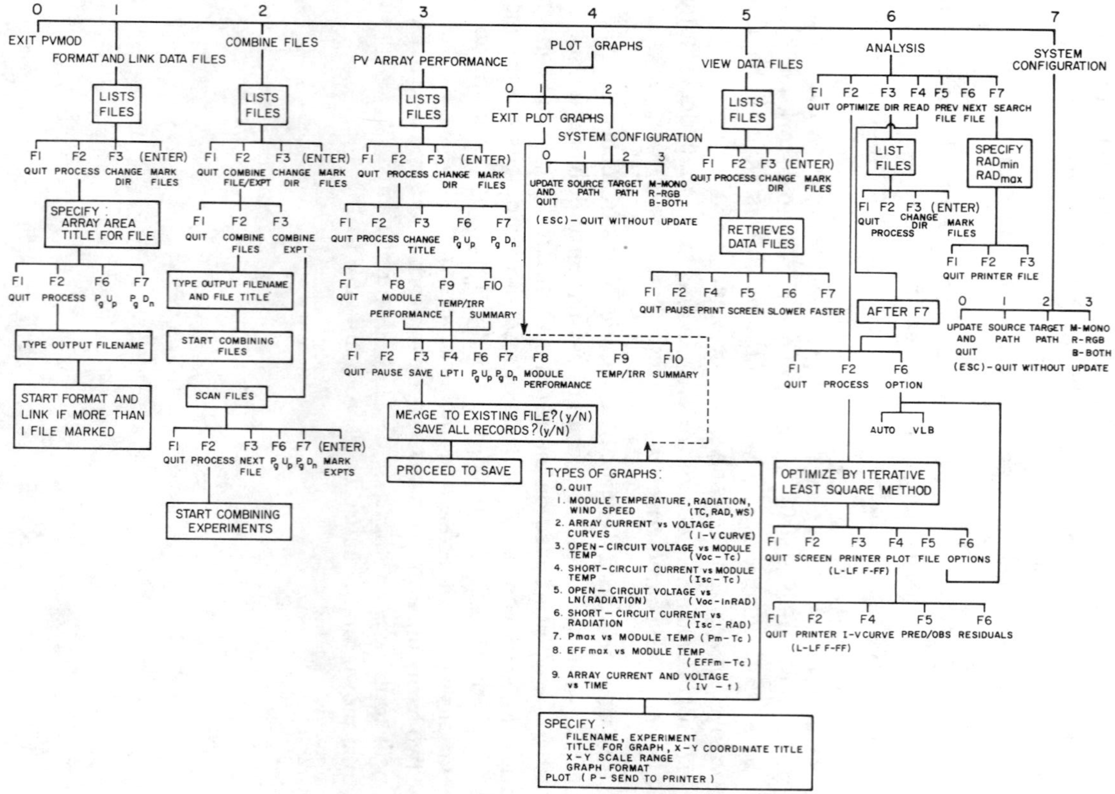

Figure 10.16 Menu tree of PVMOD 3.0.

The data-processing subprograms are the following.

Subprogram 1 Format and link data files.
Subprogram 2 Combine files.
Subprogram 5 View data files.

The analytical subprograms are the following.

Subprogram 3 PV array performance.
Subprogram 4 Plot graphs.
Subprogram 6 Analysis.

The last subprogram on the main menu is the system configuration. This is a short routine to enable the user to specify the source path for his or her data, the target path for his or her data and the types of monitor used.

A menu tree of PVMOD 3.0 is presented in figure 10.16. Extensive use of the functional keys and the directional key pad is made. The ⟨ESC⟩ key is also used to back-step to a previous menu. Whenever appropriate, a beep sound is used to indicate completion of a specific routine, while it remains in the pause state until any character keys are pressed.

10.4.1 Data file

The data being processed and analysed by PVMOD 3.0 are the module I–V characteristics curves. These I–V curves are made up of instantaneous current and voltage measurements as the PV module is forced from open circuit to short circuit by external electronic equipment. Discrete I–V measurements are taken along the curve, 80 such measurements (one experiment) within 2 s. Each measurement or data point consists of the 16 pieces of information in table 10.2. Each of these 16 pieces of information is referred to as one field in the data file. If prior to measurement the data point prompt is set to 80 data points (the maximum is 99 data points), there will be 80 records of 16 fields each in one experiment.

Figure 10.17 illustrates what a data file required by PVMOD 3.0 looks like. The first line of every data file is reserved for 'file title' or a short note or remarks for the particular file. If 80 data points were chosen prior to measurement (as in the present case), then following this first title line we have a group of records of experiments. Each experiment is composed of one identification record, one mean value record and the 80 measurement records—82 records in total for each experiment for this particular example.

The identification record and the mean value record of each experiment are useful primarily for searches required by all subprograms. Every routine needs to search for the experiment number and to size the associated subprogram variable to the number of data points. The analysis subprogram searches for the mean irradiance value as one of its options. Identification of each curve is made through the file, experiment number, module number,

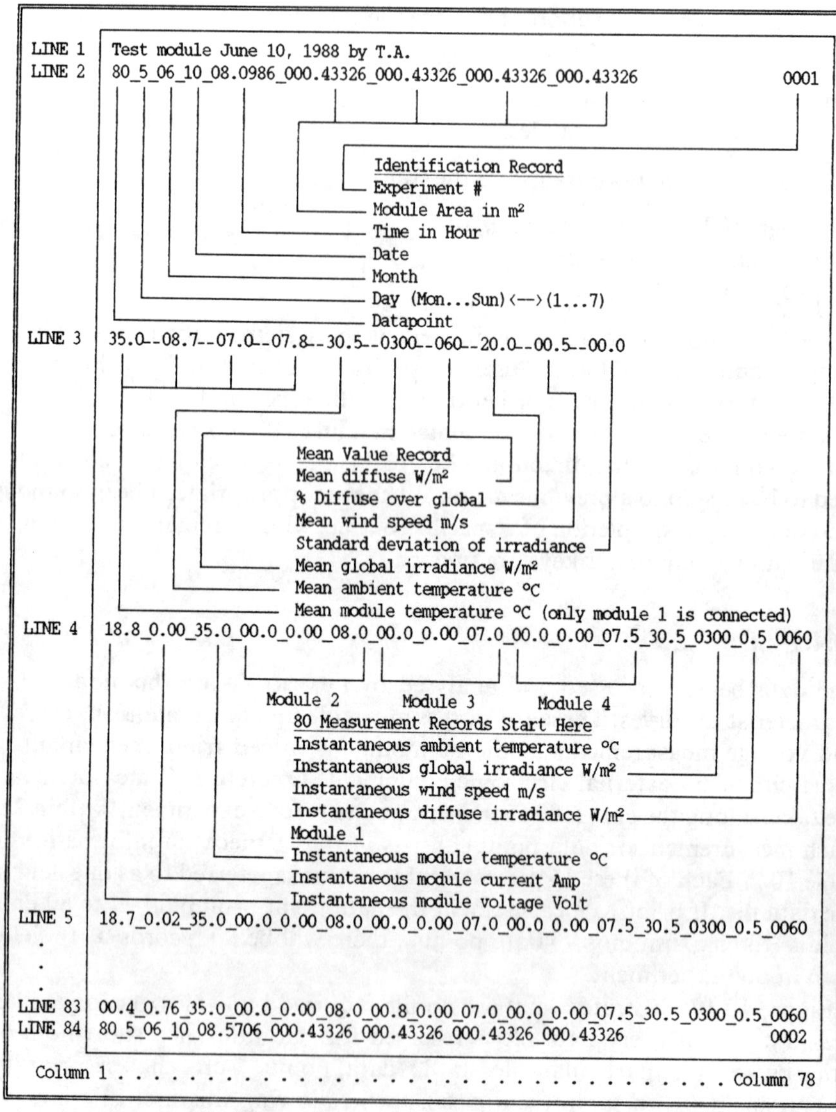

Figure 10.17 Field and record definition of a data file formatted under PVMOD 3.0 requirements.

day, date and time. Furthermore, the mean value record reduces repetitive mean value computation time.

10.4.2 Data-processing subprograms

Data directly measured manually or by data loggers in its original unit, sequence and structure are here referred to as the 'raw' data. Often, these raw data have to be recalculated for the desired unit, corrected according to the accuracy factor of the instruments and finally all raw data have to be reorganized into the structure illustrated in figure 10.17. Once raw data have been processed in this way and restructured, they are referred to as formatted data, having 'PRN' as a file extension name.

Subprogram 1 in the main menu (figure 10.15) formats raw data into the required form of figure 10.17. The source file for this subprogram is available to the user. The user can work on the source file to adapt it to his or her raw data requirements. If raw data were previously saved in several files, linking files while formatting these data is also possible.

The module *I–V* curves are measured at various irradiance levels and module temperatures. In many situations, the user will want to control one of them and to study the effect of varying the other, e.g. looking at *I–V* curves for various module temperatures at constant irradiance level or studying *I–V* curves for various irradiance levels at constant module temperature. Therefore, there will be a need for a routine to regroup experiments.

Subprogram 2 of the main menu allows the user to combine files that have already been formatted by subprogram 1. The user may want to combine daily files into weekly data files. A more powerful use of the subprogram 2 involves combining experiments rather than files. Experiments within a file can be broken down and combined with experiments in other files according to certain criteria set by the user. These criteria may be to combine all experiments of equal irradiance levels, or to combine all experiments of equal

```
                          PVMOD 3.0
                          VIEW DATA
Test module June 10 1988 by T.A.
c:\BPJUN10 .PRN
 V1   I1   T1   V2   I2   T2   V3   I3   T3   V4   I4   T4   TA  RAD  WS  DIFF
80 5  6 10  8.0986   0.43326   0.43326   0.43326   0.43326                   1
35.0   8.7   7.0   7.8  30.5   300   60  20.0   0.5   0.0
18.8 0.00 35.0  0.0 0.00  8.0  0.0 0.00  7.0  0.0 0.00  7.5 30.5  300 0.5   60
18.7 0.02 35.0  0.0 0.00  8.0  0.0 0.00  7.0  0.0 0.00  7.5 30.5  300 0.5   60
18.7 0.02 35.0  0.0 0.00  8.0  0.0 0.00  7.0  0.0 0.00  8.0 30.5  300 0.5   60
18.7 0.02 35.0  0.0 0.00  8.0  0.0 0.00  7.0  0.0 0.00  7.5 30.5  300 0.5   60
18.8 0.04 35.0  0.0 0.00  8.0  0.0 0.00  7.0  0.0 0.00  7.5 30.5  300 0.5   60
18.9 0.04 35.0  0.0 0.00  8.0  0.0 0.00  7.0  0.0 0.00  7.5 30.5  300 0.5   60
18.8 0.04 35.0  0.0 0.00  8.0  0.0 0.00  7.0  0.0 0.00  8.0 30.5  300 0.5   60
18.9 0.06 35.0  0.0 0.00  9.0  0.0 0.00  7.0  0.0 0.00  8.0 30.5  300 0.5   60
18.6 0.06 35.0  0.0 0.00  8.0  0.0 0.00  7.0  0.0 0.00  8.0 30.5  300 0.5   60
18.8 0.08 35.0  0.0 0.00  8.0  0.0 0.00  7.0  0.0 0.00  7.5 30.5  300 0.5   60
18.8 0.08 35.0  0.0 0.00  8.0  0.0 0.00  7.0  0.0 0.00  7.5 30.5  300 0.5   60
18.6 0.10 35.0  0.0 0.00  8.0  0.0 0.00  7.0  0.0 0.00  8.0 30.5  300 0.5   60
18.5 0.12 35.0  0.0 0.00  8.0  0.0 0.00  7.0  0.0 0.00  8.0 30.5  300 0.5   60
18.3 0.12 35.0  0.0 0.00  8.0  0.0 0.00  7.0  0.0 0.00  8.0 30.5  300 0.5   60

c:\BPJUN10 .PRN             EXPERIMENT (0 FOR ALL):  0
F1-QUIT    F2-PAUSE    F4-PRINT    F5-SCREEN    F6-SLOWER    F7-FASTER
```

Figure 10.18 Printer output from view data files subprogram.

module temperatures. Although these two are the most common criteria for combining files, other criteria such as experiments occurring at a certain period of the day or wind speed or ambient temperature may also be implemented. Just simply breaking down a file too huge for efficient handling may be enough reason to use this routine. Whatever reason the user has in breaking down files and combining experiments or combining whole files, the user essentially is trying to classify the huge amount of experiments that PVMOD 3.0 can handle—to classify them for ease and convenience for analysis and presentation.

Subprogram 5 allows the user to view formatted data files. Viewing can either be done on screen or on a hard-copy print-out from the printer. The speed at which data scrolls on the monitor when viewing on screen can be controlled and set slower or faster. The printing speed will have to depend on the printer setting and clock rate of the processor. Figure 10.18 is a printer output from this subprogram.

10.4.3 Photovoltaic array performance subprogram

There are three different types of report that can be generated by this subprogram. Three different emphases are behind each presentation of module performance report.

First report — A comparison of performance between the four modules at a specific irradiance and ambient condition (figure 10.19(*a*)); called by the F8 key.

PVMOD 3.0
PV ARRAY PERFORMANCE

Test module June 10 1988 by T.A.

FILENAME : BPJUN10 .PRN EXP'T 1
DATE : FRI JUN 10
TIME : 8: 5:54
DATAPTS : 80

	MOD1	MOD2	MOD3	MOD4	TA	RAD	DIFF	WS
TEMP AVE. (C)	35.0	8.7	7.0	7.8	30.5	300	60	0.5
STD DEV	0.00	0.87	0.00	0.25	0.00	0.00	0.00	0.00
Voc (V)	18.8	0.0	0.0	0.0				
Isc (A)	0.76	0.00	0.00	0.00	DIFF/RAD (%)=	20.0		
Vmax (V)	15.0	0.0	0.0	0.0				
Imax (A)	0.62	0.00	0.00	0.00				
Pmax (W)	9.3	0.0	0.0	0.0				
EFFmax (%)	7.2	0.0	0.0	0.0				
FILL FACTOR (%)	65.1	0.0	0.0	0.0				
AREA (M2)	0.4333	0.4333	0.4333	0.4333				

F1-QUIT F2-PAUSE F3-SAVE F4-LPT1: F6-PgUp F7-PgDn
F8-MODULE PERFORMANCE F9-TEMP/IRRADIANCE F10-SUMMARY

(*a*)

PVMOD 3.0
PV ARRAY PERFORMANCE

TITLE: Test module June 10 1988 by T.A.
FILENAME: c:\BPJUN10 .PRN DATE: FRI JUN 10

EXP	MOD1	MOD2	MOD3	MOD4	TA	WS	RAD	DIFF	%DIFF	TIME	STDRAD	STDDIF
1	35.0	8.7	7.0	7.8	30.5	0.5	300	60	20.0	8: 5:54	0.0	0.0
2	37.5	7.0	7.0	7.8	30.5	1.1	402	54	13.4	8:34:14	0.0	0.0
3	40.0	7.0	7.0	7.8	31.0	1.2	498	54	10.8	8:54:46	0.0	0.0
4	43.0	12.0	7.0	7.8	31.1	1.2	600	54	9.0	9:23:49	0.0	0.0
5	46.0	11.0	7.0	7.9	31.5	0.9	702	60	8.5	9:58:42	0.0	0.0
6	46.5	11.0	7.0	7.8	32.3	0.4	702	60	8.5	10:40:28	0.0	0.0
7	52.0	12.0	7.0	7.8	32.5	1.7	900	67	7.4	11:46: 9	0.0	1.8
8	51.5	12.0	7.0	7.8	33.0	1.3	900	67	7.4	11:51:14	0.0	2.2
9	51.5	12.0	7.0	7.9	33.0	1.2	900	69	7.7	11:53:44	0.0	3.0
10	52.5	12.0	7.0	7.8	33.0	1.2	906	72	7.9	13:37:14	0.7	0.0

F1-QUIT F2-PAUSE F3-SAVE F4-LPT1: F6-PgUp F7-PgDn
F8-MODULE PERFORMANCE F9-TEMP/IRRADIANCE F10-SUMMARY

(*b*)

PVMOD 3.0
PV ARRAY PERFORMANCE

TITLE: Test module June 10 1988 by T.A.
FILENAME: c:\BPJUN10 .PRN DATE: FRI JUN 10
ARRAY 1

EXP	RAD	DIF	%DIF	TC	TA	WS	VOC	ISC	VMAX	IMAX	PMX	EMAX	FF%	TIME
1	300	60	20.0	35.0	30.5	0.5	18.8	0.76	15.0	0.62	9	7.2	65.1	8: 5:54
2	402	54	13.4	37.5	30.5	1.1	19.0	1.02	14.4	0.90	13	7.4	66.9	8:34:14
3	498	54	10.8	40.0	31.0	1.2	18.9	1.26	14.3	1.10	16	7.3	66.1	8:54:46
4	600	54	9.0	43.0	31.1	1.2	18.9	1.52	14.8	1.28	19	7.3	65.9	9:23:49
5	702	60	8.5	46.0	31.5	0.9	18.7	1.78	13.8	1.56	22	7.1	64.7	9:58:42
6	702	60	8.5	46.5	32.3	0.4	18.7	1.78	13.8	1.56	22	7.1	64.7	10:40:28
7	900	67	7.4	52.0	32.5	1.7	18.4	2.22	13.3	1.92	26	6.5	62.5	11:46: 9
8	900	67	7.4	51.5	33.0	1.3	18.5	2.24	13.3	1.92	26	6.5	61.6	11:51:14
9	900	69	7.7	51.5	33.0	1.2	18.5	2.24	13.3	1.94	26	6.6	62.3	11:53:44
10	906	72	7.9	52.5	33.0	1.2	18.4	2.24	13.6	1.88	26	6.5	62.0	13:37:14

F1-QUIT F2-PAUSE F3-SAVE F4-LPT1: F6-PgUp F7-PgDn
F8-MODULE PERFORMANCE F9-TEMP/IRRADIANCE F10-SUMMARY

(*c*)

Figure 10.19 (*a*) Report 1: module performance report comparing four *I–V* curves. Only one module is connected. (*b*) Report 2: thermal stress comparison of four modules. Only one module is connected. (*c*) Report 3: summary performance report of one module.

Second report A comparison of thermal stress (module temperature) between the four modules; the relation of these stresses to irradiance and ambient condition (figure 10.19(*b*)); called by the F9 key.

Third report A summary report of each module which is useful to correlate module response and thermal stress to irradiance and ambient condition on a per module perspective (figure 10.19(*c*)); called by the F10 key.

All the three reports can be merged to existing text files or saved as a new text file both on a per record basis or as the whole report. Record in this context refers only to one screen page. The performance report can also be printed by a printer.

10.4.4 Plot graphs subprogram

It is important to check the system configuration (subprogram 7) and to make sure that the type of monitor used is properly set, when using this subprogram.

Nine types of graph can be plotted on a graphics-supported monitor and by pressing 'P' the graphics page can be dumped onto a printer. The menu tree in figure 10.16 lists the possible nine types of graph.

The user has to specify the name of the formatted data file, the experiment

```
                           PVMOD 3.0
                           PLOTGRAPH
                      GRAPH TYPE 2  IV CURVE

          0. QUIT
          1. ENTER THE DATA FILENAME : BPJUN10.PRN
          2. HOW MANY EXPERIMENTS TO BE SELECTED   : 10
             (MAXIMUM OF 10 CURVES PER GRAPH TYPE)

                                  1    2    3    4    <--MODULE NUMBER
     1 EXPERIMENT NUMBER :   1    Y    N    N    N
     2 EXPERIMENT NUMBER :   2   >Y    N    N    N
     3 EXPERIMENT NUMBER :   3    Y    N    N    N
     4 EXPERIMENT NUMBER :   4    Y    N    N    N
     5 EXPERIMENT NUMBER :   5    Y    N    N    N
     6 EXPERIMENT NUMBER :   6    Y    N    N    N
     7 EXPERIMENT NUMBER :   7    Y    N    N    N
     8 EXPERIMENT NUMBER :   8    Y    N    N    N
     9 EXPERIMENT NUMBER :   9    Y    N    N    N
    10 EXPERIMENT NUMBER :  10    Y    N    N    N

   ENTER NUMBER -->
```

(*a*)

```
                              PLOT ROUTINE

        0. QUIT
        1. MAIN TITLE: BPJUN10.PRN   13: 37: 14   905.9W/M2 STD= .67
        2.    X TITLE: V
        3.    Y TITLE: I

              4. AUTO SCALE             5. MANUAL SCALE

                X MINIMUM:   0.20        X LOWER LIMIT:  0
                  MAXIMUM:  19.00          UPPER LIMIT:  20
                Y MINIMUM:   0.00        Y LOWER LIMIT:  0
                  MAXIMUM:   2.35          UPPER LIMIT:  2.5

        6. FORMAT: S - SYMBOL  L - LINE  B - BOTH ) B

        7. PLOT

ENTER NUMBER --)
```

(*b*)

Figure 10.20 (*a*) Selecting files, modules and experiments for plotting, (*b*) Setting graph options.

number and the module number (figure 10.20(*a*)) to be plotted. Default settings for graph titles, coordinate titles, coordinate scaling and graph format can be changed. Zooming in on a graph (magnifying a graph) is possible by setting the lower and upper limit of the *x* and *y* coordinates in the manual-scale option (figure 10.20(*b*)). The default setting is the auto-scale option wherein the limits are set by the range of the data. A maximum of 10 curves is possible in one graph (figure 10.21).

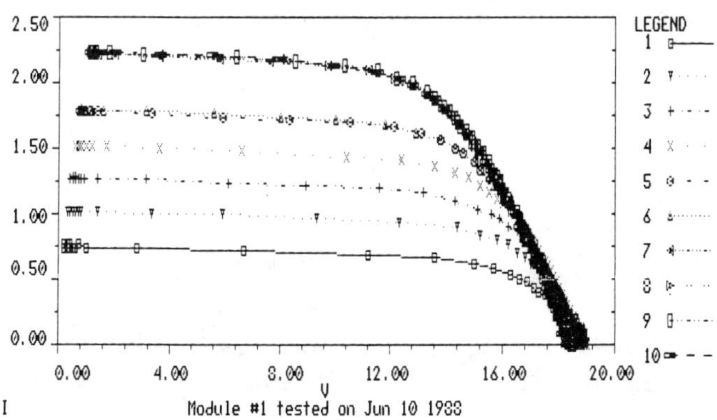

Figure 10.21 *I–V* characteristics curves from the plot graphs sub-program of PVMOD 3.0.

10.4.5 Analysis subprogram

This subprogram is essentially an iterative numerical solution to a least-squares optimization procedure to determine the five parameters (R_s, R_{sh}, A, I_{ph} and I_0) of the theoretical model for a PV module (see equation (10.1).). This is an implementation of an algorithm proposed by Braunstein *et al* (1977).

The load voltage V and the load current I of the PV module are usually given by

$$V=-IR_s+\frac{1}{Z}\ln\left(\frac{I_{ph}-BI-V/R_{sh}}{I_0}+1\right) \tag{10.1}$$

where

$$Z=\frac{q}{Ak_B T} \qquad B=1+\frac{R_s}{R_{sh}}. \tag{10.2}$$

The voltage V in equation (10.1) is an implicit function of the current I and contains five parameters: R_s, R_{sh}, A, I_{ph} and I_0. This equation is non-linear in its parameters and an approximate equation can be derived for equation (10.1). (See Braunstein *et al* (1977) for the derivation of equations (10.3)–(10.14).) The following equation is also an implicit function of the current and the voltage but it contains only three parameters: R_s, R_{sh} and Z.

$$V=V_{oc}-IR_s\frac{1}{Z}\ln[\arg(I, V)] \tag{10.3}$$

$$\arg(I, V)=\frac{B(I_{sc}-I)-V/R_{sh}}{BI_{sc}-V_{oc}/R_{sh}}+\exp[Z(I_{sc}R_s-V_{oc})] \tag{10.4}$$

where V_{oc} is the short-circuit current.

The values of V_{oc} and I_{sc} are directly measured. The remaining two parameters I_0 and I_{ph} are easily calculated from

$$I_{ph}=\frac{BI_{sc}-(V_{oc}/R_{sh})\varphi}{1-\varphi} \tag{10.5}$$

$$\varphi=\frac{\exp(ZI_{sc}R_s)-1}{\exp(ZV_{oc})-1} \tag{10.6}$$

$$I_0=\frac{BI_{sc}-V_{oc}/R_{sh}}{\exp(ZV_{oc})-\exp(ZI_{sc}R_s)} \tag{10.7}$$

after computing the two parameters R_s and μ as follows. By applying the least-squares method to equation (10.3), we obtain two analytical functions for R_s and μ:

$$R_s=\frac{a+bc/d}{be/d-d} \tag{10.8}$$

$$\mu = \frac{1}{Z} = \frac{eR_s - c}{d} \tag{10.9}$$

where

$$a = \sum_j V_j \ln[\arg(I_j, V_j^*)] - V_{oc} \ln[\arg(I_j, V_j^*)] \tag{10.10}$$

$$b = \sum_j \{\ln[\arg(I_j, V_j^*)]\}^2 \tag{10.11}$$

$$c = V_{oc} \sum_j I_j - \sum_j I_j V_j \tag{10.12}$$

$$d = \sum_j I_j \ln[\arg(I_j, V_j^*)] \tag{10.13}$$

$$e = \sum_j I_j^2 \tag{10.14}$$

where V_j^* is the calculated voltage from the previous iteration while I_j and V_j are the measured currents and voltages.

The solution procedure is outlined in the block diagram in figure 10.22. The program reads in the n measured points, determines the least error value ε and calculates the average experimental error σ_{exp}, where U_j is the voltage measurement error at each point. The subscript k denotes the step number. The shunt resistance R_{sh} is defined parametrically in steps; in each step, R_{sh} is decremented by ΔR_{sh}. The superscript L denotes the number of iterations needed for a solution.

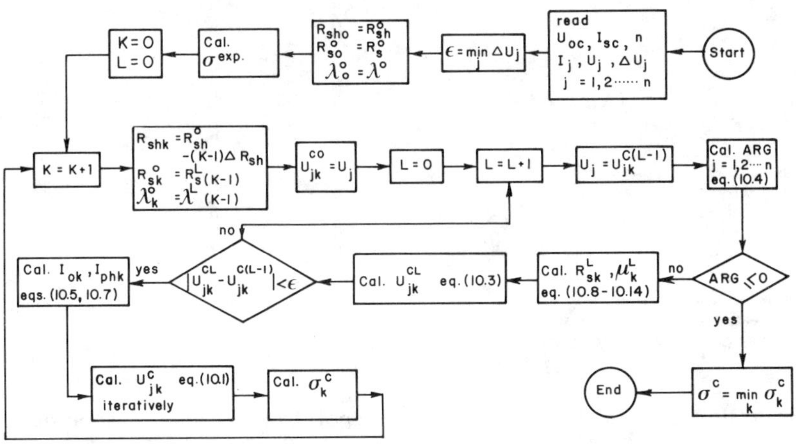

Figure 10.22 A block diagram to determine module parameters.

The analysis subprogram is capable of scanning through several formatted data files marked by the user and searching for only those experiments having the irradiance range specified by the user. The user can set the subprogram to search all experiments within the marked files having an irradiance level between and inclusive of 300 to 900 W m^{-2}, or just

900 W m^{-2}, or whatever is specified. The optimization procedure determines the best estimate for the five parameters in equation (10.1) for each experiment within the specified irradiance range. With this search option, it is possible to compare estimated parameters for all *I–V* curves at, say, 900 W m^{-2} and to see whether these parameters are sensitive to irradiance level or to module temperature.

Figure 10.23 shows a complete report from this subprogram that can be saved in a file or printed by a printer. The estimated coefficient is presented together with its 90% confidence range. An analysis of variance report is also presented. Finally, the predicted voltage is compared to observed voltage. Note that the original 80 data points are now reduced to just 40 data points. The subprogram eliminates redundant data points and any erratic jumps along the curve due in large part to signal noise.

```
                         PVMOD 3.0
              LEAST SQUARE OPTIMIZATION REPORT

DATA FILENAME: a:\JUN10   .PRN
DATE: FRI  JUN  10       TIME: 13:37
EXPERIMENT  10
MODULE  1  TOTAL CELL AREA IS  .43326  M2
                                AVE.       STD DEV
AMBIENT TEMP ('C)              33.00        0.00
MODULE TEMP  ('C)              52.50        0.00
GLOBAL RADIATION  (W/M2)      906           0.00
DIFFUSE RADIATION (W/M2)       72           0.00
WIND SPEED (M/SEC)              1.20        0.02

EXECUTION OF ROUTINE:  Date: 03-22-1989
                       Start= 14:18:56    End= 14:22:45    Duration=  0: 3: 49

  INITIAL VALUES  (lower bound)=  7.9  volt
            Rsh=  1000
             Rs=  1
              A=  36
           DRSH=  10
   EXP VOLT ACC=  .5
    EXPT. ERROR=  5.151608E-03

   FINAL RESULTS
            Rsh=  120
             Rs=  1.169788
              A=  44.86786
            Iph=  2.260567
             Io=  9.507746E-07
MIN. CAL. EPROR=  4.773978E-04
           STEP=  89

R-Squared=  .9994377
Mean Square Error= 0.2549463E-02  with  32  degrees of freedom
```

MODEL FITTING RESULTS at Rsh= 120

	COEFFICIENT	STD DEV	T-VALUE	90% C.L.	90% Confidence Range
RS	1.170	0.0170	68.755	0.0287	1.1411 - 1.1985
A	44.868	0.0145	3102.699	0.0244	44.8434 - 44.8923
IPH	2.261	0.231E-06	9807071	0.389E-06	2.2606 - 2.2606
IO	0.951E-06	0.104E-12	9141862	0.176E-12	0.951E-06 - 0.951E-06

ANALYSIS OF VARIANCE FOR THE FULL REGRESSION

SOURCE	SUM OF SQUARE	DF	MEAN SQUARE	F-RATIO
MODEL	145.0122	5	29.002	11375.898
ERROR	0.0816	32	0.003	
TOTAL	145.0938	37		

DATAPOINT	CURRENT (A)	VOLTAGE (V)	CAL. VOLTAGE (V)	CALCULATED-MEASURED VOLTAGE (V)	(%)
0	0.00	18.40	18.40	0.00	0.00
1	0.04	18.30	18.34	0.04	0.21
2	0.14	18.10	18.15	0.05	0.28
3	0.22	18.00	18.00	0.00	0.03
4	0.28	17.90	17.89	-0.01	-0.03
5	0.33	17.80	17.80	0.00	0.01
6	0.39	17.70	17.69	-0.01	-0.07
7	0.44	17.60	17.59	-0.01	-0.04
8	0.50	17.50	17.48	-0.02	-0.14
9	0.56	17.40	17.36	-0.04	-0.24
10	0.60	17.30	17.28	-0.02	-0.12
11	0.65	17.20	17.18	-0.02	-0.12
12	0.72	17.10	17.04	-0.06	-0.38
13	0.76	16.90	16.95	0.05	0.31
14	0.80	16.80	16.87	0.07	0.41
15	0.86	16.70	16.74	0.04	0.24
16	0.90	16.60	16.65	0.05	0.32
17	0.98	16.50	16.47	-0.03	-0.16
18	1.02	16.40	16.38	-0.02	-0.11
19	1.08	16.20	16.24	0.04	0.26
20	1.12	16.10	16.15	0.05	0.29
21	1.18	16.00	16.00	-0.00	-0.01
22	1.22	15.90	15.90	-0.00	-0.02
23	1.28	15.80	15.74	-0.06	-0.38
24	1.34	15.60	15.58	-0.02	-0.14
25	1.40	15.40	15.41	0.01	0.06
26	1.46	15.20	15.23	0.03	0.21
27	1.52	15.00	15.04	0.04	0.29
28	1.58	14.80	14.84	0.04	0.29
29	1.64	14.60	14.63	0.03	0.20
30	1.70	14.40	14.40	-0.00	-0.03
31	1.76	14.20	14.14	-0.06	-0.44
32	1.80	13.90	13.95	0.05	0.38
33	1.88	13.60	13.52	-0.08	-0.62
34	1.92	13.30	13.26	-0.04	-0.28
35	1.98	12.80	12.80	0.00	0.03
36	2.04	12.30	12.17	-0.13	-1.08
37	2.08	11.50	11.60	0.10	0.86
38	2.14	9.90	9.93	0.03	0.29
39	2.24	0.00	0.00	0.00	0.00

Figure 10.23 Least-squares optimization report from the analysis subprogram.

Graphics are also supported by this subprogram. Three graphs (figures 10.24(*a*)–10.24(*c*)) showing the goodness of fit between the theoretical model and the observed *I–V* curve are generated by pressing one of three functional keys. A plot of the residuals in figure 10.24(*c*) shows that the fitting error does not exceed the measurement precision of 0.1 V.

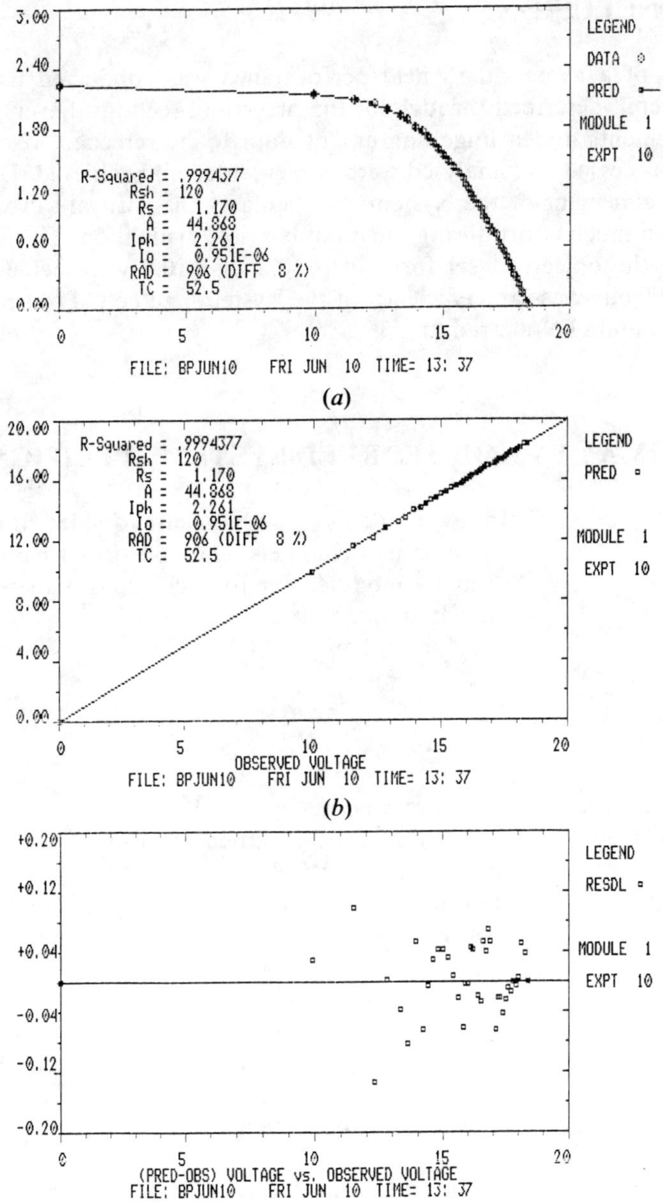

(c)

Figure 10.24 (*a*) Predicted and observed *I–V* curves. Graphical output 1 of the least-squares optimization report. (*b*) Graphical output 2 of the least-squares optimization report. (*c*) Graphical output 3 of the least-squares optimization report. Plot of the residuals between predicted and observed data.

10.5 CONCLUSION

Evaluation of a PV module's field performance was conducted using the whole 'system', described throughout the preceding sections. Several weeks of measurements and a huge amount of data to be screened, rejected or accepted, processed and analysed were implemented and achieved. Development and refinement of the 'system', particularly the software PVMOD 3.0, also required much effort during the module's field evaluation. Through this experience, the objectives set forth in §§ 10.2 and 10.4 were achieved. To gauge the effectiveness and efficiency of the 'system', an actual test report in chapter 11 should be referred to.

APPENDIX A PVMOD TEST BED SYSTEM SPECIFICATION

PV module	16 PV modules or arrays are divided into four measurement channels. Only one input per channel (four PV modules per four channels) can be measured at any one time. Resolution, 0.02 A, 0.1 V.
Input rating	PV modules or arrays. 75 W, 10 A, 100 V.
Global and diffuse tilted irradiance	Class 1 pyranometer, 15° tilt. Resolution of measurement, 6 $W m^{-2}$. Sensitivity, 4–6 $\mu V W^{-1} m^{-2}$. Non-linearity, 0.0% at 500 $W m^{-2}$, and 0.7% at 1000 $W m^{-2}$. Impedance, 1100–1500 Ω. Response time, less than 5 s (1/e), 99% value after 24 s. Spectral range, 305–2800 nm (50% transmission points). Temperature dependence, ±1% (from −10 to 40 °C). Cosine and azimuth error, 3% at 10° sun altitude.
Temperature sensor	Two-terminal integrated-circuit temperature transducer. Linear current output, 1 $\mu A K^{-1}$. Range, from −55 to 150 °C. Resolution, 0.1 °C.

Wind sensor	Three-cup anemometer. Range, 0–20 m s^{-1}. Accuracy, 0.5 m s^{-1}.
A–D converter	CMOS 16-channel multiplexer. 8 bit A–D converter. Conversion time, 100 μs per channel. Input range, 0–5.12 V. Sensitivity, 0.02 V.
Ramp generator	Slope of 0.76 V s^{-1}.
Output	*I–V* curves of four PV modules or arrays with global and diffuse irradiance, wind speed and temperature. Maximum of 99 data points (voltage, current, temperature, wind speed and irradiance per data point) per *I–V* curve. 2.5 s to generate 99 data points. 8000 bytes per irradiance level (four *I–V* curves, temperature, wind speed and irradiance).
Operation	Computer-controlled monitoring and storage. Mode, irradiance-specified monitoring and storage; time interval monitoring and storage. Constant irradiance check before and after *I–V* curve generation.
Software support	Full PVMOD 3.0 features. Graphical presentation of data. Statistical description of data. Theoretical models and numerical method of PV module characterization.

NOMENCLATURE

The symbols in parentheses indicate the parameters as seen on the computer print-outs.

A		Ideality factor in the lumped model of the solar cell
η_{max}	(EFFmax)	Efficiency at the maximum-power point ($V_{max}I_{max}$/(radiation × area of cells))
FF	(FF)	Fill factor
I	(I)	Net current of the solar cell (A)

I_D		Diode current of the solar cell (A)
I_{max}	(Imax)	Current at maximum power (A)
I_{ph}	(Iph)	Photocurrent of the solar cell (A)
I_{sc}	(Isc)	Short-circuit current (A)
I_0	(Io)	Reverse saturation current of the solar cell (A)
k_B		Boltzmann's constant, 1.381×10^{-23} J K^{-1}
P_{max}	(Pmax)	Power at the maximum-power point (W)
q		Electronic charge, 1.602×10^{-19} C
R_s	(Rs)	Series resistance in the lumped model of the solar cell (Ω)
R_{sh}	(Rsh)	Shunt resistance in the lumped model of the solar cell (Ω)
T		Absolute temperature (K)
V	(V)	Terminal voltage of the solar cell (V)
V_{max}	(Vmax)	Voltage at maximum power (V)
V_{oc}	(Voc)	Open-circuit voltage (V)
$\left.\frac{dV}{dI}\right\|_{I=I_{sc}}$		Inverse slope of I–V characteristics at the short-circuit point
$\left.\frac{dV}{dI}\right\|_{V=V_{oc}}$		Inverse slope of I–V characteristics at the open-circuit point
φ		Photon flux per unit time

REFERENCES

Braunstein A, Bany J and Appelbaum J 1977 *Energy Conversion* **17** 1–6

Treble F C 1985 *Proc. 4th Asian School on Solar Harnessing, Bangkok, 12–20 December 1985* (Bangkok: Asian Institute of Technology)

BIBLIOGRAPHY

Ashok S and Pande K P 1985 *Solar Cells* **14** 61–81

Goldstein L and Case G R 1987 *Solar Energy* **21** 37–43

International Rectifier HEXFET Databook

Kar S, Ashok S and Fonash S J 1980 *J. Appl. Phys.* **51** 3417

Lasnier F and Ang T G 1987 *Photovoltaic Test and Instrumentation Series—PVMOD 1.0* (Bangkok: Energy Technology Division, Asian Institute of Technology)

Rauschenbach H S 1980 *Solar Cell Array Design Handbook* (New York: Van Nostrand–Reinhold)

11 Testing of Photovoltaic Modules Under Natural Conditions Using the Asian Institute of Technology Photovoltaic Module Test Bed System

11.1 INTRODUCTION

Financial and technical feasibility of a photovoltaic (PV) power system has materialized into a commercial market in some specific regions and within specific applications. Among the many factors involved in making an investment decision in favour of a PV power system (PVPS), the two most likely questions are: 'How reliable is this new system?' and 'How much does it cost to produce a certain kilowatt hour?'

Let us look at reliability first. The simplest measure of reliability by a user (who is usually from a remote area) is the answer to the question: 'Will my appliances run as they used to?' Electronic equipment that goes into a PVPS has been the focus of much product development and technical refinement for various specific operating conditions for the past few years, and much improvement in terms of reliability, robustness and serviceability have been achieved. What we are interested in now is whether the amount of energy in kilowatt hours promised by the system so designed is going to be fulfilled throughout the useful lifetime of the system. By fulfilling this designed kilowatt hour energy output, any appliances connected to the system are going to operate as required; failure to do so will cause some of the connected appliances to stop running owing to insufficient energy output from the PVPS. This sizing problem due to poor reliability is related to how design engineers understand the energy output from the installed PV module, and not to oversize in order to optimize cost and not to undersize in order to optimize reliability. An accurate prediction of the energy output from a PV module subjected to the working conditions of the installation site is vital to this end.

Another common concern of a user is the cost of the system. The cost of a PVPS is always related to the energy output expected from the system.

Electronic equipment of higher power-carrying capability costs more. The overall system cost per kilowatt hour will be much more when it is underutilized if there is an unexpected drop in module output. Even PV modules are priced and compared by the user in terms of US dollars per peak watt. More often, the manufacturer's measured module power output is not duplicated in every installation site. The price per peak watt of power output will certainly shoot up when the irradiance is less and the temperature is higher in the field than in the test conditions of the manufacturer.

This chapter will try to address the sizing and the costing problem as presented above. The approach that will be taken here is to ensure an accurate prediction of the PV module performance at the site of installation. This will involve taking series of measurements of module performance subjected to characteristic environmental stress of the site. An investigation will be made of the irradiance condition of the site, the module temperature behaviour and its effects and the performance of the module as a function of the environmental stress, and finally the measured performance will be simulated by a theoretical model.

11.2 PHOTOVOLTAIC MODULE TEST

A single-crystal PV module was tested at the Energy Technology Energy Park, Asian Institute of Technology (AIT), from 8 June 1988 to 4 July 1988. The module was installed on an open frame, with 78 cm height of open space above the ground, tilted at 15° south. 36 cells of 10 cm diameter were connected in series giving a total module area of 0.433 26 m^2.

Module performance was monitored in terms of module current, voltage and module back-surface temperature. The current–voltage (I–V) characteristics of the module under constant global irradiance from 300 to 900 $W\,m^{-2}$ at 100 $W\,m^{-2}$ intervals were compiled. The following response variables were correlated to measured ambient conditions in terms of ambient temperature, global and diffuse irradiance on the plane of the module and wind speed:

(*a*) Open-circuit voltage V_{oc}

(*b*) Short-circuit current I_{sc}.

(*c*) Voltage V_{max} at maximum power.

(*d*) Current I_{max} at maximum power.

(*e*) Maximum power P_{max} and maximum efficiency η_{max}.

(*f*) Fill factor FF.

Some of the response variables were strongly dependent on the module temperature, which is a function of irradiance, ambient temperature and module thermal inertia.

11.2.1 Irradiance time window

Figure 11.1 presents the time period wherein the irradiance level is detected during the test period. Ideally, on a clear day, each irradiance level will have two time windows: one before noon and the second after noon. As the irradiance level increases, these morning and afternoon irradiance time windows will approach one another until they fuse together at 900–1000 W m^{-2}, around noon. However, the sky condition in the test site is such that cloudy days continue throughout the test period, causing the time window for each irradiance level to spread out with a specific probability density function along the hour axis. For our present purpose, further investigation was not implemented. However, further study on this concept will provide a probability density function model for each irradiance level along the time axis, which will correspond to a specific module power output. This will lead to computation of an overall probability density function of daily total energy output from the module during a particular period or season.

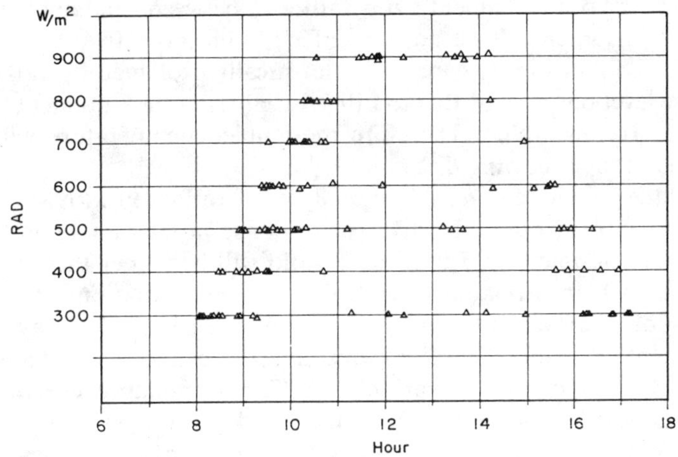

Figure 11.1 Irradiance window. The time period wherein the irradiance level E_e is detected during the test period.

For our present purpose, figure 11.1 graphically conveys the extent of cloud effects during the period of irradiance monitoring. As noted above, an ideal clear sky condition will make figure 11.1 look like an inverted ‘V’ with a peak at noon of around 900 W m^{-2}. Any data point between the two ‘legs’ of the V is due to an irradiance drop from a higher to a lower irradiance level when the sky is not clear. For instance, around noon (1200 hours), an ideal clear-sky condition will provide an irradiance level of around

900–1000 $\mathrm{W\,m^{-2}}$ for a specific location. Having an irradiance level lower than this during this period, thereby falling between the two 'legs' of the V, indicates that cloud effects have brought about this drop. Evidently, an important implication is the rise in module temperature when data points fall between the two 'legs'. This rise in temperature (particularly at low irradiance level) has a negative effect on the module performance by lowering module voltages and efficiency. When too many points are observed between the two 'legs' of the inverted V in figure 11.1 in a particular site, then the module is best mounted with sufficient clearance for natural cooling.

11.2.2 General performance of the module

Table 11.1 tabulates the response variables at each global irradiance level. They are average values computed from several sets of I–V characteristic curves at each irradiance level. The standard deviations within each irradiance level, except for the wind speed and the temperature of the module's back surface, are no more than 6% of the average value.

The variance of the ambient temperature T_A between irradiance levels and within each irradiance level do not differ significantly. In other words, as observed, variations in T_A cannot be statistically explained by variations in irradiance level but may be due to other variables or measurement precision around its average value. Therefore the ambient temperature will not be considered in the following discussions.

Table 11.2 presents the results of analyses of variance (ANOVA) on several response variables from table 11.1. The response variables in table 11.2 have a probability of below 0.05 that the F ratio will be higher than calculated. This means a high probability (high F ratio) that variations in irradiance level are more than variations within an irradiance level which may be due to measurement precision. Therefore, this indicates a significant change in the average value of the response variable at different irradiance levels compared with measurement precision. This observation also indicates that the measurement precision is small enough to justify further analyses using the collected data. It also verifies that the response variation with respect to irradiance level in figure 11.2(*a*) is acceptably free from precision errors.

Table 11.2 also shows that the temperature of the module's back surface T_c and the fill factor (FF) are highly sensitive to irradiance level (high F ratio). Figure 11.2(*a*) makes table 11.2 clearer. Only T_c, I_{sc} and I_{max} increased with increasing irradiance. The increase in the module temperature normally arises because of increasing irradiance; thus there are slight drops in V_{oc} and V_{max} in the figure. This probably explains the associated drop in FF and maximum efficiency η_{max} with increasing irradiance. The variation in FF is emphasized in figure 11.2(*b*). As the irradiance level increases, the I–V characteristics become less square (lower FF and η_{max}). Figure 11.2(*c*) is a plot of the module output power against the module voltage. The maximum-

Table 11.1 Summary of average module performance under natural conditions from 8 June 1988 to 4 July 1988. The module area is 0.433 m^2 of 36 cells in series.

E_e ($W\,m^{-2}$)	Experiment	T_c (°C)	T_a (°C)	Wind speed ($m\,s^{-1}$)	V_{oc} (V)	I_{sc} (A)	V_{max} (V)	I_{max} (A)	P_{max} (W)	η_{max} (%)	FF (%)
300	35	38.7	32.4	0.9	18.6	0.74	14.8	0.61	9.0	7.0	65.8
400	18	40.1	32.7	0.9	18.8	1.00	14.9	0.84	12.6	7.2	66.5
500	26	42.1	32.6	0.9	18.8	1·26	14.6	1.07	15.8	7.2	66.2
600	21	43.8	33.0	1.1	18.8	1.51	14.5	1.29	18.6	7.2	65.6
700	13	45.7	32.3	1.3	18.8	1.77	14.2	1.51	21.6	7.1	64.7
800	10	47.8	33.0	1.2	18.6	2.01	13.9	1.73	24.1	6.9	63.7
900	15	50.4	33.3	1.1	18.6	2.23	13.4	1.94	26.1	6.7	62.5

Table 11.2 Result of an ANOVA on some response variables from table 11.1. A high F ratio and a low probability indicate significant variations in the response variable between irradiance levels. The main effects mean square is computed as the square of the difference between the average of the response variable at each irradiance level and the overall average of the response variable; this square difference is then divided by its degrees of freedom which is 6. The residual mean square is computed as the square of the difference between each datum and its average within the irradiance level; this square difference is then divided by its degrees of freedom which is 131. If the probability of the F ratio (the ratio of main effects to residual mean square) is lower than 0.10, this indicates a high probability that the variances of the response variable within its population changed significantly, inferring a dominant presence of the irradiance level effect (main effects) on the response variable in addition to measurement effects.

Response variable	Mean square		F ratio	Probability of the F ratio
	Main effects	Residual		
T_c	328.114	9.944	32.995	0.0000
Wind speed	0.530	0.202	2.620	0.0197
V_{oc}	0.271	0.107	2.542	0.0232
V_{max}	4.780	0.233	20.560	0.0000
η_{max}	0.734	0.070	10.553	0.0000
FF	33.484	0.568	58.970	0.0000

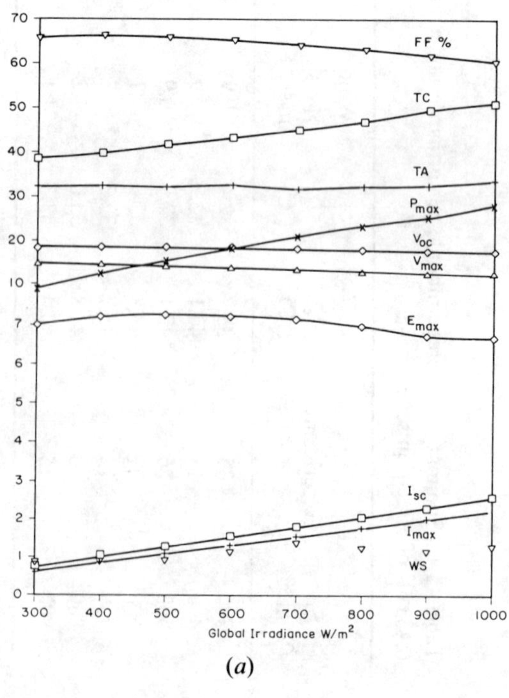

(*a*)

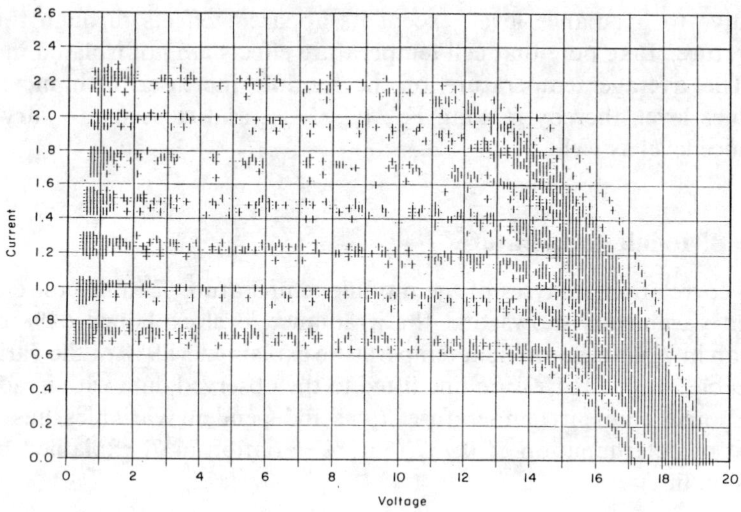

(*b*)

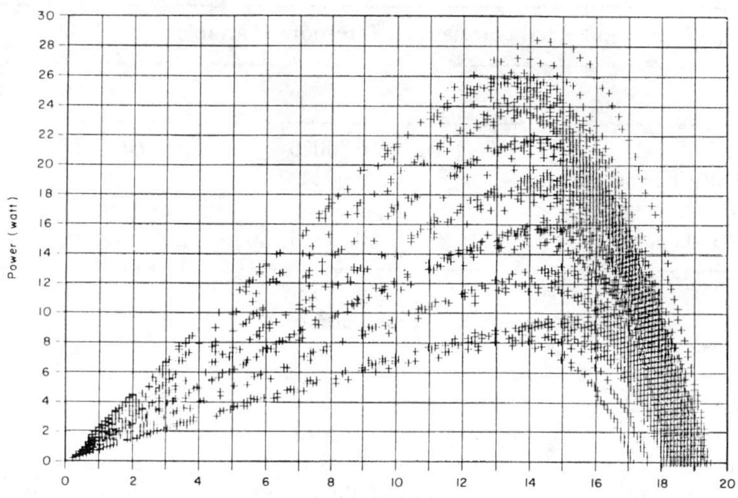

(*c*)

Figure 11.2 (*a*) Module performance as a function of global irradiance (see table 11.1). (*b*) *I–V* characteristics curves observed from 300 to 900 W m^{-2}. (*c*) Module power as a function of module voltage.

power point shifts slightly from 16 V towards 13 V as the irradiance changes from 300 to 900 W m^{-2}. In § 11.2.4 we shall show that this shift in V_{max} is due to T_c and not directly due to the irradiance level.

Theoretical considerations indicate that T_c, I_{sc}, I_{max} and subsequently P_{max} are dependent on irradiance level, and the rest of the parameters are insensitive to irradiance level, except its apparent effects through the cell temperature. Take note that cell temperature effects are not isolated in table 11.2. The average temperature of the module increases with increasing irradiance level, thereby making V_{oc}, V_{max}, η_{max} and FF seems to vary with irradiance level as well.

11.2.3 Module temperature

Partial correlation analyses of the module temperature T_c in response to the ambient parameters shows that the irradiance level explained 64% of the variation in T_c and the ambient temperature explained 34.9% of the variation in T_c (table 11.3); a regression line fitted to the observed data with irradiance level E_e and ambient temperature T_a as independent variables has a coefficient of determination of 0.737 (73.7% variation in T_c explained by the regression line).

Table 11.3 Module temperature T_c as a function of E_e and T_a.

Partial correlation (T_c response variable)		
	R	R^2 (%)
E_e	0.80020	64.0
Diffuse irradiance	0.09440	0.9
T_a	0.59085	34.9
Wind speed	0.01714	0.3

Stepwise multiple regression of $T_c=f(E_e, T_a)$	
R^2	$=0.737$
F ratio	$=189.0686$ with 133 degrees of freedom
A (constant)	$=30.006 \pm 2.085$ at 95% confidence interval
B (E_e coefficient)	$=0.0175 \pm 0.0022$ at 95% confidence interval
C (T_a coefficient)	$=1.14 \pm 0.268$ at 95% confidence interval

Therefore the empirical model for $T_c=f(E_e, T_a)$ is

$$\begin{aligned} T_c &= A+B(E_e-300)+C(T_a-25) \\ &= 30.006+0.0175(E_e-300)+1.14(T_a-25) \\ &= 30.006[1+5.832\times 10^{-4}(E_e-300)+0.038(T_a-25)] \end{aligned} \quad (11.1)$$

and these coefficients are specific to this particular module tested. A more complex model for T_c involving heat transfer analysis of the module and ambient may give us a more physically meaningful coefficient than the empirical model.

Table 11.4 lists the temperature window at 80% and 90% confidence levels. The average observed T_c from table 11.1 is included to check values predicted by equation (11.1). In every irradiance level the predicted value varies no more than 2% from the average observed T_c.

Table 11.4 Temperature ranges at 80% and 90% confidence levels of finding T_c at each irradiance level. The empirical model of $T_c = f(E_e, T_a)$ in equation (11.1) is used. See table 11.1 for average T_a values at each irradiance level. Equation (11.1) gives less than 2% error compared with average observed T_c at every irradiance level.

E_e ($W\,m^{-2}$)	Temperature range (°C) at following confidence levels		Average observed value	Calculated value (equation (11.1))
	80%	90%		
300	35–40	35–43	38.7	38.5
400	36–43	36–44	40.1	40.5
500	40–44	40–46	42.1	42.2
600	42–46	41–47	43.8	44.4
700	45–47	44–47	45.7	45.3
800	44–50	44–51	47.8	47.9
900	50–53	49–53	50.4	50.0

Figure 11.3 is a plot of the predicted against the observed temperature T_c of the module. When all the predicted points are on the full line, the multiple-determination coefficient R^2 would have been 1.0. Since R^2 is the only 0.737, the empirical model can predict approximately 73 correct T_c out of 100 trials. Points that are far from the observed full line are due mainly to the combination of the cloud effects and the thermal inertia of the module. A module may have a T_c of 52 °C at 900 $W\,m^{-2}$ but, when the clouds effects decrease the global irradiance from 900 to 300 $W\,m^{-2}$ for 3–5 min, T_c will not have decreased instantaneously, nor will T_a. Therefore, such cases are not predicted by the empirical model, since it is an instantaneous model dependent on the instantaneous value of E_e and T_a. Such cloud effects can only be predicted through a thermodynamic simulation within a day, using a heat transfer energy balance model.

The frequency distribution of T_c at each irradiance level is presented in figures 11.4(*a*)–11.4(*g*). The number above each bar is the cumulative probability value. The confidence level of finding T_c at each irradiance level is also indicated above the bar graph.

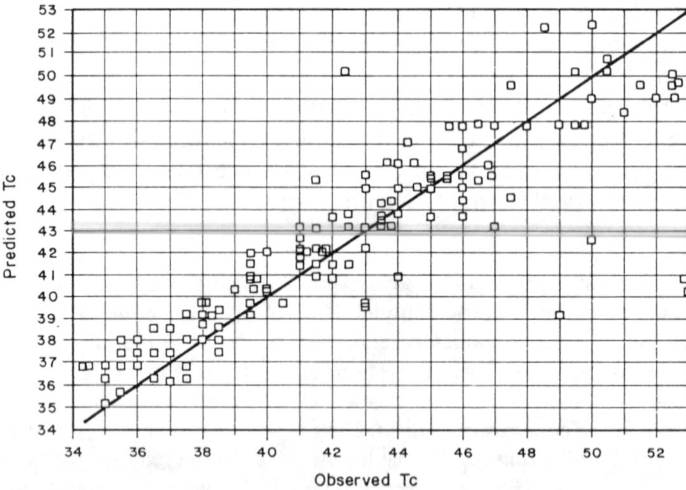

Figure 11.3 Predicted value of the module temperature T_c is plotted against the observed value showing 73.7% of total T_c variation is explained by equation (11.1). Points far from the observed line are due to the cloud effects that are not predicted by the empirical model.

11.2.4 Module response to ambient parameters

Table 11.5 is a partial correlation result of the module's response variables to ambient parameters. On the basis of the data gathered, the irradiance and the module temperature explain a high percentage of variation in all response variables of the module. The current and power of the module are sensitive to the irradiance level, while the voltage and efficiency are sensitive to the module temperature. It is interesting to note that the irradiance level induced a positive response from the module, which is not surprising; while the

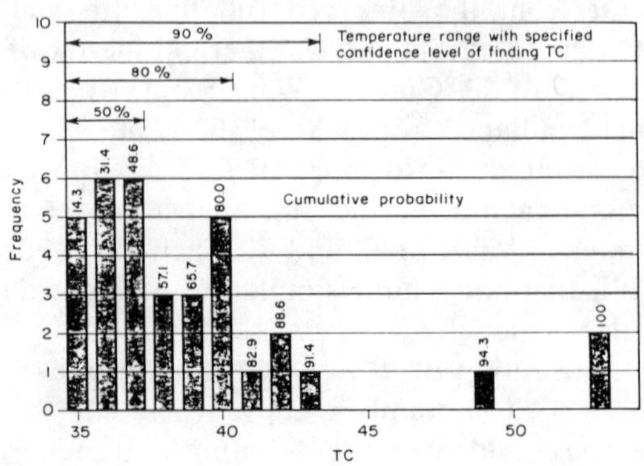

(*a*)

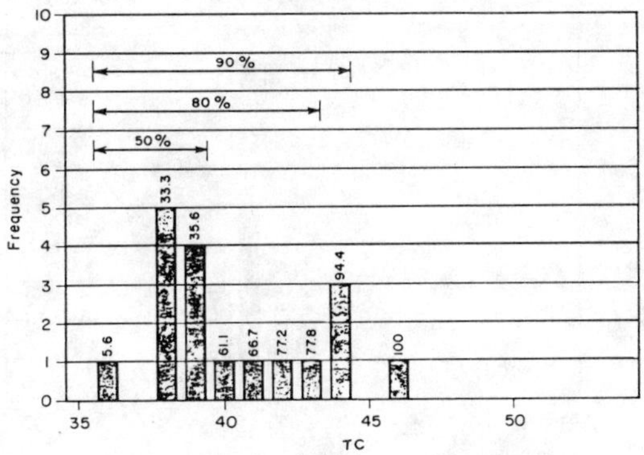
90 %
80 %
50 %
Frequency
TC
33.3
35.6
5.6
61.1
66.7
77.2
77.8
94.4
100

(*b*)

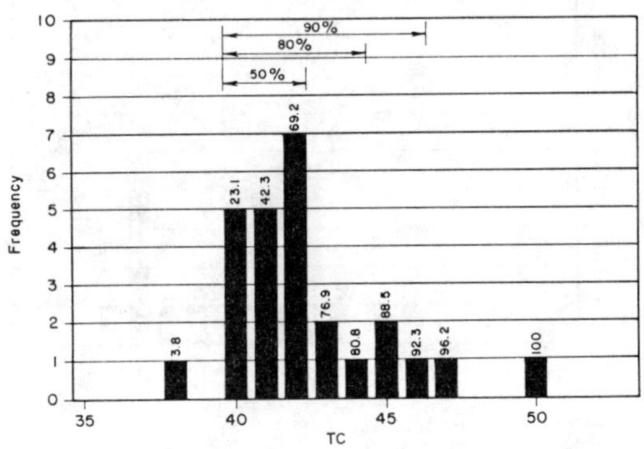
90%
80%
50%
Frequency
TC
69.2
23.1
42.3
76.9
88.5
3.8
80.8
92.3
96.2
100

(*c*)

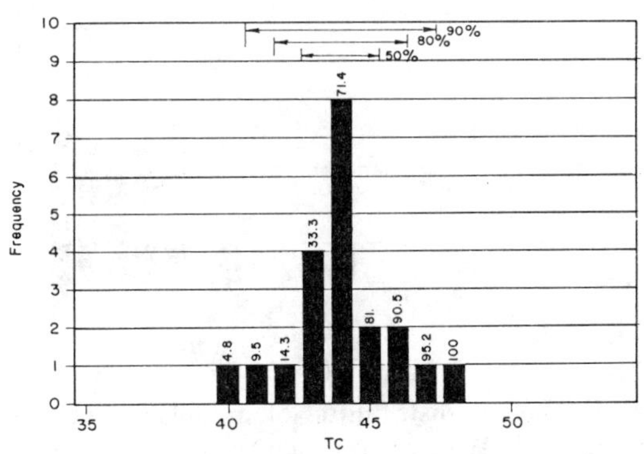
90%
80%
50%
Frequency
TC
71.4
33.3
81.
90.5
4.8
9.5
14.3
95.2
100

(*d*)

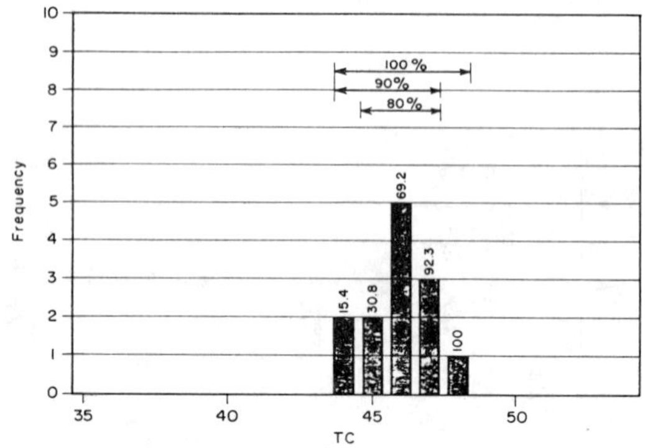

(e)

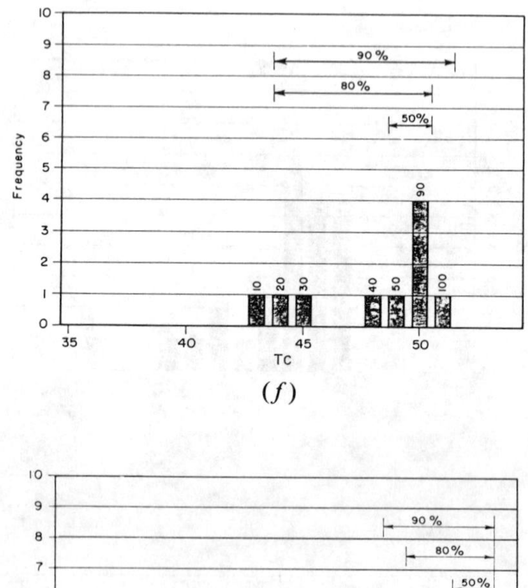

(f)

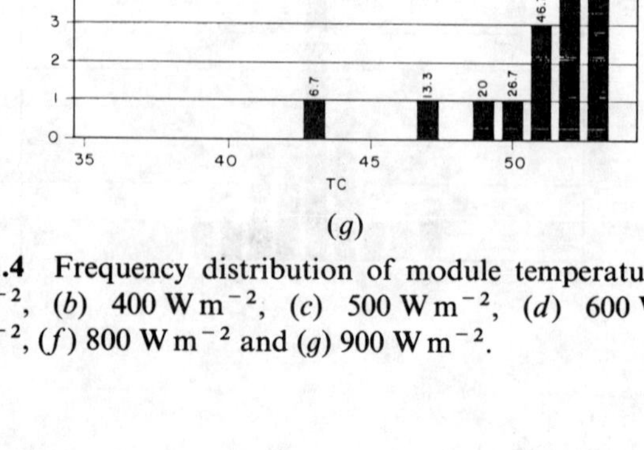

(g)

Figure 11.4 Frequency distribution of module temperature T_c at (*a*) 300 W m^{-2}, (*b*) 400 W m^{-2}, (*c*) 500 W m^{-2}, (*d*) 600 W m^{-2}, (*e*) 700 W m^{-2}, (*f*) 800 W m^{-2} and (*g*) 900 W m^{-2}.

Table 11.5 Partial correlation of response variables to ambient parameters. $R^2 \times 100\%$ refers to the percentage of the variation in the response variable explained by the corresponding ambient parameter.

	$R^2 \times 100\%$ for the following ambient parameters				
	E_e	Diffuse irradiance	T_c	T_a	Wind speed
V_{oc}	83.5	7.8	−85.8	−2.9	0.8
I_{sc}	99.5	−0.3	0.007	−34.4	−7.9
V_{max}	−0.02	0.2	−39.1	0.8	1.5
I_{max}	98.7	0.4	0.1	−12.9	−1.7
P_{max}	97.9	3.9	−22.8	−5.0	−0.3
η_{max}	19.5	2.0	−36.3	−10.8	−3.8

module temperature tends to degrade its performance. The other three ambient parameters (T_a, wind speed and diffuse irradiance) do not exhibit any direct significant effects on the response of the module.

Among the response variables, I_{sc}, I_{max} and P_{max} are almost totally controlled (explained 98–99%) by the irradiance level. P_{max} (product of V_{max} and I_{max}) which is highly sensitive to irradiance is mostly due to a higher correlation of I_{max} to irradiance compared with a lower correlation of V_{max} to T_c. On the other hand, since I_{max} is almost proportional to irradiance, η_{max} becomes highly correlated to T_c. Furthermore, the irradiance has very little effect on V_{max}, while the module temperature has very little effect on I_{sc} and I_{max}.

While the module's temperature explains a large percentage of the variations in V_{oc} (−85.8%), it does not perform as well in explaining the variations in V_{max} (−39.1%) and η_{max} (−36.3%), although it is still the highest contributor of the ambient parameters in explaining V_{max} and η_{max}. Since V_{max} and η_{max} are highly correlated to one another with a correlation coefficient R of 0.927 77 and since η_{max} is computed from V_{max}, any error in determining V_{max} will affect η_{max} substantially. Their low multiple-determination coefficient R^2 with respect to T_c might be due to slight errors introduced while searching for V_{max} along a non-continuous I–V curve composed of discrete data points.

In order to determine the rate of variation in the response variable of table 11.5 with respect to the irradiance and the module temperature, a linear model

$$Y = A + B(E_e - 300) + C(T_c - 25) \tag{11.2}$$

is fitted to the observed data. The coefficients B and C are then the rate of variation with respect to the irradiance and the module temperature

Table 11.6 Stepwise multiple linear regression. $Y = A + B(E_e - 300) + C(T_c - 25)$ for I_{sc}, V_{max}, I_{max} and P_{max}; $Y = A + B \ln(E_e - 100) + C(T_c - 25)$ for V_{oc} and η_{max}. In the MSE, V_{max} has 136 degrees of freedom and the other parameters 135. For the A value for V_{oc} the 95% confidence interval is estimated.

	A (constant)	B (E_e coefficient)	C (T_c coefficient)	R^2 (%)	MSE	F ratio	
						E_e	T_c
V_{oc}	15.518 (±0.190)	0.841 (±0.040)	−0.100 (±0.00389)	95.1	0.0056	39.52	2603.66
I_{sc}	0.797 (±0.021)	0.002 569 (±0.000 036)	−0.003 458 (±0.001 462)	99.7	0.0007	47.697	22.03
V_{max}	16.457 (±0.236)	—	−0.111 81 (±0.012 70)	69.2	0.134	—	304.89
I_{max}	0.650 (±0.026)	0.002 255 (±0.000 044)	−0.002 157 (±0.001 799)	99.5	0.001	24.628	5.64
P_{max}	11.373 (±0.425)	0.031 445 (±0.001 132)	−0.137 871 (±0.029 818)	99.1	0.297	15.209	83.43
η_{max}	5.527 (±0.462)	0.489 (±0.096)	−0.077 (±0.009 45)	66.4	0.033	8.90	264.37

respectively. A better fit is achieved for V_{oc} and η_{max} when they are expressed as a logarithmic function of irradiance:

$$Y = A + B\ln(E_e - 100) + C(T_c - 25). \tag{11.3}$$

Table 11.6 tabulates the result of a stepwise multiple linear regression. R^2 expressed in per cent is the coefficient of multiple determination indicating the percentage of response variation explained by the regression model. The mean square error (MSE) is the ratio of the error sum of squares (SS_E) to the degrees of freedom. The error sum of squares is the amount of response variation remaining after the regression is performed, which is computed from the calculated Y_i and each observed y_i value:

$$SS_E = \sum_i (y_i - Y_i)^2.$$

The degrees of freedom are the total number of observed data (138) minus the number of regression line coefficients, minus 1. The mean square error is

$$MSE = \frac{SS_E}{\text{degrees of freedom}}.$$

Since a stepwise regression is utilized, the F ratios for E_e and T_c indicate the relative effectiveness of E_e and T_c in the regression model to explain response variation when E_e and T_c are entered into the regression model in that order. Again, it is evident E_e is significant in explaining I_{sc}, I_{max} and P_{max} while T_c is significant to V_{oc}, V_{max} and η_{max}.

The numerical value under each coefficient in table 11.6 indicates the range of 95% confidence interval around the estimated coefficient. Therefore we are 95% confident that I_{sc} increased by 0.25–0.26 A for every 100 W m^{-2} rise in irradiance level, and 95% confident that V_{oc} decreased by 96–109 mV for every degree Celsius rise in the module's temperature.

The important points in table 11.6 are summarized in table 11.7. Figure 11.5(a)–11.5(f) are plots of the predicted against the observed values (table 11.8). The plot of predicted V_{max} is scattered around the observed line.

Table 11.7 Rate of variation in the response variables with respect to E_e and T_c.

With 95% confidence,

V_{oc}	decreased by 0.096 V	to 0.10 V	for every 1°C rise in T_c
V_{max}	decreased by 0.009 V	to 0.13 V	for every 1°C rise in T_c
P_{max}	decreased by 0.11 W	to 0.17 W	for every 1°C rise in T_c
η_{max}	decreased by 0.07%	to 0.09%	for every 1°C rise in T_c

while

I_{sc}	increased by 0.25 A	to 0.26 A	for every 100 W m^{-2} rise in E_e
I_{max}	increased by 0.22 A	to 0.23 A	for every 100 W m^{-2} rise in E_e
P_{max}	increased by 3.07 W	to 3.28 W	for every 100 W m^{-2} rise in E_e

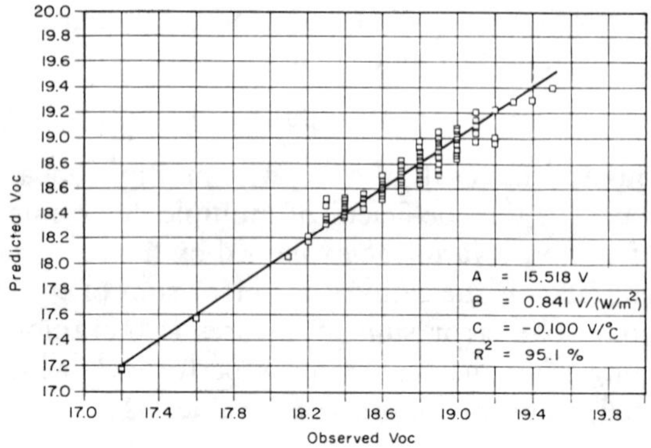

(*a*)

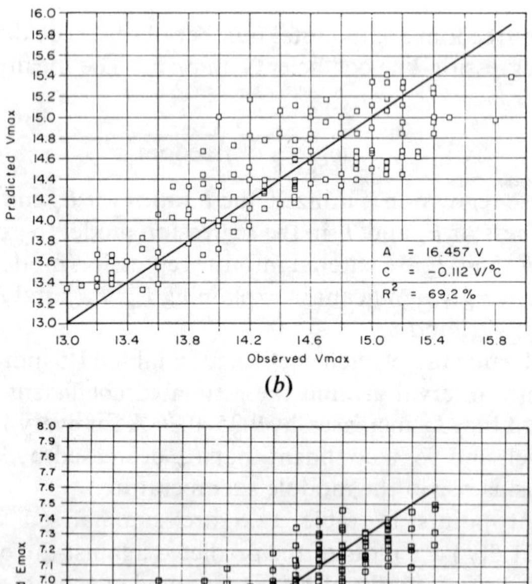

(*b*)

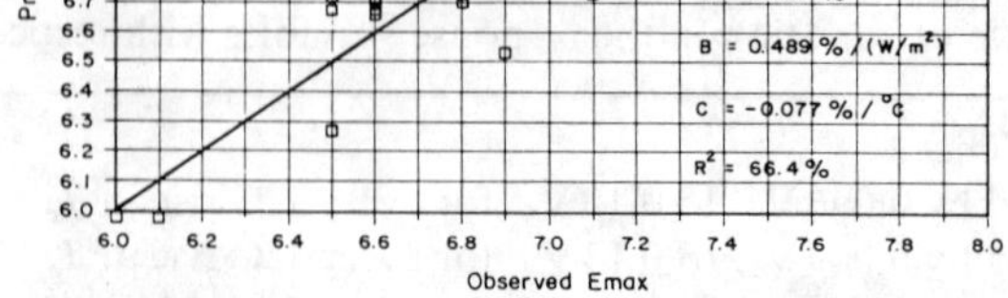

(*c*)

This is most probably due to slight errors introduced in searching for V_{max} along a non-continuous I–V curve composed of discrete data points. The plot of η_{max} also shows predicted points widely scattered and, considering a low R^2 of 66.4%, a manual computation from P_{max} and E_e instead of the empirical model may be more accurate.

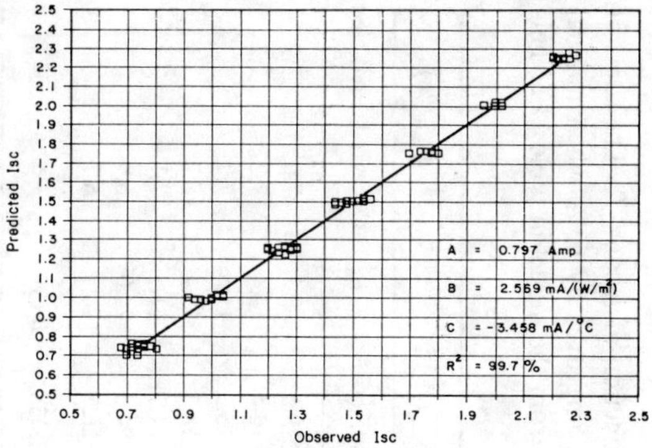

(*d*)

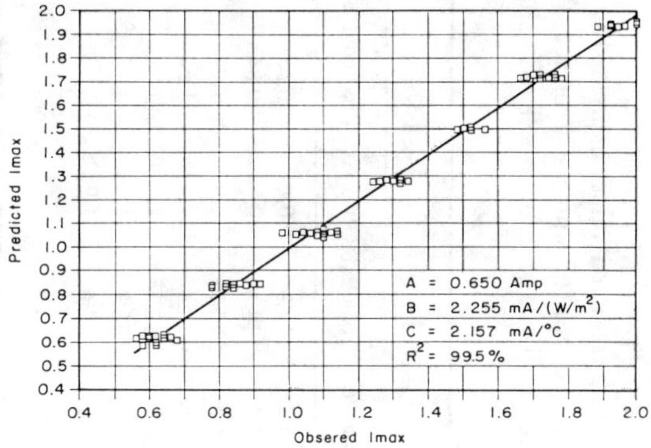

(*e*)

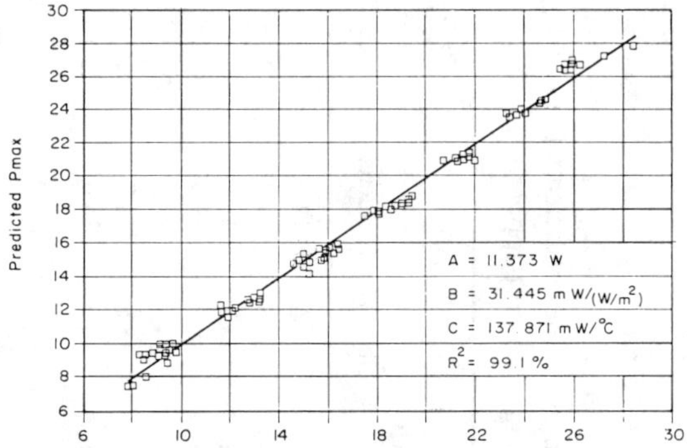

(*f*)

Figure 11.5 Plots of predicted against observed values of (*a*) V_{oc}, (*b*) V_{max}, (*c*) η_{max}, (*d*) I_{sc}, (*e*) I_{max} and (*f*) P_{max}.

Table 11.8 A comparison between observed (table 11.1) and predicted values expressed as a linear function of E_c and T_c. The sums of squares are computed as $\Sigma_{E_e=300}^{900}$ (observed–predicted)^2. The sum of squares error for $P_{max}=V_m I_m$ improves by 63% when manual computation using the predicted values of V_{max} and I_{max} is compared with the predicted value of P_{max} from the model in table 11.6. This same advantage does not apply to η_{max}.

E_e ($W m^{-2}$)	T_c (°C)	V_{oc} (V)		I_{sc} (A)		V_{max}(V)		I_{max}(A)		P_{max}(W)			η_{max}(%)	
		Observed	Predicted	Observed	Predicted	Observed	Predicted	Observed	Predicted	Observed	Predicted	$V_m I_m$	Observed	Predicted
300	38.7	18.6	18.6	0.74	0.75	14.8	14.9	0.61	0.62	9.0	9.5	9.3	7.0	7.1
400	40.1	18.8	18.8	1.00	1.00	14.9	14.8	0.84	0.84	12.6	12.4	12.4	7.2	7.2
500	42.1	18.8	18.8	1.26	1.25	14.6	14.5	1.07	1.06	15.8	15.3	15.5	7.2	7.1
600	43.8	18.8	18.9	1.51	1.50	14.5	14.4	1.29	1.28	18.6	18.2	18.4	7.2	7.1
700	45.7	18.8	18.8	1.77	1.75	14.2	14.1	1.51	1.51	21.6	21.1	21.1	7.1	7.1
800	47.8	18.6	18.7	2.01	2.00	13.9	13.9	1.73	1.73	24.1	24.0	24.0	6.9	7.0
900	50.4	18.6	18.6	2.23	2.25	13.4	13.6	1.94	1.95	26.1	26.7	26.7	6.7	6.8
Sum of squares			0.029		0.00983		0.107		0.000259		1.338	0.490		0.0431

11.3 MAXIMUM-POWER POINT MATCHING

Identification of the maximum-power point and eventually fixing the system's operating voltage at or near this point will optimize the energy production cost of the system. It is essential to investigate the behaviour of the maximum-power point of the module during the operating period of the system, the objective of which is to match the variations in the maximum-power point of the module through the day as closely as possible to the voltage requirement of the load, particularly when a tracker is not utilized.

The V_{max} window can be identified in figure 11.6. From 300 to 900 W m^{-2}, V_{max} varies within a voltage window of 13–16 V. The cluster of points at each irradiance level is due to the negative effects of T_c and is not so much an effect of irradiance level (table 11.5). Therefore, controlling T_c will help to stabilize V_{max} throughout the day.

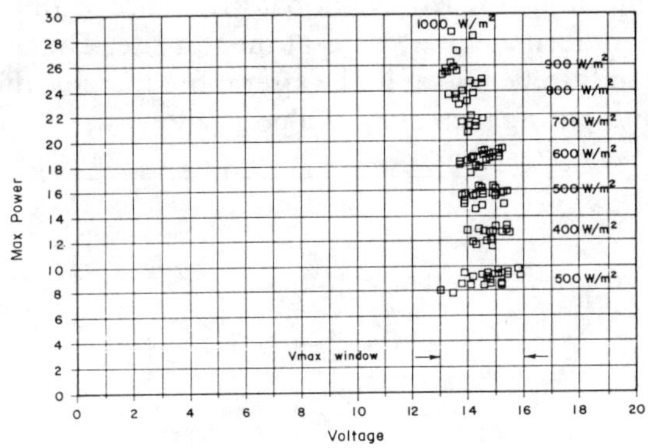

Figure 11.6 The V_{max} window.

Figure 11.7 is a plot of the frequency distribution of V_{max} within this V_{max} window. The cumulative probability at every 0.5 V interval is also indicated. There is a 44.6% probability of finding V_{max} between 14.5 and 15.5 V within a day, while there is 50.3% probability that V_{max} is between 14 and 15 V within a day. Therefore, when the system operating voltage varies between 14 and 15 V during the day, 50.3% of the time from 0800 to 1700 hours, the system is operating at the maximum-power point, and therefore maximum efficiency. Figure 11.7 can be used to measure how well the system voltage matches the maximum-power point of this particular module. The peak of the V_{max} distribution for this particular module is 14.5 V. Additional data will make the plot more symmetrical around 14.5 V, and more accurate. If a maximum-power point tracker is utilized, then figure 11.7 will still guide us to find the operating voltage during the day.

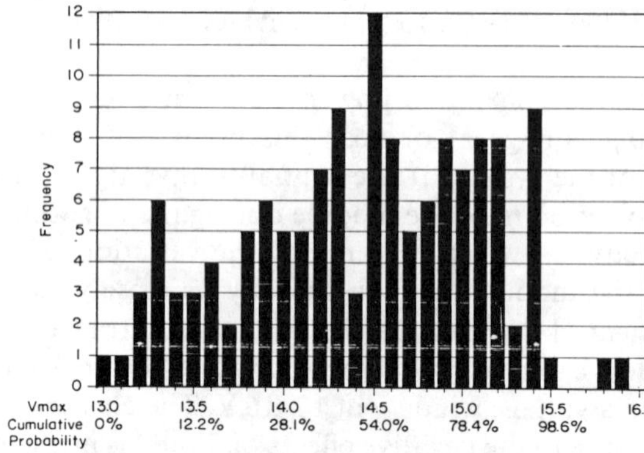

Figure 11.7 Frequency distribution of V_{max}.

Figure 11.8 shows us how the maximum efficiency varies during the day. A higher η_{max} is achieved in the morning than in the afternoon owing to the lower T_c in the morning. Generally, global irradiance is almost symmetrical around noon even during cloudy days at the test site. The module temperature, however, tends to be higher in the afternoon, thereby affecting V_{max} and η_{max} negatively by shifting their clouds of points downwards in the afternoon.

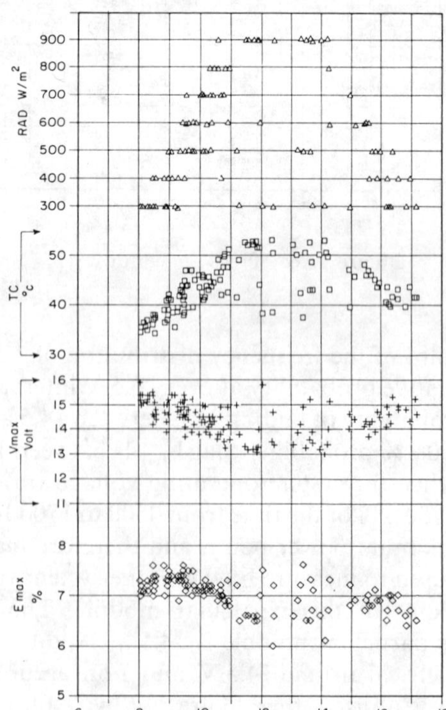

Figure 11.8 Variation in the module performance during a typical day.

11.4 THEORETICAL MODEL OF THE TEST MODULE

The output voltage V and current I of the module are usually given by

$$I = I_{ph} - \frac{V + IR_s}{R_{sh}} - I_0\left[\exp\left(\frac{q(V+IR_s)}{Ak_B T}\right) - 1\right] \tag{11.4}$$

where q is the electronic charge, k_B is Boltzmann's constant, A is the ideality factor, T (K) is the module temperature, I_{ph} is the photocurrent, I_0 is the reverse saturation current, R_s is the series resistance and R_{sh} is the shunt resistance. Following the solution technique of Braunstein *et al* (1977), the five parameters (I_{ph}, I_0, R_s, R_{sh} and A) in equation (11.4) were transformed to a three-parameter module equation. An iterative least-squares method fits the data points to the module equation.

PVMOD 3.0 has coded the above solution procedure into an interactive user-friendly personal microcomputer environment. Once test data have been properly processed by PVMOD 3.0, the numerical analyses can be directed to proceed automatically going through every data file, searching for a user-specified irradiance level, filtering data noise and saving the final report for a line printer or a user-specified output file.

11.4.1 Prediction of parameters in equation (11.4)

Experimental data are composed of the I–V characteristics curves of figure 11.2(*b*) from 300 to 900 W m^{-2}. Each curve is fitted to the theoretical model of equation (11.4) solving for the five parameters (I_{ph}, I_0, R_s, R_{sh} and A). There were 140 curves and the frequency distributions of the calculated parameters are presented in figures 11.9(*a*)–11.9(*d*). Note that most of the data beyond the 90% confidence range belong to the 300 W m^{-2} I–V characteristics.

Table 11.9 gives the analytical results from figures 11.9(*a*)–11.9(*d*), in three columns: the parameter range at the 90% confidence level; the parameter value at the peak of the distribution with the associated percentage frequency of the total observation; the mean value and its standard deviation of the

Table 11.9 Parameter range at the 90% confidence level based on 140 I–V characteristics in figure 11.2(*b*) from 300 to 900 W m^{-2}.

	90% confidence range	Peak value (% frequency)	Mean value	(Standard deviation)
R_s	0.45–1.8 Ω	1.15 Ω (10%)	1.11 Ω	(0.4 Ω)
R_{sh}	110–240 Ω	130 Ω (14.6%)	174 Ω	(64.7 Ω)
A	31–60	44 (6.4%)	45.9	(9.3)
I_0	5×10^{-10}–2×10^{-5} A	2×10^{-6} A (10%)	4.3×10^{-6} A	(0.000 02 A)

distribution. All but R_{sh} exhibit a convergence of the mean and peak values of the distribution.

When the parameter frequency distribution is analysed according to specific irradiance level, the distribution at 900 W m^{-2} irradiance level provides the best range to approximate the best parameter value. This

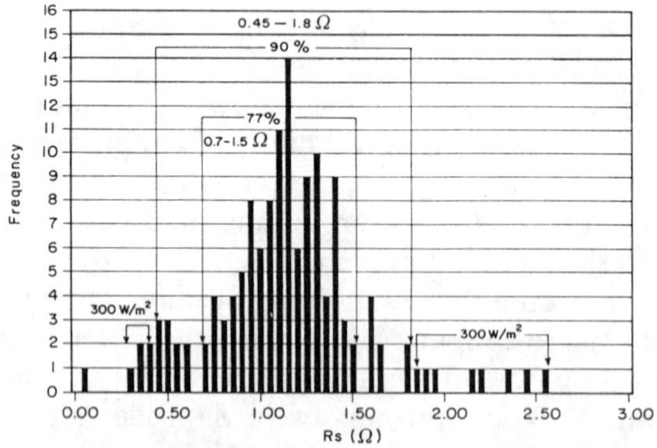

(a)

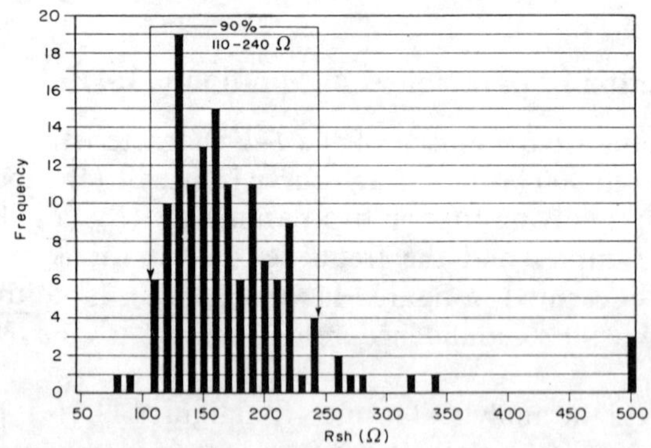

(b)

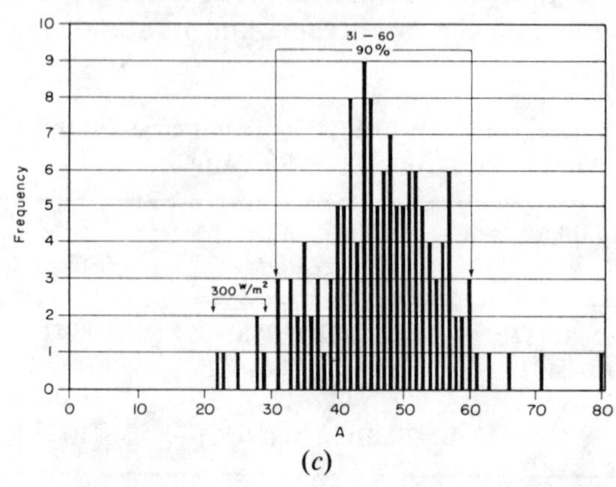

(c)

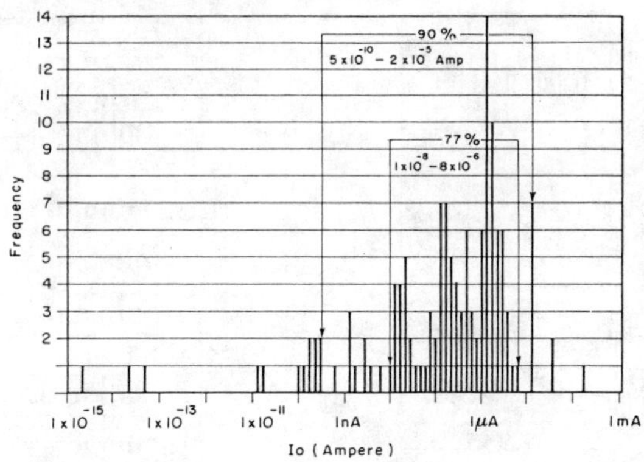

(*d*)

Figure 11.9 Frequency distributions of (*a*) R_s, (*b*) R_{sh}, (*c*) A and (*d*) I_0 (see table 11.9).

observation is based on figures 11.10(*a*)–11.10(*d*), where the overall parameter frequency distribution from 300 to 900 W m^{-2} is broken up into frequency distributions at specific irradiance levels. With reference to these figures, all the frequency distributions at 900 W m^{-2}, except R_{sh}, provide the most optimized range that covers the peak and mean value (table 11.9) of the overall frequency distribution. The distribution of R_{sh} at 900 W m^{-2} includes only the peak value of the overall distribution.

The mean value for each distribution at a specific irradiance level is denoted by an arrow head (▲) in the figures. Note that all the 900 W m^{-2} distributions have their mean values consistently close to the peak value of the overall distribution (table 11.10). It is therefore believed that the mean

Table 11.10 Mean value of each parameter at specific irradiance levels compared with the peak and mean values of the overall distribution (figures 11.10(*a*)–11.10(*d*)). The average value at 900 W m^{-2} approximates the overall peak value. For the overall average we used a total of 140 data; the module area is 0.432 m^2.

E_e (W m^{-2})	Percentage of data	R_s (Ω)	R_{sh} (Ω)	A	I_0 (μA)
300	23.6%	1.21	174	42.2	0.973
400	12.1%	1.02	192	47.6	1.503
500	17.1%	1.09	182	45.7	4.231
600	14.3%	0.95	209	51.4	17.93
700	12.9%	1.12	157	46.3	1.763
800	8.6%	1.17	164	45.1	3.189
900	11.4%	1.16	130	45.0	1.212
Overall average	100.0%	1.11	174	45.9	4.3
Overall peak		1.15	130	44.0	2.0

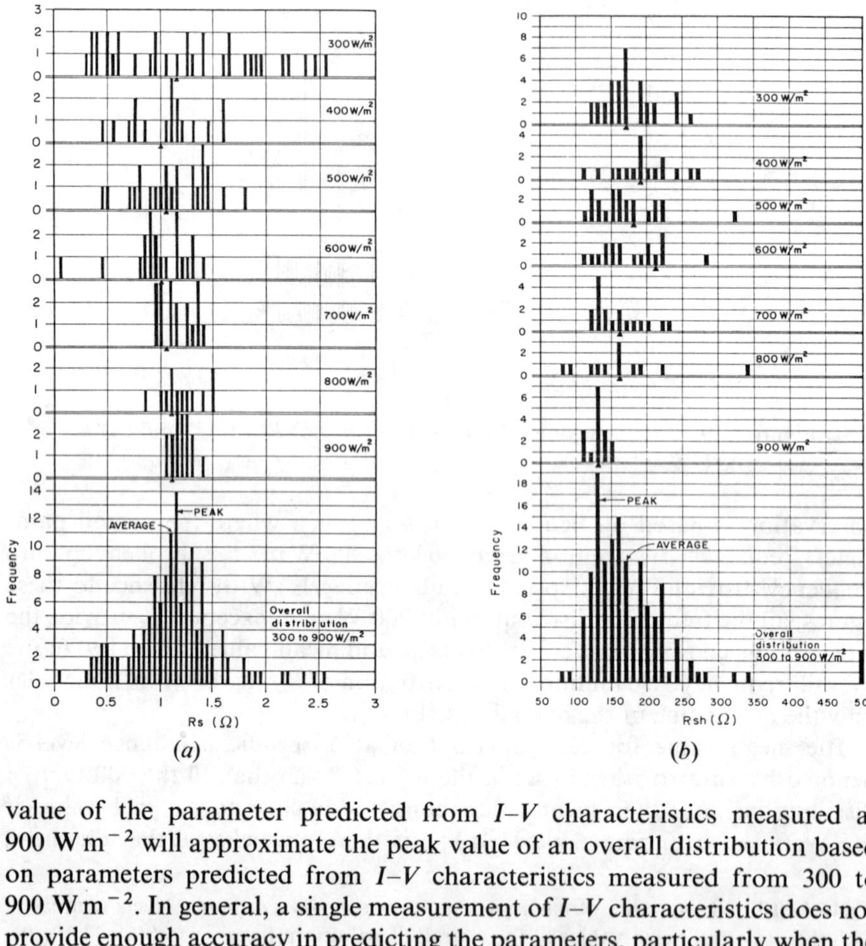

(*a*) (*b*)

value of the parameter predicted from *I–V* characteristics measured at 900 W m^{-2} will approximate the peak value of an overall distribution based on parameters predicted from *I–V* characteristics measured from 300 to 900 W m^{-2}. In general, a single measurement of *I–V* characteristics does not provide enough accuracy in predicting the parameters, particularly when the characteristics are measured at low irradiance level.

Correlation analyses (table 11.11) between the module's parameters (R_s,

Table 11.11 Correlation analyses showing no significant linear relation except I_{ph} with E_e and I_{sc}. T_c fails to have any effect on I_{ph} when subjected to a linear regression analysis.

	R_s	R_{sh}	A	I_0	I_{ph}
E_e	−0.02352	−0.19475	0.09499	0.03395	0.99791
T_c	0.02941	−0.15904	0.01528	0.02461	0.77014
V_{oc}	−0.04040	−0.00171	0.09135	0.03843	0.11254
I_{sc}	−0.02974	−0.18715	0.10338	0.03782	0.99759
FF	0.13839	0.19098	−0.15474	0.00301	−0.67085

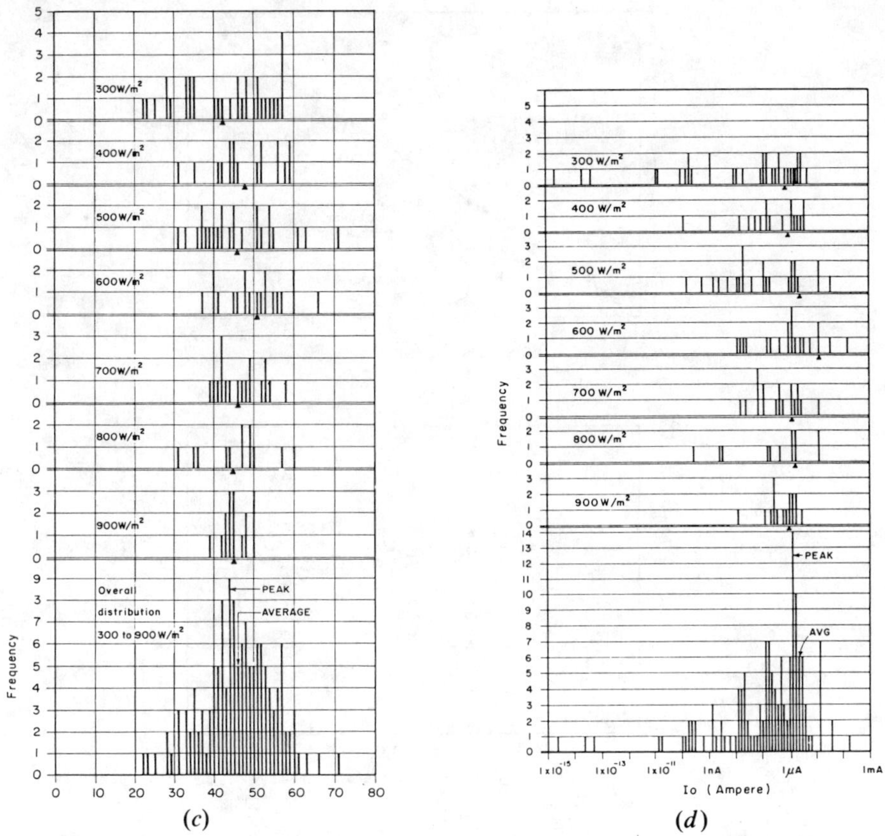

Figure 11.10 Expanded frequency distributions of (*a*) R_s, (*b*) R_{sh}, (*c*) A and (*d*) I_0 at 300–900 W m^{-2}.

R_{sh}, A, I_0 and I_{ph}) and E_e, T_c, V_{oc}, I_{sc} and FF do not show any significant linear relationship except in the case of I_{ph}. This confirms the theoretical foundation predicting a significant relation between I_{ph} and E_e and I_{sc}. When subjected to a linear regression analysis, however, the temperature coefficient of I_{ph} (-6.583×10^{-6} A K^{-1}) was much too low to cause any significant effects in I_{ph} even at hundreds of kelvins. The following equation then expresses I_{ph} as a linear function of E_e and I_{sc}, explaining 99.98% of all variations in the predicted I_{ph}, and is presented in graphical form in figure 11.11(*a*):

$$I_{ph} = A + BE_e + CI_{sc} \qquad (11.5)$$

where $A = -0.00607$, $B = 0.000671 \pm 0.0000882$ at the 95% confidence level, $C = 0.984 \pm 0.035$ at the 95% confidence level, $R^2 = 0.999814$, MSE $= 0.006986$ and the degrees of freedom are 136.

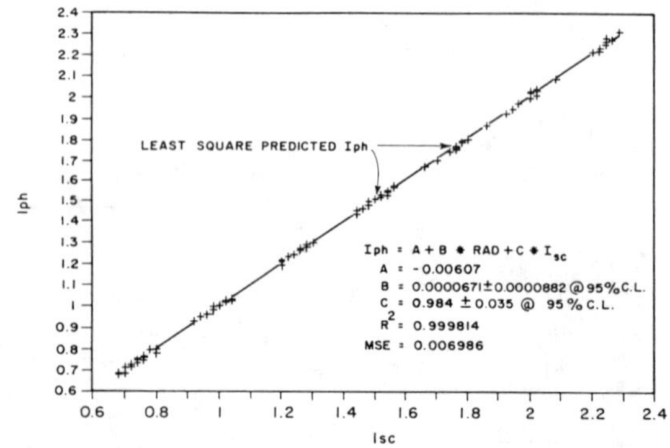

(*a*)

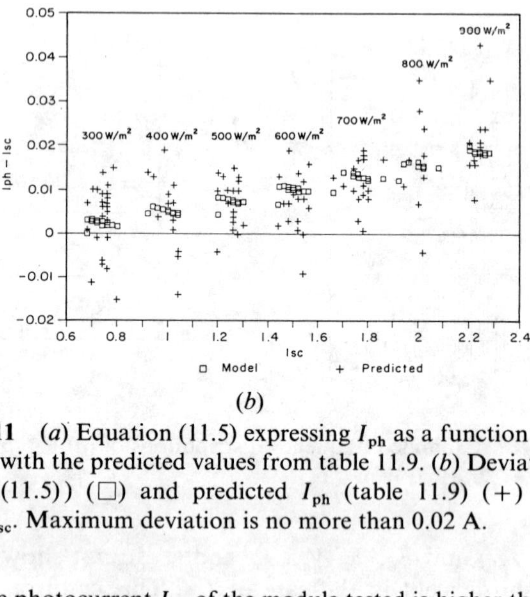

(*b*)

Figure 11.11 (*a*) Equation (11.5) expressing I_{ph} as a function of E_e and I_{sc} plotted with the predicted values from table 11.9. (*b*) Deviation of I_{ph} (equation (11.5)) (□) and predicted I_{ph} (table 11.9) (+) from the observed I_{sc}. Maximum deviation is no more than 0.02 A.

Table 11.12 The photocurrent I_{ph} of the module tested is higher than the measured I_{sc} by 2–20 mA for 300–900 W m^{-2}.

E_e (W m^{-2})	I_{sc} (table 11.1) (A)	I_{ph} (equation (11.5)) (A)	I_{ph}–I_{sc} (A)
300	0.74	0.742	0.002
400	1.00	1.005	0.005
500	1.26	1.267	0.007
600	1.51	1.520	0.010
700	1.77	1.783	0.013
800	2.01	2.025	0.015
900	2.23	2.249	0.019

Deviations from I_{sc} can be computed as equation (11.5) minus I_{sc}:

$$I_{ph} - I_{sc} = -0.006\,07 + 0.000\,067\,1E_e - (1 - 0.984)I_{sc}.$$

This is graphically presented in figure 11.11(*b*) and in table 11.12. When the measurement precision is not much better than 10 mA, I_{sc} can be a good approximation to I_{ph}.

11.4.2 Simulation of the tested module with equation (11.4)

With the empirical expression for I_{ph} and the parameter values listed in table 11.9, it is possible to simulate the *I–V* characteristics curves of the tested module at irradiance level of 300–900 W m^{-2}. The three most important regions in this process of fitting the simulated curves to the observed characteristic curves are the regions at open circuit, short circuit and maximum power. It is significant that the simulated curves fit the observed values well at these points (table 11.1) at every irradiance level.

During this simulation process, the simulated open-circuit voltage was found to be lower than the observed value in table 11.1. Partial differentiation of equation (11.4) shows the following relationship:

$$V_{oc} = f(A, -I_0) \qquad (11.6a)$$

$$I_{sc} = f(I_{ph}) \qquad (11.6b)$$

$$I_{max} = f(-I_0, I_{ph}) \qquad (11.6c)$$

$$V_{max} = f(-I_0, I_{ph}, R_s, A). \qquad (11.6d)$$

The open-circuit voltage is a positive function of A, the ideality factor, and a negative function of I_0, the reverse saturation current.

Since A and I_0 were chosen on the basis of the peak value of the overall frequency distribution of each parameter covering an irradiance range from 300 to 900 W m^{-2}, and these are close to the mean value for each distribution (table 11.9), a comparison of the coefficients of variation of A and I_0 will provide a measure of stability of their standard deviations, thereby quantifying the central tendency of the distribution:

$$\text{coefficient of variation} = \frac{\text{standard deviation}}{\text{mean value}} \times 100\%$$

coefficient of variation for $A = 20.3\%$

coefficient of variation for $I_0 = 465.1\%$.

Having the same number of *I–V* curves for A and I_0, the high coefficient of variation for I_0 denotes a poor central tendency of the I_0 distribution. This has two important implications: first, that the peak value or the mean value of the overall frequency distribution of I_0 (figure 11.10(*d*)) may not be an accurate prediction of I_0; second, an equally poor central tendency at a

Table 11.13 The additional step in adjusting I_0 fits the simulated curves to the observed I–V characteristics in the three regions well within 10% of the observed value. The values in parentheses are the percentage deviation from observed average value listed in table 11.1.

E_e (W m^{-2})	I_0 (10^{-7} A)	V_{oc} (V)	I_{sc} (A)	V_{max} (V)	I_{max} (A)	P_{max} (W)
300	0.90	18.6	0.72 (2.7%)	14.5 (2.0%)	0.58 (4.9%)	8.4 (6.6%)
400	1.20	18.8	0.92 (8.0%)	14.5 (2.7%)	0.82 (2.4%)	11.9 (5.6%)
500	1.72	18.8	1.24 (1.6%)	14.5 (0.7%)	1.05 (1.9%)	15.2 (3.6%)
600	2.34	18.8	1.49 (1.3%)	14.0 (3.4%)	1.30 (0.8%)	18.2 (2.2%)
700	2.67	18.9	1.75 (1.1%)	14.5 (2.1%)	1.47 (2.6%)	21.3 (1.3%)
800	4.12	18.7	2.00 (0.5%)	13.5 (2.9%)	1.76 (1.7%)	23.8 (1.4%)
900	5.51	18.6	2.22 (0.4%)	13.5 (0.7%)	1.92 (1.0%)	25.9 (0.7%)

Table 11.14 The final simulation model for the tested module for irradiance level from 300 to 900 W m^{-2}

$$\text{Model} \quad I = I_{ph} - \frac{V + IR_s}{R_{sh}} - I_0\left[\exp\left(\frac{V + IR_s}{Ak_BT/q}\right) - 1\right]$$

$$I_{ph} = -0.006\,07 + 0.000\,067\,1E_e + 0.984I_{sc}$$

$$R_s = 1.15\,\Omega$$

$$R_{sh} = 130\,\Omega$$

$$A = 44$$

$$I_0 = 9.01 \times 10^{-8}\text{ A at } 300\text{ W m}^{-2}$$
$$= 1.20 \times 10^{-7}\text{ A at } 400\text{ W m}^{-2}$$
$$= 1.72 \times 10^{-7}\text{ A at } 500\text{ W m}^{-2}$$
$$= 2.34 \times 10^{-7}\text{ A at } 600\text{ W m}^{-2}$$
$$= 2.67 \times 10^{-7}\text{ A at } 700\text{ W m}^{-2}$$
$$= 4.12 \times 10^{-7}\text{ A at } 800\text{ W m}^{-2}$$
$$= 5.51 \times 10^{-7}\text{ A at } 900\text{ W m}^{-2}$$

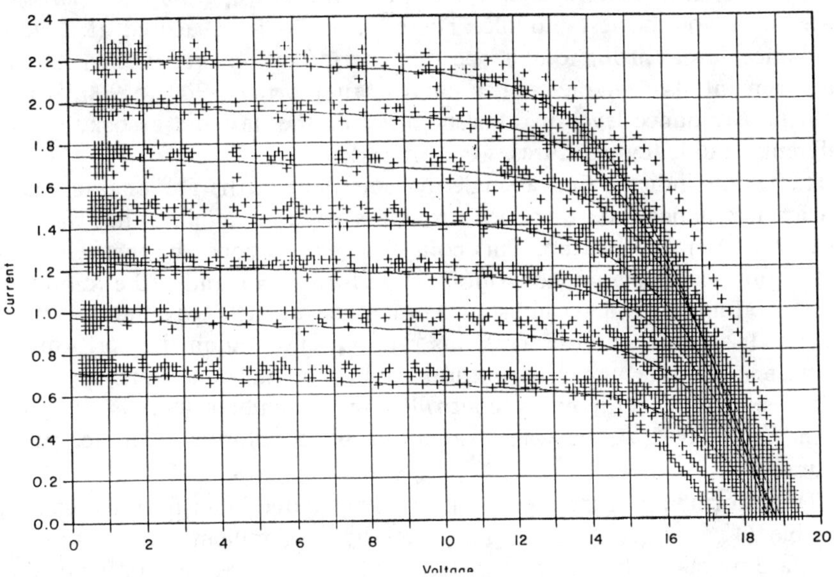

Figure 11.12 The simulated I–V curves of table 11.14 plotted with the observed I–V curves at 300–900 W m^{-2}.

specific irradiance level in figure 11.10(*d*) does not provide a good level of confidence in using their mean values as the predicted values at each irradiance level (and also their peak values), therefore compounding such inaccuracies in the overall distribution (figure 11.9(*d*)). Even though figures 11.9(*d*) and 11.10(*d*) are not completely useless, predicting I_0 from the figures by either taking the peak value, as is done for the other parameters, or using the mean value, does not provide good accuracies owing to the poor central tendency of their distribution.

To obtain a better accuracy for I_0, any curve-fitting technique can be used to determine I_0 from equation (11.4) by keeping R_s, R_{sh} and A constant, with the values presented in table 11.9, and I_{ph} expressed by equation (11.5). Table 11.13 shows that this additional step in adjusting I_0 fits the simulated curves to the observed I–V characteristics in the three regions well within 10% of the observed value. Table 11.14 presents the final simulation model for the tested module for irradiance level from 300 to 900 $W\,m^{-2}$. Figure 11.12 expresses the model graphically superimposed on top of the observed characteristics.

11.5 CONCLUSION

The foregoing performance evaluation of the module was carried out with extreme attention to accuracy and precision with which all variables are measured. Performance evaluation was performed using the I–V characteristics curves of the test module. These curves were measured at a stable irradiance level throughout a specified period of several weeks of daily measurement. 140 characteristics curves ranging from 300 to 900 $W\,m^{-2}$, with the irradiance level at different times of the day and associated with different module temperatures, were processed.

The results indicate that a single measurement of the I–V characteristics at each irradiance level does not provide a wide enough viewing frame to observe a complete picture. This complete picture concerns the behaviour of the physical phenomenon (module performance) under the stimuli of both random and deterministic parameters in its surrounding. From figures 11.2–11.10, one observes this random effect upon the performance of the test module under its natural environmental conditions at the installed site. Only with many controlled measurements (clouds of points as in figure 11.2(*b*)) can an evaluation be conclusive with acceptable accuracy.

The performance of the test module was evaluated in terms of its response variable (V_{oc}, V_{max}, I_{sc}, I_{max}, P_{max} and η_{max}). Ambient parameters that stimulate much of their responses (see table 11.5) were the irradiance level and the thermal component of the surroundings, quantified not in terms of the ambient temperature which this report was not able to characterize, but rather in terms of the module temperature measured at its back surface.

I_{sc}, I_{max} and P_{max} are highly sensitive to the irradiance level (table 11.5 and figures 11.14(*c*)–11.14(*e*)). The effect due to the thermal component of the surrounding measured through T_c, the module's temperature, becomes comparable with the effect due to the irradiance component in the case of V_{oc}, V_{max} and η_{max} (figures 11.14(*a*), 11.14(*b*) and 11.14(*f*)). V_{max} was found to be independent of the irradiance level (see table 11.5 and figure 11.14(*b*)).

The maximum power available from the test module was found to be around 26 W at 900 W m^{-2} at 50 °C (table 11.8). The module will most probably produce not more than 30 W at 1000 W m^{-2} based on projections from figure 11.14(*e*). This will be a 25% reduction from the power rating (45 $W_p \pm 10\%$ at 25 °C) reported by the manufacturer. In financial terms, a user will be paying 33% more for power that does not exist in his particular system using this test module.

However, caution must be observed here; the data available can only indicate a 25% power reduction of the module owing to differences in the operating conditions to which the module is subjected, out of the manufacturer's laboratory. They are not conclusive in indicating the quality or deterioration of any module. Moreover, only one module is tested.

Nevertheless, these results point to an important fact: if installed operating conditions are very different from those where the manufacturer's specifications were measured, a field test of the module in question will be necessary, unless financial constraint can allow an overdesign by more than 25%, or an additional capital outlay of about 33%.

According to the empirical model for the maximum efficiency, even at 25 °C and 1000 W m^{-2}, the attainable maximum module efficiency will only

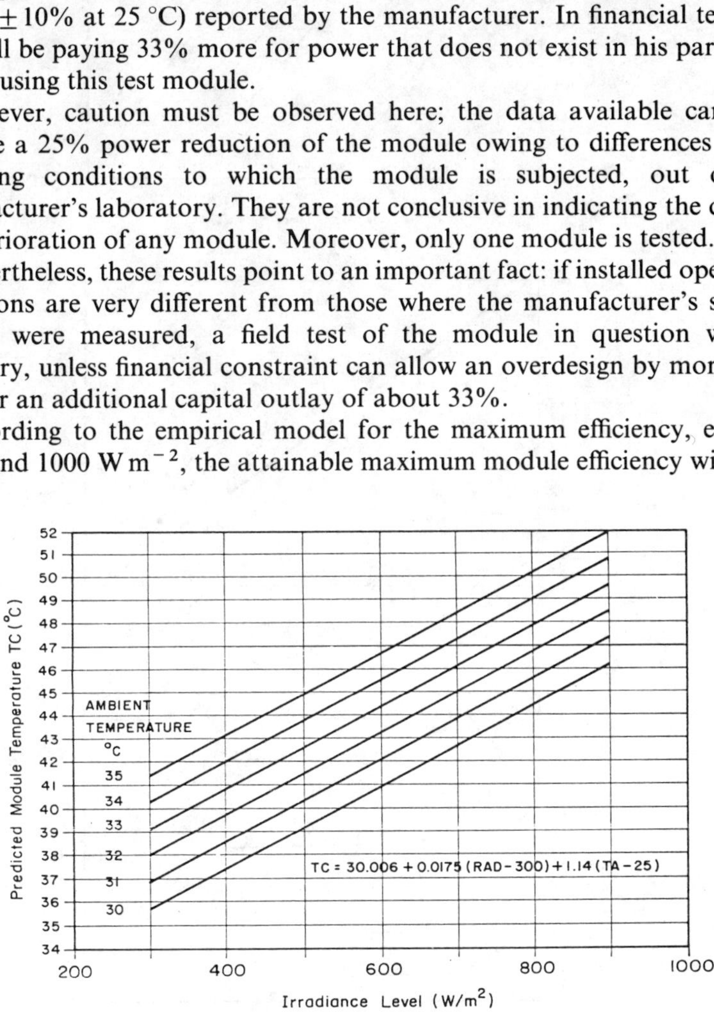

Figure 11.13 Module temperature T_c predicted from the intersection of the irradiance and the ambient temperature family of curves. The graph is plotted on the basis of equation (11.1).

be around 9%. At a more realistic module temperature of 52 °C, the module efficiency would have to drop to around 7%.

Figures 11.13 and 11.14(*a*)–11.14(*f*) are graphical tools based on analyses conducted in § 11.2. They can be used for quick prediction of the module

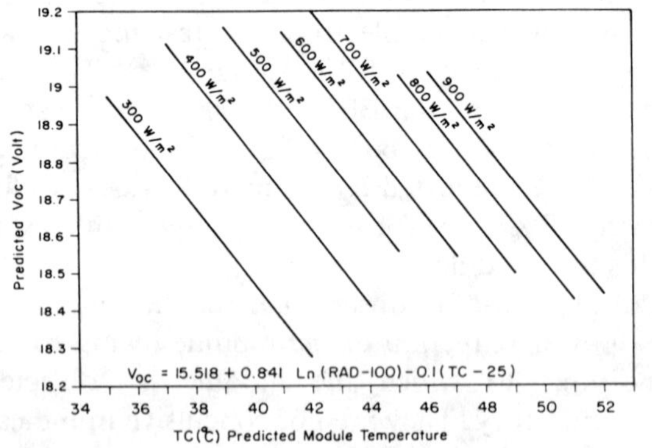

(*a*)

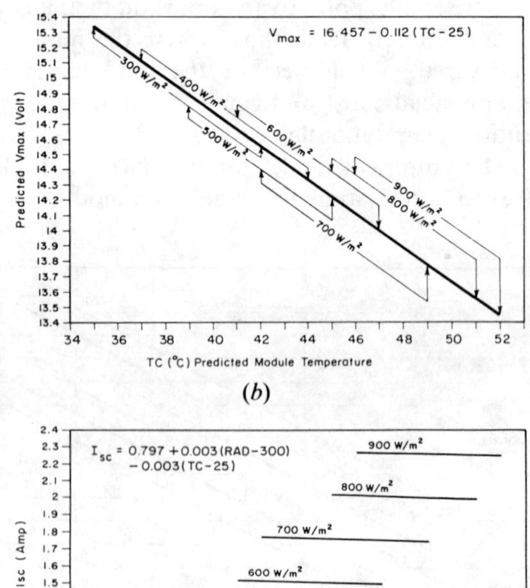

(*b*)

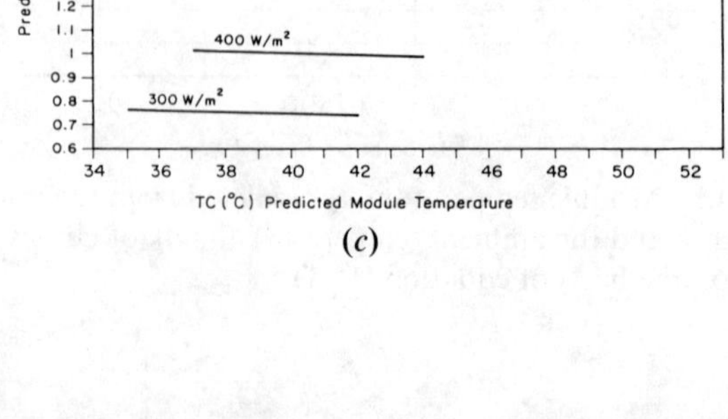

(*c*)

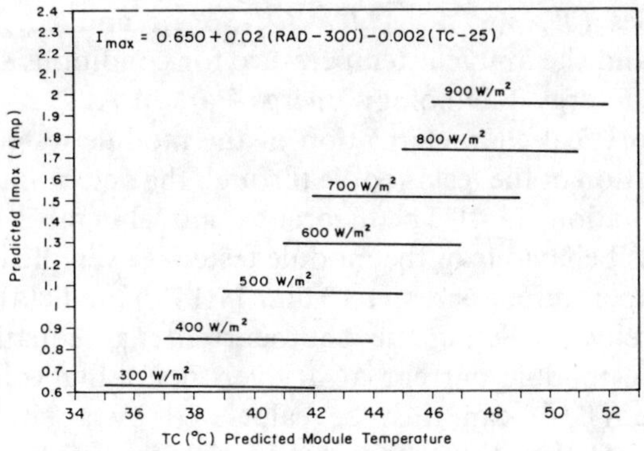

(*d*)

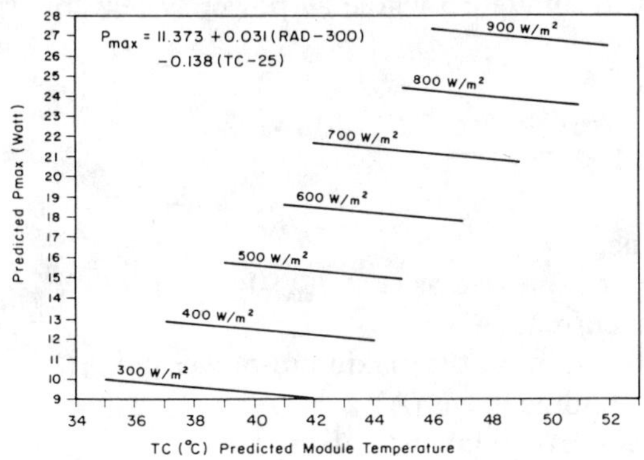

(*e*)

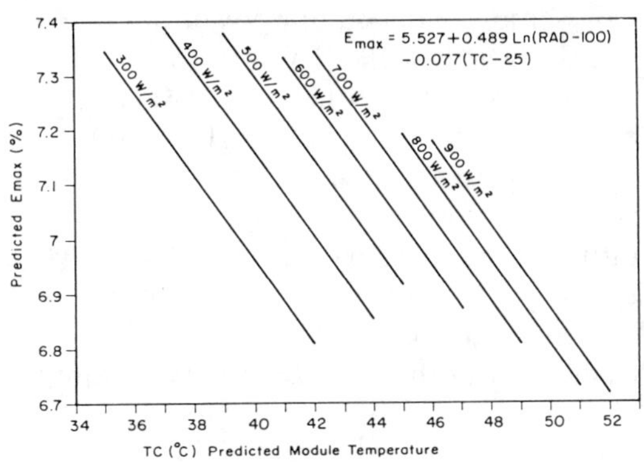

(*f*)

Figure 11.14 Prediction of (*a*) V_{oc}, (*b*) V_{max}, (*c*) I_{sc}, (*d*) $\dot{I}_{max}$, (*e*) P_{max} and (*f*) η_{max} from irradiance and module temperature.

response variables (T_c, V_{oc}, V_{max}, I_{sc}, I_{max}, P_{max} and η_{max}) knowing the irradiance level and the ambient temperature for conditions similar to those prevailing in the Energy Technology Energy Park at AIT.

Finally, a theoretical characterization of the module tested is made by a numerical simulation of the test module through the determination of the five parameters in equation (11.4). The theoretical model in table 11.14 simulates the instantaneous behaviour of the module tested. Given the irradiance level and ambient temperature, figures 11.13 and 11.14(*c*) predict the short-circuit current I_{sc}, thereby predicting the photocurrent I_{ph} (equation (11.5) and table 11.14). The module current at a given operating voltage (equation (11.4) and table 11.14) can then be calculated by an iterative process. Table 11.13 verifies that the theoretical model predicts accurately at the three most important points in the module's *I*–*V* characteristics curve. The model in table 11.14 can then be used to simulate long-term behaviour of the module tested or to simulate a whole PV power system for sizing and design purposes.

NOMENCLATURE

E_e	Irradiance ($W\ m^{-2}$)
FF	Fill factor expressed as $(V_{max}I_{max}/V_{oc}I_{sc}) \times 100\%$
I	Module current (A)
I_{max}	Module current at the maximum-power point (A)
I_{ph}	Module photocurrent (A)
I_{sc}	Module short-circuit current (A)
I_0	Module reverse saturation current (A)
k_B	Boltzmann's constant, $1.38 \times 10^{-23}\ J\ K^{-1}$
MSE	Mean square error
P_{max}	Module power at the maximum-power point (W)
q	Electronic charge, 1.6×10^{-19} C
R_s	Module series resistance (equation (11.4)) (Ω)
R_{sh}	Module shunt resistance (equation (11.4)) (Ω)
SS_E	Error sum of squares
T_A	Ambient temperature (°C)
T_c	Temperature of module's back surface (°C)
V_{max}	Module voltage at the maximum power point (V)
V_{oc}	Module open-circuit voltage (V)
η_{max}	Module efficiency at the maximum-power point (%)

REFERENCE

Braunstein A, Bany J and Appelbaum J 1977 *Energy Conversion* **17** 1–6

12 Characterization Procedure of a Photovoltaic Refrigeration System

12.1 INTRODUCTION

The experimental observation and analysis of the performance of a photovoltaic (PV) refrigeration system are classified into two sets of performance variables for discussion purposes in this chapter. These two sets are as follows.

(i) Electrical variables: these include compressor consumption, compressor duty cycle, PV array output and battery state of charge.
(ii) Thermal variables: these include the temperatures of the load and air inside the refrigerator, the coefficient of performance of the compression cycle, and the thermal heat transfer between the system and the ambient.

These two sets of performance variables are functions of the following.

(*a*) Irradiance which is a function of space and time.
(*b*) Ambient temperature which is also a function of space and time.
(*c*) Pattern or profile of system utilization.

    (1) Types and amount of loads to be refrigerated.
    (2) Frequency and duration of introducing new loads.
    (3) Cooling requirements (minimum temperature).

12.2 CHARACTERIZATION PROCEDURE FOR A PHOTOVOLTAIC REFRIGERATION SYSTEM

The following is the characterization procedure conducted at field-installed conditions.

(*a*) Electrical performance at no load and at thermally stable conditions.

    (i) Energy consumption.
    (ii) Compressor current.

(iii) Nominal voltage of the system.
(iv) 'On' period d (min) of motor–compressor unit.
(v) One duty cycle D (min) of motor–compressor unit.
(vi) d/D.
(vii) Watt–hours of energy consumed per cycle.
(viii) Average power per cycle.
(ix) Array efficiency.

(*b*) Thermal performance.

(i) Cool-down time: the time (h) to cool a load to a specified temperature.
(ii) Hold-over time: the time (h) to hold the temperature of a load at a specified temperature.
(iii) Door opening.
(iv) Loading and unloading of loads (water).

(*c*) Determination of the thermal heat transfer coefficient of the refrigerator enclosure.

(*d*) System breakdown test with computer-controlled load profile to verify the adaptability of the system to the user's loading and unloading behaviour and to the meteorological condition of the site of installation.

12.3 INSTRUMENTATION SET-UP

Figure 12.1 illustrates the instrumentation set-up used in studying some of the operating aspects of the PV refrigeration system. The PV array supplies 'photopower' to the refrigerator with monitoring instrumentation attached to record temperatures, currents and voltages at different points of the system automatically.

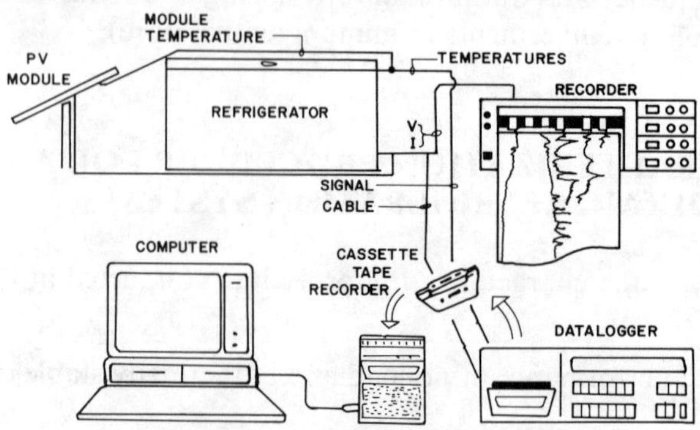

Figure 12.1 Experimental set-up.

Integrated-circuit (IC) temperature sensors of type AD590 detect the temperatures of the PV modules, the ambient temperature, and the air and wall temperatures inside the enclosure as shown in figure 12.2. These temperature measurements together with measurements of currents and voltages of the array, battery and motor–compressor unit are recorded by a data logger and then stored in a digital cassette tape. Sequential information is stored in the tape according to a user-programmable time interval provided by the data logger. This information is then fed into a microcomputer. Basic programs were developed to retrieve, process and plot these data to visualize the behaviour of the experimental system.

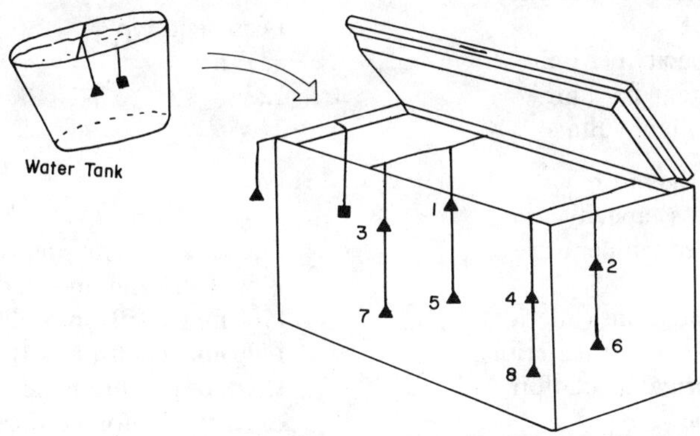

Figure 12.2 Position of IC temperature transducers inside the refrigerator enclosure: ■, surface-type flat sensor; ▲, cylindrical sensor.

12.4 EXPERIMENTAL OBSERVATION

A summarized description of the refrigeration system used in the experiment is presented in table 12.1.

Figures 12.3 and 12.4 present the temperature evolution of 34 kg of water. The temperature of the internal wall surface is an approximate measure of the temperature of the evaporator for the reason that the evaporator coil is placed behind the internal wall surface.

It takes 34 kg of water, at an initial temperature of 28 °C, 32 h to cool to 4 °C at an ambient temperature between 28 and 32 °C. This is defined as the 'cool-down' time of the refrigerator enclosure. The refrigerator enclosure can maintain the temperature of 34 kg of water, initially at 1 °C, at below 10 °C within 24.4 h at an ambient temperature at between 28 and 32 °C. This is defined as the 'hold-over' time of the refrigerator enclosure. These two terms indicate a measure of the thermal performance of the refrigeration system in a physical and practical sense to the user.

Table 12.1 The PV refrigeration system.

PV modules	
Number of modules	6
Number of modules in series	2
Number of modules in parallel	3
Peak power of module	40 W
Short-circuit current of module	2.42 A
Open-circuit voltage of module	20 V
Peak power of array	240 W
Storage	
Number of batteries	4
Type	Lead–acid type
Capacity per unit	63 A h
Voltage per unit	12 V
Nominal voltage	24 V
Refrigerator	
Gross capacity	200 l
Rated temperature	+3 °C minimum interior
	+50 °C maximum exterior
Dimensions (interior)	1100 mm × 350 mm × 520 mm
(exterior)	1400 mm × 650 mm × 1070 mm
Thermal insulation	15 cm of polyurethane
Type	Hermetic motor–compressor
Nominal voltage	24 V

The normal operating input current of the motor–compressor unit is 3.4 A at 24 V. Table 12.2 gives a summary of the duty cycle of the motor–compressor unit. From this table, it is observed that the d/D ratio approaches the no-load ratio as the load is cooled for several days. This is explained by the fact that, as the water releases all its heat energy to the evaporator coil and is cooled, the motor–compressor unit will gradually not respond to the presence of the thermal heat energy sources (loads) inside the refrigerator enclosure except that heat energy is transferred through the wall from the ambient, which is also similar to a no-load condition. Even though the physical presence of the water is evident, the 'thermal presence' of the water as seen by the system is non-evident; therefore, the system will operate close to the no-load condition.

Experimental observations are performed to investigate the effects of opening a door for 10 min, without changing loads and with 16 kg of water inside. A two-dimensional temperature stratification of air inside the refrigerator enclosure is observed. However, there is no substantial increase in the energy consumption of the compressor.

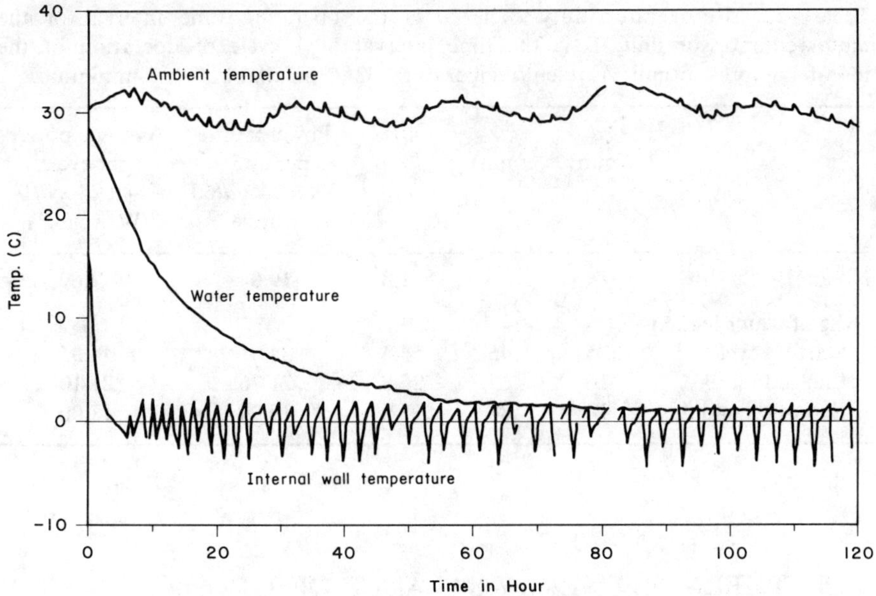

Figure 12.3 Temperature evolution in the cooling of 34 kg of water.

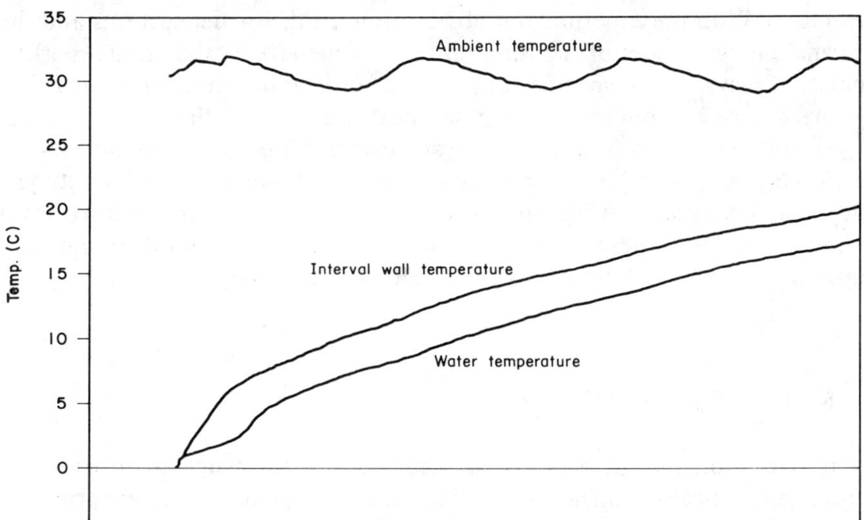

Figure 12.4 Temperature evolution of 34 kg of water when the compressor is turned off at 217.7 h.

Table 12.2 Refrigerator duty cycle. *d* is the operating time interval of the motor–compressor unit; *D* is the time interval of 1 cycle of operation of the motor–compressor unit. Ambient temperature, 32 °C maximum, 28 °C minimum.

	d (min)	*D* (min)	*d*/*D* (%)	Power × time per cycle $3.4 \times 24 \times d/60$ (W h cycle^{-1})	Average power per cycle $3.4 \times 24 \times d/D$ (W cycle^{-1})
No load	14	119	11.8	19.04	9.60
16 kg of water loaded					
Starting cycle	25	46	54.4	34.0	44.35
End of first day	16	62	25.8	21.76	21.10
End of second day	14	88	15.9	19.04	13.00

12.5 THERMAL ENERGY BALANCE OF THE REFRIGERATOR ENCLOSURE

Computer simulation of the refrigeration system will provide the user with valuable performance characteristics which are not available in short-term field tests. With the experimental observations in the earlier sections and the thermal energy model of the refrigerator in appendix A, the annual performance of a system for specific conditions of utilization can be investigated.

The determination of the thermal heat transfer coefficient of the refrigerator enclosure is required for system simulation. These coefficients are KS_a, KS_L, KS_w and MC_w. Appendix A outlines the procedure for estimating these coefficients using the observed data on load temperature from two experiments. These data are similar to the load temperature curve in figure 12.4. Table 12.3 tabulates the results of these analyses.

12.6 CONCLUSION

PV refrigeration system field test procedures were investigated in order to meet the practical technical needs of the user. Simple tests and experimentation, although not exhaustive, are conducted to characterize the performance of the PV refrigeration system. A need for investigating the annual system performance is also emphasized through system modelling and computer simulation.

Table 12.3 Determination of the thermal heat transfer coefficient (see Appendix 12A for the derivation of the mathematical model).

Model $T_L(t) = P_1 \exp(P_2 t) + P_3 \exp(P_4 t) + P_5 + P_6 \sin(bt + c) + P_7 \cos(bt + c)$

	First experiment, 32 kg of water	Second experiment, 17 kg of water
P_1	−1.810	−3.601
P_2	−0.044	−0.050
P_3	−25.441	−22.743
P_4	−0.008	−0.014
P_5	27.849	29.194
P_6	0.018	0.007
P_7	0.314	0.129
b	0.259	0.256
c	−10.111	9.008
K_1	2839.522	1399.807
K_2	148.791	89.469
S	0.908 m^2	0.453 m^2

$A = 44.471$ h $\quad Q_A = 0.939$
$B = 33.005$ h $\quad Q_E = 0.499$
$C = 0.365$
$E = 18.243$

$K_L = 9.148\ \mathrm{W\,m^{-2}\,°C^{-1}}$
$KS_w = 0.455\ \mathrm{W\,°C^{-1}}$
$KS_a = 1.249\ \mathrm{W\,°C^{-1}}$
$MC_w = 123.647\ \mathrm{kJ\,°C^{-1}}$

The following are the test procedures conducted at an ambient temperature between 28 and 32 °C and a short summary of the results.

(*a*) Electrical performance at no-load or at thermally stable conditions.

- (i) Energy consumption, 260.0 W h day^{-1}.
- (ii) Compressor current, 3.4 A.
- (iii) System's nominal voltage, 24.0 V.
- (iv) $d = 14.0$ min.
- (v) $D = 119.0$ min.
- (vi) $d/D = 11.8\%$.
- (vii) Power × time per cycle, 19.0 W h.
- (viii) Average power per cycle, 9.6 W.
- (ix) Array efficiency, 11.0%.

(*b*) Thermal performance

	Duration (h)	Water mass (kg)	Temperature range (°C)
(i) Cool-down time	32	34	28–4
(ii) Hold-over time	24	34	1–10

(iii) Door opening: induced internal temperature stratification; no significant effects on energy consumption.
(iv) Loading and unloading of loads (water).

(*c*) Determination of the thermal heat transfer coefficient of the refrigerator enclosure:

$$K_L = 9.148\ \mathrm{W\,m^{-2}\,°C^{-1}}$$
$$KS_w = 0.455\ \mathrm{W\,°C^{-1}}$$
$$KS_a = 1.249\ \mathrm{W\,°C^{-1}}$$
$$MC_w = 123.647\ \mathrm{kJ\,°C^{-1}}.$$

APPENDIX A DETERMINATION OF THE THERMAL HEAT TRANSFER COEFFICIENT OF THE REFRIGERATOR ENCLOSURE

A.1 Theoretical model of the refrigeration system

The following assumptions are made.

(i) No door opening.
(ii) Compressor off.
(iii) No condensation.
(iv) $MC_i \ll MC_w$.

The energy balance equations of the system are (figure 12.5)

$$KS_L(T_i - T_L) = MC_L \frac{\mathrm{d}T_L}{\mathrm{d}t} \qquad (12.1)$$

$$KS_a(T_a - T_w) = MC_w \frac{\mathrm{d}T_w}{\mathrm{d}t} + KS_w(T_w - T_i) \qquad (12.2)$$

$$KS_w(T_w - T_i) = KS_L(T_i - T_L). \qquad (12.3)$$

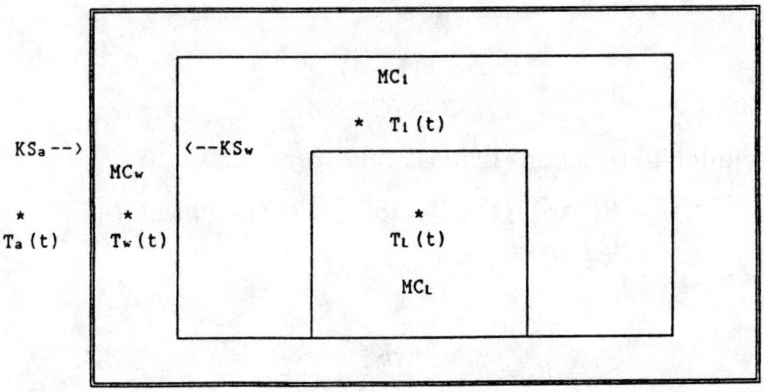

Figure 12.5 Schematic diagram of the refrigeration system (see nomenclature for notation).

Let

$$A = \frac{MC_L}{KS_L} \tag{12.4a}$$

$$C = \frac{KS_w}{KS_a} \tag{12.4b}$$

$$B = \frac{MC_w}{KS_a} \tag{12.4c}$$

$$E = \frac{KS_L}{KS_w}. \tag{12.4d}$$

Therefore, substituting equations (12.4*a*)–(12.4*d*) into equations (12.1)–(12.3) and rearranging, we have

$$A\frac{dT_L}{dt} + T_L - T_i = 0 \tag{12.5}$$

$$B\frac{dT_w}{dt} + (C+1)T_w - CT_i - T_a = 0 \tag{12.6}$$

$$(E+1)T_i - ET_L - T_w = 0. \tag{12.7}$$

Eliminating T_w and T_i from equations (12.5)–(12.7) and solving for T_L, we have

$$K_1\frac{d^2T_L}{dt^2} + K_2\frac{dT_L}{dt} + T_L = T_a(t) \tag{12.8}$$

where

$$K_1 = AB(E+1)$$
$$K_2 = B + ACE + AE + A.$$

The solution to equation (12.8) is of the form

$$T_L(t) = P_1 \exp(P_2 t) + P_3 \exp(P_4 t) + P_5 + P_6 \sin(bt+c) + P_7 \cos(bt+c) \quad (12.9)$$

where

$$P_2 = \frac{-K_2 + (K_2^2 - 4K_1)^{-1/2}}{2K_1} \quad (12.10a)$$

$$P_4 = \frac{-K_2 - (K_2^2 - 4K_1)^{-1/2}}{2K_1}. \quad (12.10b)$$

Equation (12.9) is a function of time and an implicit function of the four parameters *A*, *B*, *C* and *E*. The *P* parameters are determined by carrying out a non-linear regression analysis on equation (12.9).

A.1.1 Two experiments to determine A, B, C and E

Two experiments were conducted with $S_{L1} \neq S_{L2}$, $M_{L1} \neq M_{L2}$ while keeping other parameters constant; the symbols with the primes denote the corresponding quantities for the second experiment. The load temperature of the second experiment can also be described by equation (12.8) with a different set of parameters. Therefore we have four expressions to solve for the four unknowns: *A*, *B*, *C* and *E*. Then, with equations (12.4*a*)–(12.4*d*), the value of the coefficients can be calculated.

Therefore,

$$B = B' \qquad C = C'$$

$$A' = Q_A A \qquad Q_A = \frac{MC_{L2}}{S_{L2}} \frac{S_{L1}}{MC_{L1}}$$

$$E' = Q_E E \qquad Q_E = \frac{S_{L2}}{S_{L1}}$$

$$K_1 = AB(E+1) \quad (12.11a)$$
$$K_2 = A + B + AE(1+C) \quad (12.11b)$$
$$K_1' = Q_A AB(1+Q_E E) \quad (12.11c)$$
$$K_2' = Q_A A + B + Q_A Q_E AE(1+C). \quad (12.11d)$$

A.2 Equations to calculate the heat transfer coefficients

A.2.1 Calculation of the K values for the two experiments

From equation (12.10),

$$K_1 = \frac{P_2/P_4 - 1}{P_2^2 - P_2 P_4} \qquad (12.12)$$

$$K_2 = \frac{(K_1 P_2^2 + 1)}{P_2} \qquad (12.13)$$

where P_2 and P_4 are the time coefficient parameters of the exponential terms in equation (12.9) after a non-linear regression analysis is executed.

After K_1 and K_2 have been calculated for the two experiments from equations (12.12) and (12.13), they are substituted into equations (12.11*a*)–(12.11*d*). When these equations are rearranged, *A*, *B*, *E* and *C* can be calculated from the following equations:

$$(1-Q_E)Q_A A^2 + (Q_A Q_E K_2 - K_2')A + \frac{1-Q_A Q_E}{(1-Q_E)Q_A}(K_1' - Q_A Q_E K_1) = 0 \qquad (12.14)$$

$$B = \frac{K_1' - Q_A Q_E K_1}{(1-Q_E)Q_A A} \qquad (12.15)$$

$$E = \frac{K_1}{(K_1' - Q_A Q_E K_1)/(1 - Q_E)Q_A} - 1 \qquad (12.16)$$

$$C = \frac{K_2 - A - B}{AE} - 1. \qquad (12.17)$$

Then the heat transfer coefficient of the refrigerator enclosure can be calculated from the values of *A*, *B*, *C* and *E* using equations (12.4*a*)–(12.4*d*).

NOMENCLATURE

C_L Thermal heat capacity of the load (kJ kg^{-1} °C^{-1})
C_w Thermal heat capacity of the refrigerator enclosure (kJ kg^{-1} °C^{-1})
d Operating time interval of the motor–compressor unit (min)
D Time interval of 1 cycle of operation of the motor–compressor unit (min)
KS_a Thermal heat transfer coefficient between the wall of the refrigerator enclosure and the ambient air (W °C^{-1})
KS_L Thermal heat transfer coefficient between the load and the air inside the refrigerator enclosure (W °C^{-1})
KS_w Thermal heat transfer coefficient between the air inside the

	refrigerator enclosure and the wall of the refrigerator enclosure (W °C^{-1})
M_L	Mass of the load (kg)
M_w	Mass of the enclosure (kg)
S_a	External surface area of the vertical wall of the refrigerator enclosure (m^2)
S_L	Surface area of the load (m^2)
S_w	Internal surface area of the vertical wall of the refrigerator enclosure (m^2)
T_a	Ambient temperature (°C)
T_i	Mean temperature of the air inside the enclosure (°C)
T_L	Mean temperature of the load (°C)
T_w	Mean temperature of the wall of the enclosure (°C)

13 Field Trial Procedure for a Photovoltaic Centrifugal Pumping System

13.1 OBJECTIVE

A field trial procedure has been designed to evaluate the performance of photovoltaic (PV) water-pumping systems under natural environmental conditions. The procedure is presented in a manual format with the intention that it be used by field engineers or laboratory researchers as a reference guide. A directly connected PV water-pumping system is considered throughout the procedure although some of the test procedures may apply to other system configurations.

The procedure emphasized tests of each component of the system. These components are matched by simulating the whole system. Instantaneous system analysis and long-term performance behaviour are investigated. Transient system analysis is also performed to look into the starting conditions of the system.

Information with regard to system reliability, performance and maintentance will be a direct output of the field trial. Technical recorded data from the field trial will also serve as relevant input for economic appraisal. It is also the intention of the proposed field trial to appraise the service provided to the user, i.e. the average number of litres of water pumped daily over a representative period of time.

13.2 APPROACH

Directly connected PV water-pumping systems operate under variable conditions of energy supply. Fluctuations in solar irradiance, the main source of its energy, sets the operating point of the directly connected motor–pump system in continuous motion along its output characteristics curve. Its effects on a system working against a constant static manometric head is to induce a wide range of flow velocities and associated velocity–head losses, leading to a variable total manometric head. In contrast with the PV system, a conven-

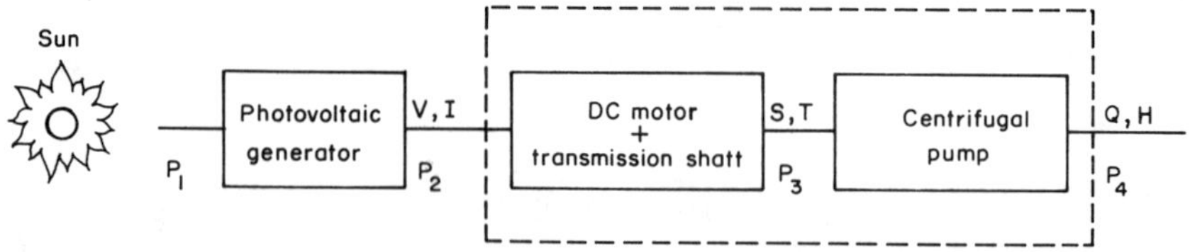

Figure 13.1 Block diagram of a PV water-pumping unit.

tional water-pumping system operates at constant speed and constant head. Therefore, the procedure will consist of a dual approach.

(i) The energy supply is maintained under steady-state operating conditions (use of an auxiliary power supply; PV array disconnected). Mechanical as well as hydraulic performances are monitored.
(ii) The system is monitored on site at constant static head. Overall system performance as well as matching conditions between the PV array and the motor–pump system are surveyed.

Table 13.1 Solar PV pumping system.

Subassembly
(*a*) Motor
(*b*) Pump
(*c*) Maximum-power point tracker
(*d*) Inverter
Components
(*a*) Motor
(i) Stator
(ii) Armature
(iii) Carbon brushes
(iv) Commutator
(*b*) Pump
(i) Impeller
(ii) Casing
(iii) Water seal
(iv) Self-priming device
(v) Foot valve
(*c*) Maximum-power point tracker
(i) Sensing circuit
(ii) Series–parallel switching device
(*d*) Inverter
(i) DC–DC conversion
(ii) Frequency variation and speed variation

Table 13.2 Parameters affecting efficiency.

(1) Motor
(*a*) Type of motor: permanent magnet, series, shunt, compound, AC
(*b*) Resistance in the windings: armature and stator
(*c*) Brush area
(*d*) Pressure of brushes on commutator
(*e*) Surface of commutator (undercutting; smoothness)
(*f*) Speed
(*g*) Air gap
(*h*) Bearings
(*i*) Weight of armature
(*j*) Cooling fan
(*k*) Balancing of armature
(2) Pump
(*a*) Type of pump
(*b*) Impeller diameter, suction eye, vanes
(*c*) Speed of rotation
(*d*) Quality of seal: mechanical, gland pack
(*e*) Casing involute, smoothness
(*f*) Clearance of impeller, especially the suction eye
(*g*) Foot valve design
(*h*) Balancing of impeller with armature

Figure 13.1 presents a schematic diagram of a PV pumping system. Table 13.1 lists the various components and parts of a typical motor–pump combination. Factors affecting the efficiency of the motor–pump system are listed in table 13.2. Each component has its intrinsic characteristics (table 13.2) over all operating conditions of the system. It is desirable that the intercept of all respective component characteristics follows the maximum-power line of the PV array generator. Depending on this internal matching, which is one of the objectives of this field trial, the efficiency of the overall system and related performances will meet an acceptable range.

System design, particularly of the PV array capacity, is reviewed to ensure that sufficient energy is produced to start the motor as early in the day as possible.

13.3 FIELD TRIAL PROCEDURE

The field trial procedure (figure 13.2) schematically presented below consists of three experimental investigations.

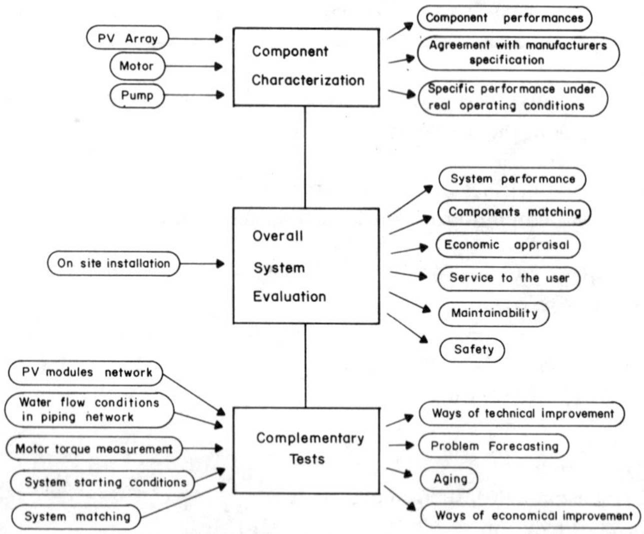

Figure 13.2 Schematic diagram of the field trial procedure.

(i) Characterization of components.
(ii) Evaluation of overall pumping system.
(iii) Complementary tests.

13.3.1 Characterization of components

A summary of the components characterization is given in table 13.3.

Table 13.4 shows the characterization of the dynamic behaviour of the different components of the system. In this procedure the static head H_s is kept constant by pumping in a closed-loop system and constitutes the significant variable used in the equation.

Depending on the purpose of the field trial, it can be interesting to simulate realistic pumping conditions in keeping the total head constant. This allows one to take into account the well and the piping as well. Therefore a programmable pressure controller using the following algorithm (for a borehole centrifugal pump) can be developed (Herrmann *et al* 1987).

At the outlet of the pump,

$$H=\frac{p_r(\dot{Q})}{\rho g}+\frac{\dot{Q}^2}{2gA^2}$$

and, in the piping,

$$H=H_s+\frac{\sum f}{2gA^2}\dot{Q}^2$$

$$P=\rho g\left(H_s+\frac{\sum f-1}{2gA^2}\dot{Q}^2\right).$$

Table 13.3 Tabular summary of characterization of components.

Component	Parameters	Variables	Sensors and auxiliary equipment	Remarks
PV array	E_e, irradiance T, temperature	I–V characteristics	Variable load Pyranometer Thermocouple, integrated circuit temperature transducer	See chapter 10 Measurement under steady-state conditions
Motor–pump system	H, effective head	I–V characteristics at constant head	Auxiliary power supply Throttling valve in the delivery pipe	PV array disconnected Motor torque[a] Constant head[b]
	S, speed	H–Q pump characteristics at constant speed	Pressure gauges Tachometer Voltmeter Ammeter	
		Motor starting conditions (I–V) at constant head	Prony brake[c] Dynamometer[d]	
System head	H_s, static head	H–Q characteristics at constant static head and variable speeds	Auxiliary power supply Water level measurement	Closed-loop system to keep the static head constant

[a]Depending on the selected motor, the torque coefficient and the armature resistance can be evaluated using a regression analysis based on classical motor equations.

[b]Constant head can be defined as the constant total manometric head (throttling controlled valve) or constant static head (closed loop) depending on the investigator's point of interest.

[c]If necessary to evaluate torque.

Table 13.4 Characterization of components' dynamic behaviour: ρ is the density of water.

Dynamic	Power	Efficiency
Electrical	$P_{elec}=IV$	$\eta_{elec}=\dfrac{IV}{AE}$,
Mechanical	$P_{mech}=C_0\tau S$	$\eta_{mech}=\dfrac{C_0\tau S}{IV}$
Hydraulic	$P_{hydr}=\rho g H_s\dot{Q}$	$\rho_{hydr}=\dfrac{\rho g H_s\dot{Q}}{C_0\tau S}$
Global		$\eta_T=\dfrac{\rho g H_s\dot{Q}}{AE_e}=\eta_{elec}\eta_{mech}\eta_{hydr}$
Match factor		$M_F=\dfrac{VI}{V_{max}I_{max}}$

Thus giving

$$p=p_0+K\dot{Q}^2$$

with

$$P_0=\rho g H_s$$

$$K=\frac{\sum f-1}{2gA^2}$$

where the coefficient f characterizes the dynamic energy losses and A is the pipe cross section.

Both static head and piping properties can be simulated to a real well by varying the two parameters p_0 and K only. By means of the measured pressure and flow rate, an analogue computer generates the set pressure p.

This kind of approach can be developed for surface pumps and other types of pump.

13.3.2 Evaluation of overall system

The evaluation of the overall system is given in tables 13.5 and 13.6.

13.3.3 Complementary tests

Six complementary tests (tests C1–C6) were carried out.

13.3.3.1 Test C1: photovoltaic array network

Step 1 Determine the current–voltage (I–V) characteristics of each module

Table 13.5 Test data and equipment for overall system evaluation.

Dynamic behaviour		Parameter	Variable	Sensors and auxiliary equipment	Remarks
Instantaneous	Steady state	H_s = constant	I, current V, voltage E_e, irradiance S, speed H, total head $\dot{Q}$, flow rate	Ammeter Voltmeter Pyranometer Tachometer Pressure gauges Flowmeter	Closed loop in order to keep the static head constant
Instantaneous	Starting conditions	H_s = constant	I_s, starting current V_s, starting voltage I_p, pumping current V_p, pumping voltage S_p, pumping speed E_{es}, starting irradiance threshold E_{ep}, pumping irradiance threshold	Ammeter Voltmeter x–t recorder Pyranometer	See § 13.3.3, tests C3 and C4
Long term		H_s = constant	G, daily global irradiance Q, daily volume of water pumped W, daily electrical energy	Solar integrator and pyranometer Water counter Watt–hour meters	Closed-loop system

Table 13.6 Energy computation: Q, daily volume (l) of pumped water; H_s static manometric head.

Long term	Energy	Daily average efficiency
Electrical	W_H	$\eta_w = \frac{W_H}{GA}$
Motor–pump system	$W_H = \rho g H Q$	$\eta_{mp} = \frac{\rho g H_s Q}{W}$
Overall system		$\eta_{os} = \frac{\rho g H_s Q}{GA} = \eta_w \eta_{mp}$

at different levels of irradiance and temperature (with and without a diode in order to investigate the hot-spot problem).

Step 2 Using the module I–V characteristics from step 1, construct the resulting I–V chracteristics of the array.

Step 3 Determine the I–V characteristics of the array at different levels of irradiance and temperature with and without a diode (experimental).

Step 4 (*a*) Compare the results from step 2 with the results from step 3.
(*b*) Compare the results from steps 2 and 3 with the manufacturer's data.

13.3.3.2 Test C2: characterization (indoor) of motor

This experiment is carried out in a laboratory environment. The motor is loaded (Prony brake or the pump itself). The speed, torque and current are recorded at a constant voltage provided by an auxiliary power supply. The torque may eventually be computed using a regression analysis on the basic equations of the motor or measured directly. Different motor characteristics may be drawn in order to verify the specifications of the motor and its adaptation to the PV generator.

Step 1 Determine motor efficiency as a function of the shaft output power.

Step 2 (*a*) Determine torque as a function of input current.
(*b*) Determine torque as a function of input voltage.
(*c*) Calculate torque coefficient of current and starting torque.

Step 2 (*a*) Determine speed as a function of voltage.
(*b*) Determine speed as a function of current.
(*c*) Calculate speed coefficient of voltage and motor internal resistances.

13.3.3.3 Test C3: matching of photovoltaic array and motor–pump system

Step 1 Determine the locus of maximum-power points (I_{max}, V_{max}) on the I–V characteristics curve of the PV array.

Step 2 Determine operating points (I, V) of the motor–pump system at constant head by superimposing the I–V characteristics of the motor–pump system onto the PV array I–V characteristics.

Step 3 (*a*) Calculate the match factor $M_F = IV/I_{max} V_{max}$.

(*b*) Determine the range of match factors at constant head for different conditions of irradiance and temperature.

13.3.3.4 Test C4: starting conditions of system

The speed is greater than zero; the flow rate is zero.

Step 1 Determine the starting current at different heads.
Step 2 Determine the corresponding irradiance threshold.

13.3.3.5 Test C5: pumping conditions of system

The speed is greater than zero; the flow rate is greater than zero.

Step 1 Determine the starting pumping conditions ($Q > 0$) of current, voltage and speed for different values of the static head.
Step 2 Determine the corresponding irradiance threshold.

13.3.3.6 Test C6: fluid flow conditions

The condition of flow (laminar or turbulent) is of considerable significance since the energy losses involved are quite different.

Parameters affecting the flow conditions are the flow velocity v, the pipe bore size D and the fluid's viscosity v_{fl}. These can be expressed in the form of a single dimensionless quantity known as the Reynolds number R_n where

$$R_n = \frac{Dv}{v_{fl}}$$

in typical engineering units where $1200 < R_n < 2000$ means laminar flow and $3000 < R_n$ means turbulent flow. In the case of turbulent flow, adjustment of pipe sizes has to be made in order to avoid pumping problems and to reduce frictional losses.

13.4 TEST FACILITIES

The technical recommendations for the test facilities are as follows.

(i) The pump should be mounted in a tank with adequate volume (so that the flow velocity near the pump is negligible in comparison with that within the pipe).
(ii) Pumping should be carried out in a closed loop.
(iii) There should be a flow-smoothing section to reduce eddy flow.
(iv) The water return should not be too close to the pump inlet in order to avoid turbulence and air bubbles pumped by the pump.
(v) Data should be recorded under steady-state conditions of irradiance and temperature.

13.4.1 Outdoor test facility

The outdoor test facilities are shown in figures 13.3 and 13.4.

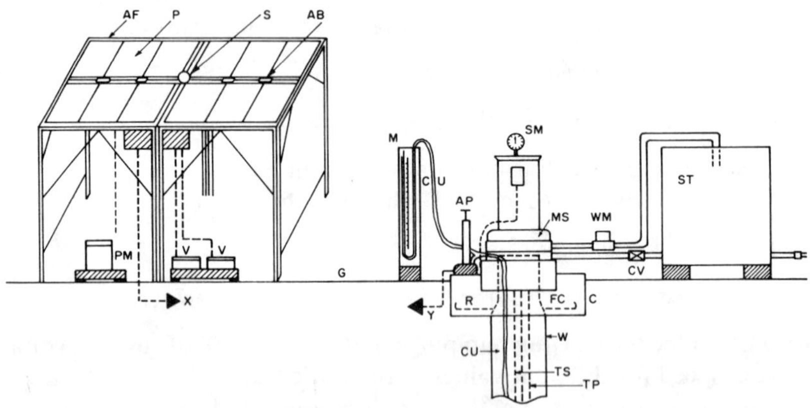

Figure 13.3 Set-up of the pumping station: AF, aluminium frame; P, solar panel; S, sensor (radiation); AB, aluminium plate; T, terminal box; PM, pyranometer; V, voltmeter; M, manometer; AP, air pump; W, well; XY, electrical wiring; G, ground level; R, steel reinforcement; MS, motor support; TS, transmission shaft; TP, transmission pipe; C, concrete platform; WM, water meter; SM, tachometer; ST, storage tank; CV, control valve; FC, funnel cover; CU, copper tube; ---, electrical wiring.

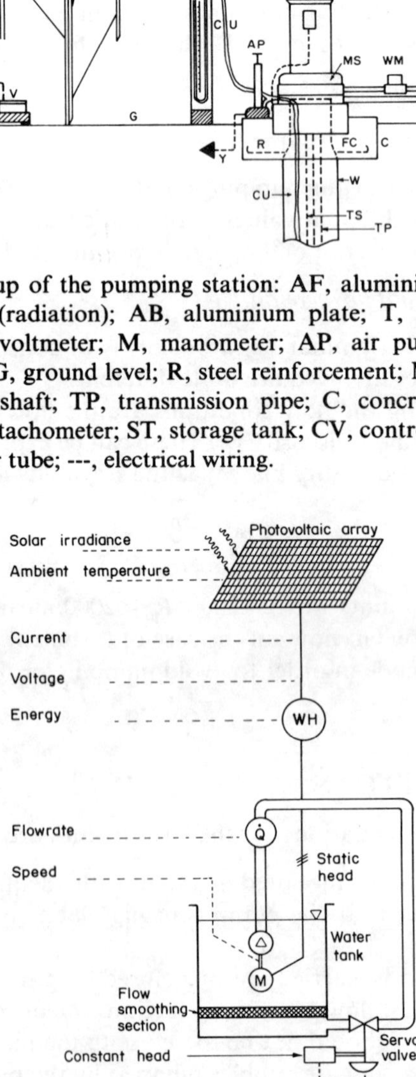

Figure 13.4 PV pumping test bed.

13.4.2 Indoor test facility

The indoor test facility is shown in figure 13.5.

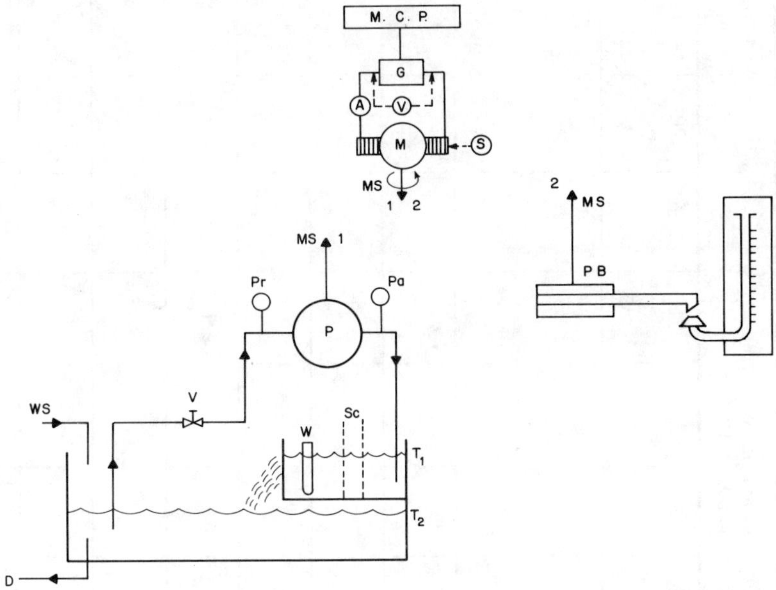

Figure 13.5 Indoor test facility for the motor–pump system: P_a (kN m^{-2}), pressure gauge, delivery side; P, test pump; V, control valve; P_r (kN m^{-2}), pressure gauge, suction side; W, water level gauge to measure flow rate (cm^3 min^{-1}); Sc, screen to reduce turbulence; T_1, flow measuring tank; MCP, monitoring and control panel; T_2, suction tank; WS, water supply pipe; D, drain; G, DC power supply; S, tachometer; PB, Prony brake; MS, motor shaft; M, motor; A, ammeter; V, voltmeter.

13.4.3 Sample data sheets

Sample data sheets are given in tables 13.7–13.9.

13.5 PRESENTATION OF RESULTS

The results are presented in figures 13.6–13.11.

13.5.1 Characterization of components

The characteristics of the system's components are shown in figures 13.12 and 13.13.

Table 13.7 Data sheet for DC motor characterization. Electrical power $P_{elec} = IV$, mechanical power $P_{mech} = (2\pi/60)\ S\tau$ or mechanical efficiency $\eta_{mech} = P_{mech}/P_{elec}$ can be calculated at any given value of current and torque.

DC motor characterization: auxiliary power supply
Load: Prony brake or the pump itself

DC power supply voltage (V)			10	13	15	17.5	20	22.5	25	27.5	30	Remarks
Current (A)	Variable	Units										
2	τ S P_{elec} P_{mech} η_{mech}	N m rev min^{-1} W W %										
3	τ S	N m rev min^{-1}										
4	τ S	N m rev min^{-1}										
5	τ S	N m rev min^{-1}										
6	τ S	N m rev min^{-1}										
7	τ S	N m rev min^{-1}										
8	τ S	N m rev min^{-1}										

Specifications	Tested by	Date	Time

Table 13.8 Data sheet for motor–pump system performance evaluation.

Motor–pump system evaluation: dynamic behaviour; auxiliary power supply												
Power supply voltage (V) Head (m)			10	13	15	17.5	20	22.5	25	27.5	30	Remarks
	Variable	Units										
1.2	I S Q	A rev min^{-1} 100 l s^{-1}										
1.5	I S Q	A rev min^{-1} 100 l s^{-1}										
1.75	I S Q	A rev min^{-1} 100 l s^{-1}										
2.00	I S Q	A rev min^{-1} 100 l s^{-1}										
2.50	I S Q	A rev min^{-1} 100 l s^{-1}										
3.00	I S Q	A rev min^{-1} 100 l s^{-1}										
3.50	I S Q	A rev min^{-1} 100 l s^{-1}										
Specifications				Tested by				Date			Time	

Table 13.9 Datasheet for overall system evaluation (dynamic behaviour).

Overall system evaluation: dynamic behaviour

Head ______ m

Solar			Electrical				Mechanical				Hydraulic						
Time (hours)	Radiation (Wm^{-2})	Power (W)	Voltage (V)	Current (A)	Power (W)	Electrical efficiency (%)	Speed (rev min^{-1})	Torque (N m)	Power (W)	Mechanical efficiency (%)	Head (m)	Time (s)	Flow (m^3s^{-1})	Power (W)	Hydraulic efficiency (%)	Global efficiency (%)	Remarks (%)

Specifications	Tested by	Date	Time

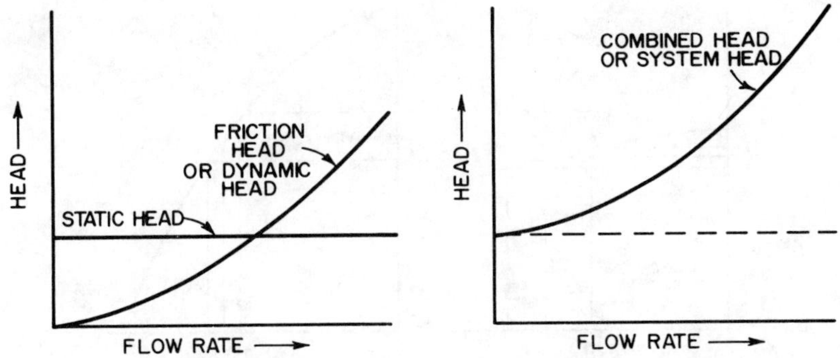

Figure 13.6 Hydraulic characteristics of a delivery pipe.

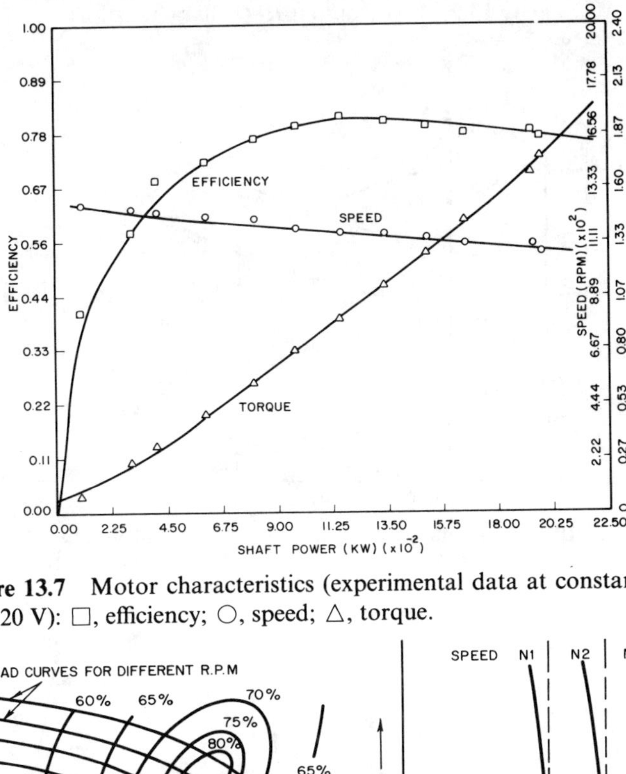

Figure 13.7 Motor characteristics (experimental data at constant voltage 20 V): □, efficiency; ○, speed; △, torque.

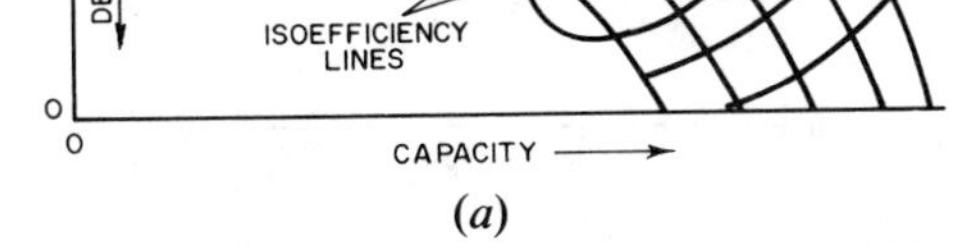

(*a*)

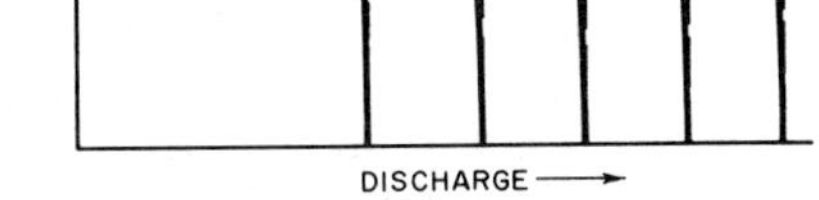

(*b*)

Figure 13.8 (*a*) Speed and isoefficiency curves (diagrammatic) for a centrifugal pump. (*b*) Speed curves for a volumetric pump.

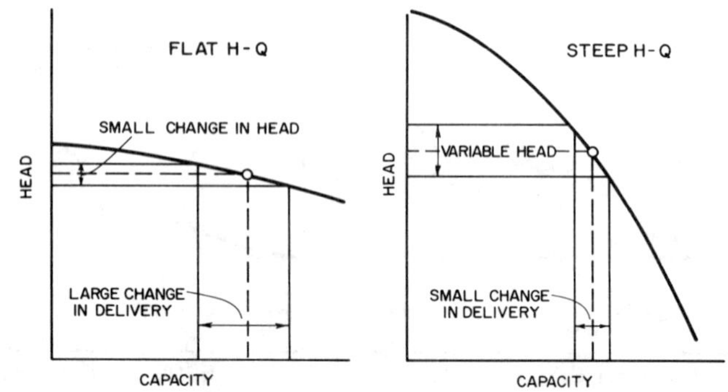

Figure 13.9 Pump output *H–Q* characteristics.

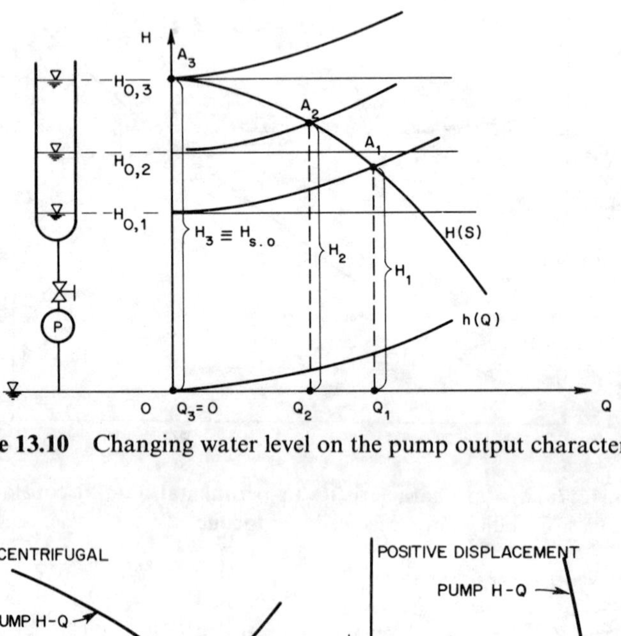

Figure 13.10 Changing water level on the pump output characteristic.

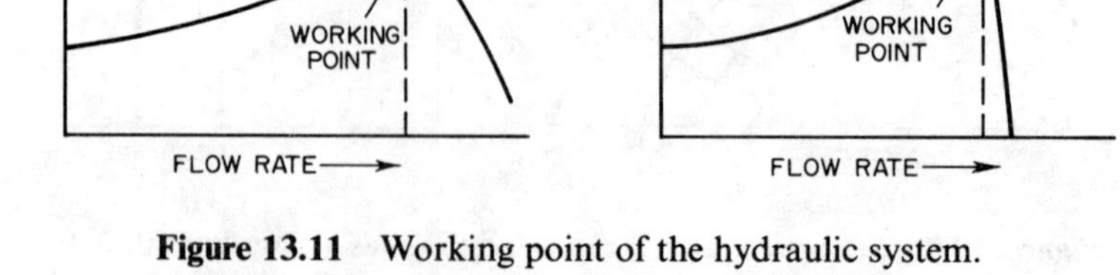

Figure 13.11 Working point of the hydraulic system.

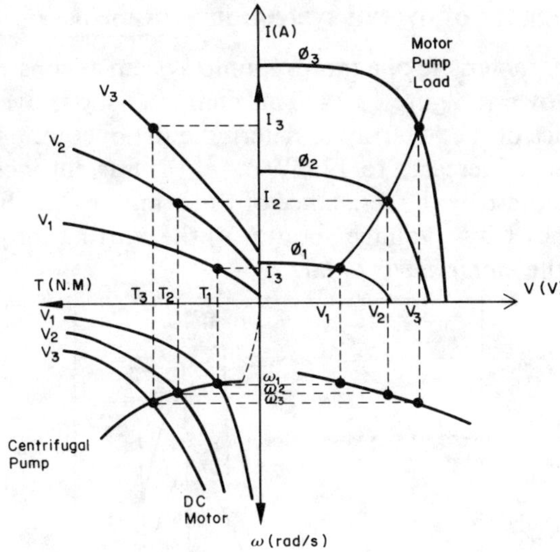

Figure 13.12 DC motor characteristics: ϕ, insolation; I, current; V, voltage; τ, torque; ω, angular speed; N, rotational speed; H, head.

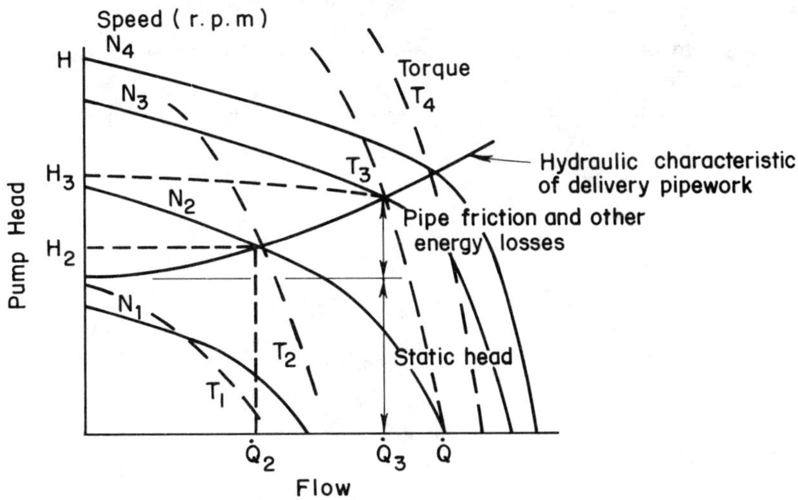

Figure 13.13 Output characteristics of a typical PV pumping system (centrifugal pump and delivery pipework characteristics).

13.5.2 Performance of overall system: instantaneous

The operating characteristic of a motor–pump system at constant static head is typically as shown in figure 13.14. This characteristic is overlaid with the I–V characteristics of the PV array to determine the operating point.

As the irradiance increases to 12.5 W m^{-2}, the current increases toward point T, which represents the resistance of the armature and brushes. When the current reaches 1.2 A (voltage about 6 V) the starting torque (0.28 N m) is achieved and the *motor starts to turn*.

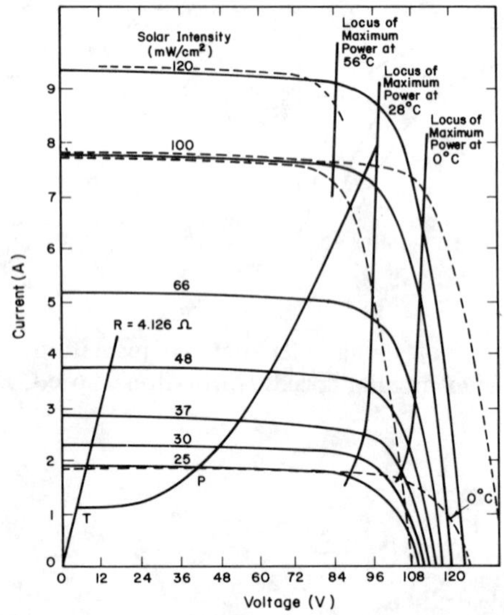

Figure 13.14 Operating characteristics of a 90 V motor with a solar array.

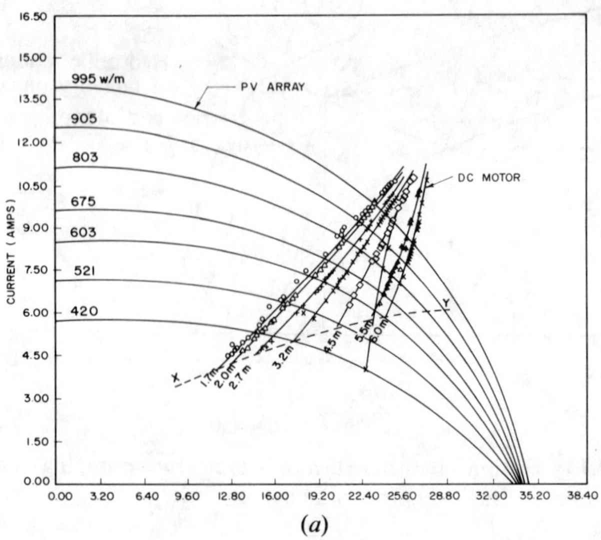

(*a*)

Rotation of the rotor generates a back electromotive force which quickly increases the array voltage. The motor speed, voltage and current all increase with increased irradiance to point P where the *pump starts to move water* (650 rev min^{-1}, 40 V and 2 A). As the irradiance increases, the operating curve passes through the maximum-power point of the array.

The most important consideration in system design is the match between the motor–pump system and the solar array; that is, the operating point of the motor under load must correspond to the locus of maximum-power points of the array over the majority of the motor speed range.

The second most important factor is the selection of the pump to match the conditions of variable motor speed, required head and maximum flow.

Figure 13.15 shows the performance of the overall system as a function of various parameters.

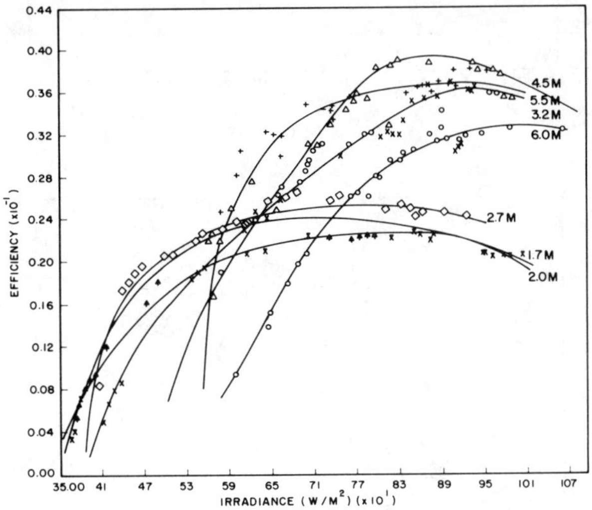

(*b*)

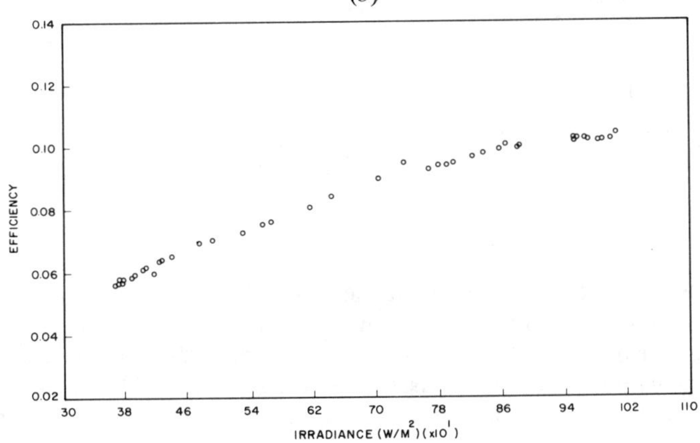

(*c*)

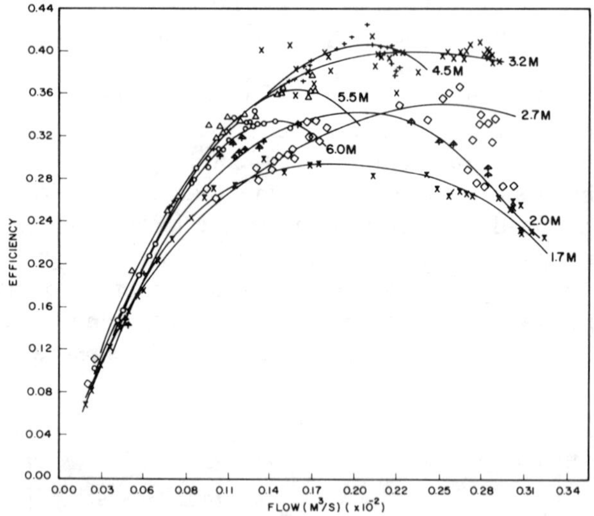

(*d*)

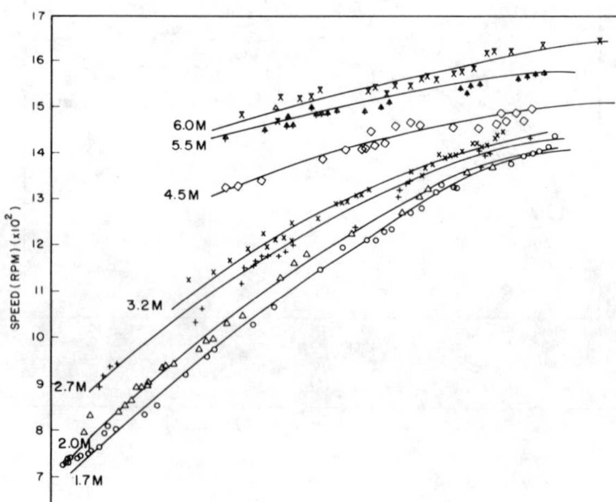

(*e*)

Figure 13.15 (*a*) *I–V* curves at various heads. (*b*) Variation in the system's efficiency with irradiance at constant heads. (*c*) Variation in the module's efficiency with irradiance. (*d*) Variation in the pump's efficiency with flow rate at constant heads. (*e*) Variation in speed with irradiance at constant heads. (*f*) Change in flow rate with irradiance at constant heads.

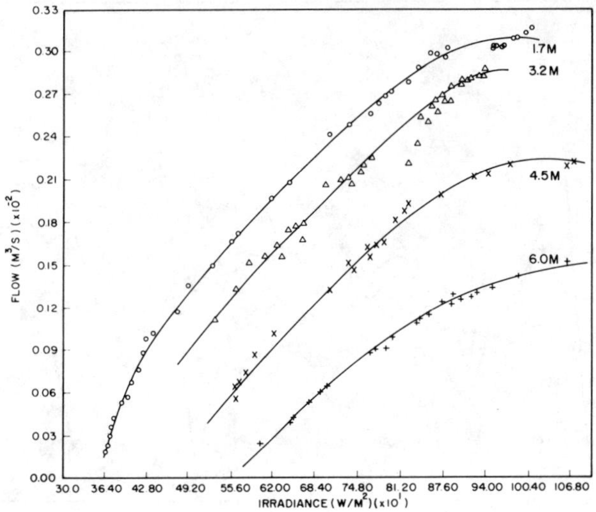

(*f*)

13.5.3 Performance of overall system: long term

Figure 3.16 shows the total daily flow for a flat-plate array over one year and figure 3.17 the input and output variations of a PV pump over one month.

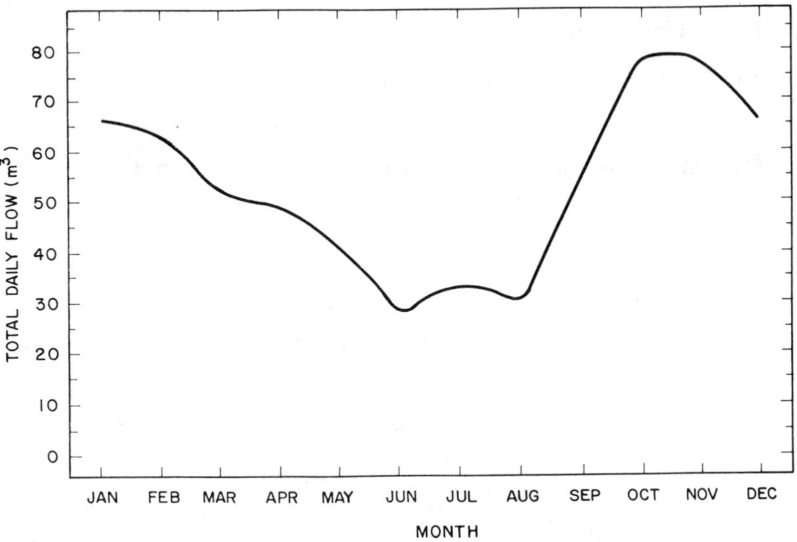

Figure 13.16 Total daily flow for a flat-plate array at a constant total manometric head over one year.

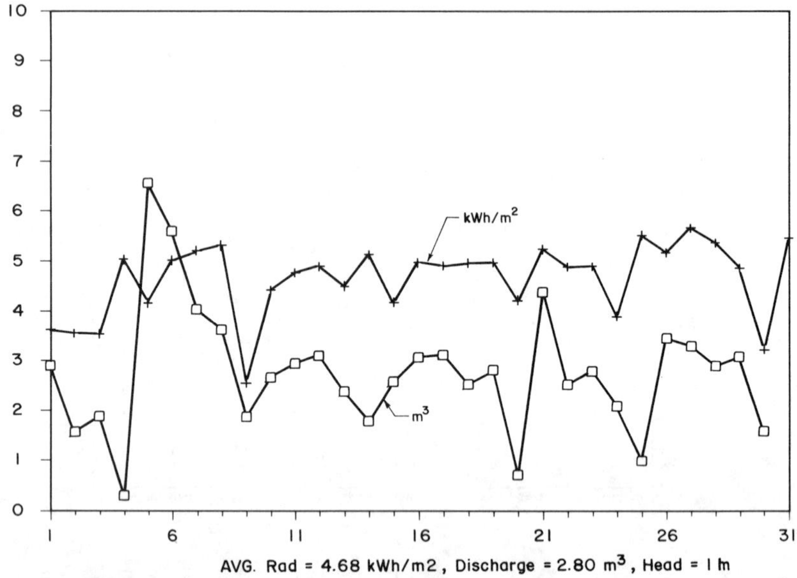

Figure 13.17 Variations in the input and output of a PV pump over 1 month (average radiation, 4.60 kW hm^{-2}; average discharge, 2.80 m^3; head, 1 m): □, discharge (m^3); +, radiation (kW h m^{-2}).

APPENDIX A PUMP BASICS

A.1 Pump classification

The pump classification is shown in figure 13.18.

A.2 PV pump configurations for low-flow deep-well applications

PV pumping system configurations are shown in figure 13.19.

A.3 Fundamentals

The total manometric head measurement of an open-tank reservoir and an open-well reservoir are shown in figures 13.20(*a*) and 13.20(*b*) respectively.

$$\text{Total head} = \text{static head} + \text{pressure head} + \text{head losses}$$

$$H = h_r(+h_a \text{ or } -h_c) + \frac{p_1 - p_0}{\rho g} + J.$$

The procedure for calculating head losses is shown in figure 13.21.

In any practical system, dynamic energy losses are the sum of the effects of

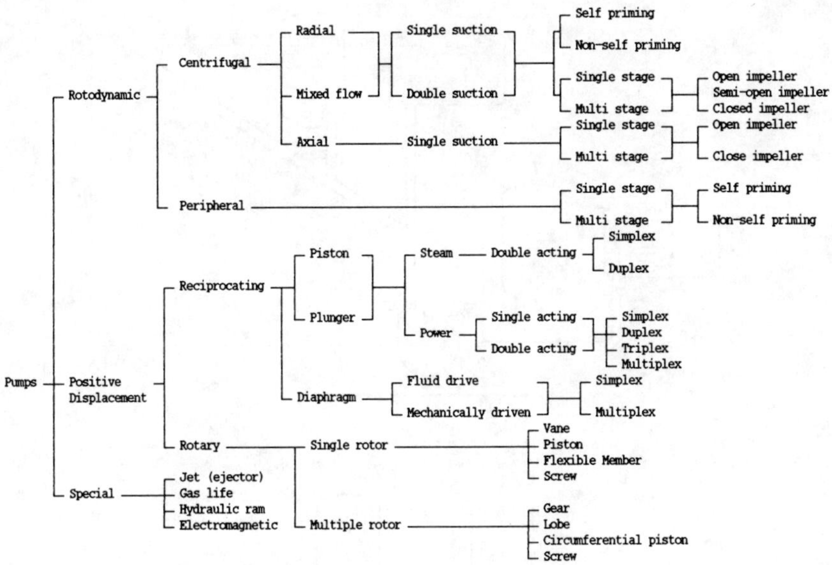

Figure 13.18 Pump classification.

friction and the result of changes in flow velocity induced by opening valves, fittings and obstructions. These have the effect of adding to the static head involved. If we consider the simple system presented in figure 13.21, hydraulic energy losses are involved at the various numerical points indicated.

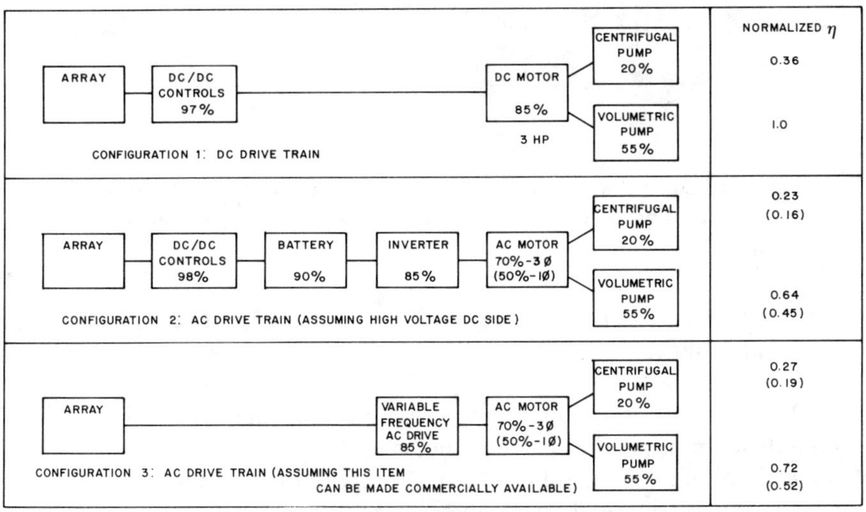

Figure 13.19 PV pumping system configurations with associated component efficiency.

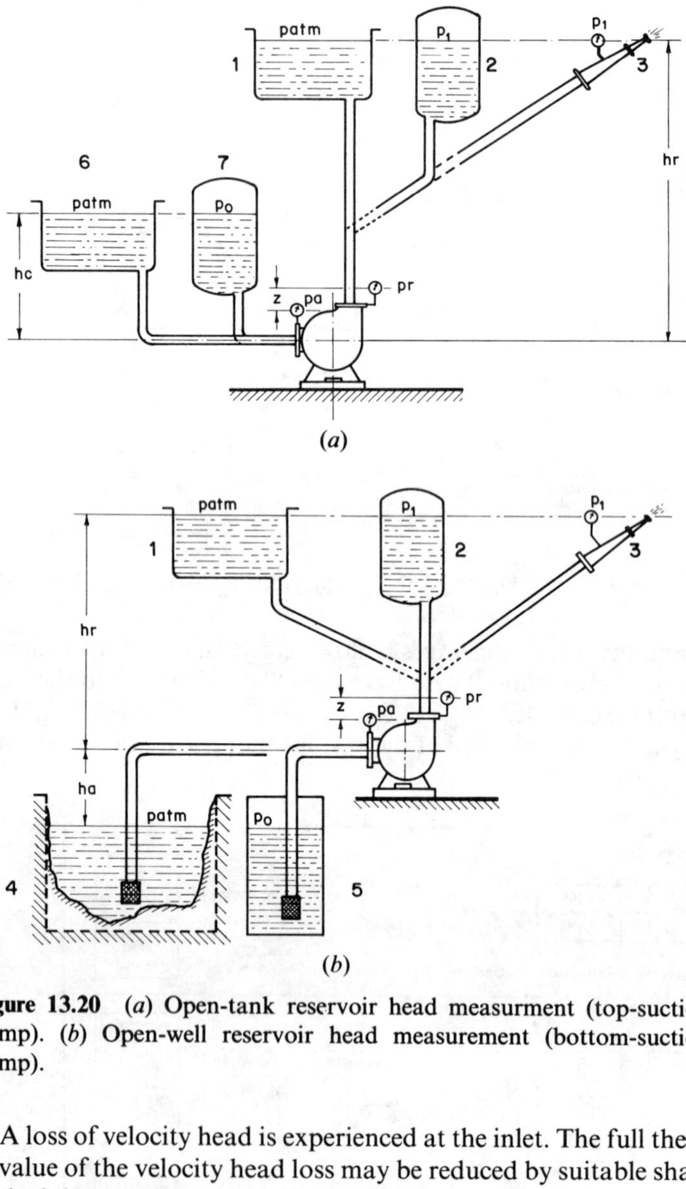

Figure 13.20 (*a*) Open-tank reservoir head measurment (top-suction pump). (*b*) Open-well reservoir head measurement (bottom-suction pump).

Point 1 A loss of velocity head is experienced at the inlet. The full theoretical value of the velocity head loss may be reduced by suitable shaping of the inlet opening.
Point 2 Frictional loss occurs in the straight run of pipe.
Point 3 Friction loss takes place at the bend.
Point 4 Frictional loss again occurs in the straight run of pipe.
Point 5 There is frictional loss in the suction fitting.

Points 2, 3, 4 and 5 taken together induce the total frictional loss on the

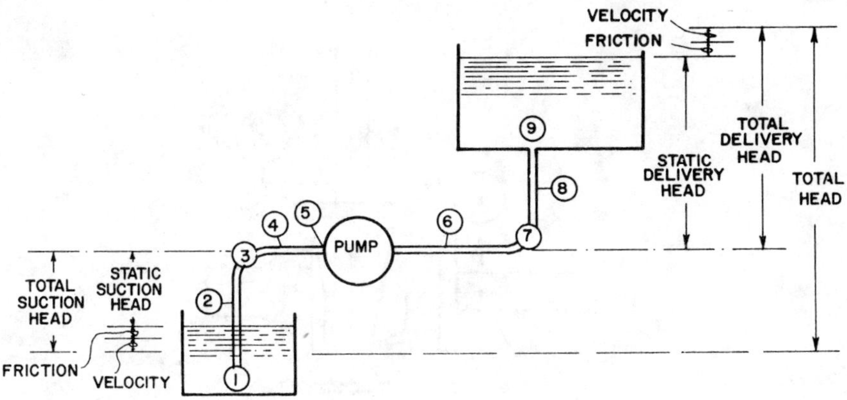

Figure 13.21 Schematic diagram for head loss calculation.

suction side or *suction friction head* J_a. The *effective suction lift* is the sum of the static suction lift, velocity head loss and the friction head.

On the delivery side, points 6, 7 and 8 represent the pipe and fitting losses; their sum is the *discharge friction head* J_r. Point 9 represents the outlet velocity head loss. Thus the effective discharge head is the sum of the static discharge head, discharge friction head and discharge velocity head:

$$J = J_a + J_r.$$

The total manometric head measurement is shown in figure 13.22. As described in this figure,

$$H = \underset{\begin{pmatrix}\text{difference between}\\ \text{pressure gauges}\end{pmatrix}}{\frac{P_r - P_a}{\rho g}} + \underset{\begin{pmatrix}\text{specific kinetic}\\ \text{energy difference}\end{pmatrix}}{\frac{V_r^2 - V_a^2}{2g}} + \underset{\begin{pmatrix}\text{pressure gauge}\\ \text{elevation}\end{pmatrix}}{z}.$$

Next we considered the flow conditions.

The condition of flow (laminar or turbulent) is of considerable significance since the energy losses involved are quite different. With laminar flow, frictional losses are predominantly due to viscous drag and are independent of the roughness of the pipe bore. With turbulent flow the shear stresses induced in the fluid are far greater than those due to viscosity alone, both in the main stream and in the region of the boundary surface. As a result of the latter, the condition of the boundary surface, i.e. whether it is smooth or rough, affects the actual frictional losses to a large extent.

Parameters affecting the flow condition are the flow velocity, pipe bore size and fluid viscosity. These can be expressed in the form of a single dimensionless quantity known as the Reynolds number R_n.

Where the Reynolds number is not greater than 1200, flow will be laminar.

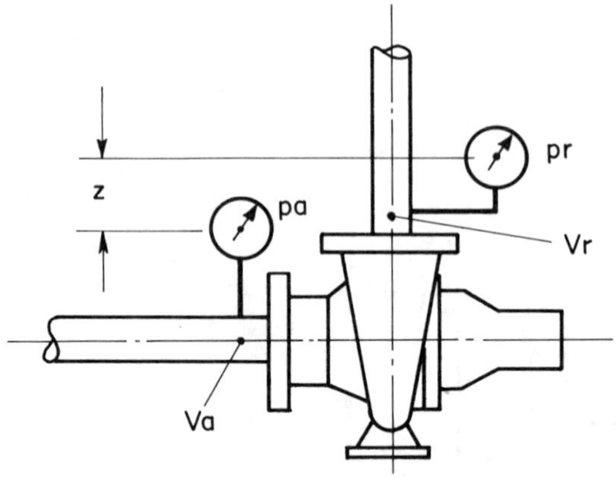

Figure 13.22 Manometric head measurement.

Laminar flow may persist up to $R_n = 2000$ under favourable conditions but, at $R_n = 3000$ and above, flow will normally be turbulent.

With most practical pumping problems the flow Reynolds number is greater than 3000 and so turbulent flow conditions apply. Low flow rates through small bore tubes may, however, involve laminar flow or Reynolds numbers approaching the transitional stage. In such cases, adjustment of pipe sizes to ensure laminar flow can result in substantial reductions in frictional losses.

A.4 Definitions

A typical pumping installation is shown in figure 13.23. The associated definitions are as follows.

Capacity or delivery — Flow rate in any suitable units.

Flow velocity v — Given by

$$\dot{Q} = 0.7854\, D^2 v.$$

Reynolds number R_n — Given by

$$R_n = \frac{Dv}{\nu_{fl}}.$$

Static or geodetic head H_s — Difference in fluid levels in the suction and delivery tanks respectively.

(Volumetric) discharge of the pump — The amount of liquid passing through the machine in unit time, usually expressed as volume per unit

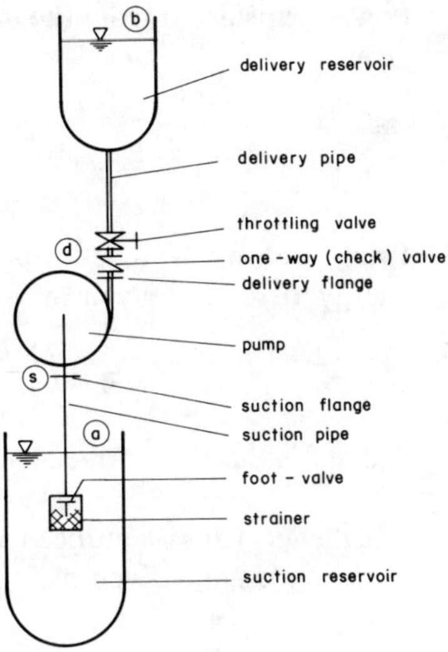

Figure 13.23 Typical pumping installation.

time. In some applications and especially with pumps, weight or mass discharge is also employed. The discharge of pumps is usually measured by orifice, diaphragm or Venturi meters, placed in the delivery line after and close to the pump.

Total manometric or effective head

Difference between the specific energy at the delivery flange and the specific energy at the suction flange. Sometimes, it is said that it is the head on the pump, or the head against which the pump is working at a given discharge. As the head is the difference between two values of specific energy, it is measured and expressed in units of length (metres, feet or inches). It is customary to have manometers connected at or near the delivery and the suction flanges of pumps.

Pump head H

Specific energy or energy per unit of weight of liquid passing through the pump, transferred to the liquid by the pump.

Hydraulic power P_{hydr}

Power transferred from the pump to the liquid and given by

$$P_{hydr} = \rho g H \dot{Q}.$$

Mechanical power P_{mech} — Power transferred from the motor to the pump and given by

$$P_{mech} = \frac{2\pi}{60} S\tau.$$

Frictional head loss h_L — Friction along stretches of pipe induce longitudinal energy losses and given by

$$h_L = f \frac{L}{D} \frac{v^2}{2g}$$

where f is the Darcy–Weisbach friction factor.

Velocity head loss h_k — Local energy losses induced by valves, fittings, and obstructions and given by

$$h_k = k \frac{v^2}{2g}$$

where k is the local loss coefficient. The velocity v relates to the undisturbed velocity upstream or downstream from the obstruction and, if a fitting connects pipes of different sizes, the velocity is usually that in the smaller pipe.

Stretch characteristic of pipe — Dynamic head as a function of the discharge $\dot{Q}$.

Total system head — Combination of static and dynamic head.

Dynamic head — Sum of frictional head losses and velocity head losses.

A.5 NOMENCLATURE FOR APPENDIX A

D	Pipe bore size
g	Acceleration due to gravity
h_a	Static suction head
h_c	Static suction head (in charge)
h_k	Velocity head loss
h_L	Friction head loss
h_r	Static discharge or delivery head
H	Total manometric head
H_s	Total static head

J	Total friction head and velocity loss
J_a	Friction head and velocity loss on suction side
J_r	Friction head and velocity loss on delivery side
L	Pipe length
P_a	Head as read on suction pressure gauge
p_{atm}	Atmospheric pressure
p_r	Head as read on delivery pressure gauge
p_0	Pressure in suction tank
p_1	Pressure in delivery tank
Q	Daily water capacity
$\dot{Q}$	Flow rate
v	Flow velocity
V_a	Average flow velocity in the suction pipe
v_r	Average flow velocity in the delivery pipe
z	Pressure gauge head
ρ	Specific or unit weight of the fluid
ν_{fl}	Fluid viscosity
1–6	Open air tank
2	1
3	Sprinkler
4	River or well
5–7	Closed tank

APPENDIX B BASIC EQUATIONS FOR A DC MOTOR

B.1 Equations

$$E = K\varphi\omega$$

$$V_a = E + R_a I_a$$

$$\tau = K\varphi I_a$$

where K is a constant for a given motor.

For a separately excited permanent-magnet motor,

$$\varphi = \text{constant}$$

$$I_a = \frac{\tau}{K\varphi}$$

$$V_a = K\varphi C_0 S + \frac{\tau R_a}{K\varphi}$$

where

$$C_0 = \frac{2\pi}{60}.$$

For a series motor, assuming that the saturation condition is not reached,

$$\varphi = K_1 I_a$$

$$I_a = \left(\frac{\tau}{KK_1}\right)^{1/2}$$

$$V_a = (KK_1 C_0 S + R_a)\left(\frac{\tau}{KK_1}\right)^{1/2}.$$

For a shunt motor,

$$\varphi = K_2 I_f = K_2 \frac{V_a}{R_f}$$

$$I_{a0} = I_a + I_f$$

$$I_{a0} = \left(\frac{\tau}{KK_2 K_3 (1 - K_3)}\right)^{1/2}$$

$$V_a = R_f K_3 I_{a0}$$

where

$$K_3 = \frac{R_a}{R_f + R_a - KK_2 C_0 S}$$

B.2 Nomenclature for Appendix B

E	Electromotive force (V)
I_a	Motor armature current (A)
I_{a0}	Motor terminal current (A)
I_f	Field current (A)
K	Motor constant
K_1	Motor coeeficient
K_2	Motor coefficient
R_a	Armature resistance (Ω)
R_f	Field resistance (Ω)
S	Speed of rotation (rev min^{-1})
V_a	Motor input voltage (V)
τ	Torque (N m)
φ	Magnetic flux (Wb)
ω	Angular velocity (rad s^{-1})

APPENDIX C SENSORS

C.1 Current–voltage characteristics of the photovoltaic modules

The procedure and developed sensor are discussed in detail in Chapters 10 and 21.

C.2 Water level measurement

A simple water level measurement is shown in figure 13.24. It consists of an air pump (or a minicompressor) equipped with a mercury manometer.

$$H = L - \frac{\rho_{Hg}}{\rho_{H_2O}} d$$

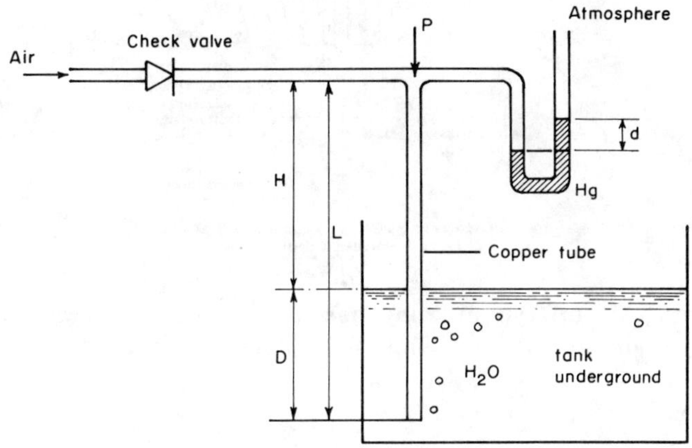

Figure 13.24 Manometer.

C.3 Torque measurement (basics)

Figure 13.25 shows the principle of torque measurement.

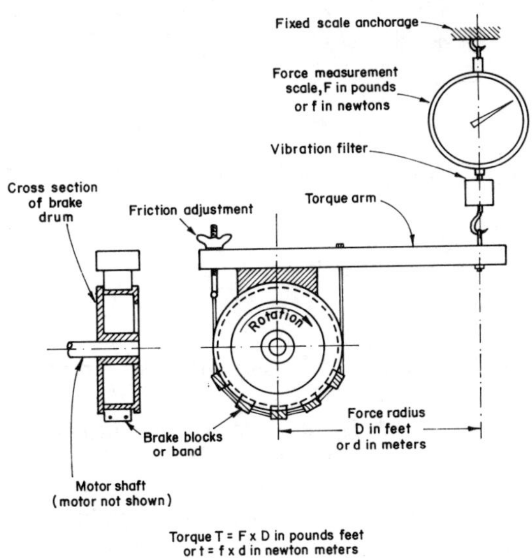

(*a*)

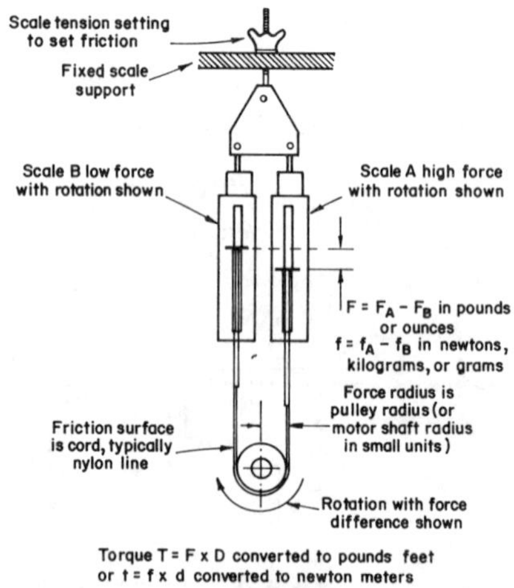

(*b*)

Figure 13.25 (*a*) Typical Prony brake for motor torque measurement. (*b*) Typical two-scale Prony brake for motor torque measurement.

C.4 Nomenclature for Appendix C

H	Water level (cm)
L	Length of copper tube (cm)
ρ_{Hg}	density of mercury, 13.6 $g\,cm^{-3}$
ρ_{H_2O}	density of water, 1 $g\,cm^{-3}$

REFERENCE

Herrmann B, Karl H, Kopf E and Lehner G 1987 *Solar Energy* **4** 275–9

BIBLIOGRAPHY

Easwarakhanthan T 1981 *Thesis No ET-81-10* Asian Institute of Technology, Bangkok

Lasnier F and Follea D 1983 *Photovoltaic Systems Engineering Lecture Notes* (Bangkok: Asian Institute of Technology)

Myles S *Lecture Notes* (Danfoss Cie)

Nir Z *Pumps and Pumping Lecture Notes* (Bangkok: Asian Institute of Technology)

Warring R H *Pumps Selection Systems and Applications* (Morden, Surrey: Trade and Technical Press)

14 Battery-testing Method for Low-water-loss and Starting, Lighting and Ignition (Automotive) Batteries

14.1 INTRODUCTION

The following battery-testing methods are recommended specifically for testing low-water-loss batteries and starting, lighting and ignition (SLI) (automotive) batteries. For testing other battery types, other testing methods should be employed. The methods described below should be used as a guide only and the battery manufacturer's instructions, if available, are preferred and should be utilized instead.

14.2 TESTING LOW-WATER-LOSS BATTERIES

The internal construction of low-water-loss batteries may include envelope separators, expanded grids, greater electrolyte reserve volume above the plates and sophisticated venting systems designed to minimize acid loss. The grids contain little or no antimony as the presence of antimony greatly increases the rate of water loss and self-discharge. The necessary mechanical strength is provided by the addition of metals such as calcium, cadmium or strontium which produce less gassing and a lower self-discharge rate. Some low-water-loss batteries are completely sealed except for the presence of small vent holes and therefore, water cannot be added to them. Testing is as follows (figure 14.1).

Step 1 Visual inspection

(*a*) Inspect for obvious damage to the case, cover or terminals and replace the battery if physical damage is discovered.

(*b*) Check the condition and size of the cables and whether the cable clamps are on tightly. If the terminals or clamps are found to be corroded, clean all corroded parts. Badly corroded cables or cables with defective terminals should be replaced.

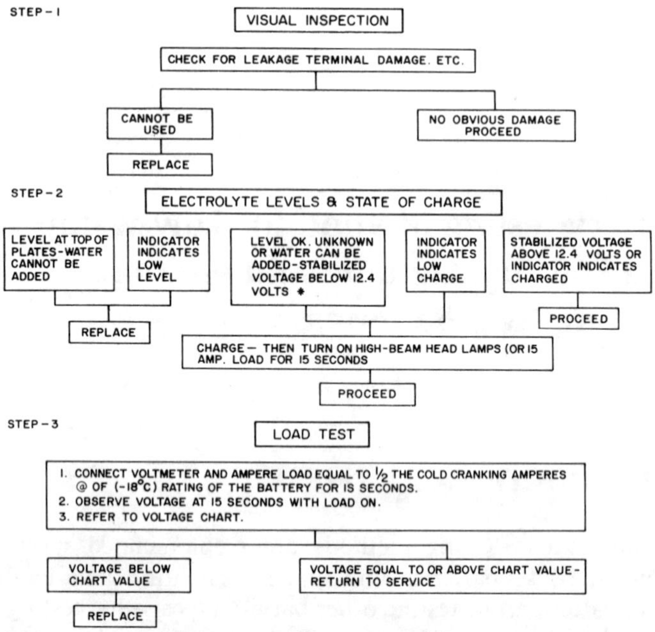

VOLTAGE CHART

ESTIMATED ELECTROLYTE TEMPERATURE		MINIMUM REQUIRED VOLTAGE UNDER 15 SEC. LOAD (USE ½ THESE VALUES FOR 6 VOLT BATTERIES)
70°F	(21°C) & ABOVE	9.6
60°F	(16°C)	9.5
50°F	(10°C)	9.4
40°F	(4°C)	9.3
30°F	(-1°C)	9.1
20°F	(-7°C)	8.9
10°F	(-12°C)	8.7
0°F	(-18 C)	8.5

* IF WATER CAN BE ADDED TO BATTERY A HYDROMETER READING OF 1.225 @ 80°F (27°C) CAN BE USED INSTEAD OF THE 12.4 VOLTAGE READING. USE 6.2 VOLTS FOR 6 VOLT BATTERIES.

Figure 14.1 Testing chart for low-water-loss batteries.

Step 2 Electrolyte levels and battery's state of charge

(*a*) In most cases, depletion of the reserve electrolyte volume above the plates signals the end of the battery life, although a simple addition of water may be all that is needed in some cases. If water cannot be added, the battery must be replaced. The electrolyte level of batteries with sealed covers with no filler caps and non-translucent cases can be checked using a built-in hydrometer which includes an electrolyte level indicator. If the electrolyte level is below the plate tops and the vents are removable, add water before proceeding further.

(*b*) The state of charge (SOC) of a low-water-loss battery can be checked with an accurate voltmeter which has been checked frequently against one of known accuracy. The battery should be kept on open circuit overnight or at least for several hours to be assured of a stabilized open-circuit voltage reading. If the stabilized open-circuit voltage is below 12.4 V, charge the battery. If a hydrometer reading can be taken, the specific gravity value should be 1.225 or above at 80 °F (26.7 °C). In the case where the SOC cannot be determined, the battery should be charged as a precaution before proceeding to step 3.

Step 3 Load procedure: the load test procedure determines whether or not the battery electrical performance is adequate and battery replacement necessary. The procedure must be carried out only if the battery is at or above the SOC specified in step 2.

(*a*) The charging guide given in table 14.1 can be used if the charging of the battery is necessary prior to the load test. It indicates only an approximation of the time involved in charging at the given rates. The charge levels indicated are those necessary to bring a battery at approximately 25% SOC to at least a level that would sustain a load test. Less charging will be needed if the battery is already at a SOC greater than 25%. If the battery still does not indicate that it has been charged after the recommended charging time, it should be charged again. The battery should be replaced if after two charges the battery's SOC is still below the required level. After the battery has been charged, attach the load test leads to the battery terminals and apply a load equal to half the cold cranking amperes at the 0 °F (−18 °C) rating of the battery for 15 s and proceed to step 3(*c*).

(*b*) If the battery is already at an adequate SOC for the load test, connect the voltmeter and load test leads to the battery

Table 14.1 12 V low-water-loss battery-charging guide.

Battery capacity (reserve min)	Slow charge	Fast charge
⩽80	10 h at 5 A 5 h at 10 A	2.5 h at 20 A 1.5 h at 30 A
>80–125	15 h at 5 A 7.5 h at 10 A	3.75 h at 20 A 1.5 h at 50 A
>125–170	20 h at 5 A 10 h at 10 A	2 h at 20 A 2 h at 50 A
>170–250	30 h at 5 A 15 h at 10 A	7.5 h at 20 A 3 h at 50 A

terminals, making sure that the load switch is in the 'off' position.

(c) Apply a test load equal to half of the cold cranking amperes at the 0 °F (−18 °C) rating of the battery and read the voltage after 15 s with the load connected. After removing the load, estimate or measure the temperature of the battery. Using the voltage chart given in figure 4.25, compare the voltage reading obtained with the minimum voltage at the electrolyte temperature. If the voltage is less than the minimum specified, the battery must be replaced. If the specified minimum is met or exceeded, return the battery to service.

14.3 TESTING STARTING, LIGHTING AND IGNITION (AUTOMOTIVE) LEAD–ACID BATTERIES

A variety of load testers, hydrometers and voltmeters are available for testing different battery types. If available, an adjustable-load tester and a hydrometer which reads in units of specific gravity, rather than a fixed-load tester and an open-circuit voltmeter or a hydrometer which reads only the battery's condition, should be used.

14.3.1 Testing with adjustable-load tester and hydrometer (figure 14.2)

Step 1 Visual inspection: visually inspect the battery for damage as in step 1 of testing low-water-loss batteries and replace the battery if obvious damage is found. Check the electrolyte level in each cell and, if necessary, fill the cells with water to just above the tops of the separators. Then charge for 15 min at 15–25 A to mix the water with the electrolyte.

Step 2 Hydrometer test: measure and record the specific gravity (corrected to 80 °F (26.7 °C)) of the electrolyte in each battery cell. If the specific gravity readings are found to range in value by 50 points (0.050 specific gravity) or more, or the lowest value is less than 1.225 specific gravity, the battery must be charged until the range is less than 50 points and the minimum value is 1.225 or greater. The battery must be replaced if these conditions cannot be attained. If these conditions are satisfied, apply a 150 A load for 15 s to remove the surface charge on the battery and proceed to step 3.

An alternative method of measuring the battery's SOC is by measuring the stabilized voltage of the battery (the voltage is stabilized if the battery has been on open circuit overnight or for several hours). The disadvantage of this method is that a lower-

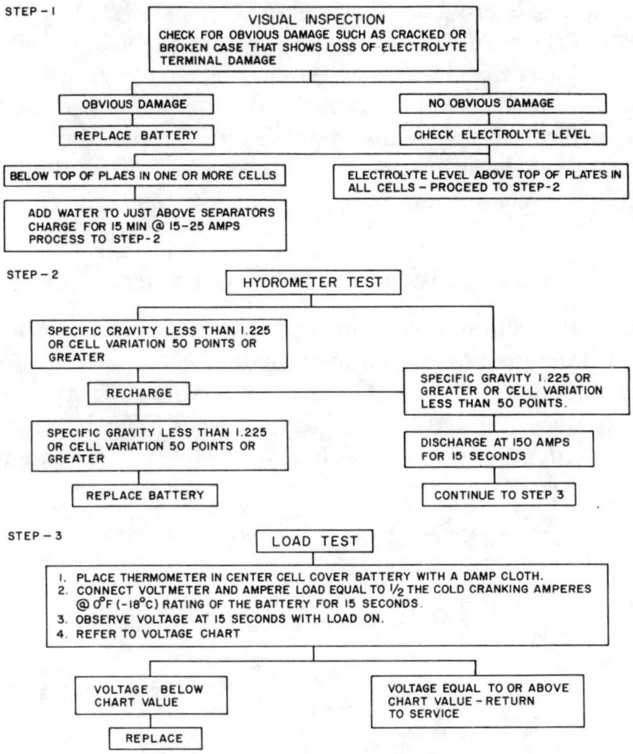

VOLTAGE CHART

ESTIMATED ELECTROLYTE TEMPERATURE		MINIMUM REQUIRED VOLTAGE UNDER 15 SEC. LOAD (USE 1/2 THESE VALUES FOR 8 VOLT BATTERIES)
70°F	(21°C) & ABOVE	9.6
60°F	(16°C)	9.5
50°F	(10°C)	9.4
40°F	(4°C)	9.3
30°F	(-4°C)	9.1
20°F	(-7°C)	8.9
10°F	(-12°C)	8.7
0°F	(-18°C)	8.6

Figure 14.2 Testing chart for SLI (automotive) lead–acid batteries with an adjustable-load tester and hydrometer.

voltage cell will not be detected. If voltage measurements can be taken for each cell, a 0.05 V range in voltage readings corresponds to a 50-point variation in specific gravity. A cell voltage of 2.07 V or a battery voltage of 12.4 V (or 6.2 for a 6 V battery) is equivalent to a specific gravity of 1.225.

Step 3 Adjustable-load test: the following instructions are meant only as guidelines and the instrument manufacturer's load testing procedure should be followed instead, if available. Firstly, disconnect battery cables starting with the ground cable. After the temperature of the centre cell has been measured, connect the voltmeter and the load test leads to the appropriate battery terminals, making certain that the terminals are free of corrosion. Next, apply a test load equal to 50% of the cranking performance rating of the battery for 15 s and then remove the load. Using the chart given in figure 4.21, determine whether the test voltage is above the minimum voltage required at the electrolyte test temperature. If it is above the minimum, the battery can be returned to service, while a test voltage below the minimum means that the battery must be replaced.

14.3.2 Testing with fixed-load tester and open circuit (figure 14.3)

Step 1 Visual inspection: follow the instructions in step 1 of testing with an adjustable-load tester and hydrometer.

Step 2 Fixed-load test: this test is used only when the available battery tester does not have an adjustable load. The battery voltage is measured instead under a heavy fixed load. In either case, the

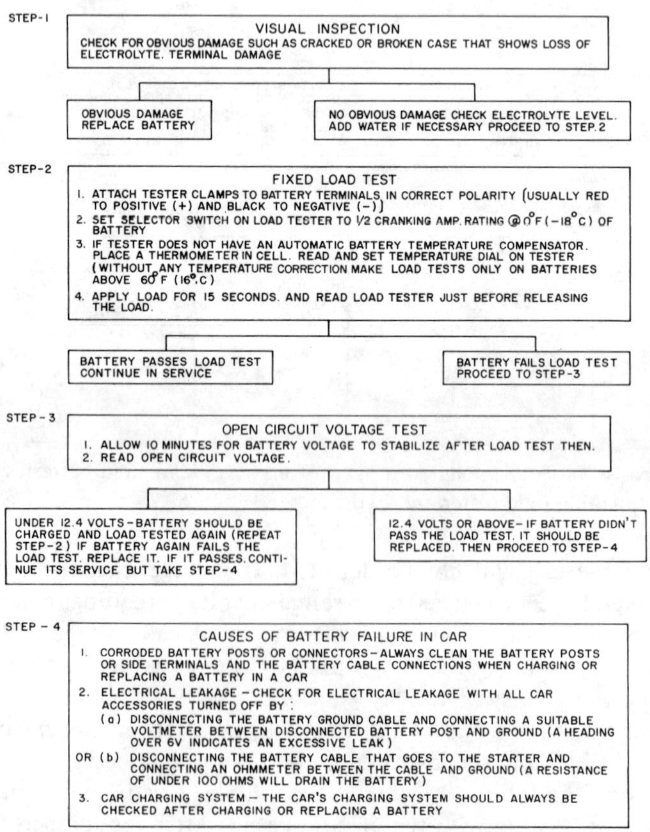

Figure 14.3 Testing chart for SLI (automotive) lead–acid batteries with a fixed-load tester and open circuit.

instrument manufacturer's instructions for load testing are preferred and should be followed when available. A selector switch or metre scale for various battery sizes and some means of correcting for the battery temperature should be included in the load tester.

(*a*) Disconnect the battery cable starting with the ground cable.
(*b*) Measure the temperature of the centre cell and set the temperature dial on the tester or insert the automatic temperature corrector probe in the centre cell. Cover the battery with a damp cloth.
(*c*) Set the selector switch on the load tester to 50% of the cranking performance rating at 0 °F (−18 °C) of the battery or three times the 20 A h capacity of the battery.
(*d*) Connect the voltmeter and load test leads to the appropriate battery terminals, making certain that the battery terminals are free of corrosion.
(*e*) Read the load tester at 15 s after application of the load test and then remove the load.
(*f*) Return the battery to service if it passes the load test; if not, proceed to step 3.

Step 3 Open-circuit voltage test: the battery's SOC is measured only if the battery fails the load test. The SOC can be determined either by measuring the specific gravity of the electrolyte with a hydrometer or by measuring the stabilized open-circuit voltage with an accurate voltmeter not less than 10 min after completing the load test. Table 14.2 can be used to determine the approximate SOC (temperature of the battery, 60–100 °F).

A digital meter or an analogue meter with an expanded scale will be needed to measure open-circuit voltage as the voltage varies slowly with the SOC. The battery should be replaced if it fails the load test and its SOC is found to be 75% or greater. If the SOC is less than 75%, the load test should be repeated after charging the battery

Table 14.2 Percentage charge as a function of open-circuit voltage.

Open-circuit voltage (V)	Charge (%)
⩾12.6	100
12.4–12.6	75–100
12.2–12.4	50–75
12.0–12.2	25–50
11.7–12.0	0–25
⩽11.7	0

according to the charging guide provided in table 14.3. If it again fails the load test, the battery should be replaced.

Table 14.3 Battery-charging guide for SLI (automotive) lead–acid battery. 6 and 12 V batteries not for use with low-water-loss batteries recommended rate and time for fully discharged condition.

Rated battery capacity (reserve min)	Slow charge	Fast charge
⩽80	10 h at 5 A 5 h at 10 A	2.5 h at 20 A 1.5 h at 30 A
>80–125	15 h at 5 A 7.5 h at 10 A	3.75 h at 20 A 1.5 h at 50 A
>125–170	20 h at 5 A 10 h at 10 A	5 h at 20 A 2 h at 50 A
>170–250	30 h at 5 A 15 h at 10 A	7.5 h at 20 A 3 h at 50 A
> 250	24 h at 10 A	4 h at 60 A

BIBLIOGRAPHY

BCI 1982 *Battery Service Manual* (Battery Council International)

Bode H 1977 *Lead–Acid Batteries*

Buresch M 1983 *PV Energy Systems: Design and Installation*

Calloway T M 1986 *Design of an Intermediate-Sized, Autonomous PV–Diesel Power Plant* (Albuquerque, NM: Sandia National Laboratories)

CIRDC 1980 *Compatible Battery for Solar Electric Power Systems Application* (Chloride India Research and Development Centre)

Macomber H L and Ruzek J B 1981 *Photovoltaic Stand-Alone Systems: Preliminary Engineering Design Handbook* (Washington, DC: US Department of Energy)

Paul T D 1981 *How to Design an Independent Power System: Best Energy Systems*

USDOE 1979 *Handbook for Battery Energy Storage for PV Power Systems* (Washington, DC: US Department of Energy)

USDOE July 1980 *Solar PV Applications Seminar: Design, Installation and Operation of Small, Stand-alone PV Power Systems* (Washington, DC: Department of Energy)

Part V

Sizing Procedure

15 The Sizing of Stand-alone Photovoltaic Power Systems: A Dynamic Simulation Approach

15.1 INTRODUCTION

One of the primary concerns in designing any photovoltaic (PV) system is the determination of the optimum relationship between the PV array and storage battery to supply a required unit of energy at a specified reliability. It is sometimes inadequate to use monthly or daily average irradiation and estimated number of continuous 'no-sun' days to determine array and battery capacities because the dynamic behaviour of the PV system and the stochastic nature of solar radiation also significantly influence the required array and storage capacity. The simulation approach adapted in this chapter uses hourly meterological data and hourly load data to simulate the energy flow in a PV system and predicts the system reliabilities under various array and battery sizes. Through numerical methods, one can formulate the discrete relationship into an analytical form. One also finds that there exist several combinations of array and battery sizes that meet a desired reliability (the substitutional nature between array and battery). The reliability–array–battery relationship is exposed to an economic environment for optimization, minimizing system cost at a specified system reliability.

The loss of load hours (LOLH) is adapted to express this reliability. In the following discussions, LOLH expresses the number of hours in a year that the PV power system is unavailable to meet the load requirements owing to energy shortage. This energy shortage does not include component breakdown or maintenance downtime. The energy shortage considered here is primarily associated with issues taken during the system sizing and design stage. These issues include the capacities of the PV array and storage battery, daily irradiation and weather prediction, peak load or load profile projection, and load-growth projection.

Once daily irradiation, weather and the load profile are accurately predicted (if possible) for a specified period (1 year), calculation is then carried out to determine the optimum combination of array and battery capacities to meet a specified reliability. Obviously, at lower reliability (higher LOLH), the overall array–battery capacity can be reduced. It can be shown that the price ratio P_w/P_B of PV arrays to battery affects the optimum array–battery combination.

However, the weather pattern and load profile can hardly be accurately predicted. In the case of a stand-alone PV power system where higher than 97% system reliability is required, the system can be extremely oversized and expensive to install. The power system has to cover those few low-probability 'no-sun' days, unexpected peak loads and changes in consumption pattern within the year. There is a diminishing marginal rate of improvement in PV system reliability per increase in investment beyond 97%. It was shown (Jones 1987) that a stand-alone PV system can be combined with the conventional generator system to obtain a hybrid system with reliabilities higher than 97% that is often more economical than either constituent alone. If properly designed, the hybrid system can overcome the major liabilities of each system. The chapter also identified the optimum energy mix for a PV–diesel hybrid system operating at 100% reliability.

15.2 MODELLING OF SYSTEM COMPONENTS

A computer program was written to simulate a system modelled by a PV array, storage battery, charge regulator and a back-up power source (figure 15.1). Meteorological data (i.e. irradiance and ambient temperature) and load consumption profile are the system's stimuli which influence each subsystem's characteristic performance and the performance of the whole system. Mathematical models for each subsystem are described in the following sections.

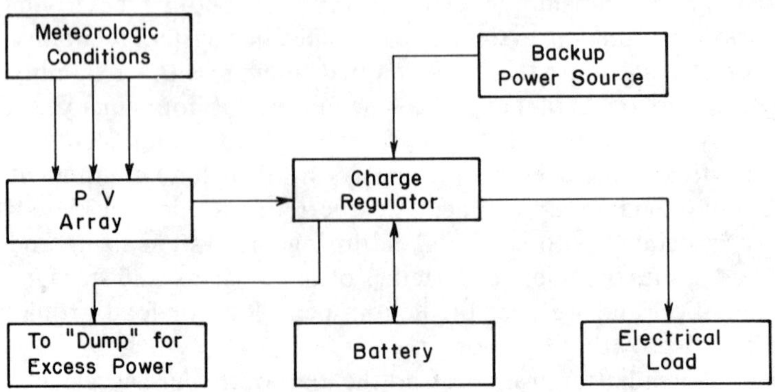

Figure 15.1 Scheme of a stand-alone PV power system.

15.2.1 Solar radiation data base

In order to reduce the required computation time, the irradiance data base used for the simulation is generated by RAMP 1.1 software found in the work of Lasnier and Ang (1986). The irradiance model (Exell 1981) constitutes the reduction of information on solar radiation in Thailand to a compact form that reproduces hourly totals and daily totals of solar radiation. The model used is expected to be applicable anywhere with a monsoon climate in southeast Asia between latitudes 25 °N and 25 °S. Irradiance on any titled plane at any specified period can be simulated.

15.2.2 Model for ambient temperature

The ambient temperature is modelled by a sinusoidal function, taking into account the daily maximum and minimum temperature and their variations within the year divided into eight periods:

$$T_a(t)=0.5\left[(T_{amax}+T_{amin})+(T_{amax}-T_{amin})\sin\left(\frac{2\pi(t-t_p)}{24}\right)\right] \quad (15.1)$$

where $T_a(t)$ is the ambient temperature at time t, T_{amax} is the maximum ambient temperature of the day and T_{amin} is the minimum ambient temperature of the day.

On the assumption that the minimum ambient temperature is at 0300 hours and the maximum ambient temperature occurs at 1500 hours, $t_p=9$ h is taken in equation (15.1). The seasonal variations of T_{amax} and T_{amin} are considered by dividing a year into eight periods during which average T_{amax} and T_{amin} are taken (Chungpaibulpatana 1981). Table 15.1 lists the temperature variations in eight periods in Thailand.

Table 15.1 Average maximum ambient temperature T_{amax} and minimum ambient temperature T_{amin} in eight periods in Thailand.

Period number	Period (months)	T_{amax} (°C)	T_{amin} (°C)
1	January 14–February 26	33.3	21.8
2	February 27–April 12	35.6	23.5
3	April 13–May 28	35.5	24.9
4	May 29–July 15	33.9	25.1
5	July 16–August 31	32.8	24.9
6	September 1–October 15	32.6	24.9
7	October 16–November 29	31.9	23.6
8	November 30–January 1	32.1	21.1

15.2.3 Model for array current–voltage

The PV array consists of modules connected in parallel. The current–voltage (I–V) characteristic of each module is represented by an equation of the form (Rauschenbach 1980)

$$I = I_{sc}\left\{1 - C_1\left[\exp\left(\frac{V}{C_2 V_{oc}}\right) - 1\right]\right\} \tag{15.2}$$

at a given cell temperature T_{c0} and irradiance G_0 and where

$$C_1 = \left(1 - \frac{I_{mp}}{I_{sc}}\right)\exp\left(-\frac{V_{mp}}{C_2 V_{oc}}\right) \tag{15.3}$$

$$C_2 = \left(\frac{V_{mp}}{V_{oc}} - 1\right)\left[\ln\left(1 - \frac{I_{mp}}{I_{sc}}\right)\right]^{-1}. \tag{15.4}$$

I (A) is the module current, I_{mp} (A) is the module current at maximum power, I_{sc} (A) is the module short-circuit current, V (V) is the module voltage, V_{mp} (V) is the module voltage at maximum power and V_{oc} (V) is the module open-circuit voltage.

To obtain the I–V curves at irradiance G and cell temperature T_c, the following transformations are applied:

$$I_{new} = I + I_{sc}\left(\frac{G}{G_0} - 1\right) + \alpha(T_c - T_{c0}) \tag{15.5}$$

$$V_{new} = V + \beta(T_c - T_{c0}) \tag{15.6}$$

where

$$\alpha = \frac{dI_{sc}}{dT} \qquad G = \text{constant}$$

$$\beta = \frac{dV_{oc}}{dT} \qquad G = \text{constant}$$

and α and β are the temperature coefficients for I_{sc} and V_{oc} respectively.

15.2.4 Model for array temperature

A simple experimental model is used to predict the array temperature from irradiance, ambient temperature and wind speed (Servant 1986):

$$T_c = T_a + \theta_G(1 + \theta_{T_a} T_a)(1 - \theta_{v_w} v_w)G \tag{15.7}$$

where T_c (°C) is the array temperature, T_a (°C) is the ambient temperature, v_w (m s^{-1}) is the wind speed and G (W m^{-2}) is the incident solar irradiance, $\theta_G = 0.0138$, $\theta_{T_a} = 0.031$ and $\theta_{v_w} = 0.042$ are constants and are determined by experimental data.

The wind speed is taken to be constant and equal to $1\,\mathrm{m\,s^{-1}}$. Further, a correction coefficient $1-1.055\eta$, where η is the array efficiency, may be added into equation (15.7) to allow for the Joule effect due to flow of electric current.

15.2.5 Model for load

Three different daily load profiles were selected to be supplied with PV electricity. These profiles are hourly defined and the yearly electric energy consumption is kept constant. Figure 15.2 shows these three profiles: constant-demand profile, constant demand with a 1 h peak and a sinusoidal profile. The simulation program allows the user to specify three parameters.

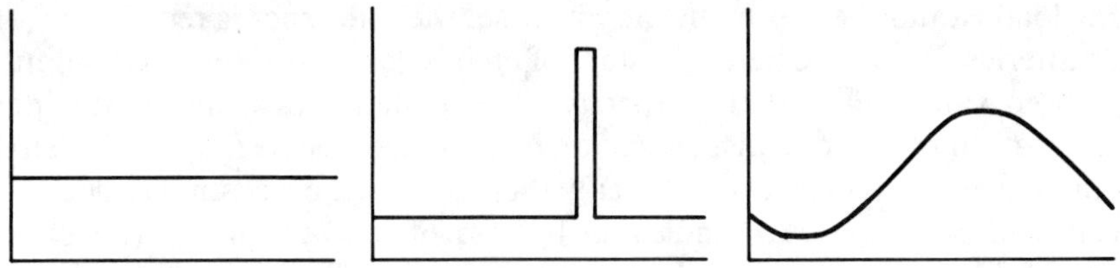

Figure 15.2 Load profiles: (*a*) Constant-demand profile; (*b*) constant-demand with a 1 h peak; (*c*) sinusoidal profile.

(i) Daily energy consumption E (kW h).
(ii) Ratio K of peak demand to the lowest demand.
(iii) The time t_p (h) of peak demand.

$E=2$ kW h, $K=10$ and $t_p=12$ h were set for the sizing case study. The mathematical formula for each profile is as follows.

The equation for the constant-demand profile (figure 15.2(*a*)) is

$$D(t)=\frac{E}{24}=D_0. \tag{15.8}$$

The equation for constant demand with 1 h peak (figure 15.2(*b*)) is

$$D(t)=\begin{cases} KE/(23+K) & \text{when } t=t_p \\ E/(23+K) & \text{when } t\neq t_p\,. \end{cases} \tag{15.9}$$

The sinusoidal profile is modelled as a cosine function (figure 15.2(*c*)):

$$D(t)=D_0+D_m\cos\left(\frac{2\pi(t-t_p)}{24}\right) \tag{15.10}$$

$$D_m=D_0\frac{K-1}{K+1} \tag{15.11}$$

where D_m is the amplitude of the cosine function.

The profile in figure 15.2(*a*) was taken as the base case for the study, and the other profiles, in figures 15.2(*b*) and (*c*) respectively, were used to investigate the effect of load patterns and the effect of time of peak demand at t_p on system reliability.

15.2.6 Charge regulator

A charge regulator is used to manage energy flow to PV arrays, batteries and loads. It also prevents batteries from overcharge and deep discharge.

The charge regulator is modelled as follows. Electrical energy from the array is used to meet the load before any excess is stored in the batteries. If batteries are full, the excess is 'dumped' by dissipation to a resistive network. If the load cannot be met by the array alone, the difference is requested from the batteries. When the battery's state of charge (SOC) drops to the minimum permitted value (30% of the capacity), the regulator disconnects the load. The duration of the load disconnection is then recorded as LOLH and such an event is a loss of load event (LOLE); these LOLES are accumulated over a specified duration (1 year) indicated by loss of load frequency (LOLF) (see figure 15.3 below).

15.2.7 Model for battery

The storage battery is modelled by the internal charge–discharge efficiencies and resistances, the battery self-discharge rate and the battery's SOC. The battery SOC $(t+\mathrm{d}t)$ can be estimated from the present SOC(t) by computing the energy input and output, and the battery self-discharge and conversion losses (Ang 1984):

$$\mathrm{SOC}(t+\mathrm{d}t)=\mathrm{SOC}(t)(1-D_s\,\mathrm{d}t)+K_1(V_B I_B-R_B I_B^2)\,\mathrm{d}t \tag{15.12}$$

where SOC(t) (W h) is the battery's SOC at time t, D_s (W h^{-1}) is the battery self-discharge rate, K_1 is the battery charge–discharge efficiency, R_B (Ω) is the battery internal charge–discharge resistance, V_B (V) is the battery voltage and I_B (A) is the battery current.

The battery charge–discharge voltage and resistances are also updated every dt according to

$$R_2=\frac{\mathrm{SOC_B}}{\mathrm{SOC_{max}}}$$

$$V_{ch}=2+0.148R_2 \tag{15.13a}$$

$$V_{dch}=1.926+0.124R_2 \tag{15.13b}$$

$$R_{ch}=\frac{0.758+0.1309(1.06-R_2)^{-1}}{\mathrm{SOC_{max}}} \tag{15.13c}$$

$$R_{dch}=\frac{0.19+0.1037(R_2-0.14)^{-1}}{\mathrm{SOC}_{max}} \tag{15.13d}$$

where V_{ch} (V) and V_{dch} (V) are the battery charge and discharge voltages, R_{ch} (Ω) and R_{dch} (Ω) are the battery charge and discharge internal resistances, SOC (W h) is the battery's SOC and SOC_{max} (W h) is the maximum SOC of the battery.

15.2.8 Back-up power–diesel generator

The operation of a diesel generator is modelled as

$$P_{gen}=\begin{cases}\text{constant} & \text{when operating}\\ 0 & \text{when stopped}\end{cases} \tag{15.14}$$

where P_{gen} is the generator power output which is determined by the generator rated power and its load factor. For the purpose of sizing, the generator model is not included in the simulation program. Instead, this model is used in the economic analysis of the back-up system.

15.2.9 Reliability index of the system

Cumulative LOLH over 1 year is chosen as the relevant index of the PV system reliability since this index can reflect the cost of load shedding to consumers (EPRI 1981). Moreover, this index has a relevant economic implication when the economics of PV hybrid systems are investigated. LOLH can be converted

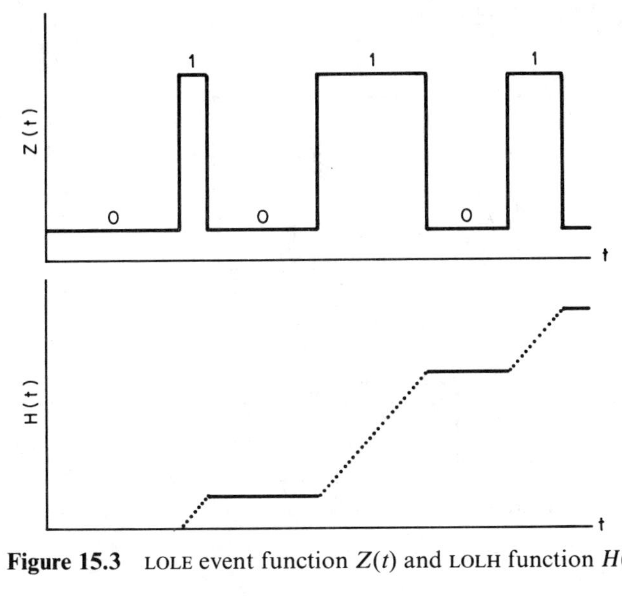

Figure 15.3 LOLE event function $Z(t)$ and LOLH function $H(t)$.

into the generator running hours, and the cost of back-up power can be calculated.

Figure 15.3 illustrates the concept of LOLH. In this figure, $Z(t)$ represents the LOLE which has the mathematical form

$$Z(t)=\begin{cases}1 & \text{if the load is shed}\\ 0 & \text{if the load is met.}\end{cases} \tag{15.15}$$

It should be noted that the system reliability here is used to describe the power supply system's availability to supply the required energy. A decrease in system reliability is due to the mismatch between battery storage capacity and the installed power rating of the array, sudden change in peak load, load profile and weather pattern not covered during the design stage. System downtime due to component lifetime and maintenance is not reflected in the reliability index. Any replacement of parts and maintenance of the system are assumed to be carried out when it is not required to supply within its normal schedule of operation. The electronics are assumed to have immediate replacement and servicing without considerable effects to the reliability index.

15.3 SIMULATION RESULTS AND SIZING PROCEDURE

The input of the simulation program includes load profile, array size (PVS) and battery capacity (BTYS). The output is the reliability index (LOLH) associated with the given input data under a given meteorological condition. The input–output arrangement and the support data base of the simulation program are shown in figure 15.4.

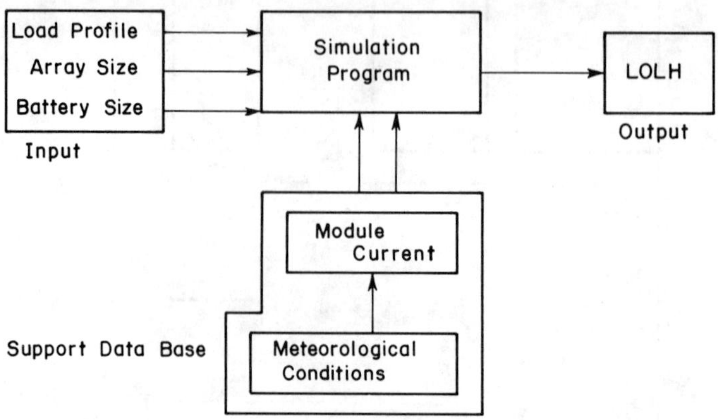

Figure 15.4 Output–input and support data base of simulation program.

The simulation results for the load profile in figure 15.2(a) at four array sizes and several battery capacities are depicted in figure 15.5 presenting the LOLH–PVS–BTYS relationship on an LOLH–BTYS map. Using a two-dimensional inverse interpolation technique (Gerald 1978), the combinations of PVS and BTYS which provide the same reliability (LOLH) can be obtained from figure 15.5, and the combinations can be depicted on a BTYS–PVS map. Figure 15.6 shows the iso-LOLH curves for six LOLH values, indicating the substitutional relationship between BTYS and PVS. The iso-LOLH curves are functions of the form

$$f(X)=\exp\left(\sum_{i=1}^{6}\frac{a_i}{X_i}\right) \tag{15.16}$$

where X_i is the PV peak wattage, $f(X)=Y$ (W h) is the battery size and a_i are polynomial constants.

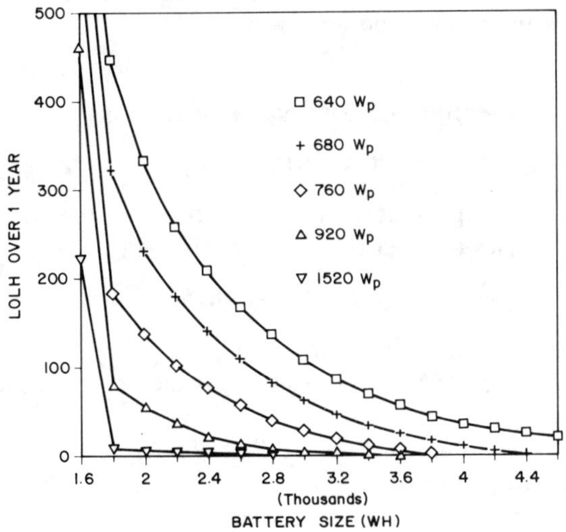

Figure 15.5 Simulation results of LOLH against BTYS at four array sizes for the load profile in figure 15.2 (a): □, 640 W_p; +, 680 W_p; ◇, 760 W_p; △, 920 W_p; ▽, 1520 W_p.

$a_0, a_1, a_2, \ldots, a_6$ are calculated by performing a polynomial regression analysis after transforming equation (15.16) into its natural logarithmic form. A good fit is important since the slope of the tangent to the iso-LOLH curves is a significant factor in the sizing analysis.

Since iso-LOLH curves indicate that there exist many BTYS–PVS combinations meeting a prescribed reliability (LOLH), it is preferable to find the combination leading to a minimum PV system cost. If we consider the

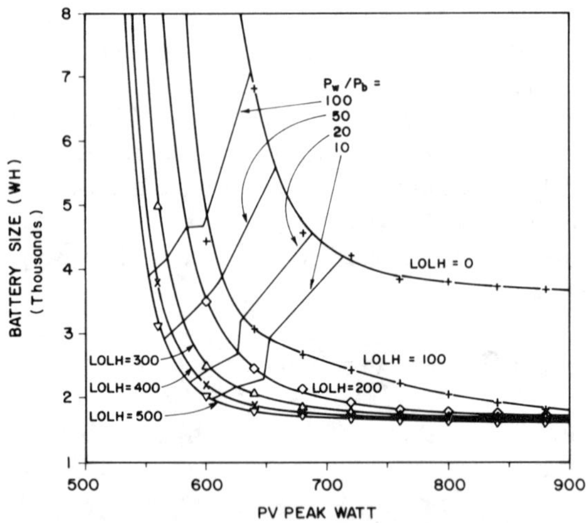

Figure 15.6 Iso-LOLH curves and optimum design points at four price ratios for a stand-alone PV power system.

iso-LOLH curve representing the LOLH value of R_0, i.e.

$$g(\text{PVS}, \text{BTYS}) = R_0 \tag{15.17}$$

where g is a function representing an iso-LOLH curve, the cost of purchasing PV arrays and batteries to meet reliability R_0 is

$$C = P_w \text{PVS} + P_B \text{BTYS} \tag{15.18}$$

where C (US \$) is the total investment, P_w (US \$ W_p^{-1}) is the array price, P_B (US \$ W h^{-1}) the battery price, PVS (W_p) is the array size, and BTYS (W h) is the battery size.

The objective is to minimize $P_w\text{PVS} + P_B\text{BTYS}$ subject to $g(\text{PVS}, \text{BTYS}) = R_0$.

The mathematical solution can be found through the method of the Lagrange multiplier (Rao 1984): the Lagrange function is

$$L = P_w \text{PVS} + P_B \text{BTYS} + m_L(g - R_0) \tag{15.19}$$

where m_L is the Lagrange multiplier. Taking the partial derivatives of equation (15.19) with respect to PVS, BTYS and m_L and setting them to zero gives

$$P_w + m_L \frac{dg}{d\text{PVS}} = 0 \tag{15.20}$$

$$P_B + m_L \frac{dg}{d\text{BTYS}} = 0 \tag{15.21}$$

$$g = R_0. \tag{15.22}$$

Equations (15.20) and (15.21) can be combined and written as

$$-\frac{d\text{BTYS}}{d\text{PVS}}=\frac{P_w}{P_B}. \tag{15.23}$$

The left-hand side of equation (15.23) is the marginal rate of substitution (MRS) between PVS and BTYS. It represents the rate at which PVS is decreased as BTYS is increased along the iso-LOLH curve. Equation (15.23) indicates that the minimum cost to meet reliability R_0 occurs at the point where the MRS equals the price ratio P_w/P_B. Thus the optimum point on an iso-LOLH curve is the point where the negative slope of the tangent is equal to the price ratio.

If the lifetimes of the array and battery are different, the price ratio must reflect the real costs to the system. For instance, the array price is about US $ 12 W_p^{-1} and the battery price is about US $120 kW h^{-1}. If arrays have a lifetime twice as long as that of batteries, the price ratio can be roughly taken as 50 (discount rate is ignored). Figure 15.6 also shows the optimum design points on each iso-LOLH curve for price ratios of 100, 50, 20 and 10.

15.4 ECONOMIC ANALYSIS OF A PHOTOVOLTAIC HYBRID SYSTEM

Table 15.2 lists the optimum BTYS–PVS combinations at various LOLHs for a stand-alone PV power system with $P_w/P_B=50$ supplying the load profile in figure 15.2(*a*). These are graphically presented in figure 15.6. Table 15.3 lists the technical and economic environments on which the cost analyses are based. By taking the cost of installing and running a diesel system from table 15.3 and using the LOLH value in table 15.2, the first-year cost of the PV–diesel hybrid system at various energy mixes can be calculated as in table 15.4 and graphically illustrated in figure 15.7. The capital cost and the first-year recurring cost are sufficient to determine the optimum energy mix

Table 15.2 Optimum sizes at price ratio $P_w/P_B=50$ for the load profile in figure 15.2(*a*).

LOLH (h)	PVS (W_p)	BTYS (W h)
0	657	5566
100	611	3809
200	593	3455
300	581	3168
400	572	3019
500	565	2902

Table 15.3 Base case data for cost analysis.

PV array	
Cost	US\$12 W_p^{-1}
Lifetime	10 years
Maintenance cost	US\$0 year^{-1}
Battery	
Cost	US\$120 kW h^{-1}
Lifetime	5 years
Maintenance cost	US\$0 year^{-1}
Diesel generator (including power distribution facilities)	
Rated power	1 kW
Load factor	50%
Efficiency	2 kW h l^{-1}
Capital cost	US\$400
Maintenance cost	5% of the capital per year
Lifetime	7000 h
Liquidation yield	US\$0
Diesel price	US\$0.75 l^{-1}
Discount rate	10%
Energy escalation rate	4%
General inflation rate	4%

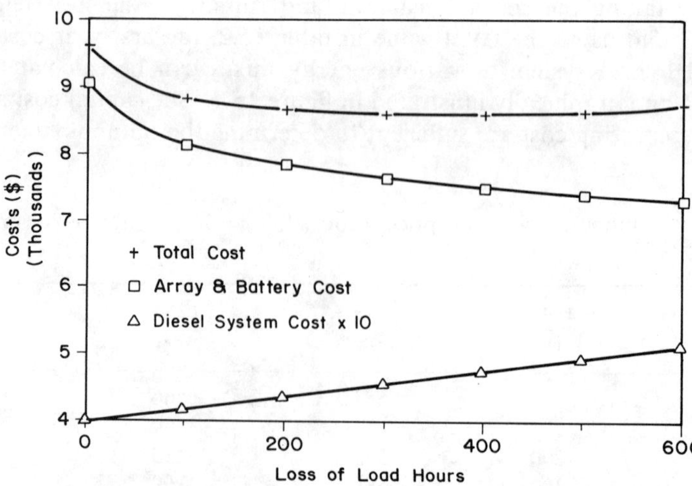

Figure 15.7 PV hybrid power system cost (+), PV system cost (□) and diesel system cost (x10) (△) as a function of PV system LOLH at 100% hybrid system reliability.

(or the optimum LOLH of the PV constituent in the hybrid system). This optimum energy mix is found to be when the diesel system is allowed to operate 354 h (table 15.4) in a year; figure 15.7 shows that a minimum hybrid system cost is achieved. Note that the LOLH of 354 h indicates the PV system's unavailability which is being taken up by the diesel system, both of which make up the hybrid system operating at 100% reliability supplying the load profile in figure 15.2(*a*). This level of LOLH corresponds to 576 W_p of array and 3093 W h of storage battery (figure 15.6 and table 15.2).

Figures 15.8 and 15.9 shows the cash flows of the constituents in the hybrid system. Figure 15.10 shows the result of a sensitivity analysis, indicating a

Table 15.4 Results of cost analysis.

LOLH	Capital costs: PV arrays and batteries	Capital costs: diesel generator	Diesel fuel cost, first year	Diesel generator maintenance cost, first year	Total costs
0	9056.5[a]	0	0	0	9056.5
0	9056.5	400	0	20	9605.3
100	8134.4	400	18.75	20	8822.7
200	7843.8	400	37.50	20	8671.7
300	7639.4	400	56.25	20	8606.7
400	7500.0	400	75.00	20	8606.9
500	7391.3	400	93.75	20	8637.8

[a] Stand-alone PV system.

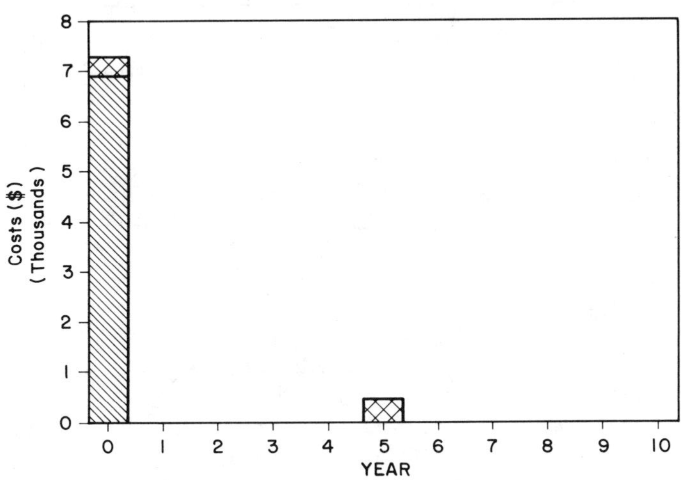

Figure 15.8 Investments of a PV system in arrays (▧) and batteries (▩).

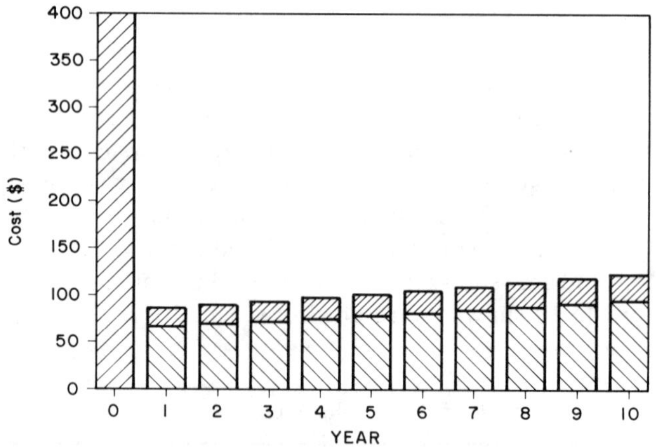

Figure 15.9 Diesel generator capital (▨), fuel (▧) and maintenance (▩) costs.

strong response to changes in capital cost and a weak sensitivity to changes in fuel cost. This low sensitivity to fuel cost makes the hybrid system adaptable to unexpected peak load or bad weather during the year. The generator can take more running hours to cover these additional load events using relatively less expensive resources (diesel fuel) than the more expensive PV module. This is particularly significant when the load and weather projection are not accurate.

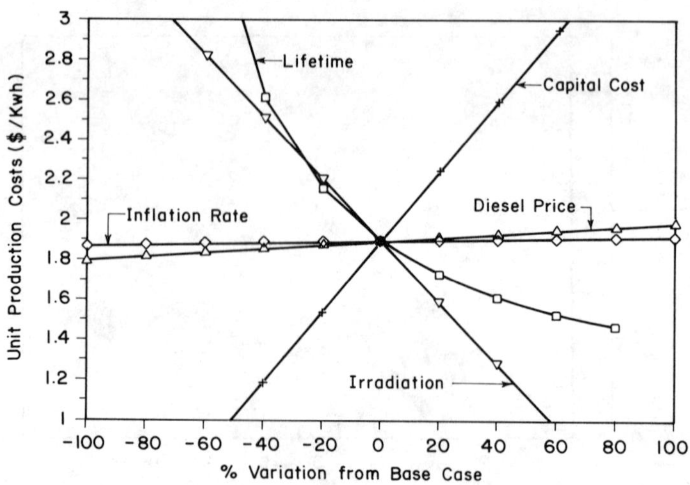

Figure 15.10 Sensitivity analysis.

15.5 DISCUSSION

15.5.1 Effect of load profiles on sizing

Besides the load profile in figure 15.2(*a*), two more load profiles (figures 15.2(*b*) and 15.2(*c*)) with the same daily energy consumption (2 kW h) as the profile in figure 15.2(*a*) were also tested to find their optimum design points as a stand-alone PV power system.

Figure 15.11 compares the sizing results (iso-LOLH curves) at LOLH = 0 for the three profiles. It shows that the required BTYS and PVS are less for the load profile that fits the solar irradiance profile well. For instance, the sinusoidal load profile with peak time at 1200 hours (load profile in figure 15.2(*c*)) fits the irradiance profile well; so the required storage capacity and array size to meet LOLH = 0 are the lowest in the three tested profiles.

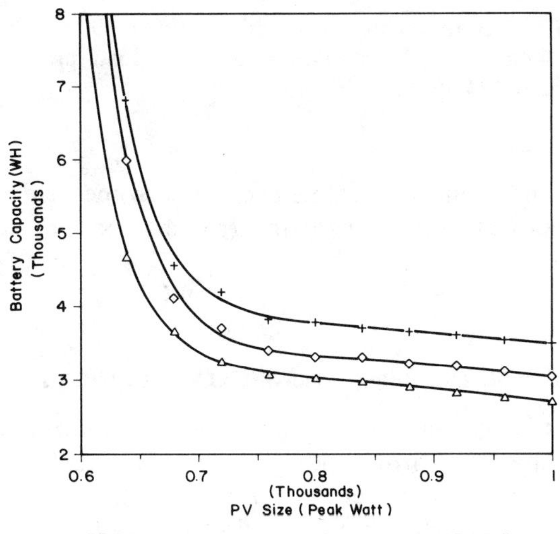

Figure 15.11 Comparison of iso-LOLH curves (LOLH = 0 h) for the load profiles in figures 15.2(*a*) (+), 15.2(*b*) (◇) and 15.2(*c*) (△).

The time t_p when peak demand occurs in a day affects system size also. If peak demand occurs at noon, the required PVS and BTYS are less. Figure 15.12 shows the variations in LOLH with peak load time for two PV systems: one is matched with the load profile in figure 15.2(*b*); the other is matched with the load profile in figure 15.2(*c*). Both systems are designed to meet the condition LOLH = 0 h at 1200 hours.

It is clear that the LOLH increases when the peak load time departs from 1200 hours. The effect of load pattern implies that load management is

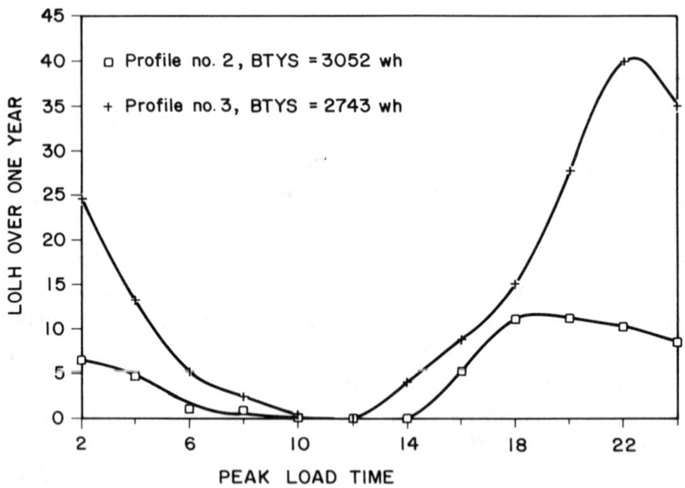

Figure 15.12 Variations in LOLH with t_p (PVS = 1000 W_p): □, load profiles in figure 15.2(*b*); BTYS = 3052 W h; +, load profile in figure 15.2(*c*), BTYS = 2743 W h.

important for stand-alone PV systems because, if some loads can be shifted from night to noon, the investment in array and battery can be reduced significantly.

15.5.2 Loss of load frequency and battery operations

Although all PVS–BTYS combinations on an iso-LOLH curve cause the same expected LOLH value, the corresponding LOLFs (the LOLEs) in a given period, say 1 year, are quite different. For systems with a large PVS and a small BTYS, the batteries will have high charge and discharge rates so that the average loss-of-load duration is short but a loss of load occurs frequently (high LOLE). On the other hand, a large BTYS and a small PVS results in low charge and discharge rates so that loss of load is of longer duration. Thus, for the same LOLH (the sum of all durations of LOLEs in a year), systems with a small PVS and a large BTYS have a low LOLF of longer duration, and the opposite designs have a high LOLF of shorter duration. The relations of LOLF against PVS for five LOLH values are shown in figure 15.13. The battery capacities matching the PVS to cause the given LOLH values are not shown in the graph but they can be obtained from figure 15.6.

In power system engineering, LOLF is an important reliability index because consumers are more sensitive to LOLF than to LOLH. However, it is not an

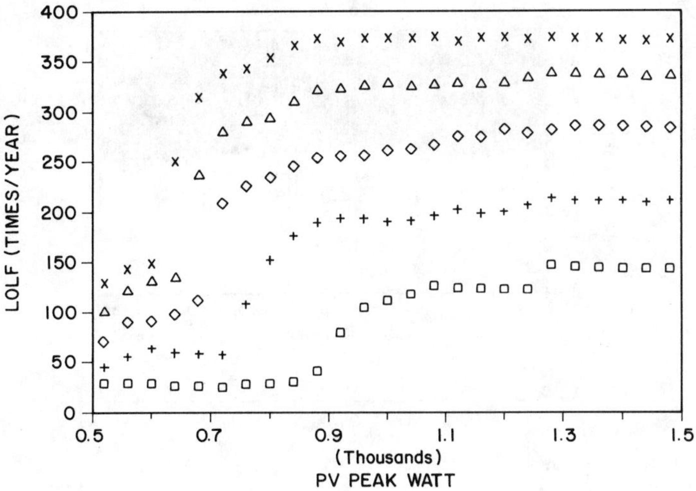

Figure 15.13 Variations in LOLF with array sizes at five LOLH values: □, LOLH = 100 h; +, LOLH = 200 h; ◇ LOLH = 300 h; △, LOLH = 400 h; ×, LOLH = 500 h.

important index here because, if a diesel generator is added to a stand-alone PV system (to charge batteries), LOLF is minimized.

The choice of battery and its cyclic operation is dependent on the PVS–BTYS combination. Systems with a small storage battery (BTYS) will experience a daily charge–discharge cycle and require a deep-cycle type of battery with long cycle life, high depth-of-discharge and high charge and discharge rates. Systems with a large BTYS will be able to discharge a lower percentage of its capacity per evening, thereby maintaining sufficient charge to supply the load during low-irradiation periods and to become fully charged during periods of high irradiation (summer months). These systems are usually designed around the low-irradiation period and require batteries of the shallow-cycle type of operation, and low self-discharge rate. Figure 15.14 shows the variation of the battery's SOC for three types of battery charge–discharge cycle: deep-cycle, shallow-cycle and medium-cycle.

Figures 15.14(*a*) and 15.14(*b*) show that a stand-alone PV power system designed for 100% reliability does not utilize the battery capacity in an optimum manner. Obviously, these two figures show that the storage batteries were of sufficient size for a few days of deep discharges to about 30% of capacity while for a higher proportion of the time the battery's SOC floats above 50% of capacity. In such cases, a back-up system might be investigated to find the possibility of offsetting battery cost to serve these few days of heavy demand on the battery by a fossil fuel system.

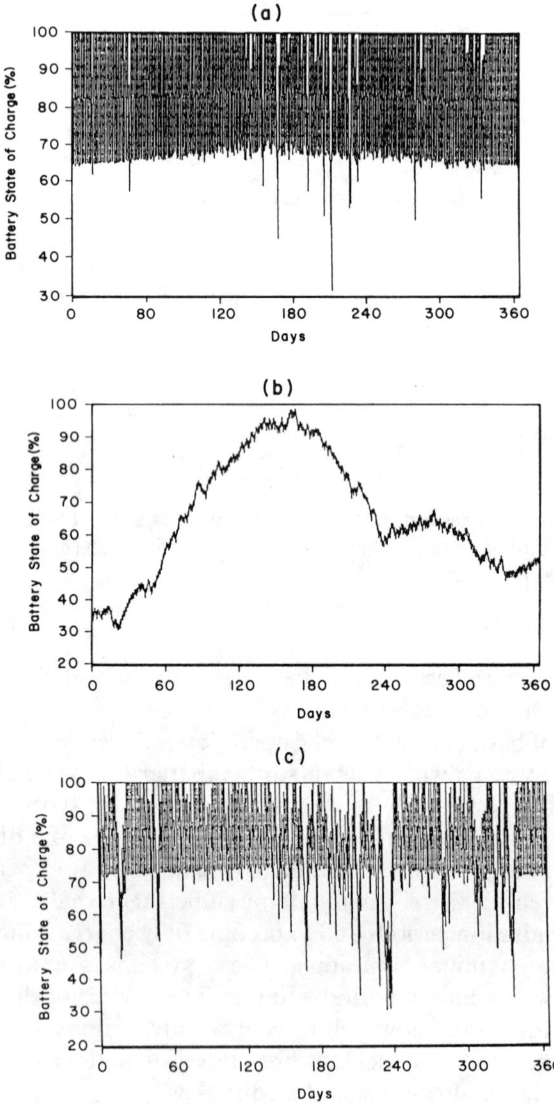

Figure 15.14 Variation in the battery's SOC over 1 year (LOLH = 0 h; load profile in figure 15.2(*a*)): (*a*) deep-cycle battery; (*b*) shallow-cycle battery; (*c*) medium-cycle battery.

15.6 SUMMARY

A dynamic simulation approach to stand-alone PV power system sizing has been developed which gives the optimum combination of array size and battery capacity to meet the reliability requirement and the possibility to incorporate a diesel back-up system, transforming the stand-alone PV power system into a PV–diesel hybrid power system. The conclusions of the study are summarized as follows.

(i) Load profile influences the size of a PV power system significantly. Systems with a load profile similar to the irradiance profile are preferable. Thus load management is important for PV systems to save initial investment and to ensure the expected reliability.
(ii) Iso-LOLH curves indicate the substitutional relationship between array and battery in a PV power system. Different PVS–BTYS combinations on an iso-LOLH curve have different LOLFS, affecting the choice of battery type and its cyclic operation (shallow cycle or deep cycle).
(iii) A stand-alone PV power system exhibits minimum system cost at a specified reliability when the marginal rate of substitution between array and battery is equal to the ratio P_w/P_B of the array price to the battery price.
(iv) If properly designed, a PV–diesel hybrid power system with the diesel system operating as the back-up unit can often be more economical than either constituent alone. From the viewpoint of future growth in demand, the PV–diesel hybrid power system has a clear advantage.

NOMENCLATURE

BTYS	Battery size (W h)
C	Cost of PV system (US \$)
D_m	Amplitude of cosine function load profile (W)
D_s	Battery self-discharge rate (W h h^{-1})
D_0	Daily average demand (W)
$D(t)$	Demand function (W)
E	Daily energy consumption (kW h)
$f(x)$	Curve-fitting function for iso-LOLH curve
g	Analytical function of iso-LOLH curve
G	Incident solar irradiance (W m^{-2})
I	Module current (A)
I_B	Battery current (A)
I_{mp}	Module current at maximum power (A)
I_{sc}	Module short-circuit current (A)
K	Ratio of peak demand to the lowest demand
K_1	Battery charge–discharge efficiency
K_2	Battery charge–discharge resistance (Ω)
P_B	Battery price (US \$ W h^{-1})
P_w	Array price (US \$ W$_p^{-1}$)

PVS	Array size (W_p)
R_{ch}	Battery charge internal resistance (Ω)
R_{dch}	Battery discharge internal resistance (Ω)
R_o	Reliability (h)
SOC_{max}	Maximum SOC of battery (W h)
$SOC(t)$	Battery's SOC at time t (W h)
t_p	Peak time (h)
$T_a(t)$	Ambient temperature at time t (°C)
T_c	Array temperature (°C)
T_{amax}	Maximum ambient temperature (°C)
T_{amin}	Minimum ambient temperature (°C)
v_W	Wind speed (m s^{-1})
V	Module voltage (V)
V_B	Battery system voltage (V)
V_{ch}	Battery charge voltage (V)
V_{dch}	Battery discharge voltage (V)
V_{mp}	Module voltage at maximum power (V)
V_{oc}	Module open-circuit voltage
$Z(t)$	Function of LOLE
α	Temperature coefficient of I_{sc} (A °C^{-1})
β	Temperature coefficient of V_{oc} (V °C)
θ_G	Experimental constant related to irradiance
θ_{T_a}	Experimental constant related to ambient temperature
θ_{v_w}	Experimental constant related to wind speed

REFERENCES

Ang T G 1984 *AIT Thesis No ET-84-10* Asian Institute of Technology, Bangkok

Chungpaibulpatana S 1981 *AIT Thesis No ET-81-9* Asian Institute of Technology, Bangkok

EPRI 1981 *Reliability Index for Power Systems EL-1773 Research Project 1353-1* (Palo Alto, CA: Electrical Power Research Institute)

Exell R H B 1981 *Solar Energy* **26** 161–8

Gerald C F 1978 *Applied Numerical Analysis* 2nd edn (Reading, MA: Addison-Wesley)

Hoover E R 1980 *Proc. 14th Photovoltaic Specialists' Conf. San Diego, CA, 7–10 January 1980* (New York: IEEE) pp 1258–61

Lasnier F and Ang T G 1986 *Computer Package for Photovoltaic System Sizing, Design and Financial Evaluation* (Bangkok: Asian Institute of Technology)

Rao S S 1984 *Optimization Theory and Applications* (New York: Wiley Eastern)

Rauschenbach H S 1980 *Solar Cell Array Design Handbook* (New York: Van Nostrand–Reinhold)

Servant J M 1986 *Proc. 9th Bienn. Congr. of the Internal Solar Energy Society* (Oxford: Pergamon)

BIBLIOGRAPHY

Wang Y J 1987 *The Sizing of Stand-Alone Photovoltaic Systems: AIT Research Study* (Bangkok: Asian Institute of Technology)

16 Prediction of Photovoltaic System Performance Using Cumulative Frequency Curves of Irradiance

16.1 INTRODUCTION

The prediction of the system performance of photovoltaic (PV) systems over longer time scales can be made possible by this type of simulation method which involves the simulation of the system over certain representative or 'specially created' days. The performance of the PV system at any given instant or during any time interval is dependent on the state of charge (SOC) of the storage media (say the battery), and on the electrical energy flow from the modules during that time interval. (The SOC of the battery is of course a reflection of the performance of the system prior to the time interval under consideration.) Owing to day-to-day fluctuations of the irradiance, the SOC of the battery at any given time of the day will also be fluctuating, even if the electrical energy consumption (of the load) is constant. However, after remaining at this transient state for a few days, the SOC will fluctuate about a daily mean value owing to the day-to-day fluctuations of the solar irradiance. This also means that the mean SOC will not be a constant but will exhibit a repetitive diurnal cyclic variation.

16.2 DAILY AND SEASONAL VARIATIONS IN THE BATTERY'S STATE OF CHARGE

Except in case of unrealistically large sizes, the batteries usually exhibit a daily cyclic variation in SOC, with the same daily energy consumption. However, when the battery size is very large, it is observed that the variation in SOC is a 'seasonal variation' rather than a daily cyclic variation. These cases are illustrated in figures 16.1, 16.2 and 16.3. These figures show the variation pattern of SOC of two batteries of different sizes: one in the range of a frequently used size for PV systems (2800 W h), and the other an extremely

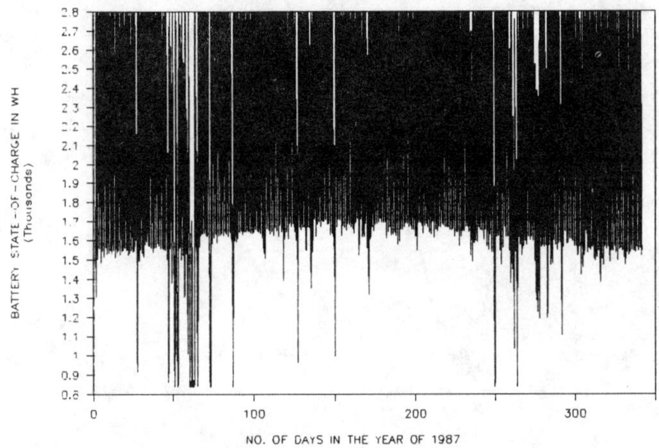

Figure 16.1 Daily cycle of SOC for a 2800 W h battery.

large size (50 000 W h). For the estimation of this battery SOC, the sizing program and battery model by Lasnier and Ang (1988) is used. It can be observed from figure 16.1 that the battery of 2800 W h capacity shows a daily cyclic variation in SOC, and this cycle repeats throughout the year. Figure 16.2 shows this daily cyclic variation in a clearer manner. In figure 16.3, the SOC variation over one year (1987) of the battery of 50 000 W h capacity is shown. It is evident from this figure that the SOC of very large batteries exhibits a seasonal variation over the year, but in actual photovoltaic systems the batteries are not of much larger size. Therefore a daily cyclic variation in SOC can be expected in batteries. Thus, to analyse the performance of the system for a period, it is sufficient if the

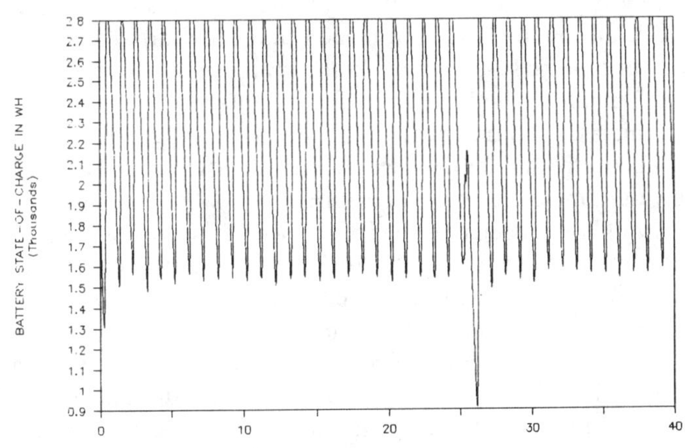

Figure 16.2 An expanded view of figure 16.1.

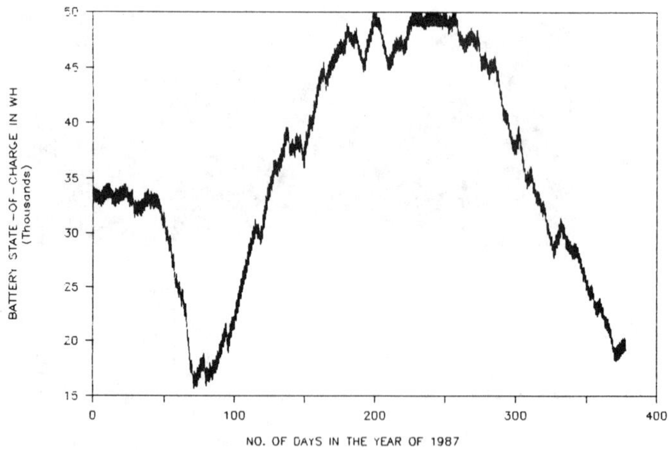

Figure 16.3 Seasonal cycle of SOC for a 50 000 W h battery.

system can be analysed for a representative day of that period provided that the representative day reflects all the statistical information (irradiance, temperature, etc) of that period.

The repetitive-day methods, one of which is discussed here, are used to determine this mean diurnal cycle, but the time scale over which this mean cyclic level is determined can be a month, a year, or even a number of years. The 'typical meteorological day (TMD)' method is a one-repetitive-day method. The prediction of long-term performance of solar energy systems may be attempted using this TMD method. The solar cells experience solar irradiance fluctuation patterns which are time dependent. All the necessary statistical information regarding this fluctuation must be taken into account while selecting the 'mean day' over which the system is to be simulated.

Most locations exhibit a variation such that the diurnal solar irradiance is symmetrical about the solar noon. So this makes a fairly realistic condition. However, the mean diurnal curve does not contain all the necessary information concerning the irradiance fluctuations, especially in terms of long-term irradiance fluctuations. The correct statistical distribution of all the hourly irradiance values incident on the aperture of the PV module over the specified time scale should be reflected in the irradiance curve constructed to represent the typical meteorological day. At the same time, the correct mean daily irradiance averaged over the time scale should not be violated.

The above two conditions can be simultaneously satisfied by probability and cumulative frequency curves (CFCs) of solar irradiance. For an idealized climate of identical days which are symmetric about solar noon, the time dependence of irradiance for half the daytime period is identical with the relation between irradiance and the cumulative time of occurrence.

Therefore, if we could generate the CFC of hourly irradiance values of all the hours over the entire time period during which the solar cells are exposed to the sun, then we can take that curve as representing the diurnal variation in the solar irradiance from early morning to noon. Also, the mirror image of the CFC about the solar noon (the maximum value of irradiance) can be said to represent the corresponding variation during the afternoon hours.

It is true that no actual day has the same diurnal irradiance pattern which the TMD exhibits. (The TMD is a fictitious day, after all.) However, the TMD is created by properly including the necessary information regarding the long-term irradiance fluctuation.

16.3 THE SIGNIFICANCE OF CUMULATIVE FREQUENCY

The absolute frequency (or frequency) is defined as the number of times that a value x occurs in a sample. The relative frequency is defined as

$$\text{relative frequency of } x = \frac{\text{frequency of the value } x}{\text{size of the sample}}.$$

If, for a certain value of x, we sum all the frequencies corresponding to the sample values which are smaller than or equal to x, we obtain the cumulative frequency corresponding to that x:

$$\text{cumulative relative frequency} = \frac{\text{cumulative frequency}}{\text{size of the sample}}.$$

Therefore the value of this cumulative relative frequency always lies between 0 and 1 including both values. The sum of all the relative frequencies in a sample is equal to 1.

The sum of all the relative frequencies of all the values that are smaller than or equal to x is called the cumulative frequency function. Therefore, we can say that the cumulative frequency represents the sum of the frequencies from the lowest to the highest class.

Since the relative frequency of a class reflects the proportion of the observations contained in that class, the cumulative relative frequency of a class is the proportion of the observations with values less than or equal to the upper limit of the class (or equivalently less than the lower limit of the next class). The cumulative distribution function represents the probability that the random variable x assumes a value less than or equal to some specified value.

The cumulative frequency distribution is obtained by plotting on the vertical axis the cumulative relative frequency of a class against the lower limit of the next class on the horizontal axis and joining all consecutive points by line segments.

The primary use of the cumulative distribution is to determine what are commonly known as quantile values. With reference to a cumulative relative frequency distribution, a 'quantile' is defined as the value below which a certain proportion of the values will fall. The quantile value is read off the horizontal axis opposite the desired proportion on the vertical axis. (The most common quantile is percentile.)

16.4 OBTAINING CUMULATIVE FREQUENCY CURVES FOR IRRADIANCE VALUES

Let us take a time period of 1 year (for example). Then there will be 13×365 values of hourly irradiances H for that year. (The number of 'solar hours' per day is taken as 13, considering the irradiance from morning (0600 hours) to evening (1800 hours).) These 13×365 values are then put in ascending order, irrespective of the chronological order in which they were recorded. Therefore the clock times at which they were recorded are totally neglected. Now these values are divided by their mean value H', and hence 13×365 values of H/H' are obtained. These values are taken as the y axis values.

16.5 FRACTIONAL TIME *F*

Here, the fractional time corresponding to a value H/H' is defined as the number of hours for which the irradiance values are less than H divided by the total number of hours. These fractional time values are plotted on the x axis for each corresponding H/H' on the y axis. It is obvious that the maximum value of fractional time F is 1 which corresponds to the highest value of H/H'. Also, it should be noted that, for each successive reading of H/H', the value of F increases by $1/(13 \times 365)$.

It is interesting to note, that in 1 year, there will be a number of hourly irradiance values which are numerically equal. When arranged in ascending order, all these numerically equal values are put together. It should be noted that each such value will correspond to a different fractional time, although the H/H' values are numerically equal. (Obviously, this will create a horizontal segment in the CFC.) Therefore, all the irradiance data obtained are contained in the CFC, although the order is altered, and hence the total solar energy obtained throughout the year is the same as the actual case.

The next step is to convert the fractional time on the x axis into clock time representation. The maximum value of the fractional time, i.e. 1, is marked as 1200 hours (noon) and the minimum value of fractional time as 0600 hours morning. That is, the full range on the x axis is equally divided from 0600 to 1200 hours. The corresponding irradiance values for these hours are obtained by multiplying by H' the corresponding H/H' value of the clock hour.

Therefore the maximum value of irradiance will be at 1200 hours. These are the irradiance values of the TMD in morning hours (figure 16.4).

It should be noted that in the case of CFCs any rare and/or highly unusual fluctuation in irradiance occurring in any day of the year will not have any significant effect on the tendency of the irradiance curve for a TMD. At the same time the cumulative irradiance energy for the year is properly weighted. This is the main advantage of CFCs over those obtained by simply averaging the hourly irradiances of the year.

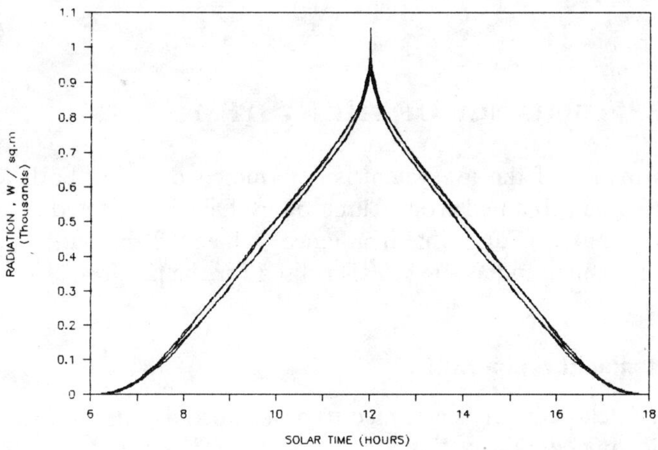

Figure 16.4 Irradiance profile of a TMD day based on CFCs for Bangkok for the years 1984–7.

The mirror image of the curves for morning hours about the axis passing through 1200 hours is taken and those values are taken as the irradiance values for afternoon hours, i.e. 0600 and 1800 hours will have the same TMD irradiances and so on.

16.6 DESCRIPTION OF PHOTOVOLTAIC SYSTEM

The sizes and ratings given below are taken for the purpose of simulating the system performance only.

16.6.1 Modules

For a module of type PW-P400, the rating is 40 W and 12 V. The number of modules is 45. Therefore the total power rating is $45 \times 40 = 1800$ W. The total module area is 45×0.3781 m^2.

16.6.2 Battery

The watt–hour capacity of the battery set is taken as 2600 W h. This is also taken as the allowable maximum SOC of the battery. The allowable minimum SOC is taken as 30% of the maximum value.

16.6.3 Load consumption

A constant load profile of 3 kW h day^{-1} is taken for the system simulation.

16.7 PERFORMANCE OF THE SYSTEM

The performance of the PV system is determined by using both the actual irradiance value (obtained from Climatology Division, Meteorological Department, Bangkok), and the irradiance values of the TMD. The sizing program of Lasnier and Wang (1987) is used for the purpose.

16.7.1 Ambient temperature

The ambient temperature is assumed to be a sinusoidal distribution, as given by the following equation:

$$T_a(t) = \tfrac{1}{2}[(T_{amax} + T_{amin}) + (T_{amax} - T_{amin}) \sin(\omega t)]$$

where T_{amax} and T_{amin} are the maximum and minimum ambient temperatures of the day.

The seasonal variations in T_{amax} and T_{amin} are considered by dividing a year into eight periods during which T_{amax} and T_{amin} are taken to be constant. These values of T_{amax} and T_{amin} in each period are obtained from the mean maximum and minimum ambient temperature for each month from 1976 to 1981 in Thailand (Wang 1987).

Although the temperature profile changes only eight times in 1 year, the first part of the simulation program should compute the module current for all 365 days of the year using the actual hourly irradiance values.

However, when the TMD irradiance values are used, it is sufficient to calculate the module current for only eight representative days of the year. This is simply because the TMD irradiance values, which are obtained from the CFCs of hourly irradiances for the year, have already incorporated the long-term irradiance fluctuation of the year. It is interesting that the time taken to compute the module current is reduced by a factor of 33 compared with the time taken for the computation using actual irradiance values.

16.7.2 Performance parameters and indicators

16.7.2.1 Loss of load hours

When the SOC of the battery falls below its minimum allowable value, the regulator should isolate the load from the system, so that the whole energy from the modules is used to charge the battery until it exceeds its minimum SOC level. The number of hours during which the load is disconnected from the system is the loss of load hours (LOLH). This is a measure of system (un)reliability.

16.7.2.2 Expected energy not served

The expected energy not served (EENS) is the energy that was expected by, but not supplied to, the load during the LOLH. This indicates that the load must look for a back-up source for this energy.

16.7.2.3 Loss of module hours

On sunny days, or days in which the energy demand is relatively low, the battery's SOC may exceed its maximum allowable value. At this stage, the battery should be isolated from the system while the modules continue to supply the load. The excess energy, which would otherwise have been absorbed by the battery, can be 'assumed' to be dissipated in a resistor. The duration of this 'dissipation' is called the loss of module hours (LOMH). This is one measure of the effectiveness with which the system is utilized.

16.7.2.4 Waste

The waste is the electrical energy converted from solar energy which is not used either by the load or by the battery. This event takes place during the LOMH. The name 'waste' can be justified by the fact that it could have been avoided or minimized by proper battery sizing and/or proper load profile.

16.7.2.5 System performance

The system performance is computed from the following equations:

$$\text{solar energy input} = \text{irradiation (W h m}^{-2}) \times \text{array area}$$

$$\text{energy requirement} = \text{daily requirement} \times \text{number of days}$$

$$\text{energy consumed by the load} = \text{energy requirement} - \text{EENS}$$

$$\text{electrical energy to the battery} = \frac{\text{electrical energy from battery}}{\text{charging efficiency} \times \text{discharging efficiency}}$$

electrical energy output from modules

= electrical energy input to battery + waste + energy direct to load

cumulative conversion efficiency of the modules

$$= \frac{\text{electrical energy output from modules}}{\text{solar energy input}}.$$

16.8 COMPARISON AND DISCUSSION OF PERFORMANCE

The conversion efficiency, i.e. the efficiency with which the solar energy is converted into electrical energy by solar cells, is computed from the simulation results obtained from both methods: the TMD and actual irradiance methods. Figures 16.5 and 16.6 show the results. In figure 16.6, the efficiency is expressed considering the cumulative performance of the module for over 48 months. A maximum difference of only 0.05% in efficiency is observed when comparing both irradiance methods. The efficiency obtained by simulation using actual irradiance is always in the neighbourhood of 9%, whereas that obtained using TMD irradiance is in the neighbourhood of 8.95%.

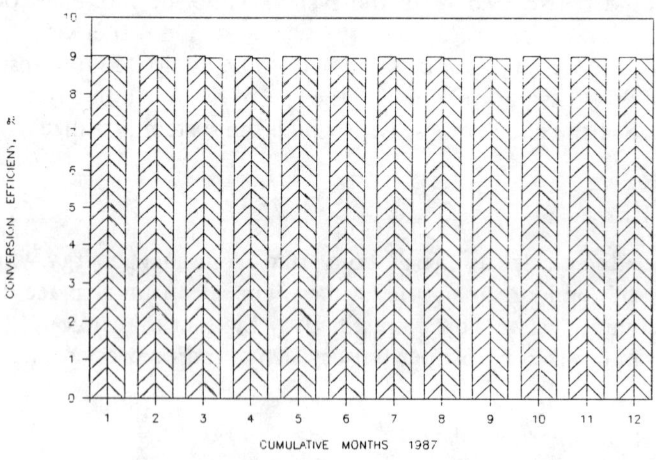

Figure 16.5 Conversion efficiency of modules from simulation using actual daily irradiance values (▨) and using TMD irradiance values (▧).

The LOLH is compared in figure 16.7. The 'solar-powered hours' and the LOLH are compared in the bar chart. The LOLH is dependent on the irradiance, module size, battery size and the load profile. When simulating using TMD irradiance, the same LOLH is recorded and cumulatively added throughout most of the period, but, when simulating using actual irradiance, any

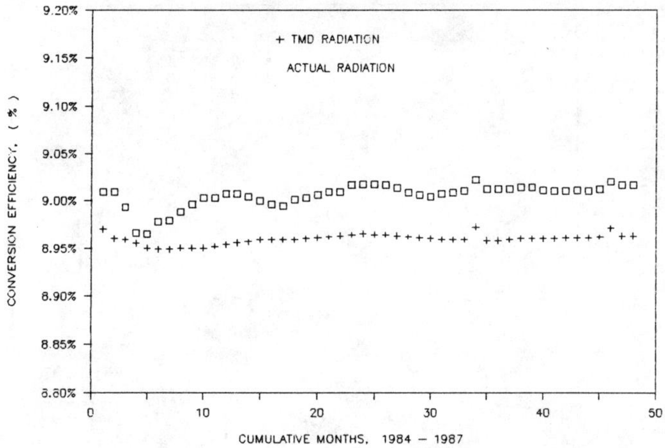

Figure 16.6 Variation over 48 months (1984–7) of conversion efficiency of modules from simulation using actual daily irradiance values (□) and using TMD irradiance values (+).

unusually high or low irradiance in a day will lead to a low or high LOLH. (Similarly an unusually high irradiance value may record a high LOMH for that 'special day'.)

In figure 16.8, the cumulative electrical energy output from the modules is compared over a period of 48 months. As the output is taken cumulatively, the difference occurring in the previous months will also be added (positively or negatively) to the difference in the next month. (In this case it is always positive.) That is why the difference in the curves in figure 16.8 seems to be increasing with increasing number of months.

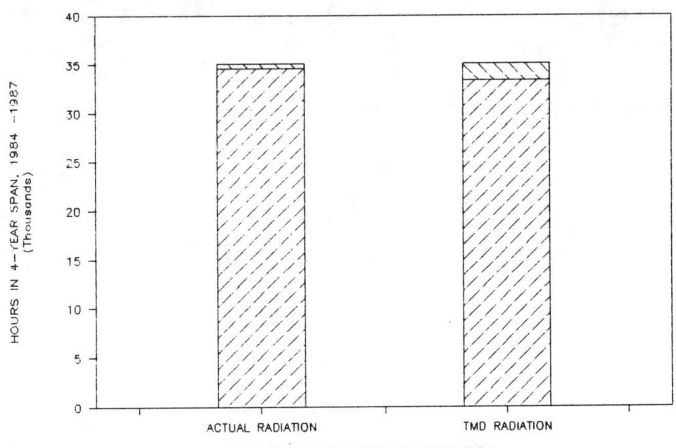

Figure 16.7 Simulation of LOLH hours from 1984 to 1987: ▨, solar-powered hours; ▧ LOLH.

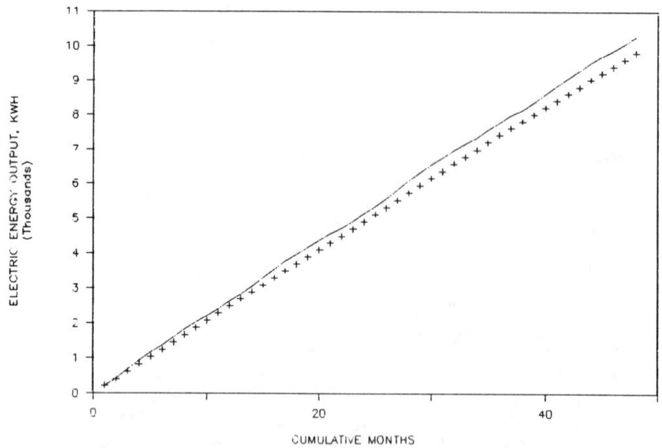

Figure 16.8 Simulation of accumulative module output over 48 months (1984–7): ——, using actual radiation; +, using TMD radiation.

REFERENCES

Lasnier F and Ang T G 1988 *Solar Photovoltaic Handbook* (Bangkok: Asian Institute of Technology)

Wang Y J 1987 The Sizing of Stand-alone Photovoltaic Systems (Bangkok: Asian Institute of Technology)

BIBLIOGRAPHY

Canavos G C 1984 *Applied Probability and Statistical Methods* (Little Brown)

Harnett D L 1982 *Statistical Methods* (Reading, MA: Addison-Wesley)

Kreyszig E 1970 *Introductory Mathematical Statistics* (New York: Wiley)

Reddy T A 1987 *The Design and Sizing of Active Solar Thermal Systems* (Oxford: Clarendon)

Part VI

Economic Analysis

17 Financial Evaluation of Renewable Energy Projects

17.1 INTRODUCTION

This chapter presents the different approaches for examining and evaluating the acceptability of investments in renewable energy projects. Procedures to carry out the different financial evaluations, their applications and limitations are presented. These procedures should be seen as part of a techno-economic decision-making process. This chapter alone will not suffice for a final evaluation since it does not encompass ecological and socio-economic effects.

The methods of calculation presented here are limited to those costs and benefits which are directly related to the renewable energy project concerned. They do not encompass the following.

(*a*) Possible external benefits.

- (i) Improvements in the physical quality of life.
- (ii) Improvements in the local economic structure.
- (iii) Reduction in employment opportunities.
- (iv) Improvement in the trade balance through the substitution of imported energy resources.
- (v) Benefits from technological training and technology transfer.
- (vi) Reduction in rural migration.
- (vii) Reduction in timber felling.
- (viii) More secure supply basis, etc.

(*b*) External costs.

- (i) Increase in environmental pollution.
- (ii) Loss of employment.
- (iii) Shortage of capital and increase in interest rates due to new investments on renewable energy projects.
- (iv) Importation and effects on currency reserves.

An investment project which is unprofitable for an individual business can be extremely worthwhile for the overall economy when considering the social benefits that it generates.

17.2 TECHNO-ECONOMIC SYSTEM OPTIMIZATION

If we regard economics as the quantitative study of the theory and practice of producing and distributing wealth and thus the basis for decision making in social affairs, then techno-economics attempts to provide the quantitative (usually monetary) basis for decision making in technical affairs. The essential point is that due attention is given to both technical and economic aspects of the problem.

A possible procedure for the technical and economic optimization of systems using renewable energy sources in developing countries is summarized in figure 17.1. It assumes that a problem (provision of energy) has been defined and a conventional energy system has been chosen.

Then the various activities in figure 17.1 can be classified into five major levels of decision making.

(i) Analysis of boundary conditions and the formulation of criteria for the preliminary selection of the alternative renewable energy system.
(ii) Formulation and technical optimization of an alternative renewable energy system.
(iii) Selection of the most favourable renewable energy system.
(iv) Economic optimization of the conventional and renewable energy system.
(v) Determination of the techno-economic optimal solution based on criteria of development policy and of environmental protection.

Techno-economic system optimization is an iterative process and takes place on various levels sometimes simultaneously as in figure 17.1. The financial evaluation methods presented in the following section will only address decision making from an individual's financial perspective.

The discussion will begin with an introduction of the different data required for the economic assessment of an investment project. They will include data reflecting the economic atmosphere where such investment is planned, and all associated expenditures and returns. This will be followed by a formal study of five approaches to financial analysis. For each approach, a quantitative formulation of the idea together with the decision-making criteria, applicability of the method and remarks on its limitations are presented. Finally, a short discussion on the importance of sensitivity analysis is reviewed.

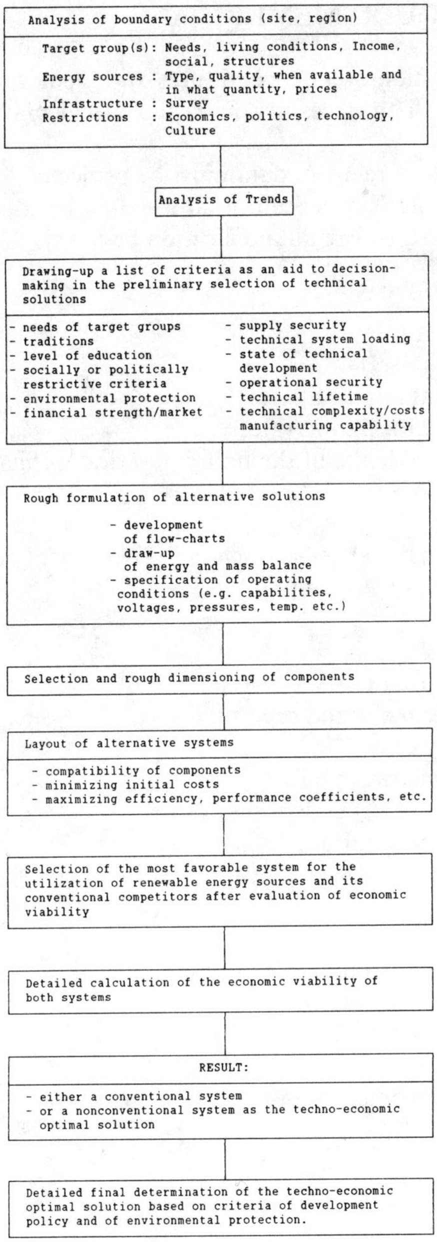

Figure 17.1 A procedure for the techno-economic optimization of systems using renewable energy sources in developing countries.

17.3 DATA REQUIREMENTS

A complete compilation of all expenditures and incomes connected with the planned investment is a necessary first step in calculating the viability of investment projects.

The following list of relevant data must be projected for every year of the investment project lifetime. Since not all the data listed may be needed, the list should be checked before data collection begins.

(*a*) Investment parameters.

(i) Interest rate.
(ii) Discount rate.
(iii) General inflation rate.
(iv) Energy escalation rate.
(v) Serviceable life of the unit per period of financial analysis.

(*b*) Equipment.

(vi) Investment expense (includes replacement costs).
(vii) Residual value of unit.

(*c*) Expenditures.

(viii) Manpower costs.
(ix) Maintenance and repair costs.
(x) Energy costs.
(xi) Administrative cost.
(xii) Taxes not dependent on profits.
(xiii) Other expenditures.
(xiv) Total operating costs (items (viii)–(xiii)).

(*d*) Income.

(xv) Revenues from energy sales.
(xvi) Savings on energy costs.
(xvii) Other incomes.
(xviii) Subsidies.
(ixx) Total incomes (items (xv)–(xviii)).

(*e*) Returns (item (ixx) minus item (xiv)).

(*f*) Depreciation.

17.3.1 Interest rate

Each type of loan—whether a mortgage, corporate bond, personal loan or whatever—carries a different price or interest rate. This interest rate, regardless of the type of loan involved, is a function of the supply of and the demand for money. Tight money supply reflects a high interest rate. Government policy is also an important determinant of the rate of interest.

Interest rate therefore is the cost of borrowing money and this must include a component that reflects the real cost of money and a component for inflation:

$$I=(1+i)(1+f)-1 \qquad (17.1)$$

where I is the market interest rate, i is the actual cost of money due to supply and demand and government policy, and f is the general inflation rate.

17.3.2 Discount rate

In the comparison of projects or the determination of the financial viability of a particular project, one must reduce the time stream of costs and benefits to a single number by computing the net present value (NPV) of a project. In the evaluation of a single project, the discount rate will affect whether the NPV is greater or less than zero. In a comparison of projects, the discount rate will affect their NPV ranking.

The correct discount rate is that rate which, when applied to future costs and benefits, yields their actual present social values. In other words, the proper rate is the rate at which society as a whole is willing to trade off the present for future costs and benefits. The choice of a proper rate is not over numerical values *per se*, although that is surely part of the issue; rather, it concerns the very concept of what the discount rate ought to measure.

In order to facilitate decision makers in the identification and investigation of various conceptual approaches to isolation of the true discount rate, a list of rate concepts are described.

17.3.2.1 Market interest rate

This is associated chiefly with corporate and government bonds and with the debt instruments of financial institutions such as commercial banks and savings and loan associations.

17.3.2.2 Marginal productivity of investment

This is the real rate of return that the economy's marginal investment projects yield. By 'real' is meant the market value of the outputs and net of all the inputs of the project, both appropriately discounted for time.

17.3.2.3 Corporate discount rate

This is the rate used by a corporation to evaluate potential investment projects. It includes both a risk premium and a 'mark-up' for corporate taxes.

17.3.2.4 Government borrowing rate

This is the rate at which the government borrows money from the private sector; thus, it is the 'cost' of government borrowing.

17.3.2.5 Own personal marginal discount rate

This is the rate at which an individual is willing to trade off his own present consumption for his own future consumption. Graphically, the own personal marginal discount rate is the slope of the consumer's intertemporal indifference curve evaluated at the current consumption distribution.

17.3.2.6 Pigouvian discount rate

A C Pigou, a noted British welfare economist, observed that individuals have faulty 'telescopic' vision concerning the future and are inclined not to make sufficient provision for it. Individuals weigh themselves too heavily and future generations too lightly. Because future generations are unrepresented now, their interests are not well regarded. Pigou suggested that the government must view itself as the trustee of the future. Thus, in evaluating public investment projects, the government should impose a lower discount rate.

A low discount rate (where 'low' means lower than market rates) would enable projects with benefits occurring farther in the future to prove more acceptable. A low discount rate extols the virtues of projects with pay-offs occurring in the more distant future.

17.3.2.7 Social opportunity cost of capital

The true opportunity costs of a government project must be considered in a project financial evaluation. By opportunity cost is meant what members of society give up now and in the future in order to accomplish the project. This premise is the basis for the arguments espousing a high discount rate. Following this opportunity cost argument, it appears that the bench mark against which to measure the returns of the public project is the returns that can be achieved in the private sector.

Following this approach, two methods can be employed. The discount rate is set higher than the market rate to account for the opportunity cost while maintaining the time costs of the project. A second method is to set the discount rate at market rate and to add the 'social opportunity cost of capital' (SOCC) to the cost of the project. Qualitatively, SOCC refers to the welfare loss borne by society when resources are transferred from the private to the public sector.

17.3.2.8 Remarks

The foregoing issues discussed are in the realm of the welfare–theoretical foundations of cost–benefit. One of three situations usually prevails regarding the choice of the discount rate. First, the rate to be employed is dictated exogenously. Second, there is no externally imposed discount rate. Here, familiarity with the underlying issues and ramifications, discussed above, comes to the fore. Third, when there is some uncertainty about the rate to be employed, a computation of the critical rate is sometimes helpful. The critical discount rate is that rate at which the NPV of the project being considered changes sign. In general, sensitivity analysis usually proves helpful. It determines how critical the discount rate really is in the project at hand.

17.3.3 Inflation rate

Inflation is defined as an increase in the prices paid for goods and services. The increase in prices can result from either cost inflation or price inflation or combinations of the two.

Cost inflation is the result of real cost increase to produce goods and services; that is, more inputs in terms of labour and capital are needed to produce a given amount of goods and services. Depletion of natural resources, natural disasters and catastrophes, social and political disruption and anything that has a negative effect on productivity tends to cause real cost increases and cost inflation.

Prices of goods and services may rise or fall even when their real costs do not change. Such a behaviour in the economy is called price inflation. One theory of price inflation holds that inflation is due mainly to increases in the amount of money held in the economy. If the supply of money increases, prices are forced to increase also. A second theory of price inflation holds that inflation is due largely to the speed and spending habits of the people. Other theories are that the price inflation responds to forces, which may be the inherent structure in the economy that resists wage and price decreases but readily allows increases.

Cost inflation and price inflation may occur at the same time, causing the money to have different values at different times.

The price indexes (PIS) are a reasonable representation of the value of the money. Governments tabulate two major categories of PIS on a routine basis. These are the Wholesale Price Indexes (WPIS) and the Consumer Price Indexes (CPIS). Each category covers specific products and has its areas of usefulness.

Normally, inflation rate is expressed as an annual percentage unless another time period is specified. The annual inflation rate is determined by

$$f(t) = \frac{\mathrm{PI}(t)}{\mathrm{PI}(t_0)} - 1 \qquad (17.2)$$

where $\mathrm{PI}(t)$ is the PI at year t and t_0 is the reference year.

17.3.4 Energy escalation rate

In the analysis of investment projects which are heavily dependent on energy it is recommended to take separate account of the development in energy prices which may not behave in sequence with the general inflation rate. Price development of various energy sources in different countries and installation sites will vary. The determination therefore of the annual energy escalation rate should be based on the PI of the particular energy source being investigated. The energy escalation rate can also be calculated from the expected real increase in the cost of energy and the general inflation rate as follows:

$$e(t)=[1+e^*(t)][1+f(t)]-1 \qquad (17.3)$$

where $e(t)$ is the annual energy escalation rate, $e^*(t)$ is the real annual rate of increase in the cost of energy and $f(t)$ is the annual inflation rate.

17.3.5 Service life of the installation

The service life of an installation is the number of years beyond which the system becomes more expensive to maintain than a new acquisition or when the efficiency of the system deteriorates such that the output product is unacceptable in quality or in quantity.

Dependable empirical values for conventional systems are available; however, it must be recognized that there is no such experience for most technologies for the utilization of renewable energy sources.

There are several different ways to estimate the useful service life of capital assets. The most sophisticated methods involve actuarial analysis; less sophisticated, but extremely useful and far more convenient, are the Internal Revenue Service (IRS) guidelines.

Use of IRS guidelines is not mandatory; they are intended solely as a guide to what might be considered reasonably normal periods of useful life. In practice, the service lives used as the basis for depreciation accounting should be determined by the asset's particular operating conditions and experience, tempered by judgment concerning the probability of technological improvements and economic changes.

Actuarial approaches to estimating service life involve construction of frequency curves, survivor curves and probable-life curves. These curves for physical properties are similar both in derivation and use to the mortality curves prepared by insurance companies to reflect human life expectancy. The survivor curve for a physical property is developed from historical data on a similar property, just as the mortality curve is developed by actuarial analysis of vital statistics on humans.

By studying the retirement history of a particular type of property, a frequency distribution of property retirement can be obtained. From this frequency distribution a cumulative frequency distribution curve—or survivor curve—can also be plotted as a probable-life curve, indicating the potential future life of an asset of any given age. Figure 17.2 shows a generalized survivor curve, with the corresponding frequency distribution and probable-life curves. The frequency curve shows the percentage of the total number of units retired during each time interval. The survivor curve shows the percentage of the total number of units of a given age still in service.

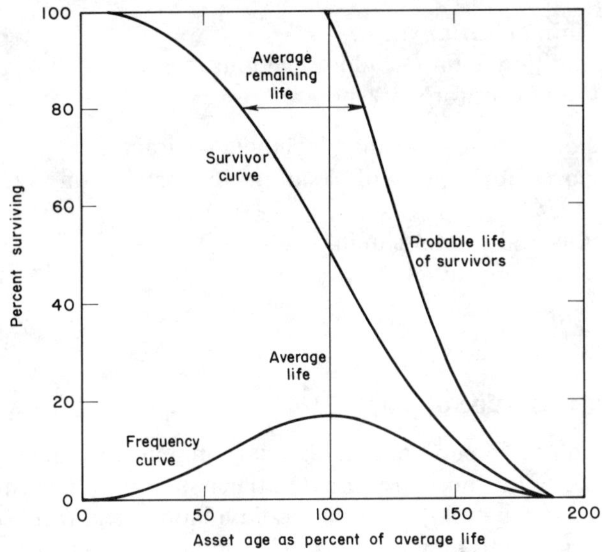

Figure 17.2 Survivor and probable-life curve.

The probable-life curve gives the total life expectancy of an asset of any given age. The horizontal distance between survivor and probable-life curves represents the remaining life of an asset of any age. The probable future life of an asset can be statistically predicted from its present age, thus providing a useful guideline in deciding at what age it should be retired or replaced, whenever the present value of its future services is less than its current replacement costs.

17.3.6 Investment costs

Various investment costs must be indexed according to their time of accrual in such a way that it is possible to order the investment expenditures (and any additional or replacement investments needed during the service life) clearly, according to a time schedule as is required for proper calculation of the financial acceptability.

The various cost groups which make up the total investment costs are as follows.

(i) Planning.
(ii) Land acquisition or leasing.
(iii) Civil works.
(iv) Plant building and structures.
(v) Connection to water, electricity, gas, sewage and traffic networks.
(vi) Machinery (ex-works).

  (*a*) Main unit equipment.
  (*b*) Additional unit A, additional unit B, etc.
  (*c*) Equipment for workshop.

(vii) Transport to port of loading including insurance.
(viii) Transport from port of loading to installation site including insurance.
(ix) Assembly and commissioning.
(x) Customs, taxes, duties, fees.
(xi) Other costs.

17.3.7 Residual value of plant

The residual value of a plant or of individual components is determined from the possibilities of its alternative use. On the assumption that equipment or parts of a plant can be sold, the expected liquidation yield from the sale at a fixed point in time is usually taken as the residual value. However, if a later sale or any further use of the parts appears improbable, then a residual value of zero would have to be employed.

There may be a high residual value from land acquisition, civil works and the erection of buildings if the site has favourable communications and varied possibilities for further use. Dismantling costs and costs for making the sale must be deducted.

17.3.8 Major operating costs

17.3.8.1 Manpower costs

When estimating trends in personnel costs over subsequent years, particularly in labour-intensive projects of long duration, it is necessary to consider carefully any increases in these costs on the basis of the anticipated general inflation rate. If opportunity costs of self-employment reach a significant level, they should also be accounted for by an appropriate sum in the financial evaluation.

17.3.8.2 Maintenance and repair costs

It is very difficult to estimate the expenses of maintaining and servicing a plant since the need for repair on a fixed unit cannot be predicted. It should also be examined whether and to what extent repair and maintenance work can be carried out by in-company staff or by the user himself.

17.3.8.3 Energy and related costs

In utilizing solar, wind and hydroenergy as well as geothermal energy, it can be assumed that the energy source will be directly available in the required quantity and quality, free of charge, after the installation of the necessary plant at any suitable selected site. On the other hand, when using biomass or conventional energy sources, it must be noted that costs will arise for the procurement, transport, processing and storage of the relevant energy sources, for feeding the plant and for discharging and discarding of residues.

Consideration should also be given to the fact that the biomass which was previously available free of charge might assume a market value owing to its potential financial value. Moreover, opportunity costs should be calculated for biomass that can be directed to alternative use such as livestock fodder.

17.3.9 Subsidies

In view of the precarious foreign trade balance of the majority of developing countries, it is likely that investments for energy supply to rural areas by utilization of renewable energy may receive subsidies from the government.

17.3.10 Depreciation

The various definitions of depreciation can be placed in one of four categories.

(i) Impaired serviceableness either physically or functionally of an asset.
(ii) Decrease in value of an asset due to time.
(iii) Appraisal concept referring to the difference between the present value of the old property and the present value of a hypothetical new (but similar) property.
(iv) Amortized cost concept in which the asset's original cost is amortized, or written off, over its useful service life.

The process of depreciation results in a consumption of the values in depreciable capital investments. Depreciation technique has the following principal purposes.

(*a*) To provide for the proper charging of depreciation to the operating expenses of an enterprise, so that profit portrayals will reflect capital consumption costs.

(*b*) To provide for recovery of capital consumption costs.

(*c*) To provide for proper recording of depreciable assets in the accounts of the company.

In determining investment feasibility, the amortized cost concept is used to meet the first two purposes mentioned above.

Once the economic life of an asset has been established, by estimate, analysis or IRS guidelines, there are four different commonly accepted approaches to calculation of annual depreciation charges.

(i) The straight-line method.
(ii) The sun-of-years digits method.
(iii) The declining-balance method.
(iv) The sinking-fund method.

17.3.10.1 The straight-line method

The straight-line method assumes a constant amount of depreciation for each year of the asset's life and is found by dividing the original cost (less anticipated residual value, if any) by the estimated service life. It offers the advantage of simplicity—an important virtue in preliminary financial studies—and may lead to understating investment profitability when income tax considerations are brought in. The annual depreciation (AD) is given by

$$\mathrm{AD}=\frac{\mathrm{CA}-\mathrm{RV}}{N} \tag{17.4}$$

where CA is the cost of the asset, RV is the present value of the residual value of the asset and N is the service life of the asset.

17.3.10.2 The sum-of-years digits method

This method offers high depreciation charges, along with corresponding low income tax liabilities during the asset's early years. This method does not require a residual value of the asset. The AD is given by

$$\mathrm{AD}=\mathrm{CA}\left(\frac{(N+1)-n}{N(N+1)/2}\right) \qquad n=1,2,\ldots,N \tag{17.5}$$

where CA is the cost of the asset, N is the service life of the asset and n is the year of depreciation charge.

17.3.10.3 The declining-balance method

The declining-balance method, similar to the sum-of-years digits method, offers accelerated depreciation charges during the early years. Depreciation is taken to be a constant percentage of the declining account balance, with the rate usually calculated at twice the corresponding straight-line rate (referred to as the double-declining-balance method). Since the asset is never 100% depreciated this way, the undepreciated portion of the account balance is written off the last year. The undepreciated balance $B(t)$ at the end of year t is given by

$$B(t)=\text{CA}(1-D)^t \tag{17.6}$$

where CA is the cost of the asset and D is the constant depreciation percentage.

For the double-declining-balance method,

$$D=2\times\frac{1}{N} \tag{17.7}$$

where N is the service life of the asset and $1/N$ is the straight-line depreciation rate.

17.3.10.4 The sinking-fund method

This is a somewhat antiquated method which assumes that a constant annual amount will be reinvested at some compound interest rate, so that the total principal and interest will accumulate at the end of the asset's life to an amount equal to the asset's original cost. This unrealistic method results in lower depreciation charges during the early years than during the later years and is seldom used except by a few public utilities who wish to maintain a high book value as their rate base. The AD is given by

$$\text{AD}=(\text{CA}-\text{RV})\left(\frac{(1+I)^N-1}{I(1+I)^N}\right)^{-1} \tag{17.8}$$

where CA is the cost of the asset, RV is the present residual value of the asset, I is the market interest rate and N is the service life of the asset.

17.4 PROCEDURE OF FINANCIAL ANALYSIS

Table 17.1 is a summary of selected methods for calculating financial feasibility. In this section, five approaches to project financial analysis will be investigated. These five approaches are as follows.

(i) NPV method.
(ii) Internal rate of return (IRR).
(iii) Annuity method.
(iv) Cost annuity comparison method.
(v) Calculation of pay-back period.

Table 17.1 Summary of selected methods for calculating the financial feasibility of a project.

Procedure for calculating economic feasibility	Criterion for the individual profitability of a project	Criterion for the relative profitability of a project	Area of application of the method	Remarks
NPV method	An investment project is only profitable when its net present value is equal to or greater than zero: $NPV \geqslant 0$ Under this condition the minimum acceptable interest on capital invested is reached or exceeded	Investment project 1 is more profitable than alternative projects 2 and 3 when its NPV is greater than those of the other investment alternatives 2 and 3: $NPV_1 > NPV_2 > NPV_3$	Reliable evaluation of the financial favourability of an investment project and comparison of investment alternatives according to capital yields anticipated to be above the minimum acceptable interest rate	This method can lead to inaccuracies in the comparison of investment alternatives with varying capital requirements and lifetimes. In such cases the investment decision should be based on the less problematic annuity method. The annuity of a project is calculated by multiplying the NPV by the CRF: $A = NPV \times CRF(DR, N)$
IRR method	An investment project is individually profitable when its IRR is equal to or greater than the cut-off discount rate which gives the minimum acceptable interest rate	A comparison of investment alternatives should not be undertaken by means of this method since it is necessary to adopt assumptions which may be unrealistic in practice and as such the danger of an incorrect decision is very great	Exact determination of the rate of interest anticipated on a capital investment when the project is executed	
Annuity method	An investment project is individually profitable	Investment project 1 is more profitable than	Reliable evaluation of the financial favourability of	When comparing investment alternatives

	when the calculated annuity is not negative: annuity $\geqslant 0$	projects 2 and 3 when its annuity is greater than those of the other alternatives 2 and 3: $\text{annuity}_1 > \text{annuity}_2 > \text{annuity}_3$	an investment project and comparison of investment alternatives from the anticipated annual return on capital over the minimum acceptable interest rate	with differing investment costs, one should consider whether investing the capital difference would influence the comparison. The NPV of a project can be calculated by multiplying the annuity by the PVF: $\text{NPV} = \text{annuity} \times \text{PVF}(\text{DR}, N)$
Cost annuity comparison method	It is not possible to determine the profitability with this procedure since there are no project-linked incomes to be set against expenditures	Investment project 1 is more profitable than other projects when its cost annuity is less than that of the other alternatives: $(\text{cost annuity})_1 < (\text{cost annuity})_2$	For exact determination of costs per year and comparison of investments with identical project goals. Applicable to the replacement problem	
Calculation of pay-back period	An investment project is individually profitable when the capital invested including a minimum acceptable interest rate will be recovered by means of the anticipated excesses of income over expenditures within the serviceable life or within a maximum acceptable pay-back period which must be shorter than the technically feasible service life	Investment project 1 is more profitable than alternative projects 2 and 3 when it has a shorter pay-back period than the other investment alternatives	Reliable evaluation of an investment project and comparison of investment alternatives based on investment risks	The length of the pay-back period does not allow conclusions to be drawn regarding the profitability of a project; therefore this method should only be applied in conjunction with others. This procedure is also of limited use for evaluating risks. It can be shown that a project with a shorter pay-back period can involve greater risks

The criteria for the individual and relative profitability of a project, calculation procedures, area of applications and limitations of each method will be presented.

Invariably, a project subjected to any financial evaluation method will have its costs and benefits spread over a number of years. In order to compare one project with another, or to determine the financial viability of a particular project, one must reduce the time stream of costs and benefits to a single number. This is most easily done by the use of a present value function (PVF). This function can be used to determine the present value of a yearly income or expenditure which escalates at some fixed percentage each year from year 1 to year N and then is discounted over N number of years.

Let

$$\mathrm{PVF}(\mathrm{DR}, B, N)=\frac{1+B}{\mathrm{DR}-B}\left[1-\left(\frac{1+B}{1+\mathrm{DR}}\right)^{N}\right] \tag{17.9}$$

$$\mathrm{PVF}(\mathrm{DR}, \mathrm{DR}, N)=1 \qquad \text{for } \mathrm{DR}=B \tag{17.10}$$

where DR is the discount rate, B is the escalation rate (general inflation rate f or energy escalation rate e) and N is the number of years.

Any yearly income or expenditure that is expected to escalate at some fixed rate B, owing to inflation f or energy price escalation e, may be multiplied by PVF(DR, B, N) in order to determine its present value.

For the special case where $B=0$,

$$\mathrm{PVF}(\mathrm{DR}, 0, N)=\frac{(1+\mathrm{DR})^{N}-1}{\mathrm{DR}(1+\mathrm{DR})^{N}}=\frac{1}{\mathrm{CRF}(\mathrm{DR}, 0, N)}. \tag{17.11}$$

The reciprocal of the PVF with $B=0$ is commonly referred to as the capital recovery factor (CRF). When multiplied by an initial sum of money, the CRF determines the periodic payment necessary to pay back that sum of money at interest DR over N periods.

17.4.1 Net present value

The NPV of an investment project at the point in time $t=0$ is the sum of present values of all the cash inflows and outflows linked to the investment. Expenditure and income are generally balanced to obtain annual returns so that the NPV of an investment project is simply the sum of the present value of its return plus the present value of its liquidation yield minus the present values of its investment costs:

NPV = annual returns − investment costs + liquidation yield

$$\mathrm{NPV}=(\mathrm{INC}-\mathrm{EXP})\mathrm{PVF}(\mathrm{DR}, f, N)-\sum_{t=0}^{N}\mathrm{IC}(t)\frac{(1+f)^{t}}{(1+\mathrm{DR})^{t}}+\mathrm{RV}\frac{(1+f)^{N}}{(1+\mathrm{DR})^{N}} \tag{17.12}$$

where INC is the annual income, EXP is the annual expenditure, IC is the investment cost, t is the time period of investment, N is the service life of investment, DR is the discount rate, f is the general inflation rate, RV is the residual value of investment project and the PVF is given by equations (17.9)–(17.11).

17.4.1.1 *Criteria*

Individual profitability
An investment project is only profitable when its NPV is greater than or equal to zero:

$$\text{NPV} \geqslant 0.$$

Relative profitability
When there are several alternative investment possibilities, the NPV of the different projects should be compared with one another and investment with the highest NPV should be selected.

17.4.1.2 *Application*

The NPV can be used for reliable evaluation of the financial favourability of an investment project and comparison of investment alternatives according to capital yields anticipated to be above the minimum acceptable discount rate.

17.4.1.3 *Remarks*

This method can lead to inaccuracies in the comparison of investment alternatives with varying capital requirements and lifetimes. In such cases, the investment decision should be based on the less problematic annuity method. The annuity of a project is calculated by multiplying the NPV by the CRF:

$$\text{annuity} = \text{NPV} \times \text{CRF}(\text{DR}, N).$$

17.4.1.4 *Example 1*

Let us determine the NPV for the diesel system (alternative 2) of Appendix A. The benefits derived from the system were not quantified into monetary values. This is a common case where energy systems are utilized. Setting INC in equation (17.12) to zero, we compute the present value of the cost of the system for a 20 year analysis period. At the twentieth year, no replacement is required. Therefore.

$$\begin{aligned} \text{INC} &= 0 \\ \text{EXP} &= \text{Operation and maintenance} + \text{fuel} \\ &= \text{US } \$400 + \text{US } \$160 \\ &= \text{US } \$560 \text{ year}^{-1} \\ \text{IC}(0) &= \text{IC}(5) = \text{IC}(10) = \text{IC}(15) = \text{US } \$2000 \\ \text{IC}(t) &= 0 \qquad \text{for all others.} \end{aligned}$$

Now

$$\text{NPV of cost} = (0-560)\text{PVF}(12\%, 5\%, 20) - \sum_{t=0}^{15} \text{IC}(t)\left(\frac{1+5\%}{1+12\%}\right)^t$$

and

$$\text{PVF}(12\%, 5\%, 20) = \frac{1+5\%}{12-5\%}\left[1-\left(\frac{1+5\%}{1+12\%}\right)^{20}\right] = 10.874.$$

Therefore

$$\begin{aligned} \text{NPV of cost} &= \text{US}\$[(0-560)10.874 - 2000(1+0.7242+0.5245+0.3798)] \\ &= -\text{US } \$6089.44 - \text{US } \$5257 \\ &= -\text{US } \$11\,346.44. \end{aligned}$$

17.4.2 Internal rate of return

Formally, the IRR is the time value of money, or the critical discount rate, that results in a zero NPV for an investment. Conceptually, the IRR is a measure of the profitability of an investment.

The formula for calculating the IRR can be established directly from the NPV formula (equation (17.12)), where NPV = 0:

$$0 = (\text{INC} - \text{EXP})\text{PVF}(\text{IRR}, f, N) - \sum_{t=0}^{N} \text{IC}(t)\frac{(1+f)^t}{(1+\text{IRR})^t} + \text{RV}\frac{(1+f)^N}{(1+\text{IRR})^N} \qquad (17.13)$$

where INC is the annual income, EXP is the annual expenditure, IC is the investment cost, t is the time period of investment, RV is the residual value, f is the general inflation rate, N is the service life of investment and the PVF is given by equations (17.9)–(17.11).

The IRR is determined by a trial-and-error procedure.

17.4.2.1 *Criteria*

Individual profitability

An investment is viewed favourably if the internal rate of return is either equal to or greater than the pre-determined discount rate:

$$\text{IRR} \geqslant \text{DR}.$$

Relative profitability
The choice should be in favour of the investment alternative with the highest IRR where this rate must be at least equal to the required discount rate.

17.4.2.2 Application

Exact determination of the rate of interest anticipated on a capital investment. The IRR indicates whether an investment of funds yields a satisfactory return.

Many situations deal with alternatives where only costs exist and the IRR of an individual alternative has no meaning. Most situations may be stated in terms of whether the extra investment required for a more expensive alternative results in enough savings or additional income to make it worthwhile.

The problem can be stated formally by letting

$$\text{NPV}_{\text{net}} = \text{NPV}_1 - \text{NPV}_2. \tag{17.14}$$

NPV_{net} is the same as the present value of the cash flows for alternative 1 less the cash flows for alternative 2 on a year-by-year basis. The IRR calculated by setting NPV_{net} to zero may be called the incremental rate of return for the more expensive alternative.

The IRR calculated by using the current value of money cash flows, accounting for the effects of inflation, is referred to as the apparent rate of return (IRR_a). The IRR calculated by using a constant value of money is referred to as the real rate of return (IRR_r). The relationship between IRR_a and IRR_r is

$$1 + \text{IRR}_r = \frac{1 + \text{IRR}_a}{1 + f}. \tag{17.15}$$

17.4.2.3 Remarks

The IRR method is not methodologically accurate when comparing two investment projects with different capital requirements and/or different service lives. A direct comparison of the IRRs in such cases can only give a correct estimation of the alternative projects if it is assumed that additional and follow-up investments can be made at an interest rate equal to the IRR of the original investment.

As the IRR of any investment departs farther from the assumed discount rate, as derived from market data, the more unlikely is the expectation to be able to invest any desired amount of capital at almost any desired interest rate. For this reason we should not base our selection simply on a comparison of the IRRs of the alternative projects.

17.4.2.4 Example 2

Let the diesel system (alternative 2) of appendix A be represented by NPV_1 and the photovoltaic (PV) system (alternative 1) of appendix A by NPV_2. Therefore we state the problem as

$$NPV_{net} = NPV_1 - NPV_2$$

as the NPV of the difference in system cost of a diesel and of a PV pumping system. When NPV_{net} is positive, the diesel system is more expensive ($NPV_1 > NPV_2$) than the PV system. To determine the IRR (table 17.2), we set $NPV_{net} = 0$ in the above problem:

$$NPV_1 = NPV_2.$$

Table 17.2 Estimate of IRR for example 2.

IRR (%)	NPV_1	NPV_2	$NPV_1 - NPV_2 = NPV_{net}$
12.0	11346	11038	308
12.1	11275	11011	264
12.2	11205	10984	221
12.3	11135	10958	177
12.4	11066	10932	134
12.5	10998	10906	92
12.6	10930	10881	49
12.7	10863	10856	7
12.8	10797	10831	−34
12.9	10732	10806	−74
13.0	10668	10782	−114

Now

$$NPV_1 = EXP \times PVF(IRR, 5\%, 20) + \sum_{t=0}^{15} IC(t)\left(\frac{1+5\%}{1+IRR}\right)^t$$

where

$$IC(t) = US\ \$2000 \qquad \text{for } t = 0, 5, 10, 15$$
$$IC(t) = 0 \qquad \text{for all others}$$
$$EXP = \text{operation and maintenance} + \text{fuel}$$
$$= US\ \$400 + US\ \$160$$
$$= US\ \$560\ \text{year}^{-1}$$

and

$$\mathrm{NPV}_2 = \mathrm{EXP} \times \mathrm{PVF}(\mathrm{IRR}, 5\%, 20) + \sum_{t=0}^{15} \mathrm{IC}(t)\left(\frac{1+5\%}{1+\mathrm{IRR}}\right)^t$$

where

$$\mathrm{IC}(0) = 630\ \mathrm{W_p} \times \mathrm{US}\ \$12\ \mathrm{W_p^{-1}} = \mathrm{US}\ \$7560$$

$$\mathrm{IC}(5) = \mathrm{IC}(10) = \mathrm{IC}(15) = 15\% \times \mathrm{US}\ \$7560 = \mathrm{US}\ \$1134$$

$$\mathrm{IC}(t) = 0 \qquad \text{for all others}$$

$$\mathrm{EXP} = \mathrm{US}\ \$150\ \mathrm{year}^{-1}.$$

17.4.3 Annuity method

The annuity method aims to convert all the net cash flows connected with an investment project into a series of annual payments of equal amounts. The conversion takes place by multiplying the NPV by CRF(DR, N) for a predetermined discount rate DR and a known planning period, N years:

$$\text{annuity} = \mathrm{NPV} \times \mathrm{CRF}(\mathrm{DR}, N). \tag{17.16}$$

17.4.3.1 Criteria

Individual profitability
An investment is considered profitable when its annuity is not negative:

$$\text{annuity} \geq 0.$$

Relative profitability
If there are several mutually exclusive investment alternatives to be compared, then the alternative with the highest annuity should be adopted so long as it is greater than or equal to zero.

17.4.3.2 Application

The annuity method can be used to compare investment alternatives even if they have very different service lifetimes, without the danger of making an incorrect decision.

17.4.3.3 Remarks

When comparing investment alternatives with differing investment costs, one should consider whether investing the capital difference would influence the comparison.

17.4.3.4 Example 3

Let us compute the annuity for examples 1 and 2. CRF(DR, N) for a 12% discount rate and a 20 year period is calculated as

$$\text{CRF}(12\%, 20) = \frac{\text{DR}(1+\text{DR})^{20}}{(1+\text{DR})^{20}-1} = \frac{0.12\times(1.12)^{20}}{(1.12)^{20}-1} = 0.133\,88.$$

From example 2, we have the following.

Example 3A

$$\text{DR} = 12\%$$

$$\text{NPV}_1 = 11\,346.44 \qquad \text{(diesel system cost)}$$

$$\text{NPV}_2 = 11\,037.82 \qquad \text{(PV system cost).}$$

Now

$$\begin{aligned}\text{NPV}_{\text{net}} &= \text{NPV}_1 - \text{NPV}_2\\ &= \text{US \$11 346.44} - \text{US \$11 037.82}\\ &= \text{US \$308.62}\end{aligned}$$

and

$$\begin{aligned}\text{annuity} &= \text{NPV}_{\text{net}} \times \text{CRF}(12\%, 20)\\ &= \text{US \$308.62} \times 0.133\,88\\ &= \text{US \$41.32.}\end{aligned}$$

Thus, US \$41.32 will be saved annually when a PV pumping system is used instead of a diesel system. The PV system will have a cost annuity (or levellized cost) of US \$1478, lower than the levellized cost of a diesel system which is US \$1519.

Example 3B
Now

$$\text{cost annuity} = \text{levellized cost} = \text{NPV}_{\text{cost}} \times \text{CRF}(\text{DR}, N)$$

and

$$\text{levellized cost of PV system} = \text{US \$11 038} \times 0.133\,88 = \text{US \$1478}$$

$$\text{levellized cost of diesel system} = \text{US \$11 346} \times 0.133\,88 = \text{US \$1519.}$$

17.4.4 Cost annuity comparison method

This is a shortened form of the annuity method without the inclusion of income in the calculation. The cost annuity is given by

$$\text{cost annuity} = \left[\sum_{t=0}^{N} \text{IC}(t)\frac{(1+f)^t}{(1+\text{DR})^t} + \text{EXP} \times \text{PVF}(\text{DR}, f, N)\right]\text{CRF}(\text{DR}, N) \qquad (17.17)$$

where IC is the investment cost, EXP is the annual expenditure, t is the time period of investment, N is the service life of investment, f is the general inflation rate, DR is the discount rate, PVF is given by equations (17.9)–(17.11) and $\text{CRF} = \text{PVF}(\text{DR}, 0, N)^{-1}$.

17.4.4.1 Criteria

Relative profitability
If there are several possible alternatives for providing a desired output, then the alternative with the lowest cost annuity should be selected.

17.4.4.2 Application

It is used to evaluate the cost per year of investment projects and for a comparison of alternatives with identical project goals. It is used in replacement problems.

17.4.4.3 Remarks

It is not possible to determine the individual profitability of an investment since there is no project-linked income to be set against expenditures.

17.4.4.4 Example 4

Using the case study in appendix A, we set up equation (17.17) as follows:

$$\text{cost annuity} = \left[\sum_{t=0}^{N} \text{IC}(t)\left(\frac{1+f}{1+\text{DR}}\right)^{t} + \text{EXP} \times \text{PVF}(\text{DR}, f, N)\right]\text{CRF}(\text{DR}, N).$$

From example 1,

$$\text{PVF}(12\%, 5\%, 20) = \frac{1.05}{0.12\text{–}0.5}\left[1 - \left(\frac{1.05}{1.12}\right)^{20}\right] = 10.874.$$

From example 3,

$$\text{CRF}(12\%, 20) = \frac{0.12 \times (1.12)^{20}}{(1.12)^{20} - 1} = 0.133\,88.$$

Therefore

cost annuity

$$= \text{US } \$0.133\,88[2000(1 + 0.7242 + 0.5245 + 0.3798) + 560 \times 10.874]$$
$$= \text{US } \$0.133\,88 \times 11\,346.44$$
$$= \text{US } \$1519 \text{ (diesel system} - \text{IC}(0, 5, 10, 15) = \text{US } \$2000;\ \text{EXP} = \text{US } \$560)$$

(compare with example 1 and example 3B). Also

$$\text{cost annuity} = \text{US}\,\$0.133\,88\left[7560\left(\frac{1.05}{1.12}\right)^0 + 1134(0.7242 + 0.5245 + 0.3798) + 150 \times 10.874\right]$$

$$= \text{US}\,\$0.133\,88 \times 11\,037.82$$

$$= \text{US}\,\$1477.74$$ (PV system

$-$IC(0) = US \$7560; IC(5, 10, 15) = US \$1134; EXP = US \$150)

(compare with example 3B).

17.4.5 Pay-back calculation

Beginning with the year of the first payment, the present values of the annual net cash flows are summed until the total reaches a value of zero. The time from commissioning up to this point is called the pay-back period.

17.4.5.1 Criteria

An individual project is favourable if the capital invested plus a minimum acceptable rate of interest is recovered by means of anticipated returns within the service life or within a maximum acceptable pay-back period, which must be shorter than the technical service life.

In evaluating the relative acceptability of alternative investments, it is assumed that the investment alternative with the shortest pay-back period is the most favourable.

17.4.5.2 Application

It can be used for evaluation of an investment project and comparison of investment alternatives based on investment risk.

17.4.5.3 Remarks

The length of the pay-back period does not allow conclusions regarding the profitability of a project; therefore, this method should only be applied in conjunction with others. This procedure is also of limited use for evaluating risk. It can be shown that a project with a shorter pay-back period can involve a greater risk.

17.5 SENSITIVITY ANALYSIS

Engineering economics deals with the future, of which we can never be certain. One or more of the variables in an economic analysis may contain considerable uncertainty, and a single evaluation of economics with some expected values of these variables seldom provides sufficient information on which to base a wise decision.

A sensitivity analysis is used to evaluate the effects of uncertainty. It is used to quantify the economic consequences of a potential but unpredictable development in important parameters.

In a sensitivity analysis the following input parameters are varied by a certain amount (percentage) from the expected value.

(i) Interest rate.
(ii) Discount rate.
(iii) Inflation rate.
(iv) Energy escalation rate.
(v) Service life.
(vi) Investment cost.
(vii) Residual value.
(viii) Operating costs.
(ix) Income.

Its effects upon the output financial parameters enumerated in § 17.4 can be investigated by constructing a figure similar to figure 17.3.

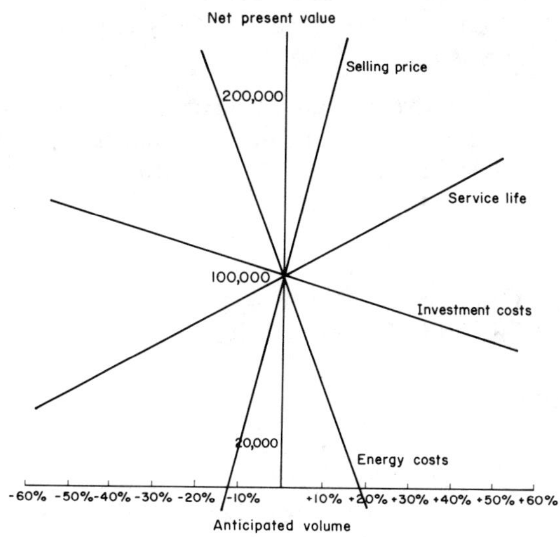

Figure 17.3 Sensitivity analysis of input parameters upon the NPV of an investment project.

A sensitivity analysis is also used in more formal risk analysis where probabilities may be assigned to possible values of variables. A graphical sample of this approach is shown in figure 17.4.

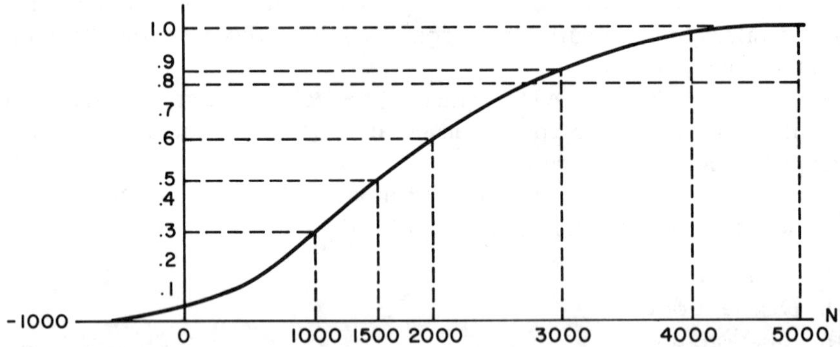

Figure 17.4 Illustrative cumulative probability density function.

The decision maker can tell at a glance that, should he choose the project, the following occur.

(*a*) The chance that its NPV turns out less than zero is about 0.05 or one in 20.
(*b*) There is no chance that it can lose more than 1000.
(*c*) The chance of a positive return is 95%.
(*d*) The chance of a return over 5000 is zero.
(*e*) The chance of a return between 0 and 3000 is about 80%.
(*f*) The expected NPV is 1500.

Clearly, the NPV cumulative probability distribution is a powerful tool for the decision maker. With this tool, the decision maker is presented with all the relevant information in a very convenient format and explicitly shown what risks a decision entails.

APPENDIX A CASE STUDY OF A THAI VILLAGE PUMPING SYSTEM FOR DRINKING WATER

A.1 Background

Consumption,	20.28 $m^3\ day^{-1}$.
Hydraulic head,	20 m.
Insolation,	5 $kW\,h\,m^{-2}\ day^{-1}$.
Nominal discount rate,	12%.
General inflation rate,	5%.
Diesel fuel escalation rate,	5%.

A.2 Alternative 1: Photovoltaic system

System cost,	US \$12 W_p^{-1}.
System size,	630 W_p.
System lifetime,	20 years.
Replacement cost,	15% system cost every 5 years for motor, pump and inverter.
Miscellaneous cost,	US \$150 $year^{-1}$.

A.3 Alternative 2: Diesel System

Diesel engine,	4 kW (minimum size available).
System cost,	US \$2000.
System lifetime,	5 years.
Operation and maintenance,	US \$400 $year^{-1}$.
Fuel consumption,	0.5 l per rated kilowatt per hour of operation.
System efficiency,	50%.
Hydraulic energy	$=0.0027 \times 20\ \mathrm{m} \times 20.28\ \mathrm{m}^3\ \mathrm{day} \times 365\ \mathrm{days}$ $=400\ \mathrm{kW\,h\,year}^{-1}$.
Hours of operation	$=2 \times 400\ \mathrm{kW\,h\,year}^{-1} \times \dfrac{1}{4\ \mathrm{kW}} = 200\ \mathrm{h\,year}^{-1}$.
Fuel cost	$=0.5\ \mathrm{l\,kW}^{-1} \times 4\ \mathrm{kW} \times 200\ \mathrm{h\ year} \times \mathrm{US}\ \$0.4\ \mathrm{l}^{-1}$ $=\mathrm{US}\ \$160\ \mathrm{year}^{-1}$.

BIBLIOGRAPHY

Barish N N and Kaplan S 1978 *Economic Analysis for Engineering and Managerial Decision Making* (New York: McGraw-Hill)

Horst F and Gerhard O 1985 *A Guide to the Financial Evaluation of Investment Projects in Energy Supply* (GTZ)

Jones B W 1982 *Inflation in Engineering Economic Analysis* (New York: Wiley)

Park W R 1973 *Cost Engineering Analysis—A Guide to the Economic Evaluation of Engineering Projects* (New York: Wiley)

Perino A M 1979 *Report No SAND 78-0931* (Albuquerque, NM: Sandia National Laboratories)

Sassone P G and Schaffer M A 1978 *Cost–Benefit Analysis—A Handbook* (New York: Academic Press)

Wright J C 1983 *Technoeconomics: Concept and Cases* (Tokyo: Asian Productivity Organization)

18 Life-cycle Cost Comparisons of Photovoltaic-powered Systems and Most Likely Conventional Alternative System

18.1 INTRODUCTION

This chapter is based on the technical report by Eskenazi *et al* (1986).

Life-cycle cost comparisons of photovoltaic (PV)-powered systems and the most likely conventional alternative system in each of the five application areas (water pumping, communications, vaccine refrigeration, lighting and home power, and multi-use) are presented. These comparisons are intended to give the reader a first-order indication of when a PV-powered system should be considered for a particular application. The financial analyses are tailored to the decision-making perspective of development bank or borrowing country project officers. Thus the financing parameters used in the analyses (amount financed, debt term, discount, and interest rate) are consistent with those for development bank loans, as opposed to those for commercial loans.

The cost comparisons are presented on two basic levels. First, detailed net present value (NPV) life-cycle cost analyses are presented as 20 year cash flows comparing the conventional alternative with PV systems using specific base-case assumptions. Next, sensitivity analyses are presented that explore system comparisons as certain base-case assumptions are varied. In each case, the sensitivity analyses cover a range of possible application loads. A graphic presentation of these sensitivity analyses allows readers to estimate the comparative viability of PV against conventional alternatives based on particular country-specific parameters.

18.2 METHODOLOGY

In each application area, conceptual designs are developed for PV-powered and conventional powered systems to meet the 'base-case' load requirements.

Each base case is intended to be representative of a typical application for which PV systems might be used. These two base-case systems are compared using a 20 year life-cycle cost analysis, whose output is the NPV of the system's total capital and operating costs over the period. The NPV cost is presented as the annualized cost for each system. In cases where the same load(s) can be used for the PV and conventional systems, only the power systems are compared (i.e. load costs are not included). This is the case for communications, multi-use and the radio load in home power systems. Where different loads are required (pumping, refrigeration and home power system lights), comparisons are made between power and load systems. The assumptions used in each analysis are detailed in the following sections.

Sensitivity analyses were performed to determine how the NPV cost varies from the base case as the following key parameters changed.

(i) Equipment capital cost.
(ii) Conventional fuel cost (diesel or kerosene).
(iii) Discount and interest rate.
(iv) Expected lifetime for conventional equipment (diesel).
(v) Irradiation.

In some applications, additional sensitivities were examined, such as the variation in refrigeration life-cycle cost as a function of the number of vaccine doses per year or the operating availability of kerosene-fuelled refrigerators. The results of these analyses are presented in graphic form in §§ 18.3–18.7. These graphs provide the reader with an initial level of comparison between PV and conventional alternatives.

Finally, graphs were constructed for each application to show overall PV best- and worst-case viabilities over the given load range. These graphs were developed by compounding the extreme sensitivity parameters as follows: for the best-case PV viability, a curve was generated using the lowest discount and interest rate, the highest fuel cost, the shortest conventional system lifetime and the highest irradiation of the sensitivity parameter range. The worst-case curve was developed using the other extremes of the ranges.

18.2.1 Common base-case assumptions

This section outlines the assumptions that are common to the base-case analyses in all five application areas. Assumptions unique to a particular application are detailed in the appropriate analysis section.

18.2.2.1 Financial assumptions

Common financial assumptions are outlined in table 18.1, and explanations of each parameter are provided below.

All costs for customs, insurance and freight (CIF) are capitalized in the total

Table 18.1 Common base-case financial assumptions.

Parameter	Assumption
Debt service	100% of system initial capital CIF cost Capitalize CIF cost 20 year term 10% per year (compounded at the end of the year)
Salvage value	Included
Installation costs	Not included
Operating labour	Not included
Diesel fuel cost	US \$0.50 l^{-1}
Kerosene fuel cost	US \$0.70 l^{-1}
General inflation	5% $year^{-1}$
Nominal discount and interest rate	10%

system cost. The CIF value is assumed to be 25% more than the free-on-board (FOB) manufacturer system cost. Installation costs are not included since these costs vary widely, according to the country and specific site. It is assumed that all systems are 100% financed for a debt term of 20 years at an interest rate of 10% $year^{-1}$, compounded at the end of each year (this is typical of a developing country loan). This assumption is made even for cases where expected equipment lifetimes may be much shorter than 20 years, such as diesel engine generator sets. Because the common analysis period of 20 years is desired, the initial capital cost in such instances is assumed to be 100% financed by the development loan; any subsequent capital replacement costs are represented as pure cash outlays at the end of each estimated equipment lifetime.

Maintenance parts costs are generally specified as a percentage of the initial capital cost. For the most part, maintenance costs appear as recurring annual costs, although some systems such as diesel generator sets also require periodic major maintenance overhaul. These overhaul costs are estimated as a percentage of the original capital cost and are assumed to occur at a specified frequency throughout the 20 year analysis. Operating labour is not included as a recurring project cost because this cost varies widely, even within the same country.

Nominal fuel costs are assumed to be US \$0.5 l^{-1} for diesel and US \$0.70 l^{-1} for kerosene. These numbers are based on values cited in the questionnaire responses (15 citations for diesel fuel costs and 13 for kerosene fuel costs), values listed in the World Bank's Domestic Petroleum Product Prices and discussions with principals from major refrigeration programmes.

All operating costs (including fuel) are escalated at a general inflation rate of 5% year^{-1} for all base-case life-cycle analyses. In addition, recurring costs are appropriately inflated for the year of their occurrence. A 10% nominal discount rate was used to obtain the NPV cost of all base-case cash flows.

18.2.1.2 Technical assumptions

Specifications such as the power system size required for the typical system in each application area were developed as part of this evaluation. The key technical assumptions common to the base-case analyses are identified in table 18.2.

Table 18.2 Common base-case technical assumptions.

Specification	System type		
	PV	Diesel	Kerosene
Component life (years)			
Array	20	Not applicable	Not applicable
Generator set	Not applicable	6	Not applicable
Power conditioning	10	10	Not applicable
Batteries	5	5	2[a]
Loads	5	5	10 (refrigeration) 3 (lights)
Major maintenance			
Engine overhaul	Not applicable	Every 3 years	Not applicable

[a] Batteries used in conventional home power systems.

The availability of power or load systems is defined as the percentage of time that the system operates within technical specifications. Potential causes of system unavailability include the following.

(i) Scheduled downtime and regular maintenance.
(ii) Downtime for repair (including unavailability of spare parts).
(iii) No fuel.
(iv) No sun.
(v) Operating outside specifications (e.g. a refrigerator operating outside the appropriate temperature range or a power system providing energy to the loads outside the voltage range at which the loads can function).

The availability values that were chosen for the PV and conventional systems in each application are outlined in table 18.3. These values are discussed in more detail in the individual analysis sections. The values are

Table 18.3 System availability.

Application	Availability (%)	
	PV	Conventional
Pumping[a]	95	95
Communications[b]	99.9	99.9 (2 diesels at 97.5)
Refrigeration[a]	95	50[c]
Home power	Not applicable[d]	Not applicable[d]
Multi-user[b]	97.5	97.5

[a] Availability of power or load system.
[b] Availability of power system.
[c] Includes unavailability of fuel.
[d] System-specific availability not determined but assumed to be equal.

based on an uninterrupted supply of fuel (except in the case of kerosene refrigeration). This is an optimistic assumption, particularly for conventional power systems. If, for example, there is no diesel fuel for 1 month of the year, diesel system availability would decrease by an additional 8%, raising the NPV cost for diesel-powered systems by approximately 9%.

The availability values also do not quantify the availability of spare parts and operation, maintenance and repair services. Clearly, all analyses depend on the reasonable availability of parts and service; otherwise, the assumptions of system lifetime and availability should be adjusted downwards accordingly.

Gasoline generator sets are not examined in the comparative cost analyses. However, the following information is provided as reference material.

(i) Gasoline generator sets are available in sizes ranging from 300 W to 100 kW, although systems at the low end of that spectrum (less than 3 kW) are predominantly used.
(ii) Gasoline generator sets in the low-size range offer the advantage of increased portability.
(iii) Gasoline generator sets have lower initial capital costs than do diesel generator sets. In the less than 3 kW range, gasoline generator sets cost US \$400–\$900 kW^{-1}. The minimum-size diesel generator set is approximately 3 kW and costs approximately US \$1800 kW^{-1}.
(iv) Gasoline fuel is more expensive (by approximately 50–100%) on a US dollar per litre basis than diesel fuel. For example, in Sudan, diesel fuel costs US \$0.51 l^{-1}, while gasoline costs US \$1.02 l^{-1}.
(v) Gasoline generator sets have a higher fuel consumption rate than do diesel generator sets. Full-load fuel consumptions were determined to be 1 $l\,kW^{-1}\,h^{-1}$ (in the 0.5–3 kW range) and 0.4 $l\,kW^{-1}\,h^{-1}$ (in the 3–25 kW range) for gasoline and diesel generator sets respectively.

(vi) Gasoline generator sets have a much shorter lifetime than do diesel generator sets (3000–5000 h compared with 20 000 h).

§§ 18.3–18.7 present the financial analyses performed for each of the applications. The models used to perform the technical sizing and life-cycle cost comparisons are described in appendix A.

18.2.1.3 *Credibility of financial analyses assumptions*

This discussion addresses whether the financial assumptions used in this evaluation result in conservative or optimistic assessments of PV system financial attractiveness. Overall, it is believed that the assumptions balance each other out, resulting in an unbiased analysis for the development loan scenario.

Conservative assessments of PV financial viability resulted from the following assumptions:

(*a*) The analyses do not include the labour costs associated with operation, maintenance and repair because they are country and site specific. This assumption is considered conservative because conventional systems generally require more of these services than do PV-powered systems.
(*b*) The supply of spare parts and fuel is assumed to be uninterrupted. Field reports indicate that the unreliable supply of these items has been a major cause of conventionally powered system unavailability.
(*c*) No price reductions are assumed for system components. Because the conventional power system represent a stabilized market, price reductions are not anticipated. However, because PV power systems involve developing technologies and little mass production, component prices are expected to decline over the years, thus making the replacement costs used in the analyses high.

Optimistic assessments of PV financial viability resulted from the following assumptions.

(*a*) No system design costs are included. PV-powered systems generally involve more costly design phases, since packaged systems and standard designs are not yet available across all applications.

For some assumptions, it is uncertain whether they have a conservative, optimistic or neutral effect on PV system financial viability. These assumptions include the following.

(*a*) Because the evaluation involved a financial analysis, macroeconomic considerations such as shadow pricing and tax revenue are not addressed.
(*b*) The implication of tax benefits to the system owner is unclear.
(*c*) Most costs are based on the purchase of one system. It is recognized that the block purchases typical of institutional procurement result in

significant system cost reductions for PV systems. Data on the effect of conventional system block purchases were unavailable.

(*d*) A 20 year PV array life is assumed. While accelerated life testing indicates that current PV arrays have lifetimes greater than 20 years, these arrays have not yet been in the field for 20 years.

The assumptions used in the financial analyses were based on development bank loans. For commercial loans, the assumptions for loan term and percentage of the loan financed are considered high. Thus, the financial results show greater PV viability than would exist under a commercial setting.

18.3 WATER-PUMPING ANALYSIS

The financial analysis presented in this section compares PV-powered water-pumping systems to diesel-powered systems for rural water supply, assuming development agency financing. The analysis shows that PV-powered systems are the least-cost option at demands of up to 25 m^3 day^{-1} at a head of 25 m (625 m^4 day^{-1}), even under unfavourable financial assumptions (figure 18.1). When the financial parameters are more favourable, PV-powered systems are competitive up to 550 m^3 day^{-1} (13 750 m^4 day^{-1}).

The graph in figure 18.1 depicts the ratio of PV- to diesel-powered pumping system NPV costs for the best-PV-case and worst-PV-case scenarios. Both scenarios assume 20 year life-cycle costing and development agency financing. At an NPV cost ratio of 1.0, PV and diesel system life-cycle costs are equal. Under the best-PV-case scenario, the five parameters shown on the graph are adjusted to 'reasonable extremes' that favour PV systems. An opposite

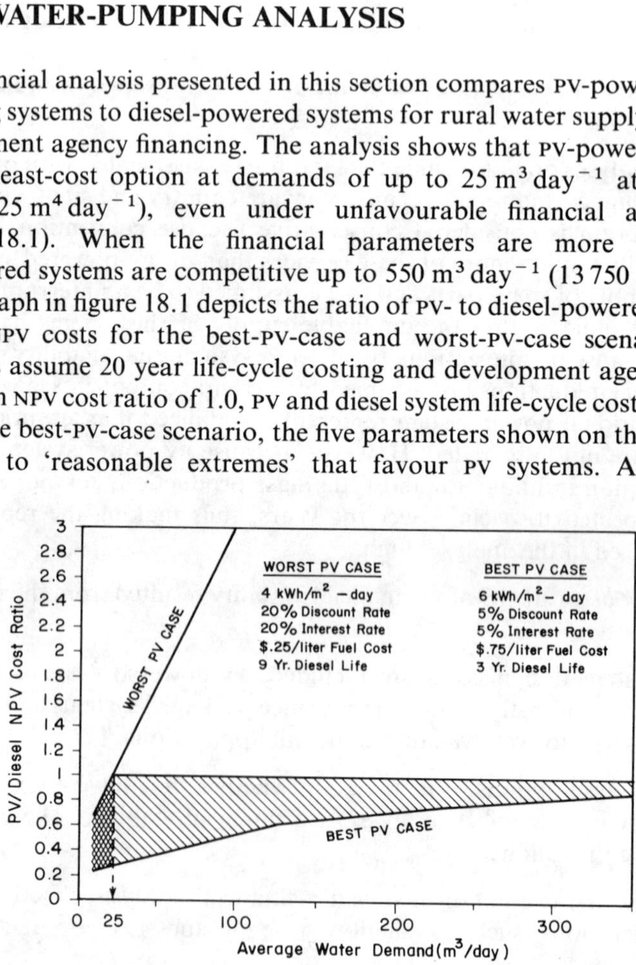

Figure 18.1 Sensitivity of water-pumping costs to best and worst conditions.

adjustment is made under the worst-PV-case scenario. The area between the two curves represents a reasonable range of financial assumptions.

This range indicates that PV-powered pumping systems are the least-cost option at loads much greater than shown by a 1983 United Nations Development Programme (UNDP)–World Bank study. It showed PV systems to be competitive up to 250 $m^4\,day^{-1}$ as opposed to the 625–13 750 $m^4\,day^{-1}$ demonstrated in this report. The major reasons for this difference are the assumption of 20 year development bank financing, the consideration of diesel system inefficiencies when the system is operated substantially below rated capacity, and recent improvements in the cost and performance of PV systems.

Figure 18.2 depicts the various cost elements of PV- and diesel-powered pumping systems. The sensitivity analyses presented in this section indicate that diesel lifetime, fuel costs, discount and interest rate, irradiation, and pumping head all have a strong impact on the cost analysis.

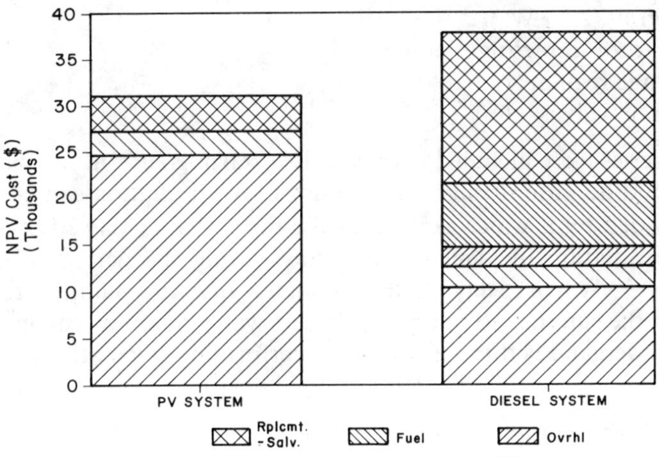

Figure 18.2 Base-case water-pumping system cost components.

18.3.1 Description of the base case

Based on field experience with pumping applications, the base-case load requirement chosen for this analysis is an average daily water demand of 50 m^3 to be pumped through a 25 m head. This load is representative of a village drinking-water system that supplies water to approximately 1250–2500 people (based on 20–40 $l\,day^{-1}$ per person). It is assumed that the load varies throughout the year, reaching a maximum that is 50% higher than the average, or 75 $m^3\,day^{-1}$. It is further assumed that the well is capable of yielding the volume and the pump flow rate demanded of it.

The conventional alternative chosen for this analysis is a diesel engine with a generator set connected to an AC motor and pump. For both the PV and conventional systems, it is assumed that water is pumped into a storage tank. Thus the total head included an assumption of some vertical distance to the tank height. Because the well and tank are considered to be the same for both systems, their costs are not included in the analysis. The PV and diesel systems are further described below, including discussions of system cost determinations.

The availability of both pumping systems is assumed to be 95%. This value is considered conservative for the PV-powered system and, since it does not include unavailability of fuel, it is optimistic for the diesel-powered system.

18.3.1.1 Photovoltaic-powered pumping system description

The PV-powered pumping system design chosen for the base-case analysis consisted of the following primary components.

(i) PV array (2.3 kW_p).
(ii) DC motor–centrifugal pump.
(iii) Voltage regulator.
(iv) Wiring, valves, etc.

No batteries were included because a low starting current for the motor is assumed and because storage for periods of low irradiation can be inexpensively provided by a water tank sized for more than 1 day of storage. The average irradiation incident on the array during the period of maximum demand is 5 $kW\,h\,m^{-2}\,day^{-1}$. It is assumed that the pump automatically starts with the sun (i.e. no operator). Excess water-pumping capacity beyond the annual demand is disregarded when performing calculations.

18.3.1.2 Diesel-powered pumping system description

The diesel-powered pumping system chosen to meet the base-case pumping requirements consists of the following primary components.

(i) Diesel generator set (6.4 kW).
(ii) AC motor–pump.
(iii) Wiring, valves, etc.

The diesel-powered pumping system is specified to operate during one period of the day at a maximum load factor of 75%. This load factor is based on user and manufacturer recommendations for optimum use to ensure maximum diesel engine life. The duration of the daily pumping period is based on the time required to pump the average daily volume requirement into the storage tank at 75% diesel load factor. It is further specified that the diesel generator set operates for at least 1 $h\,day^{-1}$. Because diesel generator sets are not commonly available in sizes less than 3 kW, a 3 kW diesel generator set must operate at less than 75% load factor in cases of small pumping demands in order to meet the 1 h requirement.

The pump has a maximum flow rate capability of 35 $m^3 h^{-1}$. This flow rate minimizes the time needed to operate the diesel over the range in demands analysed.

18.3.2 System costs

18.3.2.1 Photovoltaic-powered pumping system costs

The FOB manufacturer costs of PV-powered pumping systems have dropped dramatically from approximately US \$30 W_p^{-1} in 1978 to less than US \$10 W_p^{-1} in 1986. On the assumption that several systems are purchased at a time, current installed system costs range from about US \$11 W_p^{-1} to US \$14 W_p^{-1}. Costs vary over this range as a function of the pump selected, site-specific conditions and other options. Table 18.4 lists some of the PV-powered pumping system costs obtained during the course of this evaluation. The costs shown include CIF and installation costs.

Table 18.4 PV-powered pumping system costs (based on system prices from 1983 to 1985).

Performance requirements: Volume ($m^3 day^{-1}$)	Performance requirements: Head (m)	Pump	Motor type	Array size (W_p)	Irradiation ($kW h m^{-2}$ day)	Installed costs (US \$ W_p^{-1})
230	2	Centrifugal	Single-stage DC brushless	516	6	13.7
13	3	Centrifugal	Single-stage DC brushless floating	85	6	11.8
76	11	Centrifugal	DC motor drive surface mounted	770	5	11.7
70	27	Centrifugal	Multi-stage AC submersible	1700	—	13.3
30	65	Centrifugal	Multi-stage AC submersible	1500	—	13.3
25	34	Jack pump	DC driven	800	—	13.8
					Average	12.9

The summary of the costs assumed for the base-case PV-powered pumping system is presented in table 18.5. An initial capital cost of US $10.5 W_p^{-1} (not including installation costs) was selected for the base case. This value is considered consistent with the installed costs presented in figure 18.1. It is assumed that the motor–pump system, voltage regulator and valves will be replaced during the 20 year financial analysis period. These components are assumed to have a 5 year life and an FOB manufacturer cost of approximately US $1635 in year one. In the cash flow, this cost is escalated 5% $year^{-1}$ (the assumed general inflation rate for all the base cases) to the year of replacement. The only other cost assumed for PV-powered pumping systems is the annual maintenance and repair cost. This cost is for items not included in the regular replacement of components (e.g. lubrication of pump parts, and replacement of brushes if necessary). For the base case, these costs amount to 1% of the initial capital cost.

Table 18.5 Base-case PV-powered pumping system costs.

Specification	Cost
Initial capital costs (FOB manufacturer)	
PV array (2.3 kW_p)	US $18 149
Pump, regulator, valves	US $ 1 635
Total capital cost	US $19 784
Recurring capital costs (FOB manufacturer)	
Pump, regulator, valve replacement	US $1635 every 5 years
Other recurring costs (% initial capital costs)	
Maintenance and repair	1% $year^{-1}$ [a]

[a] Plus the appropriate escalation due to general inflation.

18.3.2.2 Diesel-powered pumping system costs

Initial capital costs (FOB manufacturer) for diesel generator sets are provided for generators ranging in size from 3 to 25 kW. (Diesel generator sets are not commonly available below 3 kW.) A regression analysis was performed on the available cost data to obtain a continuous curve of capital cost over the size range. Figure 18.3 presents this cost regression. It is assumed that the diesel generator set will be replaced every 6 years.

Initial costs (FOB manufacturer) for AC motor–pump systems to be used in conjunction with the diesel generator sets for various flow rates at different heads are provided in table 18.6. It is assumed that the AC motor–pump system and valves are replaced every 5 years. For this particular analysis, the cost for these items amounts to US $1564.

Operation, maintenance and repair costs are significant in diesel-powered

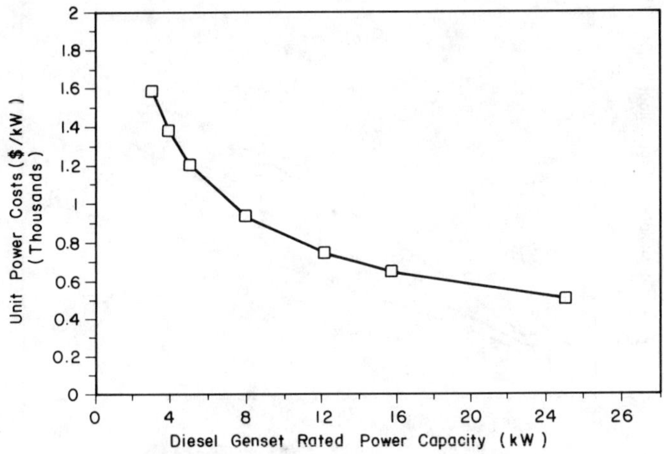

Figure 18.3 Diesel generator set costs.

pumping systems. These include the costs associated with fuel, engine, overhauls, and miscellaneous maintenance and repair items.

Fuel consumption rates vary as a function of the size of the engine, the engine's efficiency, the number of hours that the engine runs, and the load factor at which the engine operates. Figure 18.4 presents fuel consumption rates at quarter-, half- and three-quarters-load for generator sets ranging in size from 3 to 25 kW. For each comparative financial analysis, the average load factor is calculated, and then the appropriate fuel consumption rate is selected. The average number of operating hours is also computed, which allows the total annual fuel cost to be determined.

Table 18.6 AC pump costs.

AC pump costs (actual data)		
Flow rate ($m^3 h^{-1}$)	Head (m)	Cost (US $)
0.6	10	325
0.6	25	325
0.6	40	359
3.0	10	325
3.0	25	325
3.0	40	550
12.0	10	822
12.0	25	822
12.0	40	1713
35.0	10	1564
35.0	25	1564
35.0	40	1983

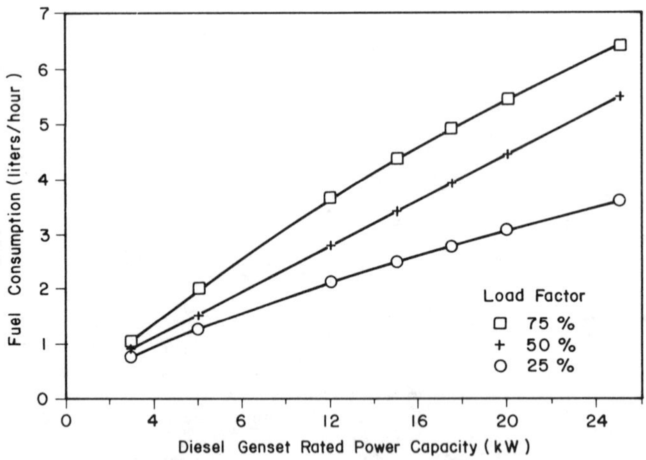

Figure 18.4 Fuel consumption rates.

Engine overhauls were assumed to be performed every 3 years. The cost associated with an overhaul was estimated to be equivalent to 15% of the initial capital cost of the engine. Miscellaneous maintenance and repair for the system are assumed to be equivalent to 2% of the system capital cost.

The operating and capital costs associated with the base-case diesel-powered pumping system are outlined in table 18.7.

Table 18.7 Base-case diesel-powered pumping system costs.

Specification	Cost
Initial capital costs (FOB manufactuer)	
Diesel generator set (6.4 kW)	US $6754
Pump, valves	US $1564
Total capital cost	US $8318
Recurring capital costs (FOB manufacturer)	
Diesel generator set replacement	US $6754 every 6 years[a]
Pump, valve replacement	US $1564 every 5 years[a]
Other recurring costs	
Engine overhaul	15% of initial generator set cost every 3 years[b]
Maintenance and repair	2% of initial system cost per year[a]
Fuel cost (1104 l at US $0.50 l^{-1})	US $522 $year^{-1}$ [a]

[a] Plus appropriate escalation due to general inflation.

[b] Engine overhauls not performed during replacement years.

18.3.3 20 year life-cycle costs

Using the base-case assumptions, 20 year cash flows are estimated. The cash flows for the PV- and diesel-powered pumping systems are shown in tables 18.8 and 18.9 respectively. The results of the cash flows are expressed as the total NPV cost for PV- and diesel-powered systems (assuming a 50 $m^3\,day^{-1}$ pumping volume and a 25 m head). For the base case, these cash flows demonstrate that the PV system is a slightly more cost-effective option: the NPV cost for the PV system was US $31 166, and the NPV cost for the diesel system was US $38 021.

The complete technical and financial model used to generate these cash flows and the sensitivity analyses presented in the following sections is provided in appendix A.

18.3.4 Sensitivity analyses

The graphs on the following pages demonstrate the projected sensitivity of the NPV life-cycle pumping cost as a function of changes in six key variables: capital cost, discount and interest rate, fuel cost, diesel system lifetime, irradiation and pumping head. The general relationship shows that diesel-powered systems become more cost effective at larger demands. This relationship is due to the decreasing costs of diesel-powered systems compared with the almost linear costs for PV-powered systems on a cost per unit volume basis.

The graphs present data in terms of the ratio between the PV- and diesel-powered system life-cycle costs. Therefore, when this ratio is less than 1.0, the PV-powered system is projected to provide lower life-cycle costs. For the capital cost sensitivity, the NPV costs are presented individually for PV and diesel systems (i.e. not in ratio form).

The base-case analysis assumes a head of 25 m. NPV life-cycle costs are considered valid for systems where the average pump–motor efficiency is 50% and pump–motor costs are similar to that used in this analysis.

18.3.4.1 Sensitivity to capital cost

This section examines the effect of changing capital cost estimates for both diesel- and PV-powered systems. The effect is modelled as a 'cost multiplier', which is examined at values of 0.75, 1.0 and 2.0, with 1.0 equalling the base-case estimate. The range of multipliers is intended to account for system capital cost variations. These variations can result from different CIF equipment costs and can account for installation costs, which are very country and site specific. This sensitivity is depicted individually for PV- and diesel-powered systems in terms of their NPV costs in figures 18.5 and 18.6.

Table 18.8 PV-powered pumping system 20 year cash flow (base case).

Year	1	2	3	4	5	6	7	8
Debt service (US $)	2905	2905	2905	2905	2905	2905	2905	2905
Operating and maintenance expenses								
Annual maintenance (US $)	208	218	229	240	252	265	278	292
Recurring capital costs								
Pump, regulator, valve cost (US $)	0	0	0	0	2608	0	0	0
Salvage value (US $)	0	0	0	0	0	0	0	0
Total cash outflow (US $)	3112	3123	3134	3145	5766	3170	3183	3197
DF $= 1/[(1+\text{DR})^{\text{years}}]$	0.9091	0.8264	0.7513	0.683	0.6209	0.5645	0.5132	0.4665
NPV stream (US $)	2829	2581	2354	2148	3580	1789	1633	1491
Total NPV of cash outflow (US $)	31 166							
Annual water pumped (m^3)	17 338	17 338	17 338	17 338	17 338	17 338	17 338	17 338

Table 18.9 Diesel-powered pumping system 20 year cash flow (base case).

Year	1	2	3	4	5	6	7	8
Debt service (US $)	1221	1221	1221	1221	1221	1221	1221	1221
Operating and maintenance expenses								
Annual maintenance (US $)	175	183	193	202	212	223	234	246
Diesel engine overhaul (US $)	0	0	1173	0	0	0	0	0
Fuel (US $)	580	609	639	671	705	740	777	816
Recurring capital costs								
Pump, regulator, valve cost (US $)	0	0	0	0	2495	0	0	0
Diesel generator replacement cost (US $)	0	0	0	0	0	11 314	0	0
Salvage value (US $)	0	0	0	0	0	0	0	0
Total cash outflow (US $)	1976	2013	3226	2095	4633	13 498	2232	2283
DF $= 1/[(1+\text{DR})^{\text{years}}]$	0.9091	0.8264	0.7513	0.683	0.6209	0.5645	0.5132	0.4665
NPV stream (US $)	1796	1664	2424	1431	2877	7619	1145	1065
Total NPV of cash outflow (US $)	38 021							
Annual water pumped (m^3)	17 338	17 338	17 338	17 338	17 338	17 338	17 338	17 338

9	10	11	12	13	14	15	16	17	18	19	20	NPV comparison
2905	2905	2905	2905	2905	2095	2905	2905	2905	2905	2905	2905	24 729
307	322	338	355	373	392	411	432	453	476	500	525	2516
0	3329	0	0	0	0	4249	0	0	0	0	0	3920
0	0	0	0	0	0	0	0	0	0	0	0	0
3212	6556	3243	3260	3278	3296	7565	3337	3358	3381	3405	3430	
0.4241	0.3855	0.3505	0.3186	0.2897	0.2633	0.2394	0.2176	0.1978	0.1799	0.1635	0.1486	
1362	2528	1137	1039	949	868	1811	726	664	608	557	510	
17 338	17 338	17 338	17 338	17 338	17 338	17 338	17 338	17 338	17 338	17 338	17 338	

9	10	11	12	13	14	15	16	17	18	19	20	NPV comparison
1221	1221	1221	1221	1221	1221	1221	1221	1221	1221	1221	1221	10 398
258	271	285	299	314	329	346	363	381	400	420	441	2116
1572	0	0	0	0	0	2106	0	0	0	0	0	2052
856	899	944	991	1041	1093	1148	1205	1265	1328	1395	1465	7020
0	3184	0	0	0	0	4064	0	0	0	0	0	3750
0	0	0	15 162	0	0	0	0	0	20 319	0	0	14 872
0	0	0	0	0	0	0	0	0	0	0	14 714	2187
3907	5576	2450	17 674	2576	2644	8885	2789	2868	23 269	3037	(11 587)	
0.4241	0.3855	0.3505	0.3186	0.2897	0.2633	0.2394	0.2176	0.1978	0.1799	0.1635	0.1486	
1657	2150	859	5631	746	696	2127	607	567	4185	497	(1722)	
17 338	17 338	17 338	17 338	17 338	17 338	17 338	17 338	17 338	17 338	17 338	17 338	

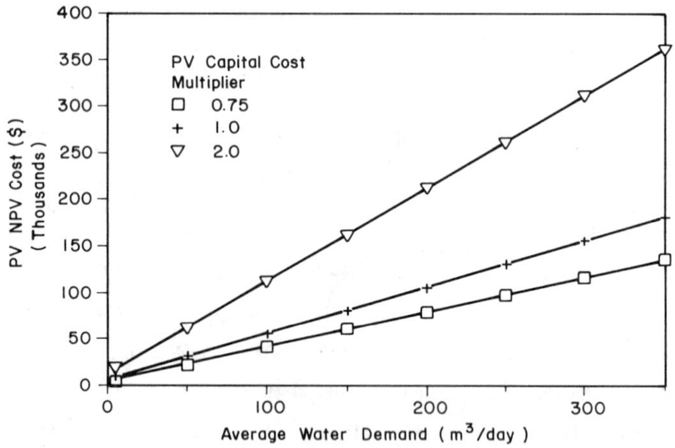

Figure 18.5 Sensitivity of PV-powered pumping costs to capital cost.

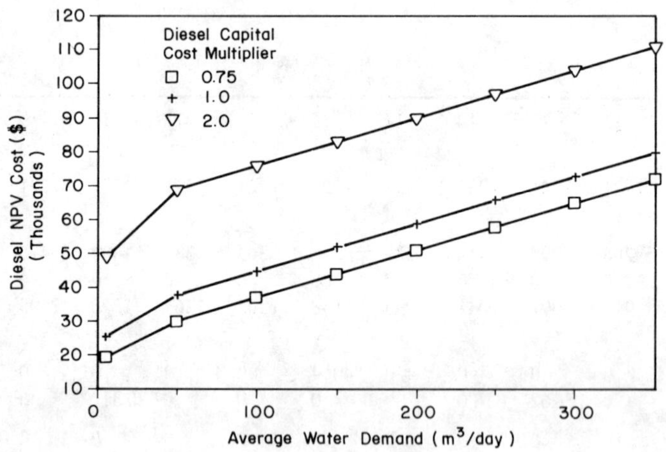

Figure 18.6 Sensitivity of diesel-powered pumping costs to capital cost.

18.3.4.2 Sensitivity to discount and interest rate

This analysis (figure 18.7) examines the changes that occur in comparative life-cycle pumping costs as the discount and interest rate applied to each cash flow is varied between 5 and 20%. As the rate increases from the 10% base-case rate, the obvious effect on both systems is a decrease in overall life-cycle costs. However, the diesel pumping cost shows a greater decrease than that of the PV system. This occurs because the diesel cash flows have a higher proportion of escalating recurring costs against the level payments for debt services. Therefore the effective reduction of this escalating diesel cost burden at higher discount and interest rates produces a greater NPV cost reduction

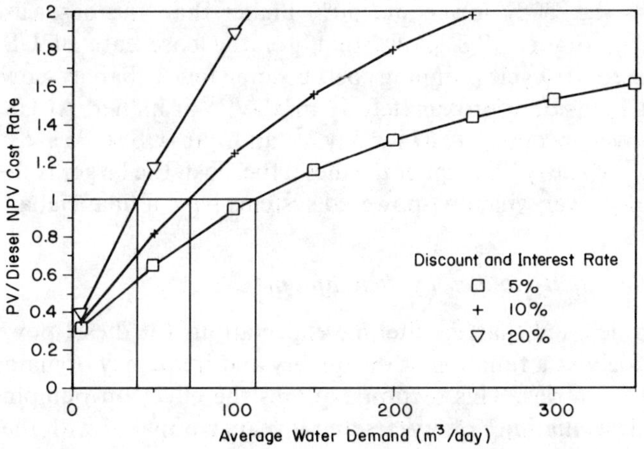

Figure 18.7 Sensitivity of pumping costs to discount and interest rate.

than in the PV cash flow, which is dominated by its large proportion of levellized debt service.

The graph in figure 18.7 demonstrates this disproportionate impact. At a 20% discount and interest rate, PV systems hold an NPV cost advantage for pumping loads only up to $40\,m^3\,day^{-1}$ whereas, at a 5% discount and interest rate, PV pumping systems appear to have an advantage up to approximately $112\,m^3\,day^{-1}$ of pumping demand.

18.3.4.3 Sensitivity to diesel fuel cost

This section examines the changes that occur in comparative life-cycle pumping costs as the cost of diesel fuel varies from US $\$0.25\,l^{-1}$ to

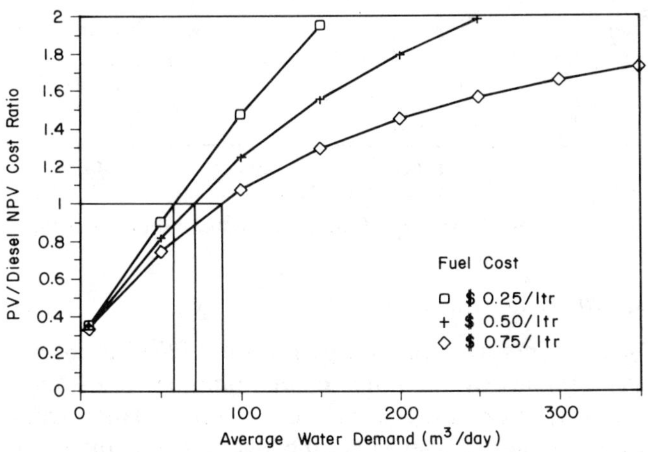

Figure 18.8 Sensitivity of pumping costs to diesel fuel cost.

US \$0.75 l^{-1} (i.e. 50% lower and 50% higher than the base-case price of US \$0.50 l^{-1}). Figure 18.8 shows that, for the base case of US \$0.50 l^{-1}, diesel-powered life-cycle pumping costs become lower than PV-powered costs at pumping loads of approximately 70 m^3 day^{-1} or higher. At US \$0.25 l^{-1}, this cross-over occurs at 57 m^3 day^{-1} and, at US \$0.75 l^{-1}, not until 90 m^3 day^{-1}. Clearly, the higher the diesel fuel cost, the larger is the pumping demand range over which PV-powered systems have a financial advantage.

18.3.4.4 Sensitivity to diesel system lifetime

It has been debated that the lifetime expectations for diesel power systems can vary widely as a function of the quality and frequency of maintenance in developing countries. This section explores the effect on pumping costs of assuming diesel lifetimes of 3 years and 9 years compared with the base-case assumption of 6 years. The financial impact of varying this assumption appears as more or less frequent diesel system replacement costs in the cash flow projections. While figure 18.9 shows that the base-case diesel system becomes less costly at pumping loads of approximately 71 m^3 day^{-1} or higher, for a diesel life of 3 years, this cross-over occurs at about 115 m^3 day^{-1}, and at 9 years it occurs at only 54 m^3 day^{-1}.

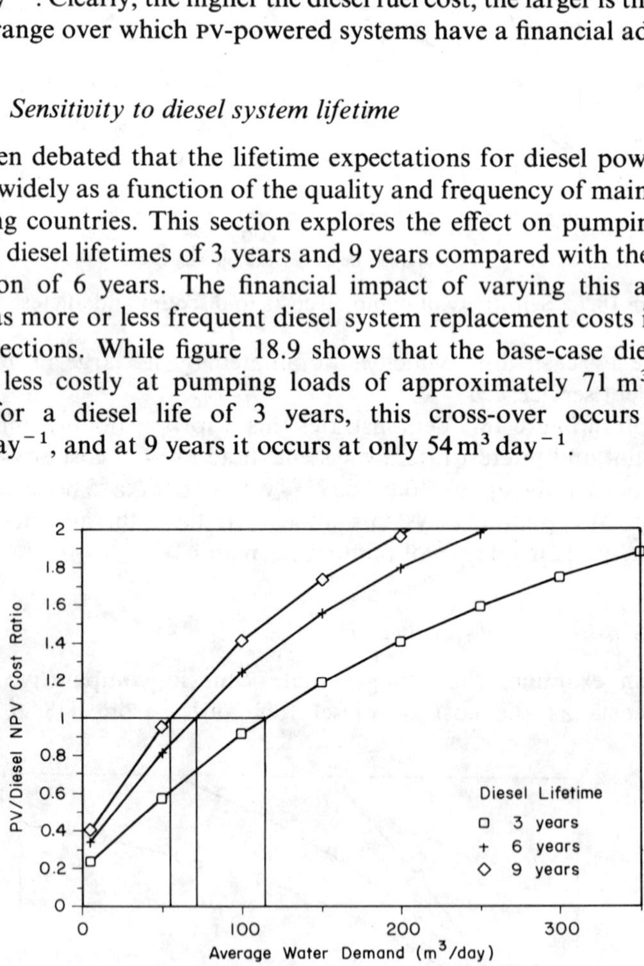

Figure 18.9 Sensitivity of pumping costs to diesel lifetime.

18.3.4.5 Sensitivity to irradiation

Figure 18.10 pressents a graph comparing the life-cycle costs of PV- and diesel-powered systems over a range of pumping demands and at varying irradiation levels. For the base-case average daily irradiation of 5 kW h m^{-2} day^{-1}, the break-even point between PV and diesel costs is about

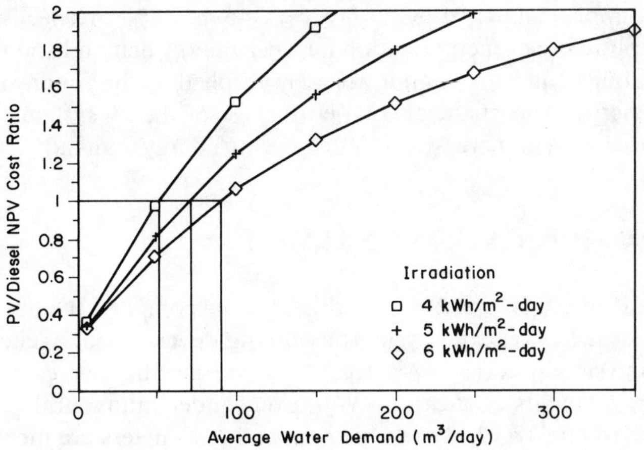

Figure 18.10 Sensitivity of pumping costs to irradiation.

70 m^3 day^{-1}. By varying the irradiation to as low as 4 kW h m^{-2} day^{-1} and as high as 6 kW h m^2 day^{-1}, the break-even points range from approximately 50 m^3 day^{-1} to 90 m^3 day^{-1} respectively.

18.3.4.6 Sensitivity to pumping head

Figure 18.11 presents a graph comparing the life-cycle costs of PV- and diesel-powered systems over a range of pumping demands and at varying pumping heads. For the base-case pumping head of 25 m, the break-even point between PV and diesel costs is approximately 70 m^3 day^{-1}. As the pumping head is varied to as low as 10 m and as high as 50 m, the break-even points

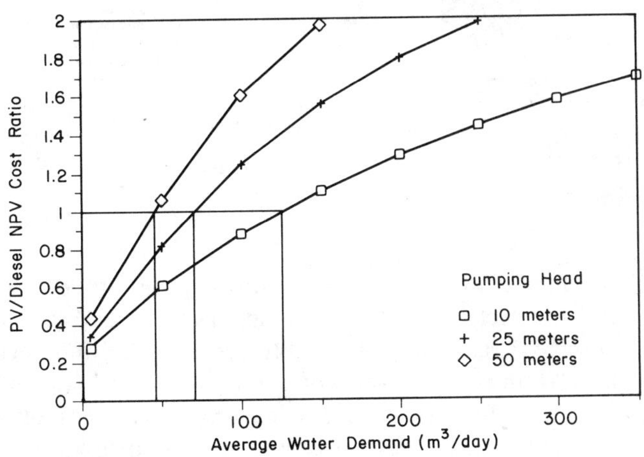

Figure 18.11 Sensitivity of pumping costs to pumping head.

range from approximately 125 $m^3\,day^{-1}$ to 45 $m^3\,day^{-1}$ respectively. This sensitivity follows the general relation between energy demand and PV system financial viability since the pumping head multiplied by the volume is directly related to energy. The financial competitiveness of the PV system compared with the diesel system increases with decreasing energy demand.

18.4 COMMUNICATION ANALYSIS

The financial analysis presented in this section compares PV- and diesel-powered microwave repeater systems assuming development agency financing. The analysis shows that PV-powered systems are the least-cost option at daily energy demands of up to 5 kW h, even under unfavourable financial assumptions (figure 18.12). When the financial parameters are more favourable, PV-powered systems are competitive up to 24 kW h day^{-1}. These results are consistent with the substantial number of commercial PV applications that are in place today.

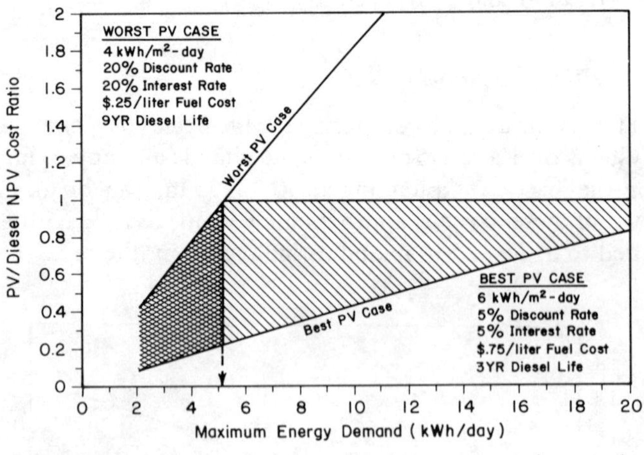

Figure 18.12 Sensitivity of communications costs to best and worst conditions.

The graph in figure 18.12 depicts the ratio of PV- to diesel-powered communications life-cycle costs for the best-PV-case and worst-PV-case scenarios. Both scenarios assume 20 year life-cycle costing and development agency financing. At an NPV cost ratio of 1.0, PV and diesel system life-cycle costs are equal. Under the best-PV-case scenario, the five parameters shown on the graph are adjusted to 'reasonable extremes' that favour PV systems. An opposite adjustment is made under the worst-PV-case scenario. The area between the two curves represents a reasonable range of financial assumptions.

Figure 18.13 depicts the various cost elements of PV- and diesel-powered microwave repeater systems. PV life-cycle costs are equally divided between debt service (initial system cost) and battery replacement, suggesting that battery life and cost are important parameters. The cost of diesel-powered systems is heavily dependent on fuel cost. Based on this information and the sensitivity analyses presented in this section, the major cost parameters in this analysis were determined to be battery life and cost, fuel cost and diesel system lifetime.

The remainder of this section discusses the analyses and assumptions leading to figures 18.12 and 18.13.

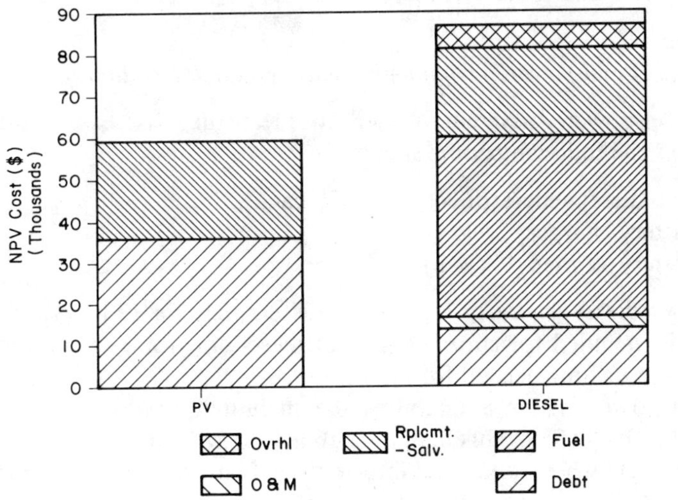

Figure 18.13 Base-case communications system cost components.

18.4.1 Description of the base case

The application considered for the base-case communications system was a DC-powered microwave repeater station. The repeater itself was assumed to be identical for both the PV and the conventional systems. Therefore the financial comparison is performed only between power systems, excluding any load costs. The base-case power requirement chosen for the typical load is a constant 7.2 kW h day^{-1}. The maximum daily load is equal to the constant daily load. Continuous 24 h operation is assumed to be a critical design parameter.

The base-case conventionally powered system is a tandem diesel generator system. It is assumed that the supply of diesel fuel and PV and diesel spare parts is never interrupted. The PV- and diesel-powered systems are discussed in detail in the following sections.

18.4.1.1 Photovoltaic-powered communications system description

The components for the base-case PV power system for communications applications include the following.

(i) PV array (2.3 kW_p).
(ii) Charge controller.
(iii) Battery storage (63 kW h).

Battery storage is sized for 7 days. The lowest-month daily irradiation incident on the array is 4 kW h m^{-2} day^{-1}. Availability of the PV-powered system is assumed to be 99.9%.

18.4.1.2 Diesel-powered communications system description

The diesel-powered system chosen to meet the base-case requirements comprises the following components.

(i) Two diesel generator sets (at 3 kW each).
(ii) Battery charger.
(iii) Battery storage (9 kW h).

The two diesel generator sets operate alternately for 12 h periods. Battery storage is included for short periods of diesel downtime (1 day) caused by dual failure or time needed for maintenance.

Diesel power system availability, not including breaks in fuel supply, is assumed to be 99.9%. This high availability is the result of having a tandem generator set, where unavailability is the square of a single generator set unavailability. Thus, on the assumption of a 97.5% availability of one generator set, the availability of a tandem set is $1-(0.025)^2$ or 99.9%.

18.4.2 System costs

18.4.2.1 Photovoltaic-powered communications system costs

The initial capital cost for the PV-powered system represents three components: the array, batteries and electronics. Recurring costs are those associated with component replacement, maintenance and repair. The PV-powered system cost breakdown for the communications base case is summarized in table 18.10. All the costs presented in this section refer to an FOB manufacturer.

PV module costs range from US \$6 W_p^{-1} to US \$8 W_p^{-1} as a function of the magnitude of the order and the potential, in the eyes of the manufacturer, for

Table 18.10 PV power system costs for communications base case.

Specification	Cost
Initial capital costs (FOB manufacturer)	
PV array (2.3 kW_p)	US \$18 182
Controller	909
Battery (63 kW h)	US \$ 9 450
Total capital cost	US \$28 541
Recurring capital costs (FOB manufacturer)	
Battery replacement	US \$9450 every 5 years[a]
Controller replacement	US \$909 every 10 years[a]
Other recurring costs (% initial capital costs)	
Maintenance and repair	0.05% $year^{-1}$

[a] Plus appropriate cost escalation due to general inflation.

future sales. Mounting hardware (e.g. support structure) and array wiring costs (area-related balance-of-system costs) amount to an additional US \$0.50–1 W_p^{-1}. For the base case, a value of US \$8 W_p^{-1} is used as the total of module and area-related balance-of-system costs.

The batteries used for communications applications are sealed deep-discharge batteries. The cost for these batteries is approximately US \$150 kW h^{-1} and the cost of the electronics is approximately US \$0.40 W_p^{-1}. The batteries are assumed to need replacement every 5 years and the control electronics every 10 years. The only additional recurring cost is for annual maintenance and repair, which is 0.05% of the total system initial capital cost.

18.4.2.2 Diesel-powered communications system costs

The capital cost of the diesel power system is based on the use of two diesel generator sets, a battery charger and a battery bank. The cost breakdown for both initial and recurring costs is detailed in table 18.11. The costs presented are for an FOB manufacturer.

Although the maximum load requirements can be met with 500 W generator sets, it is necessary to assume the use of 3 kW generator sets since diesels are not commonly available below that size. Costs are based on the same data assumed in the diesel-powered pumping system (figure 18.3 from the previous section). The engines need replacement every 6 years.

Batteries for the diesel power system are of the sealed deep-discharge variety. The battery charger price is included in the battery price, for a total

Table 18.11 Diesel-powered system costs for communications base case.

Specification	Cost
Initial capital costs (FOB manufacturer)	
Diesel generator set (2 at 3 kW)	US $9541
Battery (9 kW h)/charger	US $1350
Total capital cost	US $10 891
Recurring capital costs (FOB manufacturer)	
Diesel generator set replacement	US $9541 every 6 years[a]
Battery, charger replacement	US $1350 every 5 years[a]
Other recurring costs	
Engine overhaul	15% of engine cost every 3 years[b]
Maintenance and repair	2% of system cost per year[a]
Diesel fuel (6816 l at US $0.50 l^{-1})	US $3408 $year^{-1}$ [a]

[a] Plus appropriate escalation due to general inflation.
[b] Overhauls not performed during engine replacement years.

of US $150 $kW\,h^{-1}$ of battery capacity. These components are assumed to need replacement every 5 years.

Fuel consumption rates are based on the same values outlined in table 18.6 of the section on pumping (§ 18.3). Engine overhauls are assumed to require

Table 18.12 PV-powered communications system 20 year cash flow (base case).

Year	1	2	3	4	5	6	7	8
Debt service (US $)	4191	4191	4191	4191	4191	4191	4191	4191
Operating and maintenance expenses								
Annual maintenance (US $)	15	16	17	17	18	19	20	21
Recurring capital costs								
Battery replacement cost (US $)	0	0	0	0	15 076	0	0	0
Controller replacement cost (US $)	0	0	0	0	0	0	0	0
Salvage value (US $)	0	0	0	0	0	0	0	0
Total cash outflow (US $)	4205	4206	4207	4208	19 285	4210	4211	4212
$DF = 1/[(1+DR)^{years}]$	0.9091	0.8264	0.7513	0.683	0.6209	0.5645	0.5132	0.4665
NPV stream	3823	3476	3161	2874	11 974	2376	2161	1965
Total NPV of cash outflow (US $)	59 230							
PV system generation (kW h $year^{-1}$)	2625	2625	2625	2625	2625	2625	2625	2625

an expenditure equivalent to 15% of the engine capital cost every 3 years. Annual maintenance and repair costs are assumed to be equivalent to 2% of the capital cost of the system.

18.4.3 20 year life-cycle costs

Using the assumptions outlined in the previous sections, 20 year cash flows are presented in tables 18.12 and 18.13 for the base-case PV and diesel power systems respectively. The results are expressed as the total NPV cost for PV- and diesel-powered systems to provide an average of 7.2 kW h day^{-1} over 20 years with 99.9% reliability. This base-case value is US $59 230 for PV and US $86 529 for diesel systems. These base-case values show that, at the assumed energy demand, the cost of the PV-powered system is less than that of the diesel-powered system. Thus, for communications loads less than 7.2 kW h day^{-1} (300 W continuous), PV-powered systems are more cost effective than diesel-powered systems. (The model used to develop these NPV cost figures and the sensitivity analyses that follow is provided in appendix A.)

18.4.4 Sensitivity analyses

For communications systems, sensitivity analyses demonstrate the effect of varying the capital cost, discount and interest rate, diesel fuel cost, diesel lifetime and irradiation on the life-cycle NPV cost comparisons between PV- and diesel-powered systems. In each of the following sensitivity analyses

9	10	11	12	13	14	15	16	17	18	19	20	NPV comparison
4191	4191	4191	4191	4191	4191	4191	4191	4191	4191	4191	4191	35 676
22	23	24	26	27	28	30	31	33	34	36	38	181
0	19 241	0	0	0	0	24 557	0	0	0	0	0	22 658
0	1851	0	0	0	0	0	0	0	0	0	0	714
0	0	0	0	0	0	0	0	0	0	0	0	0
4213	25 306	4215	4216	4217	4219	28 778	4222	4223	4225	4227	4228	
0.4241	0.3855	0.3505	0.3186	0.2897	0.2633	0.2394	0.2176	0.1978	0.1799	0.1635	0.1486	
1787	9757	1477	1343	1222	1111	6889	919	836	760	691	629	
2625	2625	2625	2625	2625	2625	2625	2625	2625	2625	2625	2625	

Table 18.13 Diesel-powered communications system 20 year cash flow (base case).

Year	1	2	3	4	5	6	7	8
Debt service (US $)	1599	1599	1599	1599	1599	1599	1599	1599
Operating and maintenance expenses								
Annual maintenance (US $)	229	240	252	265	278	292	307	322
Diesel engine overhaul (US $)	0	0	1657	0	0	1918	0	0
Fuel (US $)	3578	3757	3945	4142	4350	4567	4795	5035
Recurring capital costs								
Battery replacement cost (US $)	0	0	0	0	2154	0	0	0
Diesel generator replacement cost (US $)	0	0	0	0	0	15 983	0	0
Salvage value	0	0	0	0	0	0	0	0
Total cash outflow	5406	5597	7453	6006	8380	24 359	6701	6956
$\text{DF} = 1/[(1+\text{DR})^{\text{years}}]$	0.9091	0.8264	0.7513	0.683	0.6209	0.5645	0.5132	0.4665
NPV stream (US $)	4915	4625	5600	4102	5204	13 750	3429	3245
Total NPV of cash outflow (US $)	86 529							
PV system generation (kW h year^{-1})	2626	2626	2626	2626	2626	2626	2626	2626

except capital cost, the ratios of the life-cycle NPV costs for the two systems are plotted against a range of energy demands at several values for the sensitivity parameters. In the capital cost analysis, the actual NPV costs are individually presented for the PV- and diesel-powered systems.

18.4.4.1 *Sensitivity to capital costs*

The sensitivity of the life-cycle costs of PV- and diesel-powered communi-

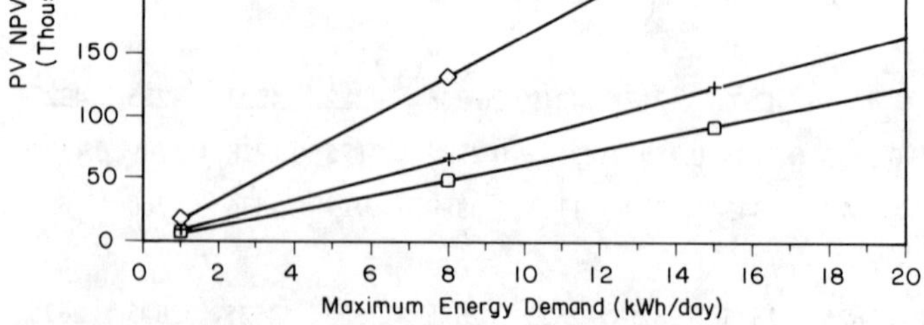

Figure 18.14 Sensitivity of communications system costs to PV-powered system capital costs.

9	10	11	12	13	14	15	16	17	18	19	20	NPV comparison
1599	1599	1599	1599	1599	1599	1599	1599	1599	1599	1599	1599	13 614
338	355	373	391	411	431	453	475	499	524	550	578	2770
2220	0	0	2570	0	0	2975	0	0	3444	0	0	5420
5287	5551	5829	6120	6426	6748	7085	7429	7811	8202	8612	9042	43 342
0	2749	0	0	0	0	3508	0	0	0	0	0	3237
0	0	0	21 419	0	0	0	0	0	28 703	0	0	21 009
0	0	0	0	0	0	0	0	0	0	0	19 265	2864
9444	10 254	7801	32 100	8436	8778	15 621	9514	9910	42 473	10 761	(8046)	
0.4241	0.3855	0.3505	0.3186	0.2897	0.2633	0.2394	0.2176	0.1978	0.1799	0.1635	0.1486	
4005	3953	2734	10 228	2444	2312	3739	2070	1961	7639	1760	(1196)	
2626	2626	2626	2626	2626	2626	2626	2626	2626	2626	2626	2626	

cations systems to the capital costs for equipment is compared using capital cost multipliers ranging from 0.75 to 2.0, where the base is 1.0. The sensitivity of communications cost to PV- and diesel-powered system capital costs is illustrated in figures 18.14 and 18.15 respectively. For the PV-powered system, NPV cost is directly proportional to energy demand as shown in figure 18.14. The NPV cost for the diesel-powered system is presented as a function of a load range. The use of a multiplier represents the effect of

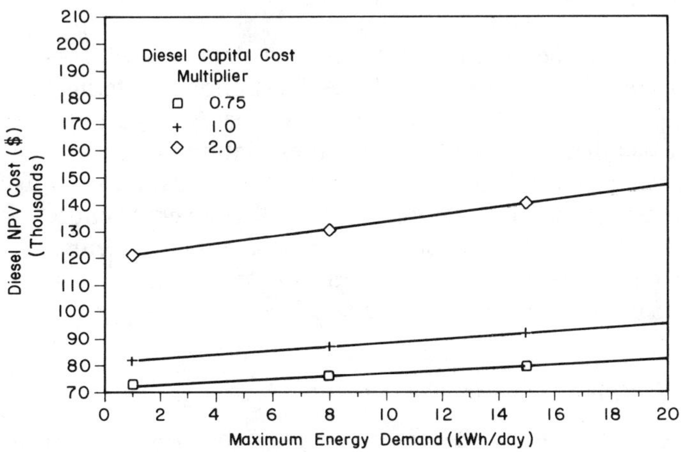

Figure 18.15 Sensitivity of communications system costs to diesel-powered system capital costs.

different CIF equipment costs and accounts for installation costs, which were not included in the base case.

18.4.4.2 Sensitivity to discount and interest rate

Figure 18.16 shows that, as the discount and interest rate increase, the demand range over which PV-powered systems are financially favourable becomes smaller. The base case, with 10% discount and interest rates, shows the cross-over between PV- and diesel-powered systems to occur at about 11.0 kW h day^{-1}. At a rate of 5%, PV is financially more attractive up to 13.4 kW h day^{-1} while, at a 20% rate, the cross-over is at 7.8 kW h day^{-1}.

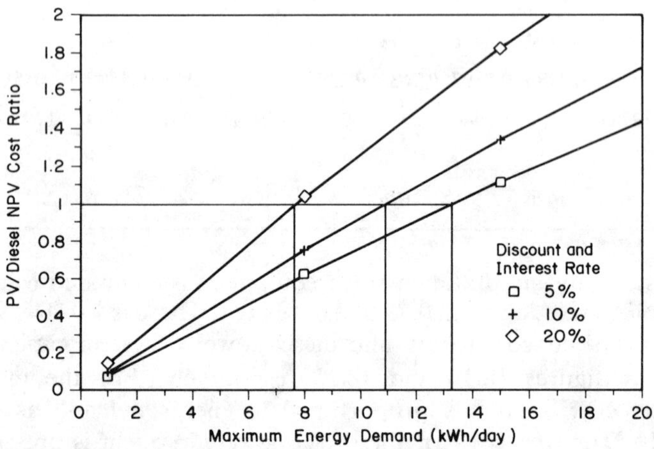

Figure 18.16 Sensitivity of communications system costs to discount and interest rate.

As discussed in the pumping analysis, this shift in the cross-over occurs because of the nature of the PV and diesel cash flows, in terms of recurring against capital costs. The diesel cash flow has a higher proportion of recurring costs that are subject to annual escalation. The base-case PV cash flow, on the other hand, is less affected by cost escalation since its proportionally larger debt service component is levellized over the entire 20 year life of the system. However, since the PV system also has large recurring costs due to battery replacement, the NPV cost ratio is less sensitive to discount and interest rate in this application than the others.

18.4.4.3 Sensitivity of diesel fuel cost

This analysis examines the effect of varying fuel cost from the base-case value of US $0.50 l^{-1}. Figure 18.17 shows the impact of using fuel costs of

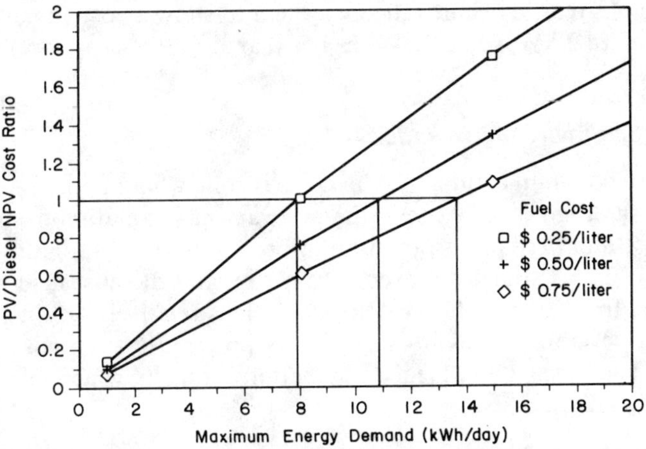

Figure 18.17 Sensitivity of communications system costs to diesel fuel cost.

US \$0.25 l^{-1}, US \$0.50 l^{-1} and US \$0.75 l^{-1}. While the base case of US \$0.50 l^{-1} shows the PV system to be more cost effective up to demands of 10.9 kW h day^{-1}, at US \$0.25 l^{-1}, the cross-over between PV and diesel systems occurs at about 7.9 kW h day^{-1}. When using the high cost of US \$0.75 l^{-1}, the system PV is the more cost-effective option up to demands of 13.8 kW h day^{-1}.

18.4.4.4 Sensitivity to diesel lifetime

Figure 18.18 demonstrates the effect of using a 3, 6 and 9 year diesel lifetime on communications system life-cycle costs (the base case uses 6 years). A

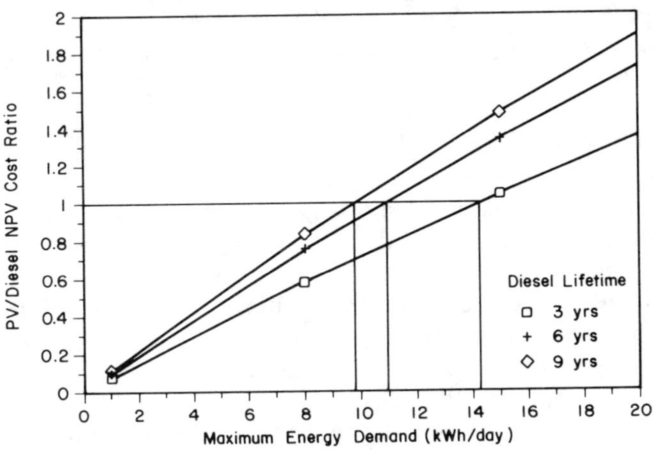

Figure 18.18 Sensitivity of communications system costs to diesel lifetime.

diesel with a 3 year life causes the PV system to show a cost advantage up to demands of 14.2 kW h day^{-1}. With a 9 year life, this cross-over occurs at 9.6 kW h day^{-1}.

18.4.4.5 Sensitivity to irradiation

The ratio of communications system costs is somewhat insensitive to change in irradiation. Figure 18.19 shows that, by varying irradiation from the base-case lowest-month daily irradiation of 4 up to an irradiation level of 6 kW h m^{-2} day^{-1}, the cross-over between PV and diesel system attractiveness ranges from 10.8 to 12.6 kW h day^{-1}. This relatively low sensitivity to irradiation variation arises because the PV-powered system costs for communications are heavily influenced by battery capacity which is not directly related to irradiation.

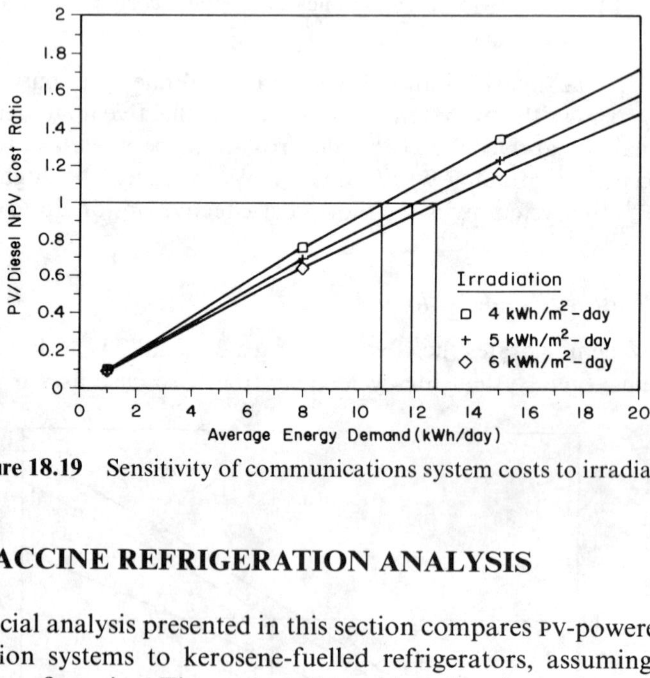

Figure 18.19 Sensitivity of communications system costs to irradiation.

18.5 VACCINE REFRIGERATION ANALYSIS

The financial analysis presented in this section compares PV-powered vaccine refrigeration systems to kerosene-fuelled refrigerators, assuming development agency financing. The analysis shows that there is no clear-cut range of viability for either PV or kerosene systems (figure 18.20). PV-powered system viability, for both small and large systems, is always in the break-even range (or very dependent on case-specific parameters).

The bar charts in figure 18.20 depict the ratio of PV- to kerosene-powered vaccine refrigerator life-cycle costs for the best-PV-case and worst-PV-case scenarios. Both scenarios assume 20 year life-cycle costing and development agency financing. At an NPV cost ratio of 1.0, PV and kerosene system life-

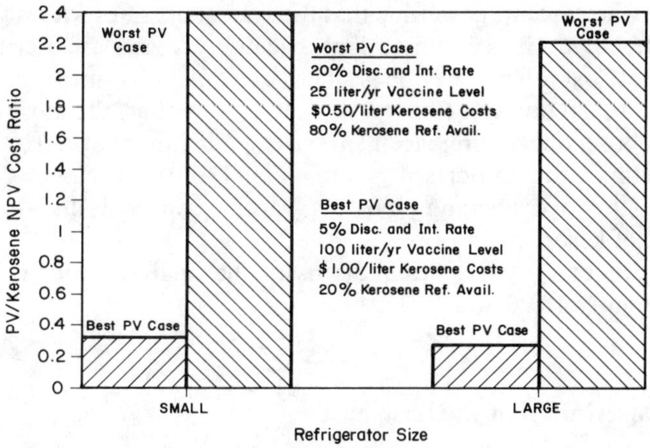

Figure 18.20 Sensitivity of refrigeration costs to best and worst conditions.

cycle costs are equal. Under the best-PV-case scenario, the five parameters shown on the graph are adjusted to 'reasonable extremes' that favour PV systems. An opposite adjustment is made under the worst-PV-case scenario.

The sensitivity analyses in this section, and the life-cycle cost elements illustrated in figure 18.21, indicate which are the most important cost parameters when comparing PV-powered and kerosene-fuelled refrigerators. The most critical assumptions are related to vaccine wastage, which is a function of the annual vaccine dose requirement and system operating

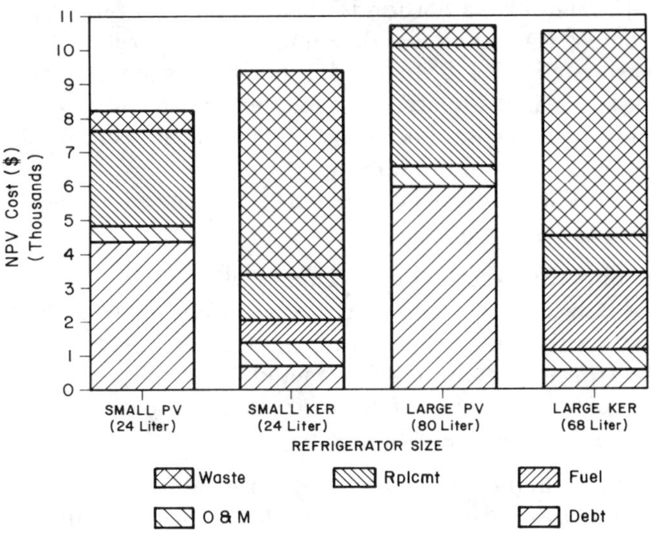

Figure 18.21 Base-case refrigeration system cost components.

availability (the percentage of time that the system operates within the proper temperature range). It is assumed that, because vaccines are a critical item, any vaccines lost owing to the system unavailability must be replaced through pure cash outlays. For the PV-powered systems, the most dominant costs are debt service and replacement costs, indicating that refrigerator and battery lifetimes are important parameters. For the kerosene-fuelled refrigerators, the overwhelming cost is vaccine wastage due to the low availability of kerosene units.

The remainder of this section discusses the analyses and assumptions leading to figures 18.20 and 18.21.

18.5.1 Description of the base case

The base-case vaccine refrigeration comparison is between PV- and kerosene-powered units. System specifications are taken directly from World Health Organization (WHO) product information sheets. The energy consumption estimates are based on test data at an ambient temperature of 32 °C, with no ice pack capable of freezing; comparable energy consumption data with freezing were not available. While system specifications are based on test conditions, they are considered valid for field operation assuming that adequate user training is provided.

An important aspect of system performance is operating availability. For vaccine refrigeration, availability is the percentage of time that the system is operating within the prescribed temperature range. When availability is less than 100%, there are both quantifiable and unquantifiable costs of lost vaccines. The quantifiable portion is the actual cost of the vaccines that are wasted. From WHO data, it was determined that an average litre of vaccine costs approximately US $19. In § 18.5.2 the impacts of this cost for both PV and diesel systems are compared. The unquantifiable impact of lost vaccines is the potential human cost of not maintaining a reliable supply of vaccines in developing country health programmes.

In the financial analysis, the cost of wasted vaccines was quantified based on the following assumptions.

(*a*) A litre of packed vaccines costs US $19.

(*b*) 50 l of vaccines are needed per year. This figure is equivalent to vaccinating the children and mothers within a village of about 45 000 people. (It is important to keep in mind that remote health centres that serve 20 000–100 000 people have been identified as having the greatest need for solar-powered vaccine refrigeration.)

The selected PV- and kerosene-powered refrigeration systems and their life-cycle cost comparisons are described in the following sections. The comparisons are made on the basis of the NPV life-cycle cost in US dollars per litre of refrigerated space.

18.5.1.1 Photovoltaic-powered refrigeration system description

The two PV-powered refrigeration systems used in the analysis are based on data provided by BP Solar Systems, for a Lec EV 5750 refrigerator, and Solarex Corporation, for a Marvel 4RTD refrigerator.

These suppliers have already sized systems for WHO vaccination programme specifications. It is assumed that the average irradiation incident on the PV array is between 5.8 and 7.0 kW h day^{-1}. The battery capacity is sized for 5 days of no sun. Both systems included the following components.

(i) PV array.
(ii) Charge controller.
(iii) Battery storage.
(iv) Refrigerator.

Technical specifications for these two systems are outlined in table 18.14. The availability (i.e. the percentage of time operating within the required temperature range) of PV-powered refrigeration systems is assumed to be 95%.

Table 18.14 Base-case PV-powered refrigeration system specifications (based on 32 °C ambient temperature, an irradiation of 5.8–7.0 kW h m^{-2} day^{-1}, 5 days no-sun security, and no freezing).

System supplier	Refrigerator vaccine storage (l)	Array size (W_p)	Battery size (kW h)	Power consumption (kW h day^{-1})
BP Solar	24	198	5.47	0.35
Solarex	80	168	3.60	0.43

18.5.1.2 Kerosene-fuelled refrigeration system description

Kerosene-powered refrigeration systems are self-contained units. The two units selected for this analysis are an Electrolux RCW 42 EK and Sibir S2325. The technical specifications for these two systems are outlined in table 18.15. The availability of these systems is assumed to be 50%.

Table 18.15 Base-case kerosene-fuelled refrigeration system specifications (based on 32 °C ambient temperature and no freezing).

System supplier	Refrigerator vaccine storage (l)	Fuel consumption (l day^{-1})
Electrolux	24	0.2
Siber	68	0.7

18.5.2 System costs

18.5.2.1 Photovoltaic-powered refrigeration system costs

The costs for the PV-powered refrigeration systems were provided by their suppliers in the WHO product information sheets and are outlined in table 18.16. The small PV-powered refrigerator unit (BP Solar) selected for analysis costs US \$3500, while the large unit (Solarex) costs US \$4781.

Table 18.16 PV-powered refrigeration system costs. System A applies to areas receiving 3.5–4.7 kW h m^{-2} day^{-1}, includes 8 days of no-sun security and assumes no ice-making during periods of no sun. System B applies to areas receiving 5.8–7.0 kW h m^{-2} day^{-1}, includes 5 days of no-sun security and assumes no ice-making during periods of no sun.

Supplier	Refrigerator vaccine storage (l)	Cost (FOB manufacturer) (US \$) (based on the purchase of 1–9 units)	
		System A	System B
AEG	90	5800	4700
BP Solar	24	4300	3500
Leroy Somer	16	9628	8278
Polar products	90	5675	5150
Solarex	(*a*) 80 (*b*) 90	(*a*) 5507 (*b*) 6458	(*a*) 4781 (*b*) 5732
Solavolt International	(*a*) 80 (*b*) 90	(*a*) 5335 (*b*) 6861	(*a*) 4386 (*b*) 5912

The PV array has a lifetime of 20 years. The batteries must be replaced every 5 years. The refrigerator and electronic controls are assumed to have a life of 10 years. To obtain replacement costs for these components, the array cost of US \$8 W_p^{-1} and battery costs of US \$150 kW h^{-1} are subtracted from the total system cost. Thus, values of US \$1096 and US \$2897 are assigned to the refrigerator and controls replacement costs for the BP Solar and Solarex systems respectively. Annual maintenance and repair costs are equivalent to 1% of the total system cost. Table 18.17 provides the cost breakdown for the PV-powered refrigeration systems.

18.5.2.2 Kerosene-fuelled refrigeration system costs

Costs for these systems are provided by their manufacturers in the WHO product information sheets. System costs are outlined in table 18.18. Based

Table 18.17 PV-powered system costs for refrigeration base case.

Specification	Cost	
	Small system BP Solar	Large system Solarex (Marvel)
Initial capital cost (FOB manufacturer)		
System cost	US $3500	US $4781
Recurring capital costs (FOB manufacturer)		
Battery replacement	US $821 every 5 years[a]	US $450 every 5 years[a]
Refrigerator, controls replacement	US $1096 every 10 years[a]	US $2897 every 10 years[a]
Other recurring costs (% initial system cost)		
Maintenance and repair	1% $year^{-1}$ [a]	1% $year^{-1}$ [a]

[a] Plus appropriate escalation due to general inflation.

Table 18.18 Kerosene-fuelled refrigerator system costs (based on the cost of one system).

Refrigerator model	Refrigerator vaccine storage (l)	Cost (FOB manufacturer) (US $)
Electrolux RC 65	142	725
Electrolux RCW 65	32	1213
Electrolux RCW 42 EK	24	552
Electrolux RCW 42 EKG	21	775
Siber S2325	68	458

on the purchase of one system, the Electrolux unit used in the analysis costs US $552 and the Sibir unit costs US $458. These entire units were assumed to need replacement every 5 years.

Fuel consumptions rates are outlined in table 18.15. A fuel cost of US $0.70 l^{-1} is used, as outlined in § 2.1. Annual maintenance and repair is equivalent to 10% of the initial system cost. Table 18.19 outlines the kerosene-fuelled system costs used in the comparative analysis.

18.5.3 20 year life-cycle costs

20 year cash flows for the small and large PV- and kerosene-powered refrigeration systems are presented in tables 18.20–18.23. In the small case,

Table 18.19 Kerosene-fuelled system costs for refrigeration base case.

Specification	Cost	
	Small system Electrolux RCW 42 EK	Large system Sibir
Initial capital costs (FOB manufacturer)		
System cost	US $552	US $458
Recurring capital costs (FOB manufacturer)		
System replacement	US $552 every 5 years[a]	US $458 every 5 years[a]
Other recurring costs		
Maintenance and repair	10% $year^{-1}$ [a]	10% $year^{-1}$ [a]
Kerosene fuel (at US $0.70 l^{-1})	US $51 $year^{-1}$ [a]	US $179 $year^{-1}$ [a]

[a] Plus appropriate escalation due to general inflation.

the NPV costs for the PV- and kerosene-powered systems are US $8252 and US $9406 respectively. The large units have NPV costs of US $10 757 for the PV system and US $10 569 for the kerosene system. Note that, for the large units, capacity is 80 l for the PV system and 68 l for the kerosene system.

Table 18.20 Small PV-powered refrigeration system 20 year cash flow (base case).

Year	1	2	3	4	5	6	7	8
Debt service (US $)	514	514	514	514	514	514	514	514
Operating and maintenance expenses								
Annual maintenance (US $)	37	39	41	43	45	47	49	52
Recurring capital costs								
Battery replacement cost (US $)	0	0	0	0	1309	0	0	0
Refrigerator, control replacement (US $)	0	0	0	0	0	0	0	0
Vaccine waste (US $)	50	52	55	58	61	64	67	70
Total cash outflow (US $)	601	605	609	614	1928	624	630	636
$DF = 1/[(1 + DR)^{years}]$	0.9091	0.8264	0.7513	0.683	0.6209	0.5645	0.5132	0.4665
NPV stream (US $)	546	500	458	419	1197	352	323	297
Total NPV of cash outflow (US $)	8252							
Refrigerator vaccine storage capacity (l)	24.0							

The significant financial impact of lost vaccines becomes clear upon examining the 'vaccine wastage' line item in each of the cash flows. For the base case, 50 l of good vaccines are required per year (i.e. for children and mothers within a village of approximately 45 000 people). If vaccines are wasted, they must be replaced at a cost of US $19 l^{-1}. In the first year, the PV systems' 95% availability results in losses of only US $50. Alternatively, the kerosene systems' 50% availability produces a first-year vaccine wastage of US $499.

It should be pointed out that the unquantifiable cost of lost vaccines (i.e. the human cost of not having vaccines when needed) is not included in this analysis. This cost could significantly alter the comparison of PV and kerosene systems. In this analysis, kerosene-fuelled refrigerators are marginally less expensive in some of the sensitivity cases presented. However, the vaccine loss from kerosene systems is substantially greater (10 times higher) than that for PV systems. Therefore, if the unquantifiable costs could be quantified, it would most likely sway many specific analyses in favour of PV systems.

18.5.4 Sensitivity analyses

Sensitivity analyses demonstrate the effect that varying certain parameters has on the life-cycle cost of refrigeration systems. The parameters include capital cost, discount and interest rate, kerosene fuel cost, litres of vaccines needed per year, and kerosene system operating availability. Sensitivity analyses are performed on the basis of NPV levellized annual cost per unit litre

9	10	11	12	13	14	15	16	17	18	19	20	NPV comparison
514	514	514	514	514	514	514	514	514	514	514	514	4375
54	57	60	63	66	69	73	76	80	84	88	93	445
0	1671	0	0	0	0	2132	0	0	0	0	0	1967
0	2232	0	0	0	0	0	0	0	0	0	0	860
74	77	81	85	90	94	99	104	109	114	120	126	604
642	4550	655	662	669	677	2818	694	703	712	722	733	
0.4241	0.3855	0.3505	0.3186	0.2897	0.2633	0.2394	0.2176	0.1978	0.1799	0.1635	0.1486	
272	1754	230	211	194	178	675	151	139	128	118	109	

Table 18.21 Small kerosene-fuelled refrigeration system 20 year cash flow (base case)

Year	1	2	3	4	5	6	7	8
Debt service (US $)	81	81	82	81	81	81	81	81
Operating and maintenance expenses								
Annual maintenance (US $)	58	61	64	67	70	74	78	82
Fuel (US $)	54	56	59	62	65	68	72	75
Recurring capital costs								
Total system replacement (US $)	0	0	0	0	881	0	0	0
Vaccine waste (US $)	499	524	550	577	606	637	668	702
Total cash outflow (US $)	691	722	754	788	1704	860	899	940
DF $=1/[(1+\text{DR})^{\text{years}}]$	0.9091	0.8264	0.7513	0.683	0.6209	0.5645	0.5132	0.4665
NPV stream (US $)	629	597	566	538	1058	485	461	438
Total NPV of cash outflow (US $)	9406							
Refrigerator storage capacity (l) 24.0 (refrigerator space)								

Table 18.22 Large PV-powered refrigeration system 20 year cash flow (base case).

Year	1	2	3	4	5	6	7	8
Debt service (US $)	702	702	702	702	702	702	702	702
Operating and maintenance expenses								
Annual maintenance (US $)	50	53	55	58	61	64	67	71
Recurring capital costs								
Battery replacement cost (US $)	0	0	0	0	861	0	0	0
Refrigerator, control replacement (US $)	0	0	0	0	0	0	0	0
Vaccine waste (US $)	50	52	55	58	61	64	67	70
Total cash outflow (US $)	802	807	812	818	1685	830	836	843
DF $=1/[(1+\text{DR})^{\text{years}}]$	0.9091	0.8264	0.7513	0.683	0.6209	0.5645	0.5132	0.4665
NPV stream (US $)	729	667	610	559	1046	468	429	393
Total NPV of cash outflow (US $)	10 757							
Refrigerator vaccine storage capacity (l)	80.0							

of capacity to account for different capacities in the large refrigerator system comparison. A sensitivity analysis was not performed on irradiation because the system specifications were based on irradiations ranging from 5.8 to 7.0 $\text{kW h m}^{-2}\,\text{day}^{-1}$.

9	10	11	12	13	14	15	16	17	18	19	20	NPV comparison
81	81	81	81	81	81	81	81	81	81	81	81	690
86	90	94	99	104	109	115	120	127	133	139	146	702
79	83	87	92	96	101	106	112	117	123	129	136	650
0	1124	0	0	0	0	1434	0	0	0	0	0	1324
737	774	812	853	896	940	987	1037	1089	1143	1200	1260	6041
983	2152	1075	1125	1177	1232	2724	1350	1413	1400	1550	1623	
0.4241	0.3855	0.3505	0.3186	0.2897	0.2633	0.2394	0.2176	0.1978	0.1799	0.1635	0.1486	
417	830	377	358	341	324	652	294	280	266	253	241	

9	10	11	12	13	14	15	16	17	18	19	20	NPV comparison
702	702	702	702	702	702	702	702	702	702	702	702	5976
74	78	82	86	90	95	99	104	110	115	121	127	608
0	1100	0	0	0	0	1403	0	0	0	0	0	1295
0	5899	0	0	0	0	0	0	0	0	0	0	2274
74	77	81	85	90	94	99	104	109	114	120	126	604
850	7855	865	873	882	891	2303	910	920	931	943	955	
0.4241	0.3855	0.3505	0.3186	0.2897	0.2633	0.2394	0.2176	0.1978	0.1799	0.1635	0.1486	
360	3029	303	278	255	235	551	198	182	168	154	142	

18.5.4.1 Sensitivity to capital cost

The effect of varying the capital cost of PV and kerosene refrigeration systems is depicted in figures 18.22 and 18.23 respectively. In each graph, the NPV cost is shown for variations in the capital cost of 0.75, 1.0 and 2.0 times the base

Table 18.23 Large kerosene-fuelled refrigeration system 20 year cash flow (base case)

Year	1	2	3	4	5	6	7	8
Debt service (US $)	67	67	67	67	67	67	67	67
Operating and maintenance expenses								
Annual maintenance (US $)	48	50	53	56	58	61	64	68
Fuel (US $)	188	197	207	217	228	240	252	264
Recurring capital costs								
Total system replacement (US $)	0	0	0	0	731	0	0	0
Vaccine waste (US $)	499	524	550	577	606	637	668	702
Total cash outflow (US $)	802	839	877	918	1691	1005	1052	1101
DF $=1/[(1+\text{DR})^{\text{years}}]$	0.9091	0.8264	0.7513	0.683	0.6209	0.5645	0.5132	0.4665
NPV stream (US $)	729	693	659	627	1050	567	540	514
Total NPV of cash outflow (US %)	10 569							
Refrigerator storage capacity (l) 68.0 (refrigerator space)								

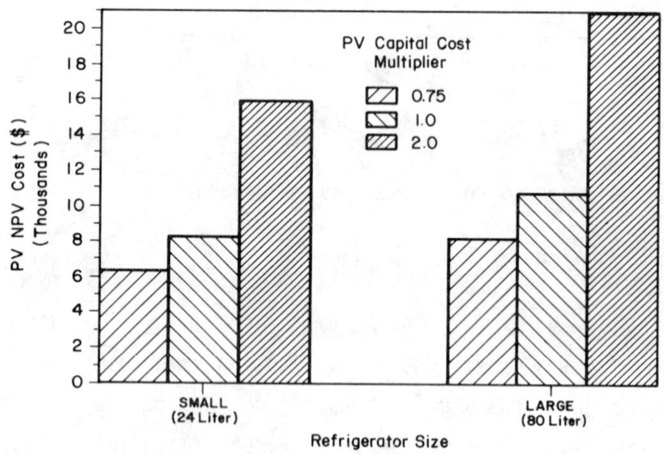

Figure 18.22 Sensitivity of PV-powered refrigeration costs to capital cost.

case for both small and large refrigerator systems. The range of multipliers is intended to account for system CIF capital cost variations. These variations could result from different CIF equipment costs and can account for installation costs.

9	10	11	12	13	14	15	16	17	18	19	20	NPV comparison
67	67	67	67	67	67	67	67	67	67	67	67	573
71	75	78	82	86	91	95	100	105	110	116	122	582
277	291	306	321	337	354	372	390	410	430	452	475	2275
0	933	0	0	0	0	1190	0	0	0	0	0	1098
737	774	812	853	896	940	987	1037	1089	1143	1200	1260	6041
1153	2139	1264	1324	1387	1453	2712	1594	1671	1751	1835	1924	
0.4241	0.3855	0.3505	0.3186	0.2897	0.2633	0.2394	0.2176	0.1978	0.1799	0.1635	0.1486	
489	825	443	422	402	382	649	347	331	315	300	286	

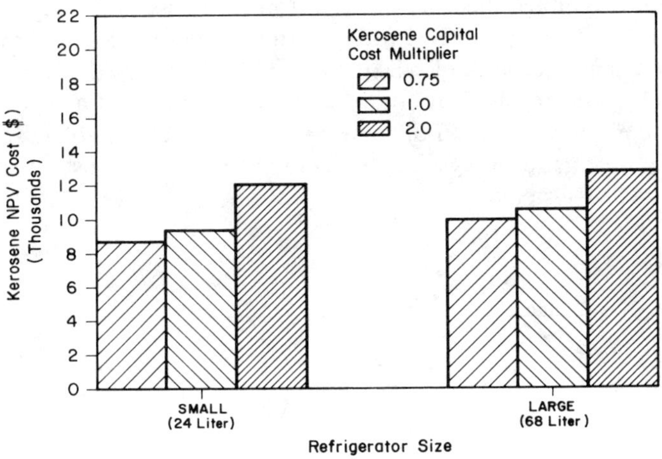

Figure 18.23 Sensitivity of kerosene fuelled refrigeration costs to capital cost.

18.5.4.2 Sensitivity to discount and interest rate

Figure 18.24 presents the sensitivity of refrigeration costs to discount and interest rate. As discount and interest rate decrease, PV-powered refrigeration systems become more cost effective than kerosene-fuelled systems beginning at rates between 10 and 20%.

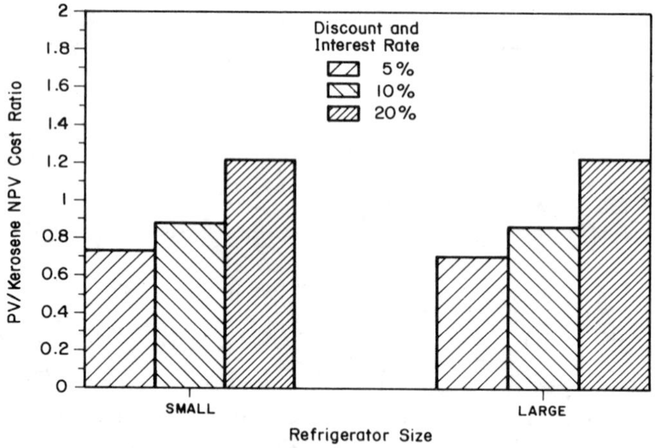

Figure 18.24 Sensitivity of refrigeration costs to discount rate.

18.5.4.3 Sensitivity to kerosene fuel cost

This section examines the effect that varying the cost of kerosene has on refrigeration costs. Figure 18.25 presents the ratio of the PV to kerosene NPV cost for kerosene fuel costs of US \$0.50 l^{-1}, US \$0.70 l^{-1} (the base case) and US \$1.00 l^{-1}. The graph shows that fuel cost does not play a significant role in the system's NPV costs. In all cases, even at a fuel cost of US \$0.50 l^{-1}, PV-powered refrigeration systems are shown to be more cost effective than kerosene-fuelled systems.

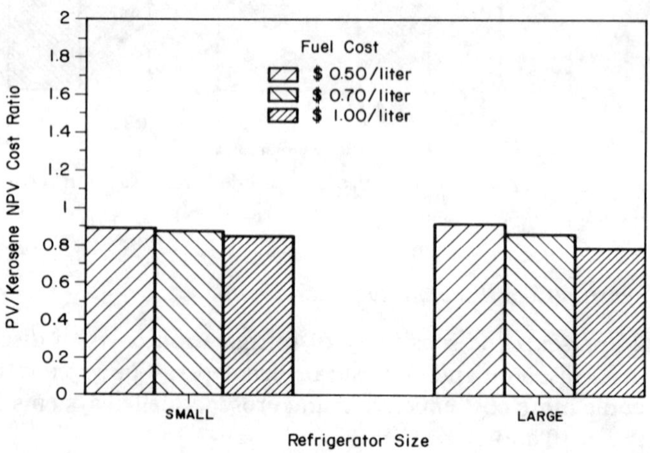

Figure 18.25 Sensitivity of refrigeration costs to kerosene fuel cost.

18.5.4.4 *Sensitivity to required litres of vaccines per year*

For the base-case analysis, it is assumed that 50 l of vaccines are needed per year. Thus, vaccines lost owing to unavailability of the system need to be replaced through pure cash outlays. Figure 18.26 shows the impact of varying the litres of vaccines required each year from 25 to 100 l, in 25 l intervals. This range corresponds to vaccinating the children and mothers of villages that have populations of 22 000–90 000 people. The graph indicates that between 25 and 50 l year^{-1}, both PV-powered systems become financially more attractive. This result is due to the lower availability of kerosene-fuelled refrigerators. Lower availability results in higher vaccine wastage and thus higher recurring costs to replace the lost vaccines. The unquantifiable cost of lost vaccines is not included in this analysis.

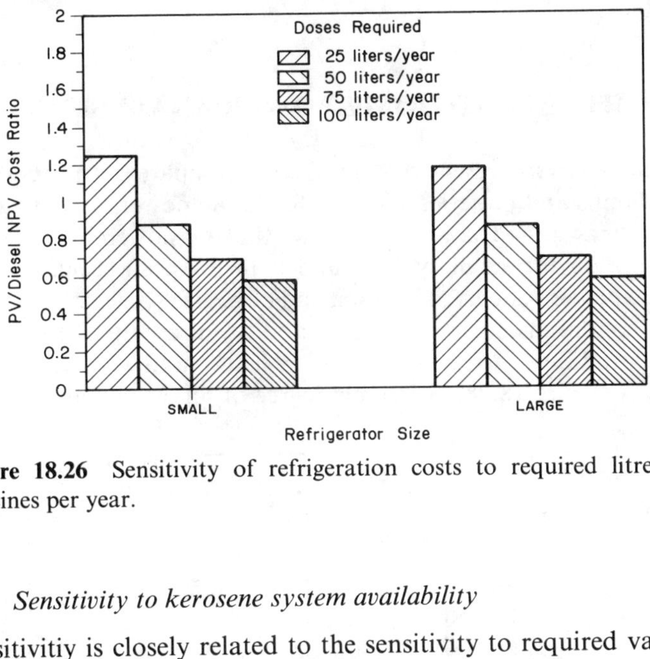

Figure 18.26 Sensitivity of refrigeration costs to required litres of vaccines per year.

18.5.4.5 *Sensitivity to kerosene system availability*

This sensitivitiy is closely related to the sensitivity to required vaccines per year (§ 18.5.4), in that it has a strong impact on vaccine wastage. Figure 18.27 demonstrates the effect of varying kerosene system operating availability from 20 to 80% on refrigeration comparative costs (the base-case operating availability is 50%). Increasing kerosene system availability from 50 to 80% has a greater impact on the PV/kerosene NPV cost ratio than decreasing system availability from 50 to 20%. Obviously, as the kerosene system availability decreases from the base case, the attractiveness of PV-powered refrigeration increases.

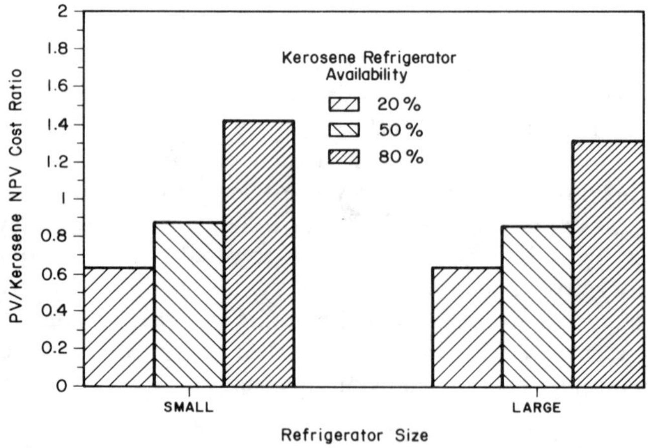

Figure 18.27 Sensitivity of refrigeration costs to kerosene system availability.

18.6 LIGHTING AND HOME POWER ANALYSIS

The financial analysis presented in this section compares PV and conventional (kerosene lamps and batteries) systems for home power, assuming development bank financing. The analysis shows that PV-powered systems are the least-cost option for small systems under all financial scenarios. Conventional home power systems of medium and large size are least cost only under the worst conditions for PV system viability: discount and interest rates between 10 and 20%, irradiation of 4 $\mathrm{kW\,h\,m^{-2}}$ and kerosene fuel cost of US $0.50 l^{-1} (figure 18.28). The high degree of financial viability of the PV-

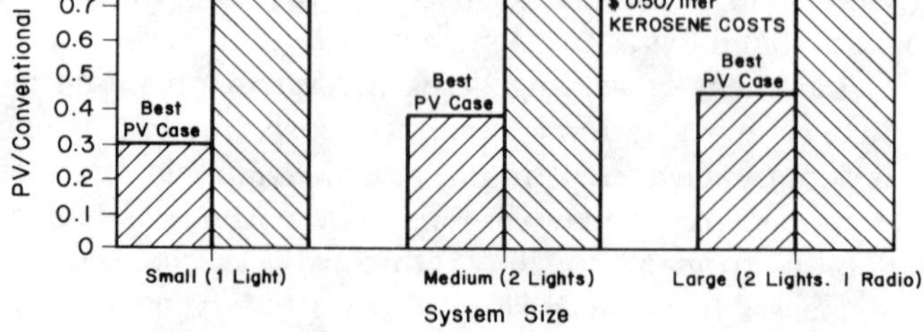

Figure 18.28 Sensitivity of lighting and home power costs to best and worst conditions.

powered systems suggests that shorter loan terms (applicable to individual private users) would still show PV system attractiveness. For example, in French Polynesia, 5 year loans to finance these types of system have resulted in a substantial expansion of the PV home power market.

The bar charts in figure 18.28 depict the ratio of PV- to kerosene-powered home power system life-cycle costs for the best-PV-case and worst-PV-case scenerios. Both scenarios assume 20 year life-cycle costing and development agency financing. At an NPV cost ratio of 1.0, PV and diesel system life-cycle costs are equal. Under the best-PV-case scenario, the four parameters shown on the graph are adjusted to 'reasonable extremes' that favour PV systems. An opposite adjustment is made under the worst-PV-case scenario.

As depicted in figure 18.29, the life-cycle cost of PV home power systems is dominated by debt service, reflecting high installed system costs. Kerosene-powered system life-cycle costs are dominated by fuel expenses. Because PV systems are the least-cost option for all the base case systems evaluated, sensitivity analysis graphs are not included in this chapter.

The remainder of this section discusses the analyses and assumptions leading to figures 18.28 and 18.29.

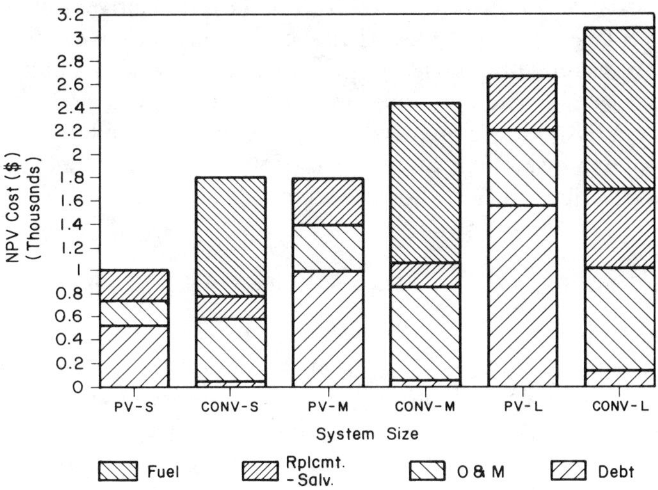

Figure 18.29 Base-case lighting and home power system cost components.

18.6.1 Description of the base case

The base-case analysis concentrates on home power systems, the major components of which are lighting. While designing comparable PV and conventional lighting and home power systems, it is difficult to select a unit for financial comparison. A common unit for comparison, in terms of equivalent lighting power cost, is an NPV levellized annual cost in US dollars per lumen, where the lumen corresponds to a source light intensity. A lux (abbreviation, lx) or a lumen (abbreviation, lm) per square metre is the standard unit of illumination (i.e. 100–200 lx is a typical level required for reading or working). While the number of lumens or lux may guide purchasing decisions in industrialized countries, it is believed that people in developing countries buy on a US dollar-per-workstation basis.

Although users generally do not compare lighting systems on the basis of lumens, the improved lighting quality (a function of higher lumens) does have an impact on purchasing decision. A 20 W fluorescent tube has been shown to provide approximately 100 lx, while a kerosene-fuelled pressure lamp will provide about 12 lx on the same surface. In developing countries, this fact is an important qualitative benefit that is not easily quantifiable in terms of purchasing decisions.

Entire home power systems often involve a mixture of conventional power sources. For example, a typical household may use kerosene for lighting and a car battery to power a radio. This combination of power sources further complicates the selection of a comparison unit. To accommodate these issues, conventional home power systems are examined and their associated costs are compared with the costs of three PV-powered systems (small, medium and large) that could be used to replace the conventional systems. Lights are replaced on a unit basis (i.e. one kerosene-fuelled lamp is replaced by one PV-powered fluorescent light). Thus the increased quality of light produced by using fluorescent lights is not quantified. For each of the three typical system designs, the financial analysis compares the life-cycle costs of conventional and PV systems on the basis of NPV cost. The conceptual system designs are described in the following sections.

18.6.1.1 Photovoltaic-powered home power system descriptions

Conceptual designs for three PV-powered systems were developed based on meeting the demand for typical small, medium and large conventional home power systems (conventional home power systems are described in §18.6.2). The PV-powered system components and operating assumptions are as follows.

Small — One 10 W fluorescent light operating for 12 h day^{-1}.

Medium — One 20 W and one 10 W fluorescent light, operating for 6 h day^{-1} and 12 h day^{-1} respectively.

Large — One 20 W fluorescent light operating for 9 h day^{-1}, one 10 W fluorescent light operating for 12 h day^{-1}, and a 12 W continuous electrical load (e.g. radio) operating for 9 h day^{-1}.

Each of the PV-powered systems comprises the following components.

(i) PV array.

(ii) Battery storage.
(iii) Loads (when calculating the cost of load devices, only the cost of lighting components is used because other devices, i.e. the radio, cost the same as those used in the conventional system).

The batteries are assumed to be a deep-discharge type that are replaced every 5 years. The battery storage is sized for 2 days. The lowest-month daily irradiation incident on the array is 4 $kW\,h\,m^{-2}\,day^{-1}$.

18.6.1.2 Conventionally powered home power system description

Three representative levels of life-style and power requirements are established for the conventional system. These levels are outlined in table 18.24.

Table 18.24 Typical conventional home power systems.

Component	Small	Medium	Large
Kerosene hurricane lamps	—	1	1
Pressurized kerosene lamps	1	1	1
12 V car battery (e.g. for a radio)	—	—	1

Usage patterns are based on the results of a study performed in Papua New Guinea. This study determined that owners typically keep one pressurized lamp lit from early evening until the morning. Those who own a second lamp (assumed to be a hurricane lamp for the base case) keep it lit from early evening until midnight.

18.6.2 System costs

18.6.2.1 Photovoltaic-powered home power system costs

The small, medium and large PV-powered systems are assumed to be one-, two- and three-module systems.

Costs for the three base-case PV home power systems are outlined in table 18.25. PV array costs are assumed to be US $8 W_p^{-1}. A simple charge controller, typical of home power applications, costs US $50. Fluorescent lights cost approximately US $40 each. The batteries are small deep-discharge batteries, such as those used in golf carts, and cost US $66 per kilowatt hour of capacity. The maintenance and repair costs include the replacement of bulbs and ballasts. Battery, controller and light replacement is assumed to occur every 5 years.

Table 18.25 Base-case PV-powered home power system costs.

Specification	Cost
Small system	
Initial capital costs (FOB manufacturer)	
PV array (39 W_p)	US $310
Battery storage (0.3 kW h)	US $20
Charge controller	US $50
Fluorescent light (1 at 10 W)	US $40
Total capital cost	US $420
Recurring capital costs (FOB manufacturer)	
Battery, controller, light replacement	US $110 every 5 years[a]
Other recurring costs (% initial capital costs)	
Maintenance and repair	4% $year^{-1}$ [a]
Medium system	
Initial capital cost (FOB manufacturer)	
PV array (78 W_p)	US $620
Battery storage (0.6 kW h)	US $40
Charge controller	US $50
Fluorescent lights (1 at 20 W; 1 at 10 W)	US $80
Total capital cost	US $790
Recurring capital costs (FOB manufacturer)	
Battery, controller, light replacement	US $170 every 5 years[a]
Other recurring costs (% initial capital costs)	
Maintenance and repair	4% $year^{-1}$ [a]
Large system	
Initial capital costs (FOB manufacturer)	
PV array (132 W_p)	US $1054
Battery storage (1.02 kW h)	US $67
Charge controller	US $50
Fluorescent lights (1 at 20 W; 1 at 10 W)	US $80
Total capital cost	US $1252
Recurring capital costs (FOB manufacturer)	
Battery, controller, light replacement	US $197 every 5 years[a]
Other recurring costs (% initial capital costs)	
Maintenance and repair	4% $year^{-1}$ [a]

[a] Plus appropriate escalation due to general inflation.

Table 18.26 Base-case conventional home power system costs.

Specification		Cost
Small System		
Initial capital costs (FOB manufacturer)		
Pressurized kerosene lamp		US $40
Recurring capital costs (FOB manufacturer)		
Lamp replacement		US $40 every 3 years[a]
Other recurring costs		
Maintenance and repair		US $40 year^{-1} [a]
Kerosene fuel (115 l at US $0.70 l^{-1})		US $81 year^{-1} [a]
Medium system		
Initial capital cost (FOB manufacturer)		
Pressurized kerosene lamp		US $40
Kerosene hurricane lamp		US $5
	Total capital cost	US $45
Recurring capital costs (FOB manufacturer)		
Lamp replacement		US $45 every 3 years[a]
Other recurring costs		
Maintenance and repair		US $60 year^{-1} [a]
Kerosene fuel (155 l at US $0.70 l^{-1})		US $109 year^{-1} [a]
Large system		
Initial capital costs (FOB manufacturer)		
Pressurized kerosene lamp		US $40
Kerosene hurricane lamp		US $5
Battery		US $65
	Total capital cost	US $110
Recurring capital costs (FOB manufacturer)		
Lamp replacement		US $45 every 3 years[a]
Battery replacement		US $65 every 2 years[a]
Other recurring costs		
Maintenance and repair		
Lamps		US $60 year^{-1} [a]
Battery		US $7 year^{-1} [a]
Kerosene fuel (155 l at US $0.70 l^{-1})		US $109 year^{-1} [a]

[a] Plus appropriate escalation due to general inflation.

18.6.2.2 Conventionally powered home power system costs

The costs of conventional systems are based on a study performed in Papua New Guinea. In this study, the initial cost of hurricane and pressurized lamps is cited at US $5 and US $40 apiece respectively. Average spare parts costs

Table 18.27 Summary of base-case life-cycle cost analysis.

System	20 year NPV cost (US $)		
	Small	Medium	Large
PV	1002	1796	2674
Conventional	1797	2447	3087

Table 18.28 Small PV-powered home system 20 year cash flow (base case).

Year	1	2	3	4	5	6	7	8
Debt service (US $)	62	62	62	62	62	62	62	62
Operating and maintenance expenses								
Annual maintenance (US $)	18	19	19	20	21	23	24	25
Recurring capital costs								
Battery, control replacement cost (US $)	0	0	0	0	111	0	0	0
Light replacement (US $)	0	0	0	0	64	0	0	0
Total cash outflow (US $)	79	80	81	82	258	84	85	86
DF $= 1/[(1 + \text{DR})^{\text{years}}]$	0.9091	0.8264	0.7513	0.683	0.6209	0.5645	0.5132	0.4665
NPV stream (US $)	72	66	61	56	160	48	44	40
Total NPV of cash outflow (US $)	1002							

Table 18.29 Small conventional home power system 20 year cash flow (base case).

Year	1	2	3	4	5	6	7	8
Debt service (US $)	6	6	6	6	6	6	6	6
Operating and maintenance expenses								
Annual maintenance (US $)	42	46	49	51	54	56	59	62
Fuel (US $)	85	89	93	98	103	108	113	119
Recurring capital costs								
Lamp replacement costs (US $)	0	0	58	0	0	67	0	0
Total cash outflow (US $)	132	141	206	155	162	237	178	187
DF $= 1/[(1 + \text{DR})^{\text{years}}]$	0.9091	0.8264	0.7513	0.683	0.6209	0.5645	0.5132	0.4665
NPV stream (US $)	120	116	155	106	101	134	91	87
Total NPV of cash outflow (US $)	1797							

are US \$20 year^{-1} for hurricane lamps and US \$40 year^{-1} for pressurized lamps. Fuel consumption (based on the usage patterns described earlier) is 40 l year^{-1} for a hurricane lamp and 115 l year^{-1} for a pressurized lamp. The lifetime of these lamps is 3 years.

The large system includes one automobile-type battery, with an initial cost of US \$65. Annual recharging costs are equivalent to 10% of the initial cost, or US \$6.5 year^{-1}. A summary of the conventionally powered system costs used for the base-case analysis is presented in table 18.26.

18.6.3 20 year life-cycle costs

Using the base-case assumptions that have been outlined, 20 year cash flows were developed for the small, medium and large PV and conventional systems. The results of these cash flows are expressed as the annualized NPV life-cycle cost in US dollars per year, as summarized in table 18.27. The actual cash flows are presented in tables 18.28–18.33.

9	10	11	12	13	14	15	16	17	18	19	20	NPV comparison
62	62	62	62	62	62	62	62	62	62	62	62	525
26	27	29	30	32	33	35	37	38	40	42	45	214
0	142	0	0	0	0	181	0	0	0	0	0	167
0	81	0	0	0	0	104	0	0	0	0	0	96
88	313	90	92	93	95	382	98	100	102	104	106	
0.4241	0.3855	0.3505	0.3186	0.2897	0.2633	0.2394	0.2176	0.1978	0.1799	0.1635	0.1486	
37	121	32	29	27	25	91	21	20	18	17	16	

9	10	11	12	13	14	15	16	17	18	19	20	NPV comparison
6	6	6	6	6	6	6	6	6	6	6	6	50
65	68	72	75	79	83	87	92	96	101	106	111	532
125	131	138	145	152	159	167	176	185	194	203	214	1024
78	0	0	90	0	0	104	0	0	121	0	0	190
274	205	215	316	237	248	365	273	287	422	315	331	
0.4241	0.3855	0.3505	0.3186	0.2897	0.2633	0.2394	0.2176	0.1978	0.1799	0.1635	0.1486	
116	79	76	101	69	65	87	59	57	76	52	49	

Table 18.30 Medium PV-powered home power system 20 year cash flow (base case).

Year	1	2	3	4	5	6	7	8
Debt service (US $)	116	116	116	116	116	116	116	116
Operating and maintenance expenses								
Annual maintenance (US $)	33	35	37	38	40	42	44	47
Recurring capital costs:								
Battery, control replacement cost (US $)	0	0	0	0	143	0	0	0
Light replacement (US $)	0	0	0	0	128	0	0	0
Total cash outflow (US $)	149	151	153	154	427	158	160	163
DF $= 1/[(1+\text{DR})^{\text{years}}]$	0.9091	0.8264	0.7513	0.683	0.6209	0.5645	0.5132	0.4665
NPV stream (US $)	136	125	115	105	265	89	82	76
Total NPV of cash outflow (US $)	1796							

Table 18.31 Medium conventional home power system 20 year cash flow (base case).

Year	1	2	3	4	5	6	7	8
Debt service (US $)	7	7	7	7	7	7	7	7
Operating and maintenance expenses								
Annual maintenance (US $)	63	69	73	77	80	84	89	93
Fuel (US $)	114	120	126	132	138	145	153	160
Recurring capital costs								
Lamp replacement costs (US $)	0	0	65	0	0	75	0	0
Total cash outflow (US $)	184	196	270	215	225	312	248	260
DF $= 1/[(1+\text{DR})^{\text{years}}]$	0.9091	0.8264	0.7513	0.683	0.6209	0.5645	0.5132	0.4665
NPV stream (US $)	167	162	203	147	140	176	127	121
Total NPV of cash outflow (US $)	2447							

9	10	11	12	13	14	15	16	17	18	19	20	NPV comparison
116	116	116	116	116	116	116	116	116	116	116	116	987
49	51	54	57	60	63	66	69	72	76	80	84	402
0	182	0	0	0	0	233	0	0	0	0	0	215
0	163	0	0	0	0	208	0	0	0	0	0	192
165	513	170	173	176	179	622	185	188	192	196	200	
0.4241	0.3855	0.3505	0.3186	0.2897	0.2633	0.2394	0.2176	0.1978	0.1799	0.1635	0.1486	
70	198	60	55	51	47	149	40	37	35	32	30	

9	10	11	12	13	14	15	16	17	18	19	20	NPV comparison
7	7	7	7	7	7	7	7	7	7	7	7	56
98	103	108	113	119	125	131	138	144	152	159	167	798
168	177	186	195	205	215	226	237	249	261	274	288	1380
87	0	0	101	0	0	117	0	0	135	0	0	213
360	286	300	415	330	346	480	381	400	554	440	462	
0.4241	0.3855	0.3505	0.3186	0.2897	0.2633	0.2394	0.2176	0.1978	0.1799	0.1635	0.1486	
153	110	105	132	96	91	115	83	79	100	72	69	

Table 18.32 Large PV-powered home power system 20 year cash flow (base case).

Year	1	2	3	4	5	6	7	8
Debt service (US $)	184	184	184	184	184	184	184	184
Operating and maintenance expenses								
Annual maintenance (US $)	53	55	58	61	64	67	70	74
Recurring capital costs								
Battery, control replacement cost (US $)	0	0	0	0	187	0	0	0
Light replacement (US $)	0	0	0	0	128	0	0	0
Total cash outflow	236	239	242	245	562	251	254	258
$DF = 1/[(1+DR)^{years}]$	0.9091	0.8264	0.7513	0.683	0.6209	0.5645	0.5132	0.4665
NPV stream (US $)	215	197	182	167	349	142	130	120
Total NPV of cash outflow (US $)	2674							

Table 18.33 Large conventional home power system 20 year cash flow (base case).

Year	1	2	3	4	5	6	7	8
Debt service (US $)	16	16	16	16	16	16	16	16
Operating and maintenance expenses								
Annual maintenance (US $)	70	77	81	85	89	94	98	103
Fuel (US $)	114	120	126	132	138	145	153	160
Recurring capital costs								
Battery, control replacement cost (US $)	0	90	0	99	0	109	0	120
Lamp replacement (US $)	0	0	65	0	0	75	0	0
Total cash outflow (US $)	200	302	288	332	244	439	267	400
$DF = 1/[(1+DR)^{years}]$	0.9091	0.8264	0.7513	0.683	0.6209	0.5645	0.5132	0.4665
NPV stream (US $)	182	250	216	227	151	248	137	186
Total NPV of cash outflow (US $)	3087							

18.6.4 Sensitivity analyses

Sensitivity analyses are provided to determine the impact of varying capital cost, discount and interest rate, kerosene fuel cost and irradiation. In the ranges of parameters selected, the PV-powered systems are always more cost effective for small systems. Medium-size conventional systems would be least costly at discount and interest rates of 17.7% and higher or at irradiation levels of 2.5 $kW\,h\,m^{-2}\,day^{-1}$ and lower. Large PV systems were found to be

9	10	11	12	13	14	15	16	17	18	19	20	NPV comparison
184	184	184	184	184	184	184	184	184	184	184	184	1564
78	82	86	90	94	99	104	109	115	120	127	133	637
0	239	0	0	0	0	305	0	0	0	0	0	281
0	163	0	0	0	0	208	0	0	0	0	0	192
261	667	269	274	278	283	801	293	299	304	310	317	
0.4241	0.3855	0.3505	0.3186	0.2897	0.2633	0.2394	0.2176	0.1978	0.1799	0.1635	0.1486	
111	257	94	87	81	74	192	64	59	55	51	47	

9	10	11	12	13	14	15	16	17	18	19	20	NPV comparison
16	16	16	16	16	16	16	16	16	16	16	16	137
108	114	119	125	132	138	145	152	160	168	176	185	885
168	177	186	195	205	215	226	237	249	261	274	288	1380
0	132	0	146	0	161	0	177	0	196	0	0	473
87	0	0	101	0	0	117	0	0	135	0	0	213
380	439	321	583	352	530	504	583	425	776	467	489	
0.4241	0.3855	0.3505	0.3186	0.2897	0.2633	0.2394	0.2176	0.1978	0.1799	0.1635	0.1486	
161	169	113	186	102	140	121	127	84	140	76	73	

more cost effective up to discount and interest rates of 13.5% at fuel costs of more than US \$0.48 l^{-1} and for irradiation levels above 3.3 $kW\,h\,m^{-2}\,day^{-1}$. Thus the approach taken was to show the extreme conditions under which conventional systems will have a cost advantage. Table 18.34 identifies those extremes.

Figures 18.30 and 18.31 show the sensitivity of home power costs to the capital costs of PV and conventional systems respectively.

Table 18.34 Assumptions necessary for conventional home power system cost-effectiveness. Any single assumption will result in conventional system cost-effectiveness.

Parameter	Small	Medium	Large
Discount and interest rate (%)	27	17.7	13.5
Kerosene cost (US $ l^{-1})	0.15	[a]	0.48
Irradiance (kW h m^{-2} day^{-1})	1.6	2.5	3.3

[a] NPV cost ratio levelled out before conventional systems showed financial attractiveness.

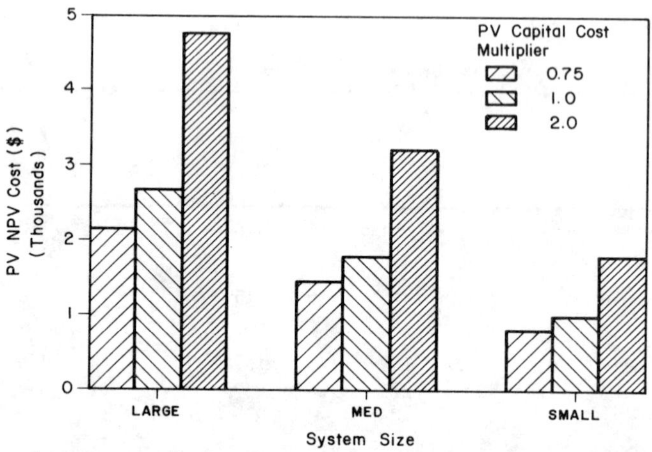

Figure 18.30 Sensitivity of PV-powered home power costs to capital cost.

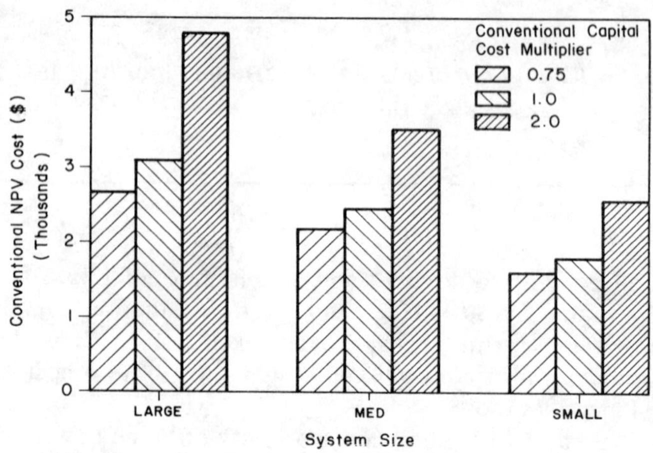

Figure 18.31 Sensitivity of conventional home power costs to capital cost.

18.7 MULTI-USE ANALYSIS

The financial analysis presented in this section compares PV- and diesel-powered multi-use systems, assuming development agency financing. The analysis shows that PV-powered systems are the least-cost option at daily energy demands of up to 2 kW h, even under unfavourable financial assumptions (figure 18.32). When the financial parameters are more favourable, PV-powered systems are competitive up to 16 kW h day^{-1}.

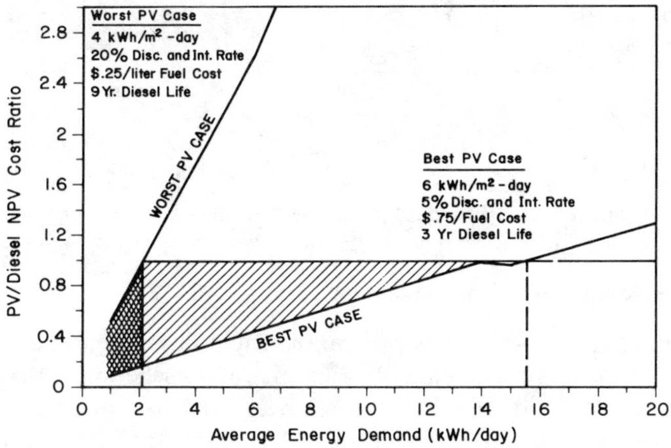

Figure 18.32 Sensitivity of multi-use system costs to best and worst conditions.

The graph in figure 18.32 depicts the ratio of PV- to diesel-powered multi-use system life-cycle costs for the best-PV-case and worst-PV-case scenarios. Both scenarios assume 20 year life-cycle costing and development agency financing. At an NPV cost ratio of 1.0, PV and diesel system life-cycle costs are equal. Under the best-PV-case scenario, the five parameters shown on the graph are adjusted to 'reasonable extremes' that favour PV systems. An opposite adjustment is made under the worst-PV-case scenario. The area between the two curves represents a reasonable range of financial assumptions.

Figure 18.33 depicts the various cost elements of PV- and diesel-powered multi-use systems. As expected, PV life-cycle costs consist primarily of debt service. The diesel life-cycle cost is dominated by fuel cost. The sensitivity analyses in this chapter indicate that discount and interest rate and fuel cost are the most sensitive parameters when comparing PV-powered multi-use systems with diesel-powered systems.

The remainder of this section discusses the analyses and assumptions leading to figures 18.32 and 18.33.

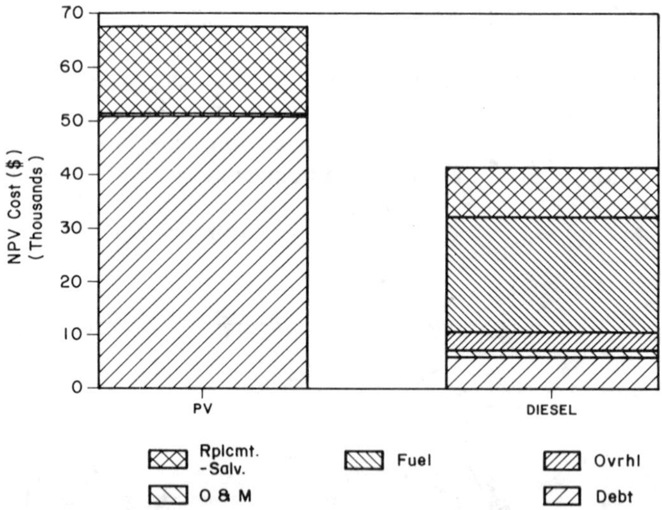

Figure 18.33 Base-case multi-use system cost components.

18.7.1 Description of the base case

The base-case multi-use load is predominantly a daytime and evening load typical of commercial, academic or administrative schedules. Load devices include AC lights and small AC appliances such as radios, televisions, microcomputers and small refrigerators. The loads are identical for the PV and conventional systems. The annual average energy demand is 10 kW h day^{-1}. The maximum energy demand is 50% higher than the average demand, or 15 kW h day^{-1}.

The financial comparison presented in this section is performed between PV- and diesel-powered systems. It is assumed that the supply of diesel fuel and PV or diesel spare parts is never interrupted.

18.7.1.1 Photovoltaic-powered system description

An AC PV power system is used for the base case. The system includes the following components.

(i) PV array (3.94 kW$_p$).
(ii) Charge controller.
(iii) Battery storage (37.5 kW h).
(iv) Inverter (2 kW).

The battery storage is sized for 2 days. The irradiation incident on the array is 5 kW h m^{-2} day^{-1}. PV power system availability is 97.5%.

18.7.1.2 Diesel-powered system description

The base-case multi-use conventional power system consists of a 3 kW diesel engine. The diesel is used for an average of 12 h day^{-1} and a maximum of 15 h day^{-1}. No back-up is provided. Diesel generator set availability is 97.6%, assuming uninterrupted fuel and spare parts supply.

18.7.2 System costs

18.7.2.1 Photovoltaic-powered system

System costs for the PV-powered multi-use base case are outlined in table 18.35. The PV-powered system cost in multi-use applications consists of the PV array, battery storage, control electronics and inverter costs. Base-case array costs are specified as US \$8 W_p^{-1}. Batteries are assumed to be of a deep-discharge type, costing US \$150 kW h^{-1} of storage capacity. The controller cost is US \$0.40 W_p^{-1}, and the inverter cost is US \$1000 kW^{-1} of power demand.

Table 18.35 Base-case PV-powered multi-use system costs.

Specification		Cost
Initial capital costs (FOB manufacturer)		
PV array (3.94 kW)		US \$31521
Battery (37.5 kW h)		US \$5625
Controller		US \$1576
Inverter (2 kW)		US \$2000
	Total capital cost	US \$40 722
Recurring capital costs (FOB manufacturer)		
Battery replacement		US \$5625 every 5 years[a]
Controller replacement		US \$1576 every 10 years[a]
Inverter replacement		US \$2000 every 10 years[a]
Other recurring costs (% initial capital cost)		
Maintenance and repair		0.1% $year^{-1}$ [a]

[a] Plus appropriate escalation due to general inflation.

18.7.2.2 Diesel-powered system

Costs associated with the base-case diesel power system are outlined in table 18.36. The capital cost breakdown for the diesel generator set is based on figure 18.3 from § 18.3. Once again, a 6 year life is assumed. Annual maintenance and repair costs are projected to be equivalent to 2% of the

Table 18.36 Base-case diesel-powered multi-use system costs.

Specification	Cost
Initial capital costs (FOB manufacturer)	
Diesel generator set (3 kW)	US $4771
Recurring capital costs (FOB manufacturer)	
Diesel generator set replacement	US $4771 every 6 years[a]
Other recurring costs	
Maintenance and repair	2% of generator set cost per year
Overhaul	15% of generator set cost every 3 years[b]
Fuel (3408 l at US $0.50 l^{-1})	US $1704 $year^{-1}$ [a]

[a] Plus appropriate escalation due to general inflation.
[b] No overhaul during diesel replacement year.

capital cost per year. Overhauls are performed every 3 years and cost an equivalent of 15% of the initial generator set cost.

Fuel requirements are taken from figure 18.4 (in § 18.3) and are based on a 3 kW diesel operating at 28% average load factor. It can be seen from the

Table 18.37 Base-base PV-powered multi-use system 20 year cash flow.

Year	1	2	3	4	5	6	7	8
Debt service (US $)	5979	5979	5979	5979	5979	5979	5979	5979
Operating and maintenance expenses								
Annual maintenance (US $)	43	45	47	49	52	55	57	60
Recurring capital costs								
Controller replacement cost (US $)	0	0	0	0	0	0	0	0
Battery replacement cost (US $)	0	0	0	0	8974	0	0	0
Inverter replacement cost (US $)	0	0	0	0	0	0	0	0
Salvage value (US $)	0	0	0	0	0	0	0	0
Total cash outflow (US $)	6022	6024	6026	6028	15 005	6034	6036	6039
DF $=1/[(1+\text{DR})^{\text{years}}]$	0.9091	0.8264	0.7513	0.683	0.6209	0.5645	0.5132	0.4665
NPV stream (US $)	5474	4978	4528	4118	9317	3406	3098	2817
Total NPV of cash outflow (US $)	67 715							
PV system generation (kW h $year^{-1}$)	3559	3559	3559	3559	3559	3559	3559	3559

figure that fuel consumption for a generator set of this size under such conditions is 0.78 l h^{-1} or 3408 l $year^{-1}$.

18.7.3 20 year life-cycle costs

The base-case 20 year life-cycle cost analyses project NPV costs of US $67 715 and US $41 486 for PV- and diesel-powered systems respectively. The cash flows used to determine these costs are presented in tables 18.37 and 18.38. The cross-over between PV- and diesel-powered system costs, for the base-case conditions, occurs at 6.2 kW h day^{-1} average energy demand.

18.7.4 Sensitivity analyses

The sensitivity of multi-use system costs to several variables, including capital cost, discount and interest rate, diesel fuel cost, diesel lifetime and irradiation is analysed. The sensitivity graphs are expressed as the ratios of the net present value life-cycle costs for PV- and diesel-powered systems over a range of electricity load demands.

18.7.4.1 Sensitivity to capital costs

To demonstrate the sensitivity of multi-use system life-cycle costs to capital costs, capital cost multipliers are applied to the PV- and diesel-powered

9	10	11	12	13	14	15	16	17	18	19	20	NPV comparison
5979	5979	5979	5979	5979	5979	5979	5979	5979	5979	5979	5979	50 902
63	66	70	73	77	81	85	89	93	98	103	108	518
0	3209	0	0	0	0	0	0	0	0	0	0	1237
0	11 453	0	0	0	0	14 617	0	0	196	0	0	13 487
0	4072	0	0	0	0	0	0	0	0	0	0	1570
0	0	0	0	0	0	0	0	0	0	0	0	0
6042	24 780	6049	6052	6056	6060	20 681	6068	6072	6077	6082	6087	
0.4241	0.3855	0.3505	0.3186	0.2897	0.2633	0.2394	0.2176	0.1978	0.1799	0.1635	0.1486	
2562	9554	2120	1928	1754	1596	4951	1321	1201	1093	994	905	
3559	3559	3559	3559	3559	3559	3559	3559	3559	3559	3559	3559	

Table 18.38 Base-case diesel-powered multi-use system 20 year cash flow.

Year	1	2	3	4	5	6	7	8
Debt service (US $)	700	700	700	700	700	700	700	700
Operating and maintenance expenses								
Annual maintenance (US $)	100	105	110	116	112	128	134	141
Diesel engine overhaul (US $)	0	0	1036	0	0	1199	0	0
Fuel (US $)	1789	1879	1973	2071	2175	2284	2398	2518
Recurring capital costs								
Diesel generator replacement cost (US $)	0	0	0	0	0	7992	0	0
Salvage value (US $)	0	0	0	0	0	0	0	0
Total cash outflow (US $)	2590	2684	3819	2888	2997	12 302	3232	3359
DF = $1/[(1+\text{DR})^{\text{years}}]$	0.9091	0.8264	0.7513	0.683	0.6209	0.5645	0.5132	0.4665
NPV stream (US $)	2354	2218	2869	1972	1861	6944	1659	1567
Total NPV of cash outflow (US $)	41 486							
Diesel system generation (kW h year^{-1})	3559	3559	3559	3559	3559	3559	3559	3559

systems. The mutlipliers vary from 0.75 to 2.0. The 1.0 multiplier corresponds to the base-case system. Variations in capital costs can result from different equipment, customs, insurance, freight or installation costs. Installation costs are not included in the base case. Figures 18.34 and 18.35 present NPV life-cycle cost for the PV and diesel multi-use power systems respectively.

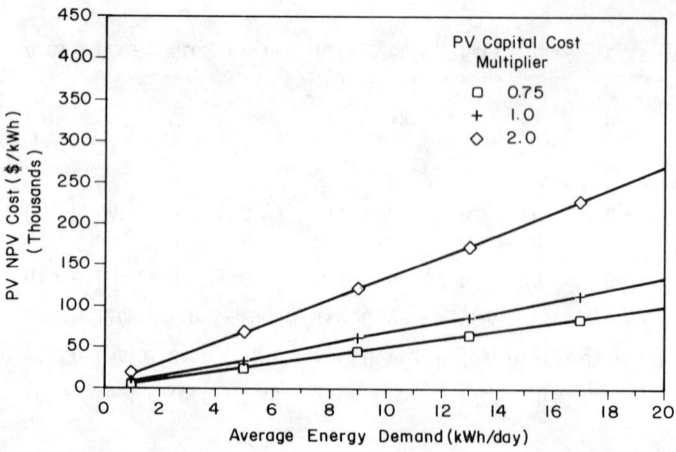

Figure 18.34 Sensitivity of PV multi-use system costs to power system capital costs.

9	10	11	12	13	14	15	16	17	18	19	20	NPV comparison
700	700	700	700	700	700	700	700	700	700	700	700	5963
148	155	163	171	180	189	198	280	219	230	241	253	1213
1388	0	0	1606	0	0	1860	0	0	2153	0	0	3387
2643	2776	2914	3060	3213	3374	3543	3720	3906	4101	4306	4521	21 671
0	0	10 709	0	0	0	0	0	0	14 352	0	0	10 505
0	0	0	0	0	0	0	0	0	0	0	8439	1254
4880	3632	3778	16 248	4094	4263	6301	4628	4825	21 535	5248	(2964)	
0.4241	0.3855	0.3505	0.3186	0.2897	0.2633	0.2394	0.2176	0.1978	0.1799	0.1635	0.1486	
2069	1400	1324	5177	1186	1123	1508	1007	955	3873	858	(441)	
3559	3559	3559	3559	3559	3559	3559	3559	3559	3559	3559	3559	

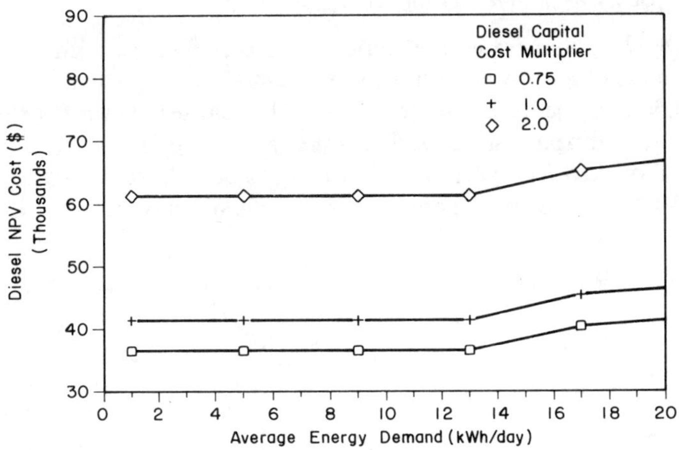

Figure 18.35 Sensitivity of diesel multi-use system costs to power system capital costs.

18.7.4.2 Sensitivity to discount and interest rate

Figure 18.36 demonstrates the effect that various discount and interest rates have on the cost of multi-use systems. If a 5% discount and interest rate is used, PV-powered systems are financially advantageous up to demands of

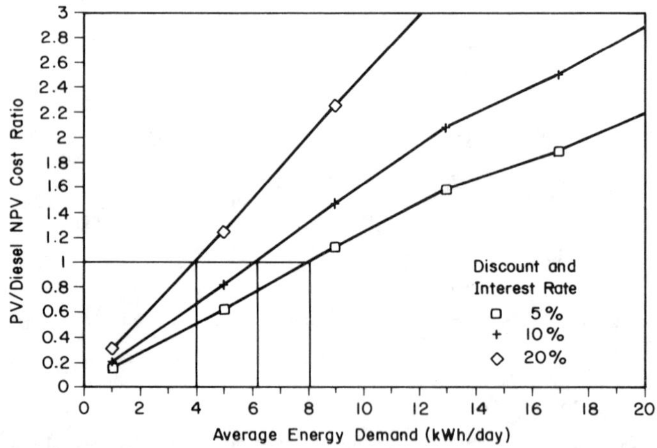

Figure 18.36 Sensitivity of multi-use system costs to discount and interest rate.

approximately 8.0 kW h day^{-1}. At a 20% discount rate, PV-powered systems are financially advantageous up to demands of only 4.0 kW h day^{-1}.

18.7.4.3 Sensitivity to diesel fuel cost

Figure 18.37 demonstrates the impact of diesel fuel price changes on the attractiveness of PV-powered multi-use systems. Varying the diesel fuel price of US \$0.50 l^{-1} (the base case) to US \$0.75 l^{-1} causes the value below which PV systems appear more favourable to move from approximately 6.2 kW h day^{-1} to 7.8 kW h day^{-1}. Moving to 50% lower than the base case (i.e. US \$0.25 l^{-1}) drops the cross-over to 4.6 kW h day^{-1}.

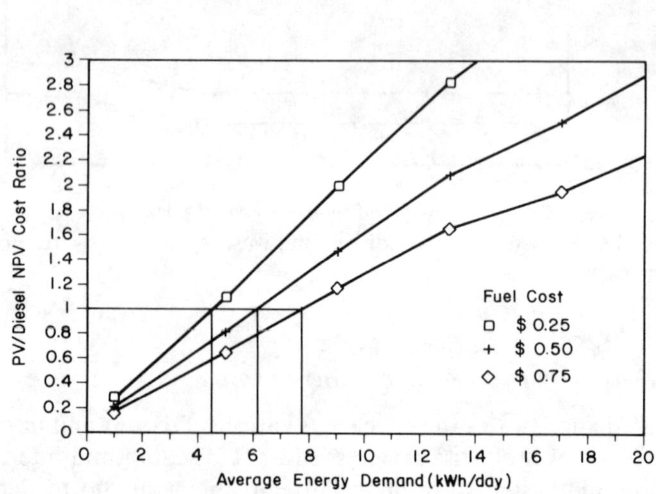

Figure 18.37 Sensitivity of multi-use system costs to diesel fuel costs.

18.7.4.4 Sensitivity to diesel lifetime

Figure 18.38 shows the impact variations in the diesel generator set life-time have on multi-use system costs. While the 6 year base-case lifetime results in PV systems having a cost advantage at energy demands less than 6.2 kW h day^{-1}, the cross-overs for 3 and 9 year lifetimes occur at 8.0 kW h day^{-1} and 5.5 kW h day^{-1} respectively.

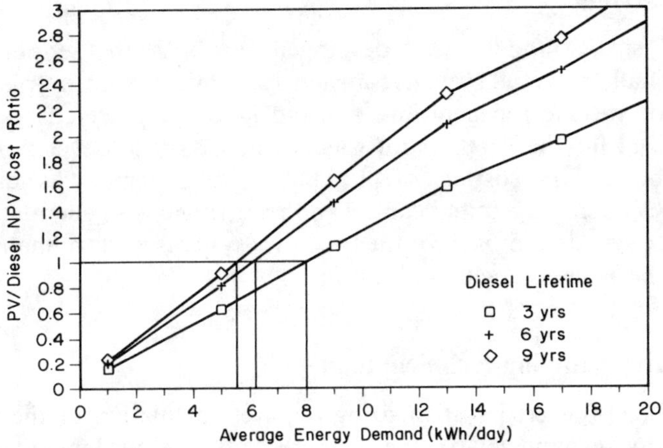

Figure 18.38 Sensitivity of multi-use system costs to diesel generator set lifetime.

18.7.4.5 Sensitivity to irradiation

Figure 18.39 shows the effect that varying irradiation has on the comparative cost of PV and diesel multi-use systems. At 4 kW h m^{-2} day^{-1}, PV systems retain a cost advantage up to daily energy demands of approximately

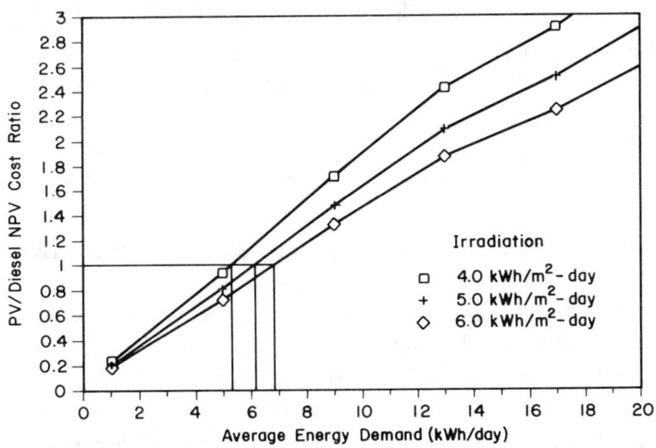

Figure 18.39 Sensitivity of multi-use system costs to irradiation.

5.4 kW h. At an irradiation of 6 kW h m^{-2} day^{-1}, PV systems are competitive up to 6.8 kW h day^{-1}.

APPENDIX A: SIZING AND FINANCIAL MODELS

A.1 Overview

Technical sizing models were developed for both the PV-powered and conventionally powered systems for each application except refrigeration. In the case of vaccine refrigeration, PV- and kerosene-powered systems were already sized for the WHO; so it was not necessary to develop a technical sizing model. The costing of the base-case systems was outlined in §§ 18.3–18.7. Using the initial capital and recurring costs generated, NPV life-cycle costs were determined for the base-case systems using a financial model that was the same for each application.

A.2 Water-pumping technical model

PV- and diesel-powered system sizing requires calculation of the maximum daily energy demand (MED), the average daily energy demand (AED), the maximum daily power demand (MPD) and the average daily power demand (APD) of a load. PV array sizing considers the maximum ratio of daily energy demand (ED) to irradiation anticipated over the year (IN). Diesel-powered system sizing relates primarily to the maximum daily power demand.

The AED over the year is used to calculate the annual fuel consumption (FC). The average water demand (AWD) (in cubic metres per day) is used to determine the total volume of water pumped in a year. The daily water demand throughout the year (in cubic metres per day, where a day equals 24 h) and the total static and dynamic head (the height that water must be pumped (in metres)) are needed as inputs to calculate these values. The ED is a direct function of the hydraulic energy demand (HED) and the motor–pump efficiency (MPE). HED can be calculated as follows:

$$\text{HED} = \frac{9.8\ \text{VH}}{3600}\ \text{kW h day}^{-1}$$

where VH (m^4 day^{-1}) is the volume of water to be pumped multiplied by the head.

The maximum HED can be calculated using the maximum VH, where the maximum VH is the maximum water demand (MWD) multiplied by the head. The average HED is calculated similarly, but using the AWD.

Because there is an inefficiency in the coupling between the motor and pump in converting electrical energy into mechanical energy, the energy required from the power system is greater than the HED. For a given MPE, the ED can be calculated as follows:

$$\text{ED} = \frac{\text{HED}}{\text{MPE}} \text{ kW h day}^{-1}.$$

Once again MED and the AED can be calculated using the appropriate HEDs. The EDs are used to perform PV array and generator set sizing and to determine the water outputs of the respective systems.

A.2.1 Photovoltaic-powered system sizing

The equation for calculating the required PV array size is as follows:

$$\text{PV array size} = \frac{\text{PPR}}{\text{SE}} \left(\frac{\text{ED}}{\text{IN}}\right)_{\text{max}} \text{ kW}_{\text{p}}$$

where PPR (kW m^{-2}) is the PV power rating, SE is the system operating efficiency (conversion efficiency of sunlight into electrical energy including temperature effects, wiring and control system losses), PPR/SE is the system sizing factor (a value of 1.1 is used in this analysis) and $(\text{ED}/\text{IN})_{\text{max}}$ is the maximum ratio of ED to IN (calculated as 1.5 times the ratio of average AED to average irradiation for this analysis).

The annual PV-powered system output is determined as follows:

$$\text{water output} = \text{AWD} \times 365 \times \text{PVAV m}^3 \text{ day}^{-1}$$

where AWD ($\text{m}^3 \text{ day}^{-1}$) is the average water demand over the year and PVAV is the PV power load system availability (where availability refers to the percentage of time that the system operates within specifications).

A.2.2 Diesel-powered system sizing

The required generator set size is calculated as follows:

$$\text{generator set size} = \frac{\text{MED}}{\text{MOH} \times \text{MLF}} \text{ kW}$$

where MED (kW h day^{-1}) is the maximum energy demand during the year, MOH are the maximum operating hours of the generator set in a day and MLF is the maximum load factor at which the generator set operates.

MOH, which is needed to calculate the generator set size as indicated in the equation above, is determined as follows:

$$\text{MOH} = \frac{\text{MWD}}{\text{PFR}}$$

where MWD ($m^3\,day^{-1}$) is the maximum water demand during the year and PFR ($m^3\,h^{-1}$) is the pump flow rate.

It is specified that the generator set must operate for at least 1 h. Thus, if MWD is low, such that the calculated MOH is less than 1 h, the generator set runs for 1 h at a load factor which is lower than MLF.

To determine the amount of fuel consumed by the diesel, it is necessary first to calculate the annual average daily operating hours (AOH):

$$\mathrm{AOH} = \frac{\mathrm{AWD}}{\mathrm{PFR}}$$

where AWD ($m^3\,day^{-1}$) is the average water demand during the year and PFR ($m^3\,h^{-1}$) is the pump flow rate.

Then the average load factor (ALF) is calculated:

$$\mathrm{ALF} = \frac{\mathrm{AED}}{\mathrm{PR} \times \mathrm{AOH}}$$

where AED ($kW\,h\,day^{-1}$) is the average energy demand during the year, PR (kW) is the generator set power rating and AOH are the average operating hours as calculated above.

The PR specified here is not necessarily the same as the generator set size calculated above. The generator set size calculation refers to the ideal generator set size for the load. The PR refers to the size being used, which is dictated by the commercial availability of generator sets. For example, diesel generator sets are not commonly available in sizes less than 3 kW. Therefore, even if the diesel generator set size turns out to less than 3 kW, a 3 kW diesel must be used, resulting in a PR equal to 3 kW. Using the AOH, the average FC is calculated as follows:

$$\mathrm{FC} = \mathrm{AOH} \times \mathrm{FCR} \times 365 \text{ l year}^{-1}$$

where AOH are the annual average daily diesel operating hours as calculated above, and FCR ($l\,h^{-1}$) is the fuel consumption rate for the particular engine at the ALF.

The annual water output is determined as follows:

$$\text{water output} = \mathrm{AWD} \times 365 \times \mathrm{GSAV} \text{ m}^3\text{ day}^{-1}$$

where AWD ($m^3\,day^{-1}$) is the average water demand over the year and GSAV is the generator set power load system availability (where availability refers to the percentage of time that the system operates within specifications).

A.3 Communications technical model

Sizing of a PV or conventional communications system begins by calculating the MED over the year. The MED is determined by summing the products of the wattage and duration (in hours) of each individual load. For many systems, such as repeater stations, the load can be assumed to be constant and continuous.

A.3.1 Photovoltaic-powered system sizing

The required PV array size is determined as follows:

$$\text{PV array size} = \frac{(\text{PPR/SE})(\text{ED/IN})_{max}}{\text{EBE}} \ \text{W}_p$$

where PPR (kW m^{-2}) is the PV power rating, SE is the system operating efficiency (conversion efficiency of sunlight to electrical energy including temperature effects, wiring and control system losses), PPR/SE is the system sizing factor (a value of 1.1 is used in this analysis), $(\text{ED/IN})_{max}$ is the maximum ratio of ED to IN (calculated as constant ED divided by the lowest-month daily plane-of-the- array irradiation; a value of 4 $\text{kW h m}^{-2}\,\text{day}^{-1}$ is used in this analysis), EBE = PDL + (1 − PDL)BE is the effective battery efficiency (where PDL is the fraction of PV array output that goes directly to the load, BE is the round-trip battery efficiency and 1 − PDL is the battery use factor).

The required battery capacity (BC) is obtained from the following equation:

$$\text{BD} = \frac{\text{MED} \times \text{SD}}{\text{MDD}}$$

where MED (kW h day^{-1}) is the maximum energy demand during the year, SD is the number of storage days required and MDD is the maximum depth of discharge over the total storage period.

The electrical output of the PV power system is as follows:

$$\text{electrical output} = \text{AED} \times 365 \times \text{PVAV} \ \text{kW h year}^{-1}$$

where AED (kW h day^{-1}) is the average energy demand and PVAV is the PV power system availability.

A.3.2 Diesel-powered system sizing

The required generator set size is calculated as follows:

$$\text{generator set size} = \frac{\text{MED}}{\text{MOH} \times \text{MLF} \times \text{BCE}} \ \text{kW}$$

where MED (kW h day^{-1}) is the maximum daily energy demand during the year, MOH are the maximum diesel operating hours per day, MLF is the maximum load factor and BCE is the battery charger efficiency.

The required BC is calculated using the equation

$$\mathrm{BC} = \frac{\mathrm{MED} \times \mathrm{SD}}{\mathrm{MDD}}\ \mathrm{kW\,h}$$

where MED (kW h day^{-1}) is the maximum energy demand during the year, SD is the number of storage days required and MDD is the maximum depth of discharge over the total storage period.

The FC is calculated as follows:

$$\mathrm{FC} = \mathrm{AOH} \times \mathrm{FCR} \times 365$$

where AOH (h day^{-1}) are the average generator set operating hours over the year and FCR (l h^{-1}) is the fuel consumption rate, which is a function of engine size and load factor.

The electrical output of the generator set system is as follows:

$$\text{electrical output} = \mathrm{AED} \times 365 \times \mathrm{GSAV}$$

where AED (kW h day^{-1}) is the average energy demand and GSAV is the generator set power system availability.

A.4 Lighting and home power technical model

A.4.1 Photovoltaic-powered system sizing

The required PV array size is determined as follows:

$$\text{PV array size} = \frac{(\mathrm{PPR}/\mathrm{SE})(\mathrm{ED}/\mathrm{IN})_{max}}{\mathrm{EBE}}\ \mathrm{W_p}$$

where PPR (kW m^{-2}) is the PV power rating, SE is the system operating efficiency (conversion efficiency of sunlight to electrical energy including temperature effects, wiring and control system losses), PPR/SE is the system sizing factor (a value of 1.1 is used in this analysis), $(\mathrm{ED}/\mathrm{IN})_{max}$ is the maximum ratio of ED to IN (calculated as constant ED divided by the lowest-month daily plane-of-the-array irradiation; a value of 4 kW h m^{-2} day^{-1} is used in this analysis), $\mathrm{EBE} = \mathrm{PDL} + (1 - \mathrm{PDL})\mathrm{BE}$ is the effective battery efficiency (where PDL is the fraction of PV array output that goes directly to the load, BE is the round-trip battery efficiency and $1 - \mathrm{PDL}$ is the battery use factor).

The required BC can be determined as follows:

$$\mathrm{BC} = \frac{\mathrm{MED} \times \mathrm{SD}}{\mathrm{MDD}}$$

where MED (kW h day^{-1}) is the maximum energy demand during the year, SD is the number of storage days required and MDD is the maximum depth of discharge over the total storage period.

In calculating the array size and BC, it is necessary to know the ED, which is calculated by summing the products of the wattage and the duration (in hours) of each individual load. Table 18.39 provides an example of this calculation.

Table 18.39 Typical loads for a home power system.

Load	Power rating (W)	Duration ($h\,day^{-1}$)	Energy demand ($W\,h\,day^{-1}$)
Fluorescent light	20	9	180
Fluorescent light	10	12	120
Radio	12	9	108
	—		—
Peak demand	42		408

A.4.2 Conventionally powered system sizing

For the conventional lighting and home power systems, no sizing model is required. Lighting requirements are specified as a number of each type of lamp. FC for each lamp is determined from field experience. Electrical loads are assumed to be powered by a battery. Batteries used in these conventional systems are not sized to meet load requirements, as they are used to provide as much power as they can produce.

A.5 Multi-use technical model

Sizing of a PV or conventional multi-use system begins by calculating the daily ED over the year. The ED is determined by summing the products of the wattage and duration (in hours) of each individual load on a daily basis. The MED and AED can then be determined. The MPD in kilowatts is specified as 1.8 times the APD where APD = AED divided by 12 h.

A.5.1 Photovoltaic-powered system sizing

$$\text{PV array size} = \frac{(\text{PPR}/\text{SE})(\text{ED}/\text{IN})_{\max}}{\text{EBE} \times \text{IE}}$$

where PPR ($kW\,m^{-2}$) is the PV power rating, SE is the system operating efficiency (conversion efficiency of sunlight to electrical energy including temperature effects, wiring and control system losses), PPR/SE is the system sizing factor (a value of 1.1 is used in this analysis), $(\text{ED}/\text{IN})_{\max}$ is the maximum ratio of ED to IN (calculated as constant ED divided by the lowest-month daily plane-of-the-array irradiation: a value of 4 $kW\,h\,m^{-2}\,day^{-1}$ is used in this analysis), EBE = PLD + (1 − PDL)BE is the effective battery efficiency (where PDL is the fraction of PV array output that goes directly to the load, BE is the round-trip battery efficiency, 1 − PDL is the battery use factor) and IE is the inverter efficiency.

The BC is determined as follows:

$$\mathrm{BC} = \frac{\mathrm{MED} \times \mathrm{SD}}{\mathrm{MDD}}\ \mathrm{kW\,h}$$

where MED (kW h day^{-1}) is the maximum energy demand during the year, SD is the number of storage days required and MDD is the maximum depth of discharge over the total storage period.

The electrical output of the PV power system is as follows:

$$\text{electrical output} = \mathrm{AED} \times 365 \times \mathrm{PVAV}\ \mathrm{kW\,h\,year}^{-1}$$

where AED (kW h day^{-1}) is the average energy demand and PVAV is the PV power system availability.

A.5.2 Diesel-powered system sizing

The required generator set size is calculated as follows:

$$\text{generator set size} = \frac{\mathrm{MED}}{\mathrm{MOH} \times \mathrm{MLF}}\ \mathrm{kW}$$

where MED (kW h day^{-1}) is the maximum energy demand during the year, MOH are the maximum operating hours per day and MLF is the maximum load factor.

The generator set must also be capable of meeting the MPD. Thus the larger value of the two (the calculated generator set size or the MPD) must be used.

The FC is calculated as follows:

$$\mathrm{FC} = \mathrm{AOH} \times \mathrm{FCR} \times 365\ \mathrm{l\,year}^{-1}$$

where AOH (h day^{-1}) are the average generator set operating hours over the year and FCR (l h^{-1}) is the fuel consumption rate, which is a function of engine size and load factor.

The electrical output of the generator set power system is calculated as follows:

$$\text{electrical output} = \mathrm{AED} \times 363 \times \mathrm{GSAV}\ \mathrm{kW\,h\,year}^{-1}$$

where AED (kW h day^{-1}) is the average energy demand and GSAV is the generator set power system availability.

A.6 Financial model

The financial model used in this analysis was designed to calculate the NPV life-cycle cost of the systems being compared. In order to calculate the NPV costs, 20 year cash flows were developed for each of the systems, as shown in the chart of accounts in table 18.40.

Table 18.40 Chart of accounts for financial analyses.

Account	PV-powered systems	Conventional systems
Initial capital cost	X	X
Equipment replacement (recurring capital cost)	X	X
Engine overhaul[a]	Not applicable	X
Maintenance and repair	X	X
Fuel	Not applicable	X
Debt service	X	X
Vaccine waste[b]	X	X

[a] Only used for generator applications (i.e. pumping, communications and multi-use).
[b] Only use for medical refrigeration.

The present value of costs for each year is determined by multiplying the future value of those costs by the discount factor (DF):

$$\mathrm{DF} = \frac{1}{(1+r)^j}$$

where r is the discount rate and j is the year.

Thus the present value of costs is calculated as follows:

$$\text{present cost} = \text{future cost}_{\text{year } j} \times \mathrm{DF}_{\text{year } j}.$$

The present value of all costs over the period of financial analysis is determined by summing the products of the future costs and discount factors for each year:

$$\text{total present costs} = \sum_{j=1}^{n} (\text{future cost}_{\text{year } j} \times \mathrm{DF}_{\text{year } j})$$

where n is the period of financial analysis.

Since each application produces an end use such as electricity, water, refrigerated volume or household expense, the total present costs can be expressed as a function of the total system output over the period of analysis. The NPV costs for each application can thus be expressed in terms of end-use product, as shown in table 18.41.

Table 18.41 Application NPV cost.

Application	NPV life-cycle unit
Water pumping	US \$ m^{-3} of water
Communications	US \$ $kW\,h^{-1}$ of electricity
Vaccine refrigeration	US \$ l^{-1} of refrigerated volume
Lighting and home power	US \$ $year^{-1}$ of household expense
Multi-use	US \$ $kW\,h^{-1}$ of electricity

REFERENCE

Eskenazi D, Kerner D and Slominski L 1986 *Technical Report No SAND85-7018/2* (Albuquerque, NM: Sandia National Laboratories)

19 A Comparative Assessment of Photovoltaics and Handpumps for Rural Water Supply: A Case Study for Thailand

19.1 INTRODUCTION

The purpose of this study is to determine under what circumstances photovoltaic (PV) pumping systems can compete technically and economically with handpumps for supplying water to rural communities. Economically competitive ranges of PV relative to handpumps for supplying water to rural communities as a function of water source depth, village size (population), solar energy resource, level of service (litres per capita per day), cost and performance characteristics will be determined.

19.2 TECHNOLOGY CONFIGURATIONS

The choice of technology will depend on a number of parameters including village size, per capita water demand, well characteristics, and cost and technical performance of the equipment. The following two systems are compared.

(i) Handpumps: each handpump installed on a well serves a certain number of persons. The number of persons served depends on the per capita water demand, the number of hours that a handpump is used daily, and the amount of time that a person is willing to spend gathering water. Handpumps are typically used for water demands up to about 40 $l\,day^{-1}$ per capita. If the village has more people than can be supported by one handpump, then two or more handpumps are used. The principal handpump system components are the handpump and the well.

(ii) PV pumping systems: each PV pumping system provides water to one or more standpipes (public faucets). The number of persons served at a faucet is determined as in the handpump case. If the village has more people than can be supported by one standpipe, then two or more standpipes are used. One well with a PV pump is used to supply all standpipes, unless well yield limitations prevent an adequate amount of water to be withdrawn from the well. In such cases, two or more wells are needed, or batteries are used to operate the pump over a longer period at a lower pumping rate. The principal PV pumping system components are the PV array, motor and pump, optional controls, battery and power-conditioning equipment, well, water storage tank, distribution network and standpipes.

The selected equipment configurations are based on the recommended application ranges of commercially available motor–pump systems. These limits are a function of water table depth and daily water demand. For example, surface-mounted centrifugal pumps are limited to applications where the water table depth does not exceed 7 m. Beyond 7 m, the centrifugal pump will lose suction. For intermediate- and deep-well applications, the choice is a jackpump or a submersible motor–pump system. The determinant in these cases is required daily flow. At a water depth of 20 m, jackpumps are a good choice if water demand is less than 30 m^3 day^{-1}. For demand more than 30 m^3 day^{-1}, a submersible motor–pump system is preferred. The representative pumping configurations should be viewed as examples only. Depending on the specific circumstances, other pumping technologies or configurations may be better.

19.3 METHODOLOGY

The analyses follows the methology presented by Cabraal *et al* (1987) which is depicted in figure 19.1.

The technologies are compared when they are providing the same level and the same quality of service. Service quality and level are defined as follows.

(i) The amount of water supplied daily per person is the same in both water supply systems being evaluated. Therefore the benefit accruing to an individual, based on availability of water, is the same in both technologies. For example, both technologies would supply 20 l day^{-1} per capita to a given village population.

(ii) The time spent by villagers gathering water is the same in both supply systems. Therefore the cost of collecting water is the same in both technologies. Water-gathering time is composed of time spent walking to and from the water point, queueing time at the water point, and water collection time. The water collection time in the case of the handpump is the time spent pumping the water. In the case of PV

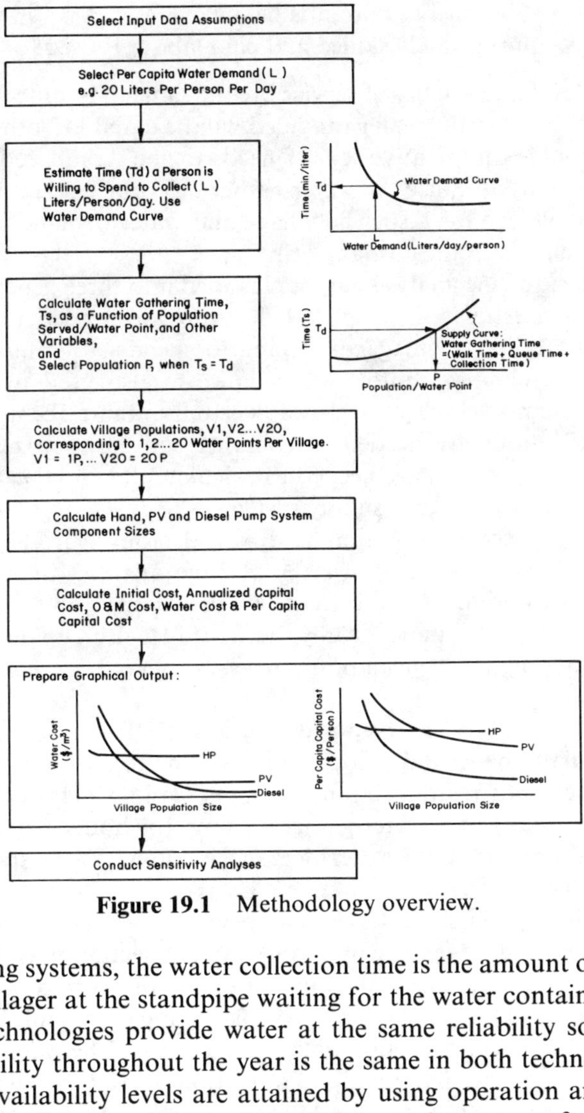

Figure 19.1 Methodology overview.

pumping systems, the water collection time is the amount of time spent by a villager at the standpipe waiting for the water container to fill.

(iii) The technologies provide water at the same reliability so that water availability throughout the year is the same in both technologies. The same availability levels are attained by using operation and maintenance practices consistent with reliable equipment performance and using adequate water storage in the case of PV systems. For example, handpump usage has been limited to 6 h day^{-1}, and pumps are assumed to be maintained regularly. Handpump maintenance is estimated based on the concept of 'village level operation and maintenance (VLOM)' promoted by the United Nations Development Programme–World Bank (1986). The handpump project has developed preliminary designs for VLOM pumps for lifts up to 25 m. In the case of PV systems, adequate storage is provided to account for cloudy days when the solar energy output is below average. Also, in the case of PV systems, appropriately skilled and paid labour is used for maintenance.

Since the level and quality of service are the same in both technologies, the benefits derived from the water provided will be equal in both technologies. Therefore, only the relative costs need to be compared using cost-effectiveness analysis. Since the water-gathering cost ('haul cost') is the same in both technologies for a specific village and water demand level, the haul cost is not included in the cost analysis.

Data needed for the analysis can be classified into three general categories. Village characteristics and people's behaviour are needed to estimate the amount of time that a villager is willing to spend gathering water. These together with technology information data are also used to calculate the number of persons served at each water supply point. Technology performance specifications are needed to compute component sizes and system configurations. Cost data are needed to calculate the initial capital and life-cycle costs of the three water supply systems.

Per capita water consumption is a key decision variable. In African countries, 20 l day^{-1} per capita is the average quantity of water used for domestic consumption in rural areas. 40 l day^{-1} per capita is the level of water consumption recommended by the World Health Organization. It also represents the higher consumption levels observed in Asia and Latin America.

A water demand curve developed by the UNDP–World Bank (1986) is used to calculate the time that a person is willing to spend gathering water. The curve shows the relationship between per capita water demand (in litres per day per capita) and water-gathering time (in hours per cubic metre). Therefore, when per capita water demand is known, time spent gathering water can be calculated.

Water-gathering time has three components: time spent walking to and from the water point, queueing time, and water collection time. Both walking and queueing time depend on the number of persons served per water point and other factors. Queueing time is modelled using commonly used waiting-line models. The water collection time is a function of the water delivery rate of the standpipe and handpump. Population served per water point is estimated by determining the population at which the water-gathering time equals the time that a person is willing to spend (from the demand curve calculation).

Village population sizes are set at multiples of the number of persons served per water point. For example, if one water point serves 200 persons, two water points will be required for a village of 400, and 10 water points will be needed for a village of 2000. In the case of a handpump system, each water point consists of one well and a handpump, except for a shallow well where two handpumps can be installed in a 10 m well of 1.2 m diameter. For PV systems, a number of water points will require one or more wells, distribution piping and storage tanks. Each well has a pump and an associated power generator. Standard engineering analyses are used to calculate system component sizes.

19.4 BASE-CASE ASSUMPTION

Base-case assumptions were set for the case study and these are depicted in tables 19.1–19.3. Village characteristics, information on system technology and costs were surveyed to approximate as close as possible situations prevailing in a typical 'up-country' village in Thailand where handpumps are commonly introduced by government ministries. The two levels of water consumption characterize water needs in these villages. It is also a common sight to find villagers with 12 plastic containers of 10 l capacity fitted to a three-wheeled cart traversing to water points within the village. This equipment has become a family possession necessary for daily water needs.

Table 19.1 Village characteristics.

Water consumption	20 and 45 l day^{-1} per capita
Number of persons per family	6
Minimum water need	10 l day^{-1} per capita
Maximum load per trip	120 l
Water point use[a]	6 h

[a] One water point is one handpump; for PV systems, one water point is one shallow well or one standpipe for a deep well.

Table 19.2 Handpump costs and technology information.

	Shallow well (depth less than 10 m)	Deep well (depth greater than 10 m)
Handpump	Suction type	Force type
	US $32	US $227 + US $2.40 × depth (m)
	15 l min^{-1}	15 l min^{-1}
	3 years	5 years
Well	US $400	US $2500
	20 years	20 years
Operation and maintenance	5% of capital cost	

Table 19.3 PV system costs and technology information.

	Shallow well	Shallow well	Deep well
Depth (m)	3	7	20
PV array cost	US \$8 W_p^{-1}	US \$8 W_p^{-1}	US \$8 W_p^{-1}
Motor–pump system	Table 19.4	Table 19.5	Table 19.6
Inverter cost	—	US \$0.75 W_p^{-1}	US \$0.75 W_p^{-1}
Inverter efficiency		80%	80%
Well	US \$400	US \$400	US \$2500
Yield	2 $m^3 h^{-1}$	2 $m^3 h^{-1}$	10 $m^3 h^{-1}$
Standpipe cost	—	—	US \$26
Storage cost	US \$20 per 2 m^3, ferro-cement jar		US \$27 m^{-3}, ferro-cement tank
Operation and maintenance			
Array cost[a]	1%	1%	1%
Motor–pump system	10%	10%	10%
Civil works	1%	1%	1%
Array life	20 years	20 years	20 years
Motor–pump system	3 years	5 years	10 years
Civil works	20 years	20 years	20 years
Storage[b]	1 day	1 day	1 day

[a] Operation and maintenance cost is computed as a percentage of the capital cost of the device.

[b] Setting water storage to 1 day consumption level is equivalent to 97% availability over the year.

Reports quoted from Thai government sources were the basis for the information on handpumps in table 19.2. Suction-type handpumps are used for shallow wells of less than 7 m suction head while force-type handpumps were used for deep-well applications. Determining the life of a handpump can be a difficult undertaking. Different parts of a handpump have differing lifetimes and these can be easily replaced within the country, where they are locally manufactured. The lifetimes listed were estimates based on yearly cost of replacement taken as cost annuity and computing the number of years when these annuities are equivalent to the present cost of a handpump. An additional 5% of capital cost was assumed for the operation and maintenance of the handpump, the platform and the well. An average pumping rate taken over the operating life of the handpump is taken to be 15 l min^{-1}.

Three different PV pumping systems were considered (table 19.3) based on existing commercial products in the country. 12 V DC submersible motor–pump systems (table 19.4) with plastic casing were found to be suitable for low discharge heads up to 3 m. Brushes have to be replaced every 4 months and operating 8–10 h day^{-1} they last for about 3 years, provided that proper operation and maintenance requirements are followed. Pumping applications with a static head exceeding 3 m can be served by a locally manufactured AC motor and surface-type centrifugal pump (table 19.5). AC

Table 19.4 12 V DC submersible motor–pump system with 25% efficiency for shallow-well 3 m head application.

Size (W)	Current (A)	Flow rate ($m^3 h^{-1}$)	Cost (US $)
24	2.0	1.0	12
42	3.5	1.3	16
60	5.0	2.3	20
96	8.0	5.3	24
156	13.0	8.7	47

Table 19.5 Surface-mounted Thai-manufactured 25% efficient AC motor–pump system for a shallow well with a head of less than 7 m.

Motor (less than 1 hp, single phase)			Centrifugal pump cost (US $)	Motor–pump cost (US $)
Power (hp)	Power (W)	Cost (US $)		
$\frac{1}{4}$	187	34	22	55
0.33	246	41	22	63
$\frac{1}{2}$	373	72	22	94
1	746	77	22	99
2	1492	94	22	116
3	2238	116	22	138
5	3730	161	25	187
7	5222	230	35	265
10	7460	279	55	334

submersible motor–pump systems (table 19.6) were configured for deep-well applications. Inverters specially designed for PV applications were used with the two AC systems. A storage system for a shallow well takes advantage of cheap ferro-cement water jars of 2 m^3 capacity at US $20 each, while ferro-cement water tanks of 12 m^3 capacity made up of six 2 m^3 sections costing US $27 m^{-3} were used for deep-well applications. Figure 19.2 depicts the shallow-well system for handpumps and PV systems. Pipings and standpipes, which are surveyed to be more expensive than the motor–pump unit used, are

Table 19.6 35% efficient AC submersible motor–pump unit for deep wells (imported).

Flow rate ($m^3 h^{-1}$)	Head (m)	Cost (US $)
3	30	855
3	35	961
3	40	1020
3	55	1202
3	70	1551
6	20	1039
6	25	1078
6	30	1153
6	35	1314
6	40	1431
6	50	1745
11	20	1235
11	28	1522
11	35	1631
11	40	1961
11	50	2075
11	60	2235
20	40	2102
20	60	3573
20	85	4145
20	105	4776
36	30	3490
36	38	4000
36	50	5059
36	60	5314
36	68	6000

not introduced for shallow-well applications. Instead, a ferro-cement water jar, used as storage, is filled through a plastic hose directly from the pump. Water is taken directly from the water jar located near the well. Therefore the number of people served per shallow well is computed from the well yield and the water consumption per person. The storage is sized to provide sufficient water to serve the 1 day requirement of the number of people served per well. Standpipes with two taps are introduced to deep-well PV systems (figure 19.3). These are located adjacent to the ferro-cement water tank near the well to minimize distribution piping cost. Well cost also includes piping cost from well to tank, while standpipe cost includes piping cost from tank to nearby standpipes. Civil works include standpipes, well and platform.

A nominal discount rate of 12% $year^{-1}$ and a general inflation rate of 5%

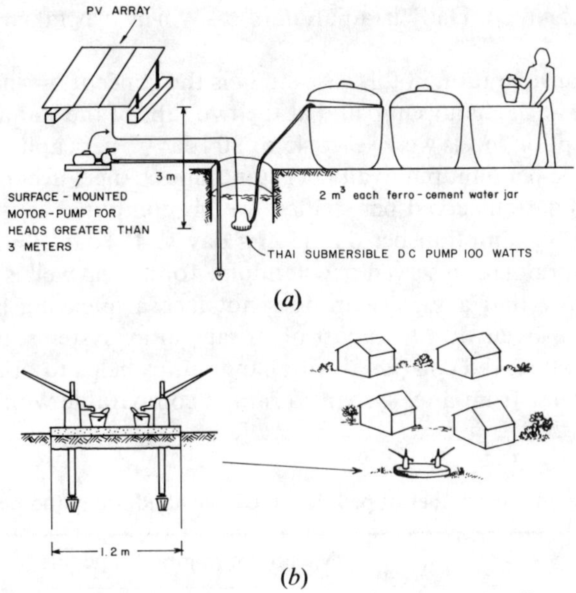

Figure 19.2 (*a*) Shallow-well PV system. (*b*) Shallow-well system with two handpumps per well.

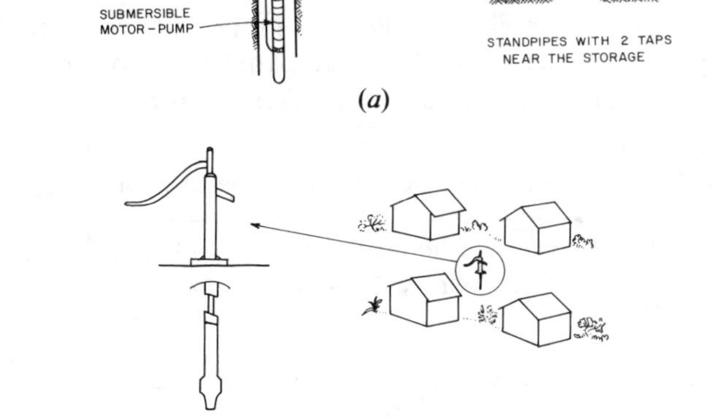

Figure 19.3 (*a*) Deep-well PV system. (*b*) Deep-well handpump system.

year^{-1} were assumed. Daily irradiation of 5 kW h m^{-2} is taken for the base case.

A unique consideration in this case study is the concept of waiting time or queueing time at the handpump and at the two taps of the standpipe. As the water consumption levels were set to 45 or 20 l day^{-1} per capita, the number of people served per handpump and per standpipe changes according to table 19.7. The population served per shallow well is computed as well yield per day by water consumption per day (10 m^3 day^{-1}/45 l day^{-1} = 222 people). Note that the population served per standpipe for a deep well is close to this value, indicating that a PV system does not face a queueing limitation as handpumps do. Even with high cost of storage in PV systems, the relatively large population served compared with handpumps helps to bring down the unit cost of water from a PV system to ranges comparable with that for the handpump.

Table 19.7 Maximum number of people served per handpump and per standpipe.

Rate (l day^{-1} per capita)	Maximum number of people		
	Per handpump (15 l/min)	PV deep well[a]	PV shallow well[b]
45	90	210	222
20	190	450	500

[a] With standpipe and two taps, 15 l min^{-1} per tap.
[b] Yield per day for every litre per day per capita: 10 m^3 day^{-1}/45 l day^{-1} per capita = 222 people.

The handpump is represented as a single-server model with Poisson input and exponential service. Since each standpipe has two taps, the standpipe is depicted as a two-server model with Poisson input and exponential service.

Handpump and PV technology were compared based on their unit water cost over a 20 year analysis period. It is further assumed that a decision can

Table 19.8 Maximum population served per well by two handpumps.

Rate (l day^{-1} per capita)	Handpumps per well	Population served per well	Population served per well for PV system	
			Deep well	Shallow well
45	2	180	210	222
20	2	380	450	500

not be made in favour of one technology or the other technology when the difference in unit water cost (in US dollars per cubic metre) over 20 years is within US \$0.01 m^{-3} of water. Within this threshold, it is assumed that handpumps and PV systems are likely to be economically and technologically comparable and substitutional. The cost of human power associated with the handpump as well as the macroeconomic impact of substituting one existing industry with another 'semi-imported' industry were not within the scope of the case study.

19.5 CASE STUDY FOR A SHALLOW WELL (3 m HEAD)

A system with three 45 W_p PV modules of 7% module efficiency installed with a 96 W US \$24 DC motor–pump system from table 19.4 is compared with a system with two handpumps per well for a shallow-well 3 m head application. In this case, the well diameter is 1.2 m and it is 10 m deep.

Table 19.9 depicts the outcome of the comparison between handpump and PV pumping at 3 m head from a shallow well. There is a 1.2 US¢ m^{-3} difference in unit water cost between the two alternatives. Since there is no

Table 19.9 Unit water cost of a shallow-well 3 m head base case at 20 and 45 l day^{-1} per capita.

	PV system
Population served per well	500 (20 l day^{-1} per capita) 222 (45 l day^{-1} per capita)
Daily water used (well yield)	10 m^3
Pump cost	US \$24 each at 96 W
PV array cost	US \$1080 at 135 W_p (3 modules)
Storage cost	US \$98 for 10 m^3
Well cost	US \$400, shallow, less than 10 m
Life-cycle cost (array)	US \$1161 (64%)
Life-cycle cost (pump)	US \$117 (6%)
Life-cycle cost (civil work)	US \$535 (30%)
Total life-cycle cost	US \$1813 (100%)
Investment	US \$3 per capita
Unit water cost over 20 years	US \$0.025 m^{-3}
	Handpump system
Population served per well	380 per well with 2 handpumps
Life-cycle cost (handpump)	US \$318
Life-cycle cost (well)	US \$430
Total life-cycle cost	US \$748
Investment	US \$1 per capita
Unit water cost over 20 years	US \$0.013 m^{-3}

queueing limitation in the PV system, water consumption level does not affect the unit water cost. The population served per well changes as consumption level changes. In the case of the handpump, water consumption affects the daily water use per well (7.5 m^3 for 20 l day^{-1} per capita and 8.1 m^3 for 45 l day^{-1} per capita), which are below the maximum well yield per day.

The array cost takes up 64% of total system life-cycle cost. By having a more efficient PV module of 8%, only two modules are required instead of three while, at the same time, the motor–pump cost can increase to four times the base-case cost and still maintain a US $0.025 m^{-3} water cost. This indicates that the efficiency of the module can be a very sensitive factor for competitiveness and also provides enough cost margin for a wider choice of motor–pump system. This still assumes a 25% efficient motor–pump unit.

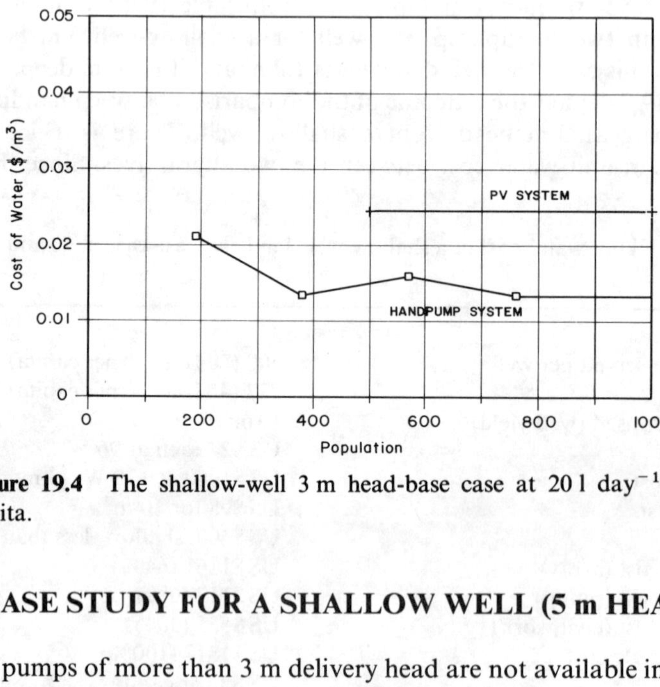

Figure 19.4 The shallow-well 3 m head-base case at 20 l day^{-1} per capita.

19.6 CASE STUDY FOR A SHALLOW WELL (5 m HEAD)

Since DC pumps of more than 3 m delivery head are not available in the Thai commercial market, a locally manufactured single-phase AC motor and surface-mounted centrifugal pump (table 19.5) can be connected to an inverter designed for solar applications with at least 80% efficiency. The single-phase AC motor–pump system has a minimum rating of $\frac{1}{4}$ hp (or 187 W) at US $55 each, with an average lifetime of 7 years. Five PV modules of 45 W_p are needed for this application. In the base case, the PV system is not competitive with the handpump technology available. The same handpump system at 3 m depth is being compared with a bulkier, more expensive PV system operating in an AC mode with two additional modules, additional cost of inverters, additional energy loss to the inverters, and a more expensive motor–pump system whose minimum power rating is almost twice as large as required.

From table 19.10, 75% of the total system life-cycle cost is for the PV array.

Table 19.10 Unit water cost of a shallow-well 5 m head base case at 20 and 45 l day^{-1} per capita.

PV system	
Population served per well	500 (20 l day^{-1} per capita) 222 (45 l day^{-1} per capita)
Daily water used (well yield)	10 m^3
Pump cost	US $55 each at 187 W
PV array cost	US $1969 at 225 W_p (5 modules)
Storage cost	US $98 for 10 m^3
Well cost	US $400, shallow, less than 10 m
Life-cycle cost (array)	US $2116 (75%)
Life-cycle cost (pump)	US $178 (6%)
Life-cycle cost (civil work)	US $535 (19%)
Total life-cycle cost	US $2829 (100%)
Investment	US $5 per capita
Unit water cost over 20 years	US $0.039 m^{-3}
Handpump system	
Population served per well	380 per well with 2 handpumps
Life-cycle cost (handpump)	US $318
Life-cycle cost (well)	US $430
Total life-cycle cost	US $748
Investment	US $1 per capita
Unit water cost over 20 years	US $0.013 m^{-3}

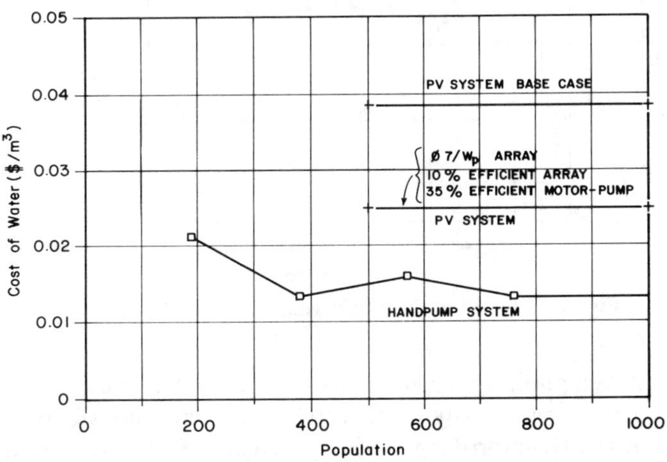

Figure 19.5 Shallow-well 5 m head case at 20 l day^{-1} per capita.

By having a 10% efficient PV module, only four modules are required instead of five. On top of this, if a 35% efficient motor–pump system is used, at the same cost, only three PV modules will be needed. However, it is only with a further reduction in PV array cost from US \$8 W_p^{-1} to US \$7 W_p^{-1} that PV systems will be competitive with the handpump system pumping at 5 m head (figure 19.5).

19.7 CASE STUDY FOR A DEEP WELL (20 m HEAD)

A system with an AC submersible motor–pump unit (table 19.6) connected through an 80% efficient US \$0.75 W_p^{-1} solar inverter to a 12×45 W_p PV array is compared with a US \$275 deep-well handpump. In the base case, the PV system is cost competitive with the handpump system when serving more than 900 persons at 20 l day^{-1} per capita (figure 19.6) or 420 persons at 45 l day^{-1} per capita (figure 19.7). The PV array cost takes up 50% of the total system life-cycle cost. In the same manner, savings in the number of PV modules, and still maintaining the required service, can be attained by higher module efficiency. Since the motor–pump efficiency assumed in the base case was based on an actual field test of a similar well known imported AC submersible pump, this efficiency (35%) will not be varied because it is unlikely that any improvement on this type of motor–pump system can be made within the country.

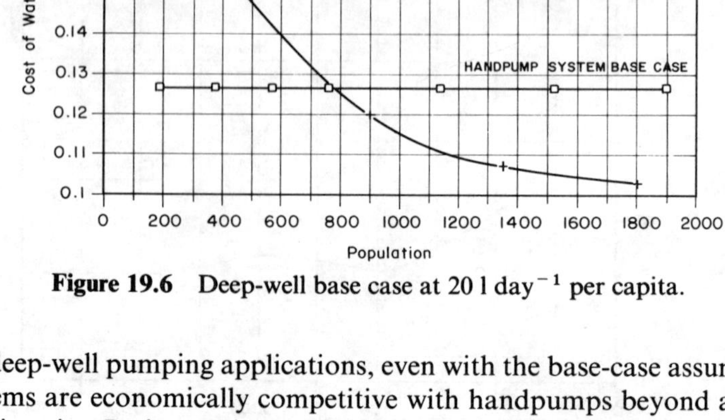

Figure 19.6 Deep-well base case at 20 l day^{-1} per capita.

For deep-well pumping applications, even with the base-case assumptions, PV systems are economically competitive with handpumps beyond a certain population size. By improving module efficiency from 7 to 10% and with an

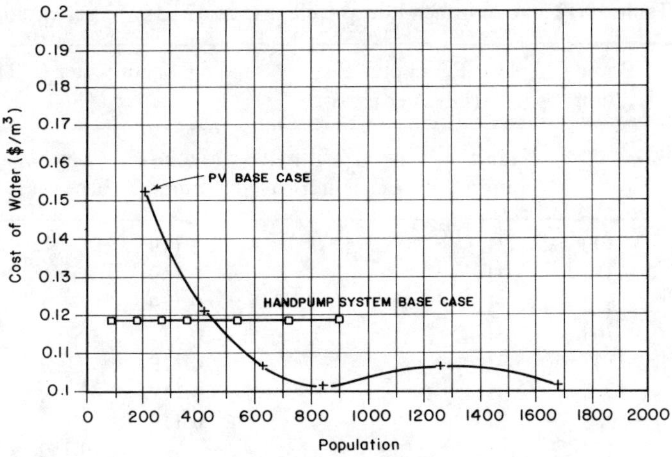

Figure 19.7 Deep-well base case at 45 l day^{-1} per capita.

array installed cost of US $8 W_p^{-1}, PV systems will be able to compete with handpumps on deep-well pumping applications at any population size (figures 19.8 and 19.9).

Table 19.11 A 20 m head deep-well case at 20 l day^{-1} per capita.

Number of water points	Population	Per capita cost (US$ per person)			Cost of water at (US $ m^{-3})[a]		
		Handpump	PV, base case	PV, improved[b]	Handpump	PV, base case	PV, improved[b]
1	190	15			0.127		
2	380	15			0.127		
3	570	15			0.127		
4	760	15			0.127		
6	1140	15			0.127		
8	1520	15			0.127		
10	1900	15			0.127		
1	450		19	15		0.152	0.126
2	900		15	12		0.120	0.101
3	1350		14	11		0.107	0.086
4	1800		13	10		0.103	0.081
6	2700		14	11		0.107	0.086
8	3600		13	10		0.103	0.081
10	4500		12	10		0.096	0.076

[a] Life-cycle cost per cubic metre of water provided during 20 years.
[b] PV array efficiency improved from 7% base case to 10%.

Table 19.12 A 20 m head deep-well case at 45 l day^{-1} per capita.

Number of water points	Population	Per capita cost (US$ per person)			Cost of water (US $ m^{-3})[a]		
		Hand-pump	PV, base case	PV, improved[b]	Hand-pump	PV, base case	PV, improved[b]
1	190	31			0.119		
2	380	31			0.119		
3	570	31			0.119		
4	760	31			0.119		
6	1140	31			0.119		
8	1520	31			0.119		
10	1900	31			0.119		
1	450		42	35		0.152	0.128
2	900		34	27		0.121	0.097
3	1350		31	23		0.107	0.082
4	1800		29	22		0.102	0.077
6	2700		31	23		0.107	0.082
8	3600		29	22		0.102	0.077
10	4500		28	21		0.097	0.075

[a] Life-cycle cost per cubic metre of water provided during 20 years.
[b] PV array efficiency improved from 7% base case to 10%.

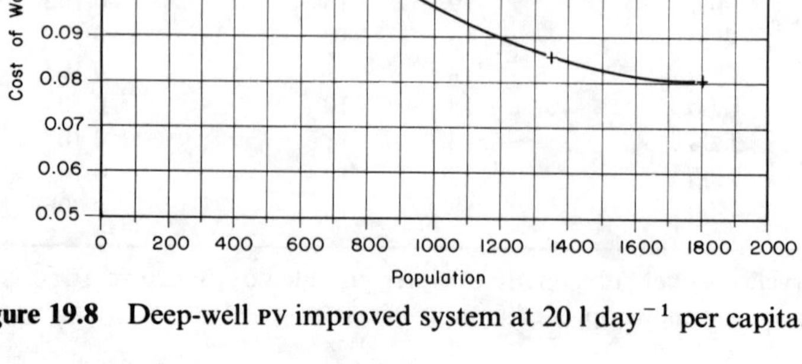

Figure 19.8 Deep-well PV improved system at 20 l day^{-1} per capita.

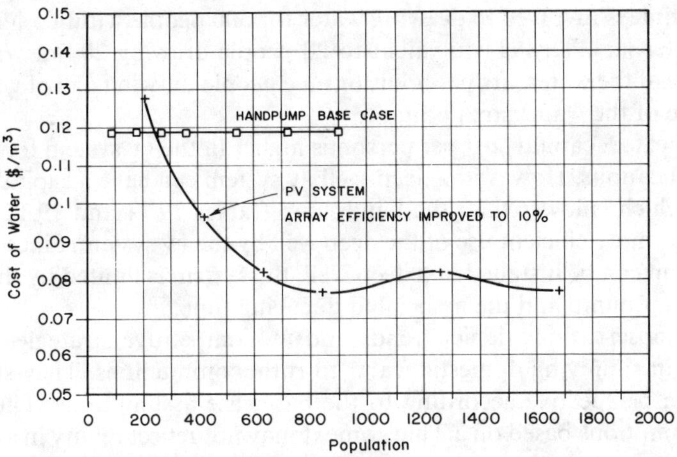

Figure 19.9 Deep-well PV improved system at 45 l day^{-1} per capita.

19.8 CONCLUSION

The case study was conducted with careful attention to the reality of component costs and to avoid partiality to the PV system. In the three areas of application discussed, there are always strategies for PV competitiveness even when base-case assumptions were not competitive. It can be fairly noted that, for pumping applications involving a static head of less than 3 m, a PV system of three modules and a DC pump in the range of less than 100 W is already cost competitive even at the base-case level. For larger systems, strategies to improve PV system competitiveness from the base case is achieved through improvement in the techno-economic performance (in cost and efficiency) of the PV array and motor–pump system. It is interesting to note that, provided that the cost data used in the case study do not differ too far from actual purchase cost, the available technology and cost structure in Thailand is conducive to the possibility of replacing handpumps with PV pumps in shallow-well and deep-well applications.

This case study also emphasized the necessity to include the social and physical characteristics of the village. Village characteristics makes the computation of the waiting time possible. This is the main weakness of the handpump system compared with the PV system. Handpumps would have out-performed PV systems, in cost-competitiveness, if queueing time was not considered, thus making the handpump serve the same number of persons in an hour as one standpipe connected to a PV system for deep wells. For shallow-well application, the base-case PV system is similar to drawing water from an open pond. The difference is the covered well and safe storage for sanitation and hygienic safeguards. In this case, no queueing limitation or waiting time is involved in drawing water for 500 people within 5 h from five 2 m^3 water jars. This is equivalent to 20 people drawing 20 l of water each from one of the water jars per hour, or nine people drawing 45 l of water each from one of the water jars per hour.

As expected, capital cost per person is higher in the PV system for shallow-well applications. However, a deep-well PV system can have a capital cost per person which is lower than the handpump (tables 19.11 and 19.12). This is due to the more efficient use of the deep well by the PV system. Since only one handpump can be installed in a deep well, the system is limited by the output of one handpump and the associated queueing time.

These observations depict trends and PV competitive strategies that are possible in supplying domestic water to rural communities. They should be viewed in perspective according to the base-case assumptions. These base-case assumptions based on a Thai context may not reflect reality in a different cultural and economic environment. Even then, an economic case study may fail in reality when capitalistic market forces change as they have been noted to behave unexpectedly sometimes.

REFERENCES

Cabraal A, Siess A, Slominski L, Buresch M and Kenna J 1987 *Technical Report No SAND87-7015* (Albuquerque, NM: Sandia National Laboratories)

Lasnier F 1989 *AIT Report on Motor–Pump Market Survey* (Bangkok: Asian Institute of Technology)

United Nations Development Programme–World Bank 1986 *Report No INT/81/026* (Bangkok: Ministry of Public Health)

BIBLIOGRAPHY

Arlosoroff S, Tschannerl G, Grey D, Journey W, Karp A, Langeneffe O and Roche R 1987 *Community Water Supply: The Handpump Option* (Washington, DC: UNDP–World Bank)

20 Life-cycle Cost Comparison of an Alternative Power Supply for a Portable Pocket-sized Stereo Cassette Tape Recorder

20.1 INTRODUCTION

This chapter attempts to investigate the financial advantage of alternative power supply for a portable pocket-sized stereo cassette tape recorder. The commercial awareness of the possibility of introducing a small photovoltaic (PV) battery charger for electronics consumer products is uppermost in mind. A case study is presented here and table 20.1 lists a typical consumption scenario.

Table 20.1 Typical consumption scenario of a pocket-sized stereo cassette tape recorder.

Power	100 mW
Voltage	3 V DC
Battery	2 × size AA
Consumption	1 h per evening
	30 h month^{-1}
	3 W h month^{-1}
Lifetime	3 years
Annual inflation rate	5.0%
Monthly inflation rate	0.4%
Annual discount rate	10.0%
Monthly discount rate	0.8%
Period of analysis	1 month–2 years

In the following sections, three conceptual alternative power supplies for a cassette tape recorder will be described, financially evaluated and compared. These three alternative power supply are as follows.

(i) Disposable alkaline primary cell.

(ii) Rechargeable nickel–cadmium secondary cell with grid power.
(iii) Rechargeable nickel–cadmium secondary cell with PV battery charger.

A comparison of relevant information needed in the case study between an alkaline primary cell and a nickel–cadmium secondary cell is presented in table 20.2.

Since the load is the same for each alternative, the case study will only investigate the direct cost of the power supply of each alternative.

Table 20.2 Specifications of alkaline and nickel–cadmium cells.

	Alkaline	Nickel–cadmium
Size	AA	AA
Amphere–hour capacity	0.5 A h	0.5 A h
Volt	1.5 V	1.2 V
Life	Not applicable[a]	500–1000 cycles
Cost	US $0.60	US $6.00

[a] Its life terminates once the chemical reaction cannot supply the energy demanded of it.

20.2 DESCRIPTION OF THE THREE ALTERNATIVE POWER SUPPLIES

20.2.1 Disposable alkaline primary cell

It is assumed here that, below 20% of the battery capacity, power output will be too low to make the cassette tape recorder audible to a comfortable level. From tables 20.1 and 20.2, table 20.3 lists the relevant cost information.

Table 20.3 Input data for a disposable alkaline primary cell.

Battery capacity	$0.5\ \text{A h} \times 1.5\ \text{V} = 0.75\ \text{W h}$
Batteries per month	$3\ \text{W h}/(0.75\ \text{W h} \times 0.8) = 6$ pieces
Cost of battery	US $0.60 each
Replacement cost	US $3.60 for 3 sets or 6 pieces
Discount rate	0.8% month^{-1}
General inflation rate	0.4% month^{-1}

Six batteries were needed as replacement per month. In normal use, two batteries are required for each replacement. Initially, one set of batteries (or two batteries per set) is needed to start the case study. One set of batteries discharged 1.2 W h (2×0.5 A h $\times$ 1.5 V $\times$ 80%) within 12 days (1.2 W h/100 mW h per evening) at a consumption level of 100 mW h per evening. The second set of batteries has to be replaced after the twelfth day of the month and the third set after the twenty-fourth day of the month.

20.2.2 Rechargeable nickel–cadmium secondary cell with grid electricity

A battery charger connected to the conventional 220 V AC grid power is used to recharge two nickel–cadmium batteries while another two nickel–cadmium batteries are in use. The cost of electricity was assumed to be the approximate rate for an average consumer in the Bangkok metropolis. Table 20.4 presents the input data for this alternative.

Table 20.4 Input data for a rechargeable nickel–cadmium secondary cell with grid power.

Batteries required	2 in use 2 for charging
Batteries required during system's lifetime	4 54 charge–discharge cycles for each battery in 3 years
Cost of battery	US \$6.00 per piece 4 pieces $\times$ US \$6.00 = US \$24.00
Rate of grid electricity	US \$0.10 kW h^{-1}
Monthly grid energy consumed	3 charge cycles $\times$ 1.2 W h per set = 0.0036 kW h month
Cost of grid electricity consumed	0.0036 kW h $\times$ US \$0.10 kW h^{-1} = US \$0.00036 month^{-1}
Cost of AC–DC battery charger	US \$20.00 each
Initial investment	US \$44.00
Operation and maintenance	US \$0.00036 month^{-1}
Discount rate	0.8% month^{-1}
General inflation rate	0.4% month^{-1}

20.2.3 Rechargeable nickel–cadmium secondary cell with photovoltaic battery charger

Table 20.5 presents the input data for this alternative.

Table 20.5 Input data for a rechargeable nickel–cadmium secondary cell with a PV battery charger.

Batteries required	2 in use 2 for charging
Batteries required during system's lifetime	4 54 charge–discharge cycles for each battery in 3 years
Cost of battery	US $6.00 per piece
PV battery-charging system	US $7 W_p^{-1} including PV module and charger
Cost of PV charging system	US $7.00 for 1 W_p system
Initial investment	US $31.00
Discount rate	0.8% $month^{-1}$
General inflation rate	0.4% $month^{-1}$

20.3 LIFE-CYCLE COST COMPARISON

Figure 20.1 presents data in terms of the ratios between the life-cycle costs of the three power supply alternatives as a function of the utilization period of the system in the following manner.

(*a*) Grid charging to alkaline battery.
(*b*) PV charging to alkaline battery.
(*c*) PV charging to grid charging.

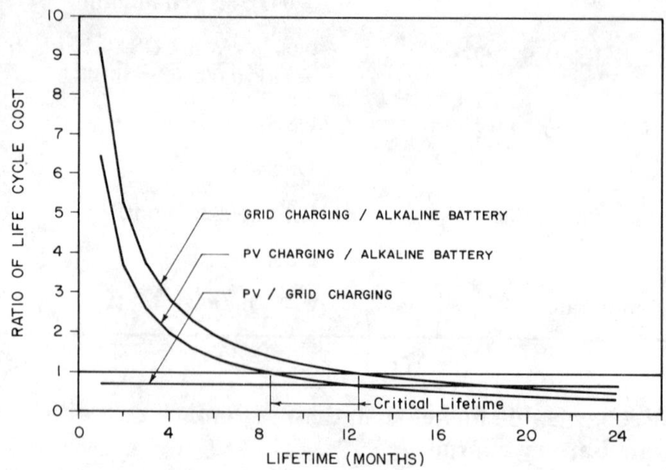

Figure 20.1 Comparison of the three power supply alternatives.

When this ratio is less than 1.0, grid charging provides a lower life-cycle cost than the alkaline battery, PV charging provides a lower life-cycle cost than the alkaline battery, and PV charging provides a lower life-cycle cost than grid charging.

The case study compares the disposable primary-cell alternative, which has a lower proportion in investment cost but a higher proportion in recurrent cost, to two alternative battery-charging systems (PV charged and grid charged) with negligible proportion in recurrent costs. As the comparison period increases, recurrent costs from each additional month included in the analysis add to the present value of the life-cycle cost. This particular difference in the cost component of the primary- and the secondary-battery alternatives highlights the significant effects of the comparison period or more appropriately the utilization period. Increasing the utilization period makes the primary-battery alternative for this particular case study less cost effective. Figure 20.1 shows that, by $8\frac{1}{2}$ months and by $12\frac{1}{2}$ months, the PV-charging system and the grid-charging system respectively will be more cost effective than the primary-cell alternative. The monthly recurrent costs of the disposable primary-cell alternative gradually catch up with the higher capital investment of the two alternative charging systems.

Since they are a direct comparison of the cost of alternative power systems providing exactly the same load requirements, the critical lifetimes of figure 20.1 can also be viewed as the pay-back period of investing in a rechargeable secondary-cell system. Savings made from not having to purchase disposable batteries are treated as income.

Increasing the rate of consumption within a utilization period does not affect the outcome of the cost comparison. It further raises the monthly recurrent cost of the disposable primary-cell alternative, while increasing the charging cycle of secondary cells without additional cost for the PV charging system and negligible increase in grid electricity cost for a grid charging system.

The major cost component of the grid-charging system is the cost of the AC–DC charger. The cost of electricity used for charging is very small compared with other costs. When comparing the two alternative charging systems, the AC–DC battery charger is essentially being compared with the total PV charging system that is usually supplied as a complete package with PV modules, wiring and charger. The PV system will only be more cost effective than the grid system, if and only if the total cost of the PV system is lower than the market price of an AC–DC battery charger. As far as these two alternatives are concerned, the economic environment is not a major factor owing to the low proportion of recurrent costs.

However, in situations where grid power is not available, PV-charged secondary cells will be the more cost-effective means of supplying DC power for small home appliances in place of disposable primary batteries.

If financial resources are scarce, then there is a non-negligible opportunity cost of the resource invested in the higher capital requirement of the charging system. This scarce resource may be allocated to other investment of a higher financial or social benefit to the owner. An argument can also be made concerning the presence of an opportunity gain of the resource invested in a charging system when other charging applications also benefit from the investment. Opportunity cost and gain are not directly treated in the case study. If a high discount rate reflects the inclusion of an opportunity cost to the analysis (sensitivity to discount and interest rate), it is still lacking in its treatment of the social cost incurred in investing a scarce resource. In specific cases where this incomplete treatment of opportunity cost is not acceptable (particularly in poor remote areas), then conclusions drawn from the present case study should be regarded with constraint considering that other non-quantifiable economic or social parameters may take precedence.

20.3.1 Sensitivity to capital cost

This section examines the effect of changing capital cost estimates for the three alternatives. The effect is modelled as a 'cost multiplier', which is examined at 0.75, 1.0 and 1.25, with 1.0 equivalent to cost estimates in tables 20.3–20.5. The range of multipliers is intended to account for system capital cost variations as a result of different cost, insurance and freight equipment and other costs which are country or site specific. These sensitivities are depicted in terms of their net present life-cycle cost in figure 20.2 as a function of their utilization period.

For the worst PV case, when the alkaline capital cost is reduced by 25% and the PV capital cost is increased by 25%, the PV system will still be cost effective when the system utilization period exceeds $14\frac{1}{2}$ months.

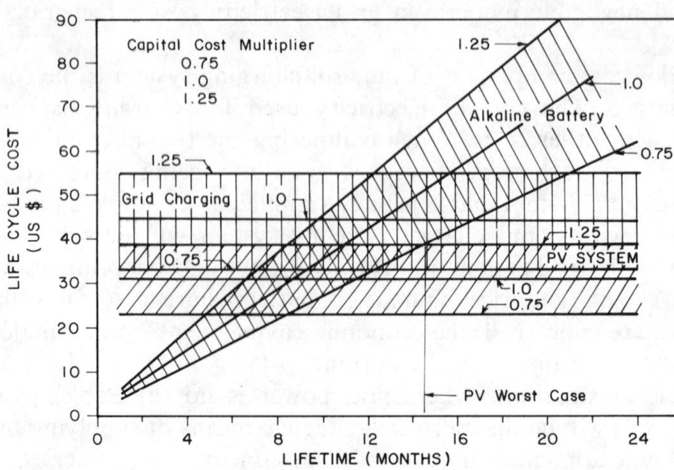

Figure 20.2 Sensitivity of the present life-cycle cost to capital cost.

20.3.2 Sensitivity to discount and inflation rate

This section examines the effects of varying the discount rate and inflation rate at half and twice the value assumed in table 20.1. The points of interest are the points of intersection of the three curves; these are the regions where decision of a choice is critical. These regions are marked by broken circles in figures 20.3 and 20.4

These two figures depict three observations.

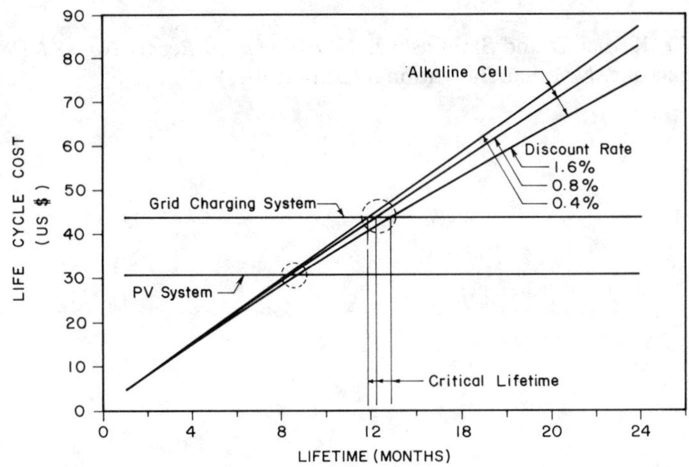

Figure 20.3 Sensitivity of the present life-cycle cost to discount rate.

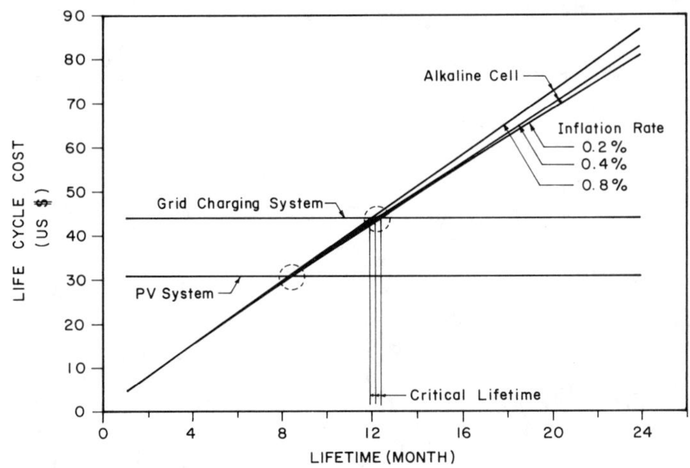

Figure 20.4 Sensitivity of the present life-cycle cost to inflation rate.

(i) The grid and PV systems are not affected by the discount and inflation rate.
(ii) The present life-cycle cost of the primary-cell alternative has a low sensitivity to the discount and inflation rate within the region enclosed by the broken circles.
(iii) The critical utilization period has a low sensitivity to the discount rate and inflation rate.

BIBLIOGRAPHY

Eskenazi D, Kerner D and Slominski L 1986 *Technical Report No SAND85-7018/2* (Albuquerque, NM: Sandia National Laboratories)

Part VII

Instrumentation

21 A Simple Metal–Oxide–Semiconductor Field-effect Transistor Electronic Variable Load

21.1 INTRODUCTION

The physical properties and the performance characteristics of a photovoltaic (PV) module depend on its basic materials, manufacturing technology and operating environmental conditions. The current–voltage (I–V) curves derived from PV modules reflect to a large extent the physical properties and their expected performance. If the curves are obtained with an acceptable accuracy under natural conditions, they will provide valuable information concerning the characteristics of the module to both the manufacturer and the system designer.

The basic principle for obtaining an experimental I–V curve is to control the load current of the PV module. The load current must be controlled in such a way as to resemble the curve in figure 21.1. This is done by varying the load current from zero, at the open-circuit condition, to its maximum value of short-circuit current. Under the short-circuit condition the voltage is ideally zero. In order to reflect an accurate characteristic of the module, the variation in the load current must be very smooth.

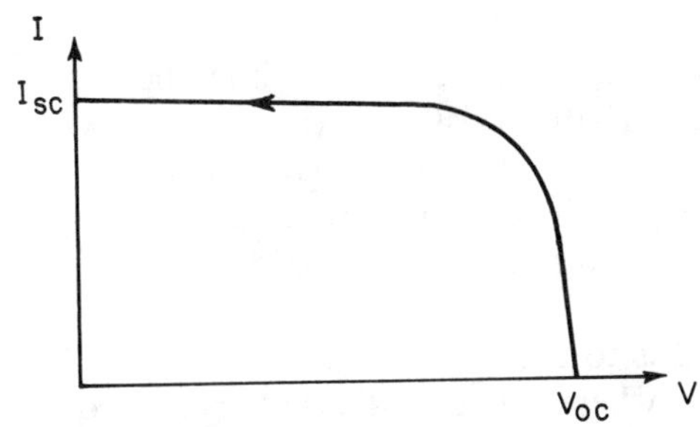

Figure 21.1 I–V curve: I_{sc}, short-circuit current; V_{oc}, open-circuit voltage.

Figure 21.2 shows the block diagram of a system used to measure the *I–V* curve and to characterize the performance parameters of a PV module.

The following are the major parts of the system.

(i) Metal–oxide–semiconductor field-effect transistor (MOSFET) electronic variable load.
(ii) Ramp generator.
(iii) Data acquisition interface.
(iv) Data processor and storage.

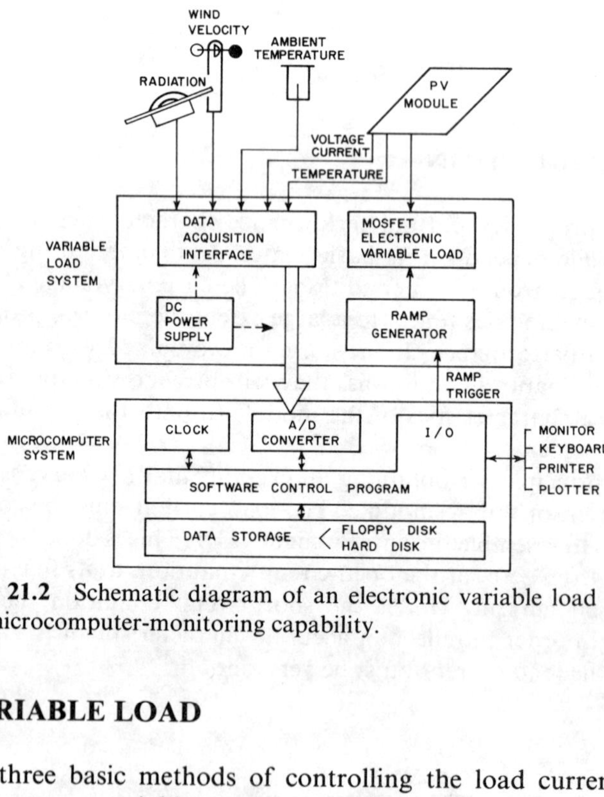

Figure 21.2 Schematic diagram of an electronic variable load system with microcomputer-monitoring capability.

21.2 VARIABLE LOAD

There are three basic methods of controlling the load current. The first method is to use a variable resistor as the load. By varying the resistance value from maximum to zero, the current can be made to flow from minimum to maximum. The variation in the resistance is usually done manually, and in order to have a smooth *I–V* curve the resistance should be varied uniformly.

A second method is to use a capacitor as the load. This is schematically shown in figure 21.3, where a PV module is connected to a capacitor, which is short-circuited at the start of a measurement. At this time the short-circuit current of the module can be measured. When switch S_1 is closed while S_2

and S_3 are open, the capacitor begins to charge. While the charge of the capacitor increases, the voltage across the capacitor increases and its current decreases. When the capacitor is fully charged, no current flows into the capacitor and an open-circuit condition across the capacitor is reached. To obtain a smooth and accurate *I–V* curve, a well dimensioned and good-quality capacitor without leakage is required. In addition, the three switches S_1–S_3 should be switched in a proper sequence, so that neither the contact nor the capacitor will be damaged.

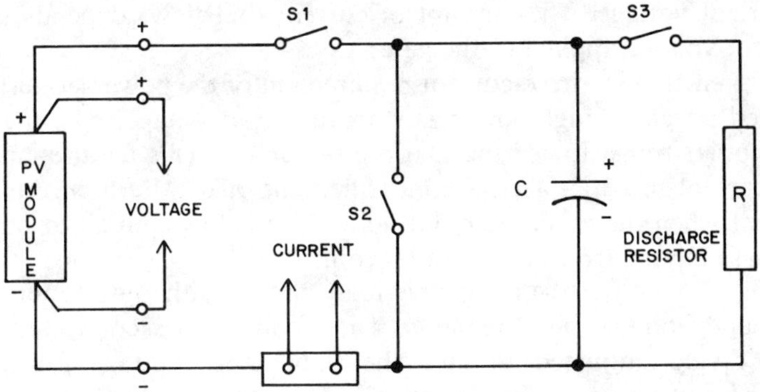

Figure 21.3 Capacitor as load to measure *I–V* curve of the PV module.

The third method is to use a transistor as the load. Figure 21.4 shows an ordinary n–p–n transistor and an FET as load. By varying the base current of the transistor and by varying the gate voltage of the FET, the apparent impedance between the collector C and the emitter E of the transistor and between the drain D and source S of the FET can be modulated. In this way, the load current which is measured across the resistor R is controlled.

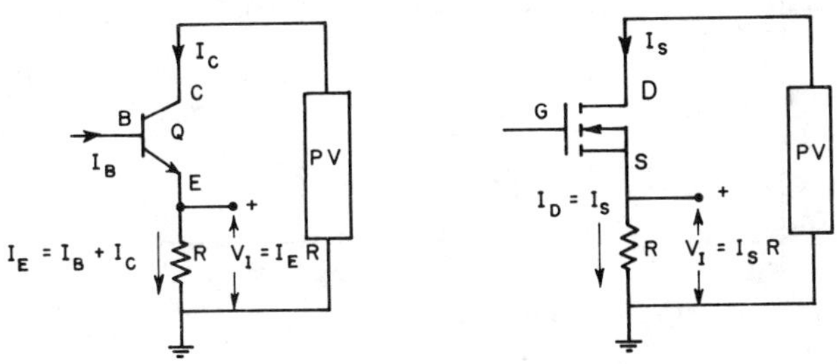

Figure 21.4 Transistor as load to measure *I–V* curve of the PV module.

The FET in figure 21.4 is an n-type channel MOSFET device. The MOSFET is a voltage-controlled device. A voltage must be applied between the gate G and source S terminals to produce a flow of current in the drain D. The gate is isolated electrically from the source by a layer of silicon oxide. Theoretically no current flows into the gate when a DC voltage is applied to it—although in practice there will be an extremely small leakage current, of the order of nanoamperes.

With no voltage applied between the gate and source electrodes, the impedance between the drain and source terminals is very high, and only a small leakage current flows in the drain until the applied voltage exceeds the drain-to-source avalanche voltage, setting up an electrical field within the MOSFET. This field modulates the resistance between the drain and source terminals and permits a current to flow in the drain in response to the applied drain circuit voltage. The amount of current that flows depends upon the amount of voltage applied to the gate.

The extremely low-drive-current requirement of the power MOSFET and the associated extremely high power gain are major advantages over the conventional bipolar transistor or the Darlington circuit. This feature enables the source current to be an accurate measurement of the drain current. This is significant when the PV module output current is low enough to be affected by the base input current of a bipolar transistor.

Figure 21.5 is a MOSFET variable-load circuit with output terminal for current and voltage measurements. The PV module is connected across a voltage divider adjusted to give the right input voltage range to any measuring instrument connected to V_1. The capacitance C_1 must be small enough not to affect the I–V curve. The current of the PV module is measured across the resistance R_8 and amplified by the operational amplifier A_1 whose gain is adjusted by R_7 to give the right input voltage range to any measuring instrument connected to V_2. R_8 must be as small as possible to be able to determine the short-circuit current accurately.

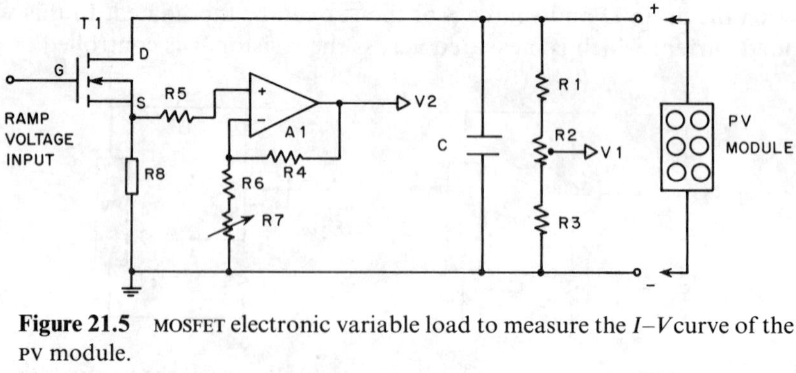

Figure 21.5 MOSFET electronic variable load to measure the I–V curve of the PV module.

21.3 RAMP GENERATOR

A voltage input at the gate terminal (figure 21.5) modulates the impedance across the drain and source terminals which in turn modulates the drain current. If the gate voltage rises linearly, then the apparent drain-to-source impedance (and in effect the drain current) changes linearly.

When a capacitor is charged from a constant-step-voltage pulse input, the voltage across the capacitor rises exponentially with increasing time at a time constant *RC*. Since the capacitor voltage cannot change instantaneously, the output starts from zero and rises toward a steady-state value dependent on the charging circuit. If the time constant is very large compared with the time frame wherein the measurement is taken, the output consists of a section from the exponential curve which is essentially a linear ramp voltage. Figure 21.6 is a simple circuit diagram of a ramp generator, using the capacitor-charging concept described.

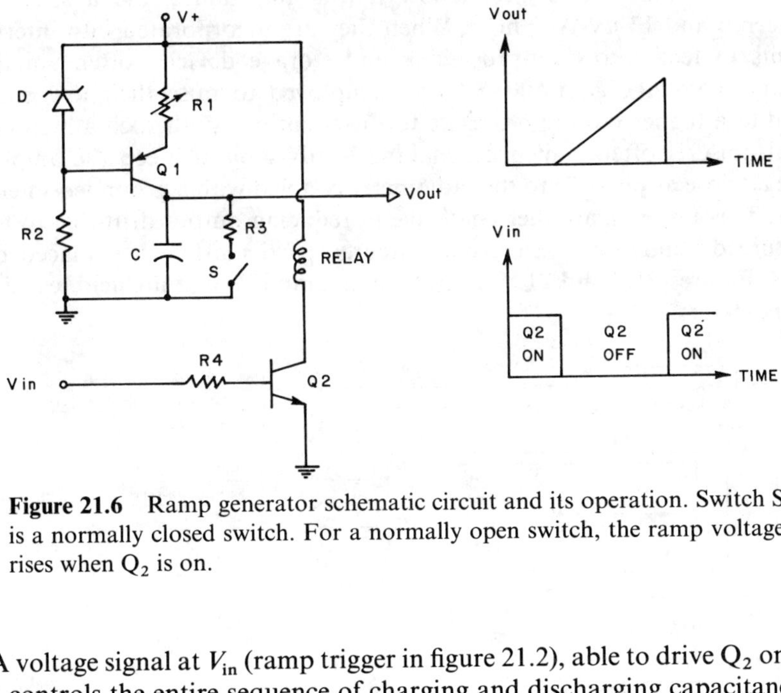

Figure 21.6 Ramp generator schematic circuit and its operation. Switch S is a normally closed switch. For a normally open switch, the ramp voltage rises when Q_2 is on.

A voltage signal at V_{in} (ramp trigger in figure 21.2), able to drive Q_2 on and off, controls the entire sequence of charging and discharging capacitance *C*. R_1 is adjusted with capacitance *C* to obtain a linear charging characteristic within a time frame required to generate the *I–V* curve at the MOSFET circuit (figure 21.5). When V_{in} drives Q_2 on, current flows through the relay coil, activating and closing switch S. With this, the capacitor is grounded through a small resistor R_3 and V_{out} is nearly zero. When V_{in} drives Q_2 off, no current flows through the relay and switch S opens, therefore allowing the capacitor to charge, and V_{out} rises in voltage linearly with time to produce a ramp voltage. The ramp voltage output is then connected to the gate terminal of the MOSFET in figure 21.5.

21.4 DATA ACQUISITION INTERFACE

The data acquisition interface consists of electronic circuits to convert the sensor input voltage to the appropriate voltage range that can be read by measuring instruments connected to it. These sensor voltages can be output voltages from an irradiance-measuring sensor, a temperature sensor, a wind sensor and a humidity sensor. The use of shielded cable is recommended to protect signals from noise interference.

A well calibrated radiometer must be used to measure solar radiation. The wavelength range of the radiometer is preferably 0.3–3 μm, the range of short-wave solar radiation useful for the PV effect. These are commercially available, having good precision, accuracy, linearity and response time. Most of the best radiometers have a low output impedance and a sensitivity between 6 and 13 $\mu V\, W^{-1}\, m^{-2}$. When they are incorporated into interface circuits to feed into data-processing and storage devices, often amplifier circuits of voltage gain 500–1000 are employed to raise their low output signal to a higher level appropriate to the situation. With such a high gain, signal noise is often a menace, making it advisable to keep the amplifier stage as close as possible to the radiometer coupled with a grounded shielded cable. This presents another challenge in reducing output drift due to temperature dependence of electronic circuits, particularly when placed outdoors. Figures 21.7 and 21.8 are two signal amplifier circuits field tested for pyranometers.

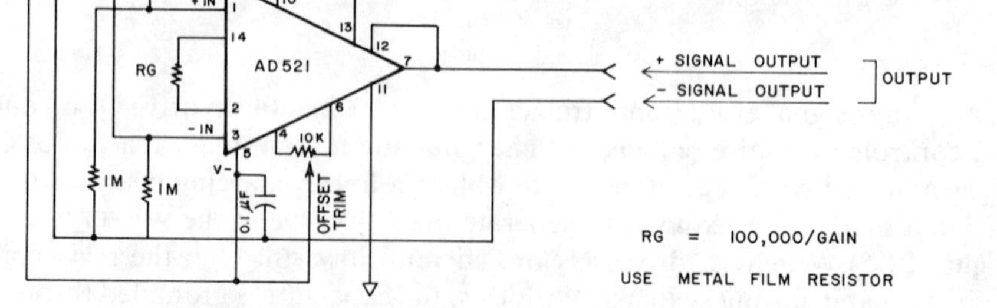

Figure 21.7 Signal amplifier for precision pyranometer.

Thermocouples or integrated-circuit temperature sensors can be used to measure module temperature and ambient temperature. Temperature sensors usually come in either a cylindrical pack or a flat pack (ideal for surface

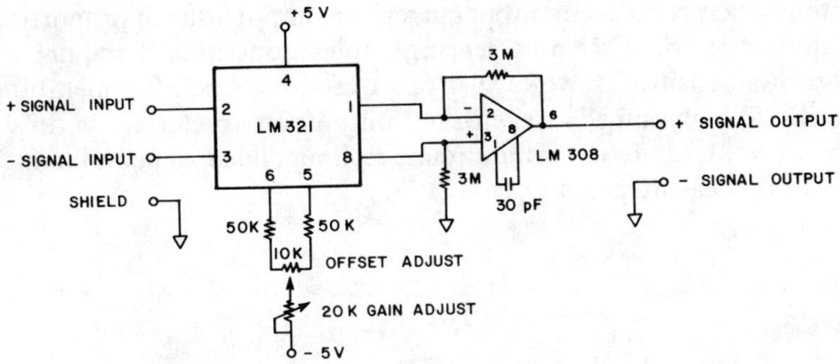

Figure 21.8 Instrument amplifier for precision pyranometer.

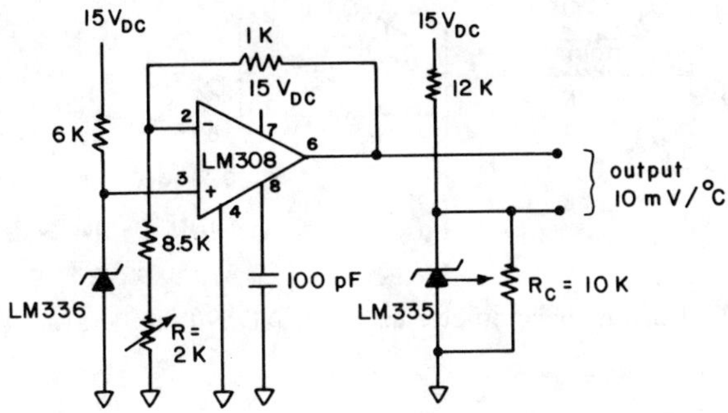

Figure 21.9 Celsius thermometer using an LM335 temperature sensor.

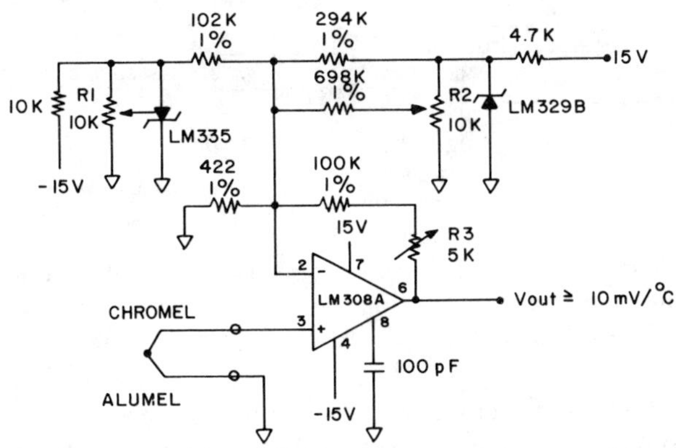

Figure 21.10 Celsius calibrated thermocouple thermometer. Terminate thermocouple reference junction in close proximity to LM335. The following adjustments should be made: apply signal in place of thermocouple and adjust R_3 for a gain of 245.7; short non-inverting input of LM308A and output of LM329B to ground; adjust R_1 so that V_{out} = 2.982 V at 25 °C; remove short across LM329B and adjust R_2 so that V_{out} = 246 mV at 25 °C; remove short across thermocouple.

mounting) and produce an output current or output voltage proportional to absolute temperature. Remote sensing applications usually require devices that are not sensitive to voltage drops over long lines. Temperature sensors with a high impedance current output are preferred in this case. Figures 21.9–21.12 present some circuits recommended by manufacturers for temperature measurements.

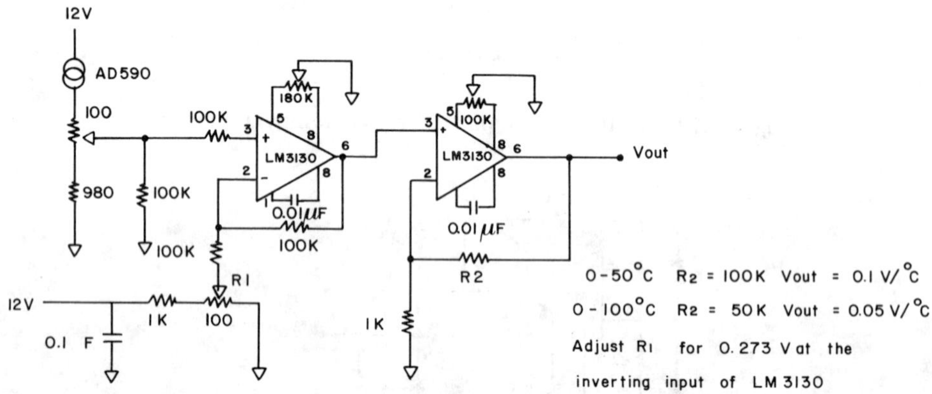

Figure 21.11 Celsius thermometer using an AD590 temperature transducer.

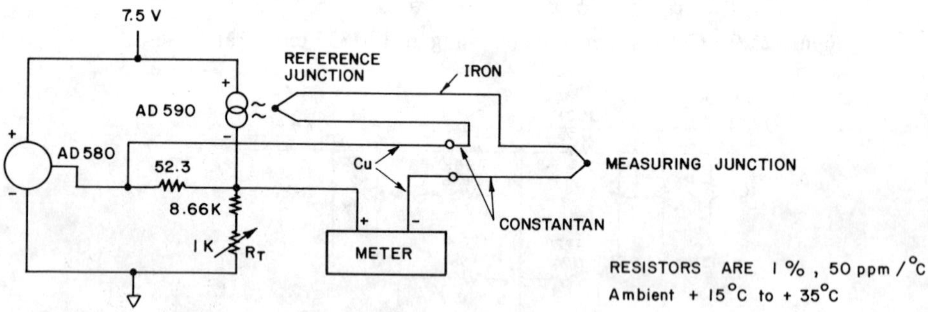

Figure 21.12 Cold-junction compensation circuit for type J thermocouple using an AD590 to monitor the reference junction temperature. Compensation accuracy will be within ± 0.5 °C for a circuit temperature between +15 and +35 °C.

Wind speed measurement and calibration can be more involved than radiation and temperature measurement. Moreover, wind tunnel calibration is required. The most convenient solution is to acquire a set of wind speed measuring devices with a meter read-out and signal output. Otherwise, figure 21.13 illustrates a field-tested circuit to measure wind speed using a suitable 'wind blade' or 'wind cup' to 'chop' a beam of light into light pulses. These light pulses are counted and accumulated electronically every 5 s and calibrated in a wind tunnel with a standard anemometer.

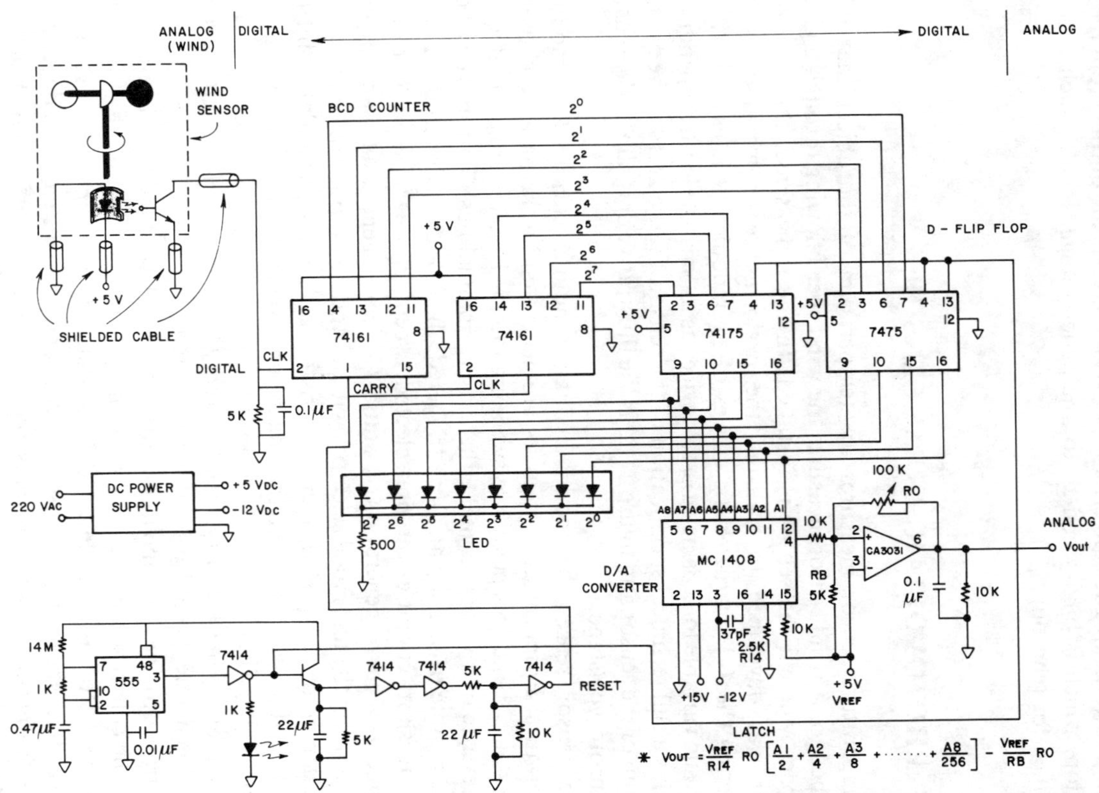

Figure 21.13 An analogue-to-digital and digital-to-analogue converter to measure wind speed: LED, light-emitting diode.

The binory-coded decimal (BCD) counter counts these wind-induced light pulses and is reset every 5 s. The output of the counter goes to the D flip–flop which stores the accumulated number of light pulses counted during the previous 5 s period. Its output is updated every 5 s before the counter is reset. The digital-to-analogue converter (MC1408) converts the output from the D flip–flop, which is binary coded, to analogue voltage signal, which is then amplified for read-out.

21.5 PHOTOVOLTAIC ARRAY SENSOR

A temperature sensor measures the temperature of the object that it is attached to. A wind sensor measures the wind speed flowing through a given air space wherein the sensor is installed. In the same sense, the PV array sensor was developed to measure the performance of PV modules by extracting the *I–V* characteristic curves of PV modules. It is essentially an electronic variable load using MOSFET devices as its variable load. Appendix A contains a full technical specification of the sensor. Figure 21.14 is the circuit diagram of the sensor and figure 21.15 is the illustrative diagram of the sensor in its prototype stage.

The sensor senses the module voltage, module current and module temperature. From these, *I–V* curves are generated by the parallel TMOSFET devices at 2, 5 and 15 s per curve-sweeping rate. *I–V* curves are generated by the presence of a voltage signal either from a manual push button trigger or from a remote trigger signal. Once triggered, the module is driven from open circuit to short circuit at the pre-selected sweeping rate. In addition to this dynamic *I–V* curve generation, a static *I–V* curve generation can be pre-selected using the test mode selector switch. Module output can be fixed at a particular point along its *I–V* characteristic curve. This particular point of interest to the user is often the maximum-power point or the point of nominal voltage of a system. By turning the static sweep dial, from V_{oc} to I_{sc}, any point along the *I–V* curve can be fixed. One full turn of the static sweep dial will generate, point by point, an *I–V* curve.

All transistor devices, including the TMOSFET, have a static on-resistance across the collector and emitter (for an n–p–n transistor) or across the drain and source (for an FET device). This static on-resistance (0.04 Ω for a TMOSFET) fixes the voltage across the drain and source to a value dependent on the drain current and prevents the real short-circuit current from the module passing through the device. By turning the test mode selector switch and selecting short instead of dynamic or static, the TMOSFET devices are shorted, drain to source, allowing a maximum current (30 A rated maximum) to flow through the 0.1 Ω current shunt. The actual short-circuit current can now be measured.

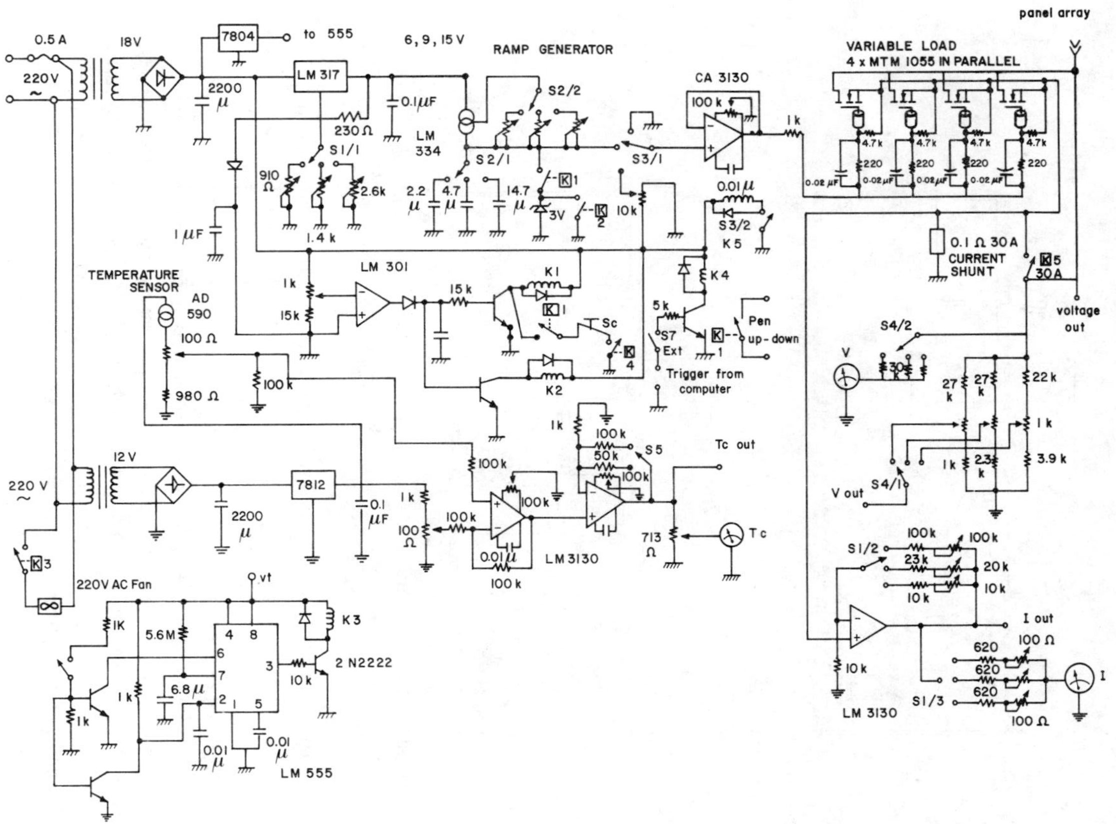

Figure 21.14 Circuit of PV array sensor. Switches: S_1, current switch; S_2, timer switch; S_3, ramp, static, short; S_4, voltage selector; S_5, temperature selector; S_6, trigger; S_7, manual external trigger. Relays: K_1, ramp; K_2, zero ramp; K_3, fan; K_4, external trigger; K_5, short output.

When a PV array field is to be tested, the temperature input must reflect as much as possible the average temperature of the modules in the field. Voltage and current ranges must also be set accordingly.

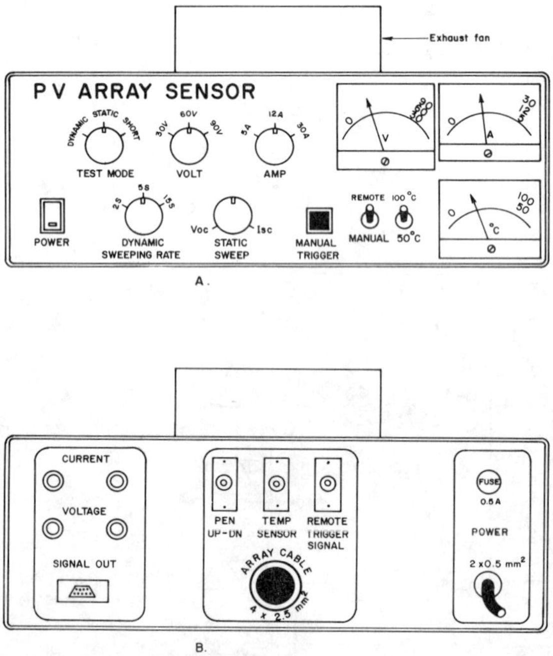

Figure 21.15 PV array sensor: (*a*) front panel; (*b*) back panel.

Along with meter read-out for voltage, current and temperature, the sensor also provides a nine-pin plug for output signals, current and voltage terminals for the *X–Y* plotter, and a 'pen up–down' signal for the *X–Y* plotter. When connected to any mechanical devices such as an *X–Y* plotter, the response time of the mechanically driven pen arm must be considered. *I–V* curve generation times can be set to 15 s per curve-sweeping rate in this case. Data loggers and microprocessors will respond faster; so a sweeping rate of 2 or 5 s will be enough. Stability of the irradiance level must be maintained when choosing the sweeping rate.

APPENDIX A PHOTOVOLTAIC ARRAY SENSOR SPECIFICATIONS

The specifications of the PV array sensor are given in table 21.1.

Table 21.1 Specifications of the PV array sensor.

Input	
Maximum input voltage	100 V DC
Maximum input current	30 A DC
Total input power dissipation at 25 °C	1000 W
Derate above 25 °C	$2\ W\ °C^{-1}$
Temperature range of PV module	100 °C (maximum)
Output	
Voltage	
Voltmeter ranges	30, 60, 90 V
9-pin plug output	$0.15\ V_{out}/V_{in}$, $0.075\ V_{out}/V_{in}$, $0.05\ V_{out}/V_{in}$
Output terminal for X–Y plotter	$1\ V_{out}/V_{in}$, $1\ V_{out}/V_{in}$, $1\ V_{out}/V_{in}$
Current	
Ammeter ranges	3, 12, 30 A
9-pin plug output and output terminal for X–Y plotter	$1.5\ V_{out}\ A^{-1}$, $0.375\ V_{out}\ A^{-1}$, $0.15\ V_{out}\ A^{-1}$
Temperature meter ranges	50, 100 °C
9-pin plug output	$100,\ 50\ mV\ °C^{-1}$
Control panel	
Trigger selector switch	Remote, trigger signal from remote station Manual, activates trigger push button
Temperature selector switch	50, 100 °C
Voltage selector switch	30, 60, 90 V
Current selector switch	3, 12, 30 A
Dynamic sweeping rate selector switch	2, 5, 15 s
Test mode selector switch	Dynamic I–V curve generation Static I–V curve generation Short-circuit current measurement
Trigger button	Press to initiate manual I–V curve generation activated in static mode.
Static sweep dial	Sweeps the I–V curve point by point as the dial is turned from V_{oc} to I_{sc}
Power supply	50–60 Hz; 220 V AC
Fuse	0.5 A
Dimension	31 mm × 11.7 mm × 25 mm

BIBLIOGRAPHY

Millman J and Taub H 1965 *Pulse, Digital and Switching Waveforms* (New York: McGraw-Hill)

22 Equipment Accommodation

22.1 INTRODUCTION

In rural and remote regions where no reliable commercial power supply exists, on-site primary power generation becomes a necessity. To keep the cost of on-site power generation to a minimum, operating equipment in these locations should not be dependent on mechanical cooling facilities. The use of passively cooled or low-energy-consuming buildings in remote areas becomes a necessity. Where on-site construction costs are prohibitive, functional, transportable or pre-fabricated buildings must be designed.

The range of transportable, small and fixed buildings in use in rural and remote areas in Australia is described below. The buildings have been designed with features to promote the natural transfer of heat from the equipment room to its surroundings and to limit the room-temperature rise to an acceptable level.

Many transportable, small and fixed buildings have been specially designed to satisfy the particular requirements of installed equipment for radio and long-line transmission and television translation. The equipment buildings designed in Australia generally belong to two classes: thermally thin light-weight construction or heavy-weight construction with high thermal inertia. Both classes of buildings have been in use in Australia for at least 15 years.

22.2 EQUIPMENT ACCOMMODATION FOR REMOTE AND RURAL AREAS

The most important parameter in the design of passively cooled buildings is the room-temperature upper limit for reliable operation of equipment. The type and size of buildings required are determined by this temperature limit along with design maximum outside shade temperature. Modern line transmission equipment and broad-band radio repeaters have an upper operating limit of 50 °C and at 55 °C the equipment must remain operational without suffering damage or permanent degradation of performance. Therefore, to satisfy an upper room-temperature limit of 55 °C, passive heat transfer design techniques must achieve a room temperature rise of not more than 5 °C for a 50 °C outside shade temperature.

22.2.1 Heavy-weight buildings

22.2.1.1 Above-ground buildings

The main feature of a heavy-weight type of building is the ability to attenuate the diurnal changes in outside air temperature. The large thermal storage effect of the building itself serves to provide passive temperature limiting to minimize the effects of heating by solar radiation. These buildings developed around 1960 for the coaxial cable telephone system were designed to limit daily room-temperature variations to under 2 °C.

The buildings constructed out of pre-fabricated concerete panels have double walls with an air gap in between, the outer and inner panels being 150 mm and 100 mm thick respectively. A 230 mm clear air space through which outside air can flow separates the ceiling from the roof which consists of galvanized steel decking, 75 mm of insulation and a concrete slab 100 mm thick. The ceiling which is 100 mm thick is covered externally by 25 mm of insulation. Improved thermal conductance from the floor slab to the natural ground is provided by 32 copper rods each of 1475 mm in length and 25 mm in diameter.

22.2.1.2 Underground buildings

The underground shelters for radio repeater equipment are steel cylinders 2.4 m in diameter and 2.6 m high. A conning tower which at certain solar-powered sites reaches up to the floor of the above-ground power and battery shelter provides entry to the shelter. Convection is the primary mechanism by which heat released from the equipment is dissipated at the shelter walls to the surrounding earth.

Temperature variations in the soil over the course of a year depend on soil depth. At shelter floor level which is 3 m below the surface, soil temperatures vary only by 5 °C while 0.5 m below the surface at the top of the shelter the soil temperature varies by as much as 15 °C over a year. Therefore, on a daily basis, the temperature of the shelter is essentially constant at floor level while the temperature at ceiling level undergoes an attenuated diurnal swing. The shelter can limit internal temperature so that it does not exceed 40 °C provided that the manhole entrance remains shaded for internal loads of up to 500 W.

22.2.2 Thermally thin buildings

The thermally transparent shelter type has the property of maintaining a small temperature difference between the internal and external temperatures for high equipment heat loads without additional passive heat-removing devices. The internal temperature therefore experiences the same diurnal temperature swing as the outside temperature.

Figure 22.1 is a cross section through a typical naturally cooled transportable shelter. The thermally transparent shelters are usually designed to provide an equipment room midheight temperature not exceeding 55 °C on a 50 °C day. Their thermal performance determined by the size of the shelter and the internal heat load will be discussed in § 22.3.

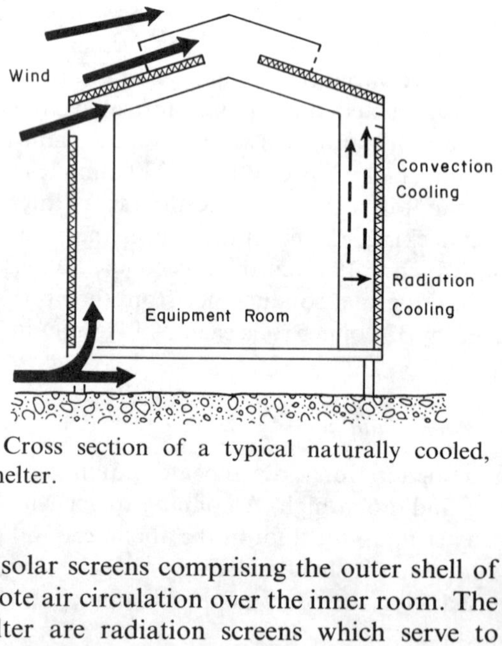

Figure 22.1 Cross section of a typical naturally cooled, thermally transparent shelter.

Radiation and solar screens comprising the outer shell of the shelter are arranged to promote air circulation over the inner room. The roof and outer walls of the shelter are radiation screens which serve to prevent solar radiation from entering the equipment room. A heat sink close to ambient temperature on the inside surface of the radiation screens removes heat radiated from the inner room while convection cooling removes heat from the radiation screens. The screens are made up of an external cladding gloss white in colour, a 25 mm air gap, reflective foil, 50 mm mineral wool insulation and an internal cladding.

22.2.3 Cabinets

Simple low-cost cabinets to be set up at over 1200 repeater sites belonging to the Digital Radio Concentrator System Network have been designed in Australia. Each cabinet designed to provide an upper room-temperature limit of 55 °C on a 50 °C day houses one repeater. The cabinet is divided into two separate compartments: an equipment compartment and a battery compartment accessed separately from opposite sides. The compartments are protected from solar radiation by sheet metal screens placed around the walls and the roof. Figure 22.2 illustrates the cabinet construction and the relative sizes of the equipment and cabinets.

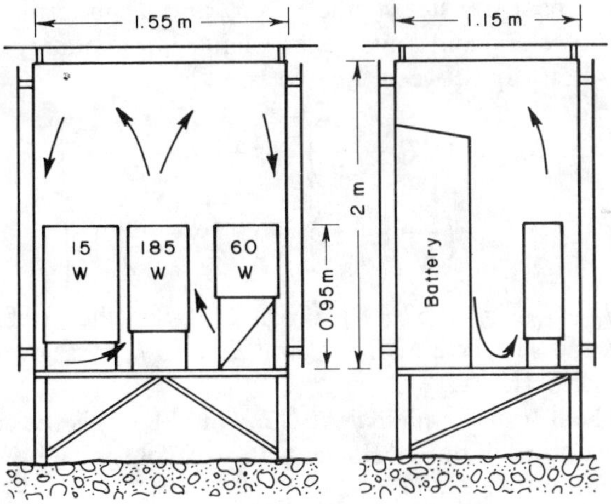

Figure 22.2 Repeater cabinets.

Heat-dissipating equipment inside the cabinets has been placed in such a way as to enhance the natural internal convection air currents. There must be sufficient space provided above the tops of the equipment to allow for heat transfer to the outside air at the walls and ceiling. A minor radiated heat inflow of around 20 W is expected on certain high-temperature days.

22.3 THERMAL PERFORMANCE

A mathematical model of the thermal performance of thermally transparent shelters has been constructed and used for the design of new shelters. Field tests have been able to validate the model up to an outside shade temperature of 43 °C as hotter temperatures have not occurred while testing was being conducted.

The model is built on the simplified convection and radiation heat transfer coefficients for each side of the equipment room walls, floor and ceiling. The convection coefficient h_c for the worst-case situation of laminar flow is

$$h_c = a\left(\frac{\theta}{l}\right)^{0.25} \quad \mathrm{W\,m^{-2}\,K^{-1}} \tag{22.1}$$

where l is the surface characteristic dimension and θ is the temperature difference over the dimension l. The constant a is 1.42 for walls, 1.32 for the ceiling and 0.59 for the floor.

Under calm conditions, the internal value h_{ci} and the external value h_{ce} of the convection coefficient are taken to be equal to h_c given in the above equation.

The radiation heat transfer coefficients h_{ri} and h_{re} in terms of the average room temperature T_R and outside ambient T_0, assuming that the wall temperature is halfway between the two, are

$$h_{ri}=\varepsilon\sigma\left(2T_R-\frac{\delta T}{2}\right)\left[T_R^2+\left(T_R-\frac{\delta T}{2}\right)^2\right] \tag{22.2}$$

$$h_{re}=\varepsilon\sigma\left(2T_0+\frac{\delta T}{2}\right)\left[T_0^2+\left(T_0+\frac{\delta T}{2}\right)^2\right] \tag{22.3}$$

where $\delta T=T_R-T_0$, $\sigma=5.67\times10^{-8}\ \mathrm{W\,m^{-2}\,K^{-4}}$ (the Stefan–Boltzmann constant), ε is the surface emissivity, and h_{ri} and h_{re} are given in watts per square metre per kelvin.

The overall heat transfer coefficient U ($\mathrm{W\,m^{-2}\,K^{-1}}$) is the combination of the convection and radiation coefficients in parallel for each side of the wall. On the assumption that $\theta=\delta T$,

$$U=\left(\frac{1}{h_{ci}+h_{ri}}+\frac{1}{h_{ce}+h_{re}}\right)^{-1}. \tag{22.4}$$

The average room temperature may be found for a given internal heat dissipation Q from

$$UA(T_R-T_0)=Q \tag{22.5}$$

where A is the free surface area of the thermally transparent equipment room containing Q.

Where necessary the empirically derived natural convection relationships involving the non-dimensional Rayleigh and Nusselt numbers can be used instead of equation (22.1).

The free-convection heat transfer coefficients may be adjusted by a suitable multiplier to account for forced convection by the design day breezes. The multipliers produce values of h_{ce} which are approximately two thirds of that suggested by the formula of the American Society for Heating, Refrigerating, and Air-Conditioning Engineers (ASHRAE) for forced-air convection:

$$h=7.43v^{0.8}\ \mathrm{W\,m^{-2}\,K^{-1}} \tag{22.6}$$

for a shelter wall cavity air speed of $v\ \mathrm{m\,s^{-1}}$. This should allow for the fact that not all surfaces will necessarily experience forced-air movement at any one time. At 50 °C ambient a wall cavity draught of 3 $\mathrm{m\,s^{-1}}$ is assumed and a multiplier of 10 is used. Similarly, at 45 °C the draught and multiplier are

1.2 $m\,s^{-1}$ and 4 respectively. At 40 °C and near calm a multiplier of 1.7 is assumed.

Heat transfer coefficients for forced convection may be derived from empirical relationships involving the Reynolds and Nusselt numbers although multipliers determined from field measurements may still be necessary for certain shelter types. This model which takes into account the effect of the prevailing breezes provides a useful estimate of the maximum room temperature for a given outside ambient.

BIBLIOGRAPHY

Heber G and Seufert C 1979 *Solar Cooling of Buildings, German Appropriate Technology Exchange*

Wishart A J 1985 *Telecommun. J. Aust.* **35**

23 Terminology

23.1 PHOTOVOLTAIC SYSTEMS TERMINOLOGY

Absorptance The ratio of absorbed to incident solar radiation. Absorptivity is the property of absorbing radiation possessed by all materials to varying extents.

Absorptivity The ratio of the incident radiant energy absorbed by a surface to the total radiant energy falling on the surface.

Air mass (AM) The length of the path through the earth's atmosphere traversed by direct solar radiation, expressed as a multiple of the path length with the sun at zenith (overhead).

Ambient conditions The conditions of the surroundings.

Albedo The ratio of the light reflected by a surface to the light falling on it.

Ambient temperature The prevailing temperature outside a building.

Ångström (abbreviation, Å) A unit of length, 10^{-10} m.

ASHRAE The acronym denoting the American Society of Heating, Refrigerating, and Air-Conditioning Engineers. ASHRAE handbooks are sources of basic data on heating and air conditioning.

Auxiliary system A system that acts as a back-up to the solar system during extended periods of extremely cold and/or cloudy weather.

Azimuth The angle between the south–north line at a given location and the projection of the earth–sun line in tthe horizontal plane.

Battery An electrical energy storage system frequently using reversible chemical reactions.

Black body An ideal surface that absorbs all the radiation incident on it and emits none; alternatively, a body that emits the maximum possible radiation at a given temperature.

British thermal unit (abbreviation, Btu) A unit of energy which is equal to the amount of heat required to raise the temperature of 1 lb of water (at 4 °C) 1 °F.

Cadmium sulphide (CdS) A yellow–orange chemical compound produced from cadmium metal in large quantities and used as a pigment. As a semiconductor, CdS is always n type.

Cathodic protection The protection of metal against the damage resulting from electrolytic reactions.
Celsius temperature scale A temperature scale based on the freezing and boiling points of water (0 °C and 100 °C respectively). The name Celsius has replaced the name Centrigrade in most scientific usage.
Collector Any device gathering the sun's radiation and converting it to a useful energy form.
Collector efficiency The ratio of the energy collected by a solar collector to the radiant energy incident on the collector.
Collector tilt angle The angle at which the collector aperture is slanted upwards from the horizontal plane.
Collimated light Parallel rays of light; the direct or beam component of solar radiation.
Concentration ratio (concentration factor) The ratio of radiant energy intensity at the hot spot of a focusing collector to the intensity of unconcentrated direct sunshine at the collector site.
Concentrator A solar collector that focuses or funnels solar radiation onto a relatively small absorber at the focal point or line.
Conduction The heat transfer through matter by exchange of kinetic energy from particle to particle.
Declination The angular position of the sun north (+) or south (−) of the plane of the earth's equator; it is a function of the time of year.
Degree day 1 day with the average ambient temperature 1 °F colder than 65 °F. For example, if the average temperature is 55 °F for 3 days, the number of degree days is $(65-55) \times 3$, or 30.
Diffuse insolation The sunlight scattered by atmospheric particulates that arrives from a direction other than the direction of direct sunlight. The blue colour of the sky is an example of diffuse solar radiation.
Direct conversion The conversion of sunlight directly into electric power, instead of collecting sunlight as heat and using the heat to produce power. Solar cells are direct conversion devices.
Direct radiation The radiation from the sun received from a narrow solid angle measured from a point on the earth's surface.
Efficiency The ratio of the measure of a desired effect to the measure of the input causing the effect, both expressed in the same units of measure.
Emittance The ratio of radiation emitted by a real surface to the radiation emitted by a perfect radiator at the same temperature. Normal emittance is the value measured at 90° to the plane of the sample; hemispheric emittance is the total amount emitted in all directions.
Energy gap The range of energies in a semiconductor forbidden to electrons; photons are readily absorbed if their energy exceeds the gap. The electron states below the gap (in energy) consitute the valence band (normally full); above the gap is the conduction band, normally empty, into which valence band electrons can be excited by light. Impurities and atomic displacements produce allowed states in the otherwise forbidden gap.

Faceted concentrator A focusing collector using many flat reflecting elements to concentrate sunlight at a point or along a line.

Fresnel lens A thin piece of plastic or glass containing tiny, carefully shaped grooves whose surfaces refract incoming light as desired, usually to a point or line focus.

Fuel cell A device, somewhat like a battery, that uses a chemical reaction to produce electricity directly, such as the reaction of hydrogen and oxygen to produce electric power with water as a product.

Heliodon A solar angle simulator useful in conducting shading assessments on buildings or solar collector arrays.

Heliostat An electro-optical–mechanical device that orients a mirror so that sunlight is reflected from the mirror in a fixed specific direction, regardless of the sun's position in the sky.

Heliothermal process A process that uses the sun's radiation to produce heat.

Hole A vacant electron state in the valence band; behaves like a positively charged electron.

Hot spot The location on a focusing collector at which the concentrated sunlight is focused and the highest temperatures are produced. If the heat is to be collected, a heat exchanger is heated.

Incidence angle The angle between the direction of the sun and the perpendicular to the surface on which sunlight is falling.

Irradiation (irradiance) The radiant energy falling per unit area on a plane surface per unit time, normally stated in watts per square metre (British thermal units per square foot per hour).

ISES The acronym for the International Solar Energy Society. A world-wide society composed of professionals, 'tinkerers' and enthusiasts interested in the utilization of solar energy.

***I–V* curves** The current–voltage characteristic curves of a solar cell.

Joule (abbreviation J) An energy unit equal to 1 N m.

Kelvin temperature scale A temperature scale whose degrees are equal to degrees on the Celsius scale but whose zero point is at the 'absolute zero' of temperature, at which all molecular motion theoretically ceases. On the Kelvin scale, the freezing point of water is about 273 K and the boiling point about 373 K.

Kinetic energy The energy possessed by virtue of an object's motion.

Knot A measure of wind speed; equal to 1 nautical mile h^{-1}.

Langley A unit of solar radiation intensity equivalent to 1.0 g cal cm^{-2}.

Latitude The angular distance north (+) or south (−) of the equator, measured in degrees.

Linear concentrator A solar concentrator which focuses sunlight along a line, such as the parabolic trough concentrator and the fixed-mirror concentrator.

Load factor The ratio of the average power output of an energy system to the rated output.

Local solar time A system of astronomical time in which the sun always crosses the true north–south meridian at 1200 hours (noon). This system of time differs from local time according to longitude, time zone and equation of time.
Megawatt, electrical (abbreviation, MWe) A million watts of electric power.
Megawatt, thermal (abbreviation MWt) A million watts of thermal (heat) energy.
Micrometre (micron) (abbreviation, μm) A unit of measure equal to one millionth of a metre, a common unit for measuring the wavelength of light. Ultraviolet light has wavelengths less than 0.4 μm, visible light covers the wavelength range 0.38–0.76 μm, and infrared radiation has wavelengths longer than 0.76 μm.
Microscale data Refers to data on insolation and weather parameters that can vary considerably over distances of a few miles, e.g. there may be more cloudiness and haze near a lake or in a city than a few miles away. Microscale data can be collected by satellites.
Normal Perpendicular, in the geometric sense.
n type A semiconductor doped with impurities so as to have free electrons in the conduction band.
Ohmic contact Areas comprising the electrical contacts for solar cells for connection to external circuitry.
Open-circuit voltage A photovoltage developed on an open circuit, which is the maximum available at a given irradiance.
Optical coating A very thin coating applied to glass or other transparent materials to increase the transmission (to reduce the reflection) of sunlight. A coating can also be used to reflect back to the heat exchanger infrared radiation emitted from it.
Parabola The locus of a point moving in a plane so that its distances from a fixed point (focus) and a fixed straight line (directrix) are equal; equation in Cartesian coordinates is $y = x^2/4f$, where f is the focal length.
Pascal (abbreviation, Pa) A unit of pressure equal to $1\ \mathrm{N\,m^{-2}}$.
Passive system A system using the sun's energy without mechanical systems, achieved by properly placed storage masses, overhangs and fenestration, use of vegetation, proper building materials and energy-conserving design.
Pay-back period The length of time required to recover the investment of a project by benefits accruing from the investment; pay-back period is an incomplete economic index.
Peak watt (abbreviation, W_p) A unit used for the performance rating of solar electric power systems. A system rated at 1 W_p will deliver 1 W at the specified working voltage under peak solar irradiation.
Photon A quantum of electromagnetic radiation; its energy E_v is related to the frequency v of the radiation by the equation $E_v = hv$ where h is Planck's constant.

Photovoltaic cell (solar cell) A semiconducting device that converts sunlight directly into electric power. The conversion process is called the photovoltaic effect.

Photovoltaic effect The generation of an electromotive force when radiant energy falls on the boundary between certain dissimilar substances in close contact.

p–n junction A junction of dissimilar semiconductor materials, where electrons move from one type to another, under specific conditions.

Potential energy The energy stored in matter because of its position or because of the arrangement of its parts, e.g. water stored behind a dam.

Power The rate at which work is done.

Power coefficient The ratio of the power extracted by a wind machine to the power available in the wind stream.

Power density The amount of power per unit of cross-sectional area of a wind stream.

Present value The value of a future cash flow discounted to the present; based on the premise that a dollar today is worth more than a dollar received in the future by virtue of the amount of interest (or return) that it earns.

p type A semiconductor doped with impurities so as to have vacancies (holes in the valence band).

Pyranometer A solar radiometer that measures total diffuse and direct radiation.

Pyrheliometer An instrument used to measure the sun's direct radiation.

Quad (abbreviation, Q) A unit of energy equivalent to 10^{15} Btu (occasionally defined at 10^{18} Btu).

Quantum A discrete unit of energy. A quantum of light is called a photon.

Quantum yield The ratio of the number of photon-induced reactions occurring to the total number of absorbed photons (or, less commonly, to the total number of incident photons).

Radian (abbreviation, rad) A unit of angular measure; $1 \text{ rad} = 57.296°$. The sunshine has an angular diameter of 0.009 rad or $\frac{1}{2}°$.

Radiometer An instrument used to measure the intensity of any kind of radiation.

Reflectivity (reflectance) The ratio of light reflected from a surface to the light falling on the surface. The reflectivity plus the absorptivity equals unity, since the incident sunlight is either reflected or absorbed.

Retrofit The installation of solar energy systems in already existing structures.

Selective coating An optical coating for heat exchangers that has a high absorptivity (low reflectivity) for incident sunlight (wavelengths less than 1 μm) and high reflectivity (low absorptivity) for infrared heat (wavelengths greater than 1 μm). The low infrared absorptivity (low emissivity) results in reduced radiant heat loss; so the collection efficiency is improved, and higher temperatures can be achieved.

Semiconductor A material whose atomic structure is such that it has an electrical conductivity intermediate between an insulator, such as glass, and a metal, such as copper. Its conductivity can be greatly influenced by the addition of minute amounts of certain elements called doping agents. Depending on the agent used, the doped semiconductor will have either an excess (n type) or a deficiency (p type) of energy-conducting electrons.

Silicon cell A photovoltaic cell made principally of silicon, a semiconductor.

Solar-altitude angle The angle between the line joining the centre of the solar disc to the point of observation at any given instant and the horizontal plane through that point in observation.

Solar cell See Photovoltaic cell.

Solar concentrator A device using lenses or reflecting surfaces to concentrate sunlight.

Solar constant The intensity of solar radiation beyond the earth's atmosphere, at the average earth–sun distance, on a surface perpendicular to the sun's rays. The value for the solar constant is 1353 $\mathrm{W\,m^{-2}}$ ($\pm 1.6\%$).

Solar distillation The process in which the sun's energy is utilized for the purification of brackish or poor-quality water. The greenhouse effect is utilized to trap heat to evaporate the liquid. The vapour so formed then condenses on the cover plate and can be collected for use.

Solar energy The energy, in the form of radiation, emitted from the sun and generated by means of a fusion reaction within the sun.

Solar farm A large array of solar collectors for generating large amounts of electric power.

Solar furnace A solar concentrator for producing very high temperatures. (Installations in France, Russia and Japan produce temperatures as high as 7000 °F.)

Solar ponds A pond of stratified water that collects and retains heat. Convection, normally present in ponds, is suppressed by imposing a stable density gradient of dissolved salts.

Solar radiation The radiant energy received from the sun both directly as a beam component and diffusely by scattering from the sky and reflection from the ground.

Solar–thermal electric conversion The conversion of solar energy to thermal energy, which powers turboelectric generators.

Specific heat capacity The amount of heat required to raise the temperature of unit mass of material one degree, usually measured in British thermal units per pound per degree Fahrenheit or joules per kilogram per kelvin.

Spectral energy distribution A curve showing the variation in spectral irradiance with wavelength.

Spectral irradiance The monochromatic irradiance of a surface per unit band width at a particular wavelength. Units often in watts per square metre per nanometer band width.

Spectral pyranometer An instrument for measuring total insolation over a restricted wavelength range.

Spectral reflectance The ratio of the energy reflected from a plane surface in a given defined wave band to the energy incident in the waveband.

Specular reflection A mirror-like reflection in which incident and reflected angles are equal.

Stationary-reflector tracking-absorber concentrating collector A spherical dish with movable absorber rod that tracks the movement of the sun.

Suntime The time measured on a basis of the sun's virtual motion. See Local solar time.

Temperature, dry-bulb The temperature of the air as indicated by an ordinary thermometer.

Temperature, wet-bulb The temperature of the air as indicated by a thermometer whose bulb is surrounded by a wet gauze and past which air is blown rapidly. The difference between the wet-bulb and dry-bulb temperatures is used to find the atmospheric humidity. When the atmosphere is completely saturated with water vapour (humidity of 100%), the wet-bulb temperature equals the dry-bulb temperature.

Thermal conductivity The amount of heat that can be transferred by conduction through a material of unit area and thickness per unit temperature difference.

Thermocouple A thermoelectric device consisting of two dissimilar wires with their ends connected together. A small voltage is generated when two junctions are at different temperatures; if one junction is kept at a reference temperature, the voltage generated in the other is a measure of the temperature of the other junction above the reference.

Thermopile A large number of thermocouples connected in series.

Thermostat A temperature sensor used to monitor the temperature in a space; provides a signal to a building's thermal control system.

Tool energy system A system for providing all energy requirements, including heat, air conditioning and electric power.

Tracking collector A collector that can rotate about one or two axes to face the sun; usually restricted to high-temperature concentrating collectors because of the complexity and cost of a tracking system.

U Overall heat loss coefficient; frequently called the *U* value.

Ultraviolet radiation The radiation having wavelengths longer than those of x-rays.

Watt (abbreviation, W) The energy rate of $1\ \mathrm{J\,s^{-1}}$.

Wind energy The kinetic energy resulting from air motion caused by the interaction of earth's rotation and solar-driven atmospheric currents.

Work The energy transfer causing a force to move through a distance with measure equal to the dot product of the force and displacement vectors.

23.2 BATTERY TERMINOLOGY

Definitions of terminology common to batteries are presented in alphabetical order in this section.

Activation polarization The polarization resulting from the rate-determining step of the electrode reaction. See also Polarization.

Active material The constituents of a cell that participate in the electrochemical charge–discharge reaction.

Ampere–hour efficiency See Efficiency, ampere–hour.

Anode The electrode in an electrochemical cell at which chemical oxidation takes place. During charge, the positive terminal of the cell is the anode. During discharge, the situation reverses and the negative terminal of the cell is the anode. However, battery manufacturers sometimes refer to the negative terminal as the anode during both charge and discharge. See also Cathode.

Anolyte The electrolyte associated with the anode in an electrochemical cell. In a secondary battery, the anolyte is associated with the negative terminal.

Available capacity See Capacity, available.

Battery Two or more electrochemical cells electrically interconnected in an appropriate series or parallel arrangement to provide the required operating voltage and current levels. Under common usage, the term battery also applies to a single cell if it constitutes the entire electrochemical storage system.

Capacity C Generally, the total number of ampere hours that can be withdrawn from a fully charged cell or battery.

Capacity, available The total number of ampere hours that can be withdrawn from a cell or battery for a specific set of operating conditions including discharge rate, temperature, initial state of charge, age and cut-off voltage.

Capacity, energy The total number of watt hours (kilowatt hours) that can be withdrawn from a fully charged cell or battery. The energy capacity of a given cell varies with temperature, rate, age and cut-off voltage. This term is more common to system designers than it is to the battery industry where the capacity usually refers to ampere hours.

Capacity, installed The total number of ampere hours that can be withdrawn from a new cell or battery when discharged to the system specified cut-off voltage at the system design rate and temperature (i.e. discharge to the system design specified maximum depth of discharge).

Capacity, rated The manufacturer's conservative estimate of the total number of ampere hours that can be withdrawn from a new cell or battery for a specified discharge rate, temperature and cut-off voltage.

Cathode The electrode in an electrochemical cell at which chemical reduction takes place. During charge, the negative terminal of the cell is the cathode. During discharge, the situation reverses and the positive terminal of the cell is the cathode. However, battery manufacturers sometimes refer to the positive terminal as the cathode during both discharge and charge. See also Anode.

Catholyte The electrolyte associated with the cathode of an electrochemical cell in a secondary battery. The catholyte is usually associated with the positive terminal.

Cell The basic electrochemical unit used to store electrical energy.

Charge rate The current applied to a cell or battery to restore its available capacity. This rate is commonly normalized with respect to the rated capacity of the cell or battery. For example, the 10 h charge rate of a 500 A h cell or battery is expressed as

$$\frac{\text{rated capacity}}{\text{charge time}}=\frac{500\ \text{A h}}{10\ \text{h}}=50\ \text{A}=\frac{C}{10}\ \text{rate.}$$

For the same cell or battery, a 5 h charge rate would be designated the $C/5$ rate and would result in a charging current of 100 A. Unfortunately, the ampere hour efficiency of many secondary batteries is less than 100%. Therefore the time required to recharge one cell completely, from 100% depth of discharge, is often longer than indicated by the so-called 'hour rate'.

Charging The conversion of electrical energy into chemical potential energy within a cell by the passage of a direct current in the direction opposite to that of discharge.

Coefficient of use The percentage of total active material contained within a cell that is converted during the discharge of 100% of rated capacity. (Also called utilization factor.)

Concentration polarization Polarization caused by the depletion of ions in the electrolyte at the surface of the electrode. See also Polarization.

Cut-off voltage The cell or battery voltage at which discharge is terminated. The cut-off voltage is specified by the cell manufacturer and is generally a function of discharge rate. Discharge beyond the specified cut-off voltage usually results in increasingly rapid decreases in cell voltage and energy output, it may permanently damage the cell and it may void a manufacturer's warranty. (Also called final voltage.)

Cycle One discharge–charge sequence to a specified depth of discharge.

Cycle life The number of cycles, to a specified depth of discharge, that a cell or battery can undergo before failing to meet its specified capacity or efficiency performance criteria.

Cycle service A duty cycle characterized by frequent and usually deep discharge–charge sequences, such as motive power applications.

Deep discharge The withdrawal of a significant percentage of rated capacity (50% or more).

Depth of discharge (DOD) The ampere hours removed from a fully charged cell or battery, expressed as a percentage of rated capacity. For example, the removal of 25 A h from a fully charged 100 A h rated cell results in a 25% depth of discharge. Under certain conditions, such as discharge rates lower than that used to rate the cell, the depth of discharge can exceed 100%.

Discharge The process of withdrawing current from a cell or battery by the conversion of chemical energy into electrical energy.

Discharge rate The current removed from a cell or battery. This rate can be expressed in amperes, but more commonly, it is normalized to rated capacity C and is expressed as C/X. For example, drawing 20 A from a cell with a rated capacity of 100 A h is referred to as the $C/5$ discharge rate (100 A h/20 A). Similarly, discharge currents of 5 A, 10 A and 33.3 A would be designated as the $C/20$, $C/10$ and $C/3$ rates respectively. See also Hour rate.

Efficiency The ratio of the useful output to the input.

Efficiency, ampere hour (Coulombic) The ratio of the ampere hours removed from a cell or battery during a discharge to the ampere hours required to restore the initial capacity:

$$\frac{\text{ampere hours discharged}}{\text{ampere hours of charge}} = \frac{\int_0^{t_d} i_d \, dt}{\int_0^{t_c} i_c \, dt}$$

where i_d and i_c are the discharging and charging currents respectively, and t_d and t_c are the discharging and charging times respectively.

Efficiency, voltage The ratio of the average discharge voltage of a cell or battery to the average charge voltage during the subsequent restoration of an equivalent capacity.

Efficiency, energy (watt hour) The ratio of the energy delivered by a cell or battery during a discharge to the total energy required to restore the initial state of charge. The watt hour efficiency is approximately equal to the product of the voltage and ampere hour efficiencies. This is sometimes referred to as the round-trip efficiency. Round-trip energy efficiencies usually do not include energy losses resulting from self-discharge, auxiliary equipment (parasitic losses) or battery equalization.

Electrochemical couple The system of active materials within a cell that provides electrical energy storage through an electrochemical reaction.

Electrolyte The medium which provides the ion transport mechanism between the positive and negative electrodes of a cell. In some cells, such as the lead–acid type, the electrolyte may also participate directly in electrochemical charge–discharge reactions.

End-of-charge voltage The cell or battery voltage at which the finishing charge is normally terminated by the charging source.

Energy capacity See Capacity, energy.

Energy density The ratio of rated energy available from a cell or battery, normalized to its weight or volume.

Equalization The process of restoring all cells in a battery to an equal state of charge. For lead–acid batteries this is a charging process designed to bring all cells to 100% state of charge. Some battery types may require a complete discharge as a part of the equalization process.

Equalizing charge A continuation of normal battery charging, at a voltage level slightly higher than the normal end-of-charge voltage, in order to provide cell equalization within a battery.

Final voltage See Cut-off voltage.

Finishing charge rate The rate specified for charging a cell or battery that is nearing 100% state of charge.

Float service A duty cycle characterized by long periods of standby operation, at full charge and constant voltage, with only occasional discharge–charge sequences.

Formation The process, during manufacture or installation, by which the electrode materials are transformed into the compositions required for cell operation. For example, the active materials of a fully charged lead–acid cell are pure lead for the negative and lead dioxide for the positive plates. However, both plates are manufactured using lead oxide. During a specially designed charging cycle, the lead oxide on the negative plates is reduced and the lead oxide on the positive plates is oxidized, thereby forming the active materials.

Gassing The evolution of gas from one or more of the electrodes in a cell. Gassing commonly results from local action (self-discharge) or from the electrolysis of water in the electrolyte during charging.

Grid An electrically conductive structure that supports the active material of a plate but does not chemically participate in the normal charge–discharge reaction.

Hour rate The discharge rate of a cell or battery expressed in terms of the length of time that a new fully charged cell or battery can be discharged at a specific current before reaching a specified cut-off voltage:

$$\text{hour rate} = C/i$$

where C is the rated capacity of the cell or battery and i is the specified discharge current. For example, if a new, fully charged cell rated at 100 A h can be discharged at 20 A for a period of 5 h before reaching the cut-off voltage, discharge of the battery at 20 A is referred to as the 5 h rate ($C/i = 100$ A h/20 A). Unfortunately, for most cells, the available capacity is not constant with discharge rate. Therefore the discharge current cannot always be calculated by dividing discharge time into rated capacity. For example, a lead–acid battery with a rated capacity of 100 A h at the 5 h rate may only deliver 88 A h at the 3 h rate before reaching the cut-off voltage. The discharge current at the 3 h rate would therefore be 29.3 A rather than 33 A as might otherwise be expected and would be designed as the $C/3.4$ rate. See also Discharge rate.

Hours of capacity The total number of hours for which a fully charged cell or battery is capable of supplying the system load demand before reaching the specified maximum depth of discharge.
Installed capacity See Capacity, installed.
Life The period during which a cell or battery is capable of operating above a specified capacity or efficiency performance level. For example, with lead–acid batteries, the end of life is generally taken as the point in time when a fully charged cell can only deliver 80% of its rated capacity. Beyond this state of aging, deterioration and loss of capacity begins to accelerate rapidly. Life may be measured in cycles and/or years, depending on the type of service for which the cell or battery is intended.
Local action The chemical reactions within a cell that convert the active materials to a discharged state without supplying useful energy through the battery terminals (self-discharge).
Motive power cell or battery A cell or battery that is intended to power electrically operated mobile equipment (fork-lift trucks) and is designed to be operated in a daily deep-cycle regime at moderate ($C/6$) discharge rates.
Nominal operating voltage The average terminal voltage of a cell or battery discharging at a specified rate and at a specified temperature.
Open-circuit voltage The terminal voltage of a cell or battery at a specified state of charge and temperature under no-load conditions.
Overcharge The forcing of current through a cell after all the active material has been converted to the charged state. In other words, charging continued after 100% state of charge is achieved. Overcharging does not increase the energy stored in a cell and usually results in gassing and/or excessive heat generation, both of which reduce battery life.
Plate An assembly of active material and, in some cases, a support grid. The plates form the positive and negative electrodes of the cell.
Polarization The deviation from open-circuit voltage caused by the flow of current in an electrochemical cell.
Power density The ratio of the rated power available from a cell or battery, normalized to its weight or volume.
Primary cell or battery A cell or battery whose initial capacity cannot be significantly restored by charging and is therefore limited to a single discharge.
Rated capacity See Capacity, rated.
Reversal The continued discharge of a cell to the point that the cell's electrical terminals change polarity.
Reversible potential The open-circuit (zero-current) potential of a reversible chemical reaction.
Secondary cell or battery A cell or battery that is capable of being charged repeatedly.
Self-discharge The process by which the available capacity of a cell is reduced by internal chemical reactions (local action).

Self-discharge rate The amount of capacity reduction occurring per unit of time in a cell as the result of self-discharge.
Separator An electrically non-conductive and ionic permeable material used to prevent the direct contact of individual plates within a cell.
Starting, lighting and ignition (SLI battery) A battery (standard automotive battery) designed primarily for high rate, low depth-of-discharge operation typical of automobile engine starting.
Stationary cell or battery A cell or battery that is generally intended for float-service usually in a fixed location.
State of charge (SOC) The available capacity in a cell or battery expressed as a percentage of rated capacity. For example, if 25 A h have been removed from a fully charged 100 A h cell, the new state of charge is 75%.
Sulphation The formation of lead sulphate crystals on the plates of a lead–acid battery. Commonly used to indicate the large crystals which form in partially discharged cells as the result of temperature cycling. These large crystals are more difficult to reduce by the charging current than are the smaller crystals that result from normal and self-discharge reactions. Sulphation can be caused by leaving the battery in a discharged state for long periods of time.
Traction See Motive power cell or battery.
Utilization factor See Coefficient of use.
Voltage efficiency See Efficiency, voltage.
Watt hour efficiency See Efficiency, energy (watt hour).

23.3 ECONOMIC TERMINOLOGY

Actual interest rate The market interest rate corrected by the inflation rate.
Amortization (i) In banking, the repayment of debts. (ii) In investment calculation, the recovery of investment expenditure by means of annual returns.
Amortization period The period needed to recover the capital invested in an investment project from the returns discounted at an appropriate rate. Also pay-back period.
Assumed interest (or discount) rate The interest (discount rate) which is assumed or required as the minimum acceptable on the capital investment for the duration of the project. Also called the cut-off discount rate.
Capital cost The cost of construction, including design costs, land costs and other costs necessary to build a facility. This does not include operating costs.
Capital recovery factor (CRF) The factor used to calculate the amount of regular payments needed to recover a present value at a given interest rate over a specified time period.
Capital servicing The interest and principal on credits.

Cash flow The flow of capital from turn-over. Here the balance of cash inflows and outflows during a specified period within the service life of an investment project.
Compounding The increase in the value of a past payment by regularly compounded interest to establish its present value.
Constant dollars The dollars (or other currency) expressed, net of inflation, in terms of today's dollars.
Continuous amortization The repayment of a debt in equal periodic amounts.
Credit duration A period agreed with the lender as to the time during which a certain debt is to be repaid.
Depreciation The periodic reduction in the value of the capital goods assets connected with an investment.
Discount rate The opportunity cost of making an investment.
Discounting The reduction in the value of a future payment calculated at a given discount rate to establish its present value.
Discounting factor The factor for calculating the present value of a single future payment accounting for the time and at a given discount rate.
External financing The securing of financial resources from outside a company to cover its capital requirement, e.g. for investment projects.
Inflation The process of general price expansion.
Inflation rate The increase in the general price level as a percentage of the previous year's prices.
Interest rate The cost of borrowing money. See Discount rate.
Internal financing The provision of financial resources within a company to permit the adoption of an investment project.
Internal rate of return (IRR) The profitability (in per cent) of an investment calculated from the net present value.
Lifetime The maximum period during which an investment facility can be used for either production or other purposes.
Linear depreciation The initial equipment costs, minus the amount anticipated as the liquidation yield, are divided by the number of years of foreseeable service life. This gives a constant annual depreciation charge. In other words, the book value of the plant decreases linearly.
Liquidation yield The income expected by the sale of assets at a fixed point in time, usually at the end of the service life.
Loss of purchasing power The reduction in value of a certain capital sum due to inflation.
Loss of value The depreciation in the value of assets due to utilization.
Market interest rate The normal rate of interest on the current capital market for the investment of internal capital or the loan of external capital taking account of duration and risk aspects.
Maximum acceptable pay-back period The maximum period within which an investment should pay itself off.

Minimum acceptable profitability The minimum profitability required by the investor on the average capital tied up.

Net present value (NPV) The sum of the present values of all cash flows connected with a particular investment project over its lifetime.

Opportunity costs The fictitious costs which correspond to the income which might be expected from an alternative application of scarce resources. They are sometimes also referred to as 'shadow prices'.

Pay-back period The amortization period. Also called the pay-off or pay-out period.

Payments (i) The payments by the investor to third parties in connection with an investment project. (ii) The payments in connection with an investment project on the part of third parties to the benefit of the investor.

Planning period The period between the initiation of an investment project and its conclusion. The planning period is equal to or longer than the service life.

Present value The value of a future amount discounted or a past amount compounded over a specified time period at a specified discount rate.

Present value factor (PVF) The factor used to calculate the present value of a sequence of equal payments or cash flows occurring at regular intervals over a specified time period and at a given interest rate.

Profitability The return on investment.

Profitability threshold This describes a profit–loss situation from which point onwards an investment project is economically favourable. Indicators e.g. the cut-off discount rate or the break-even point.

Recovery factor, capital See Capital recovery factor.

Remaining service life The difference between the planned or actual service life of the plant and its technical lifetime. If the service life and the technical lifetime are equal, then the remaining service if zero.

Renewable-energy plant Plant for the utilization of renewable energies.

Renewable-energy technologies The technologies for the utilization of renewable energies.

Residual value The residual value of material assets connected with an investment project consists of their initial investment costs minus the depreciation up to a certain point in time (usually the service life).

Return The excess of the income in a period (usually one calendar year) over the running costs.

Risk supplement The increase in the interest rate to be used to take account of any particular risk of loss.

Sensitivity analysis An investigation during an investment evaluation with the aim of determining the sensitivity in a financial indicator following a given variation in a project input parameter.

Service life The period during which an investment facility should or will be used economically. The service life can be shorter than the technical lifetime; however, it is usually assumed that the technical lifetime and service life are equal.

Servicing of debt Equal to capital servicing or financial costs.
Technical lifetime The maximum period during which plant or components are technically operational.

23.4 SOLAR GEOMETRY

Angle of incidence (of direct solar radiation) The angle between the direct solar radiation beam and the normal drawn outwards from the plane considered (e.g. from the collector aperture).
Aphelion The point in the earth's orbit when it is farthest from the sun, approximately 152×10^6 km.
Azimuth (angle) The arc of the horizon (in degrees) intercepted between a given point and an adopted reference direction: usually the true north, and measured in a clockwise direction, 0–360°.
Heliodon A solar angle simulator, useful in conducting shading assessments on buildings or collector arrays, usually having a model table, tilting for the latitude, rotating for the hour of day, using a lamp to represent the sun, at some distance away, mounted on a vertical rail, allowing adjustment for the date (for declination).
Orientation (angle) The direction that a collector (or a building) faces, expressed as the aximuth angle of the horizontal projection of the surface normal.
Perihelion The point in the earth's orbit when it is closest to the sun, approximately 147×10^6 km.
Solar altitude (angle) The angle between a straight line from the apparent position of sun to the point of observation and the horizontal plane through that point of observation.
Solar azimuth (angle) The projected angle between a straight line from the apparent position of the sun to the point of observation and due north. The angle is measured clockwise using the projections on the local horizontal plane, 0–360°.
Solar declination The angle subtended between the earth–sun line and the plane of the equator (north positive), zero on equinox dates, varying between +23.45° (December 22) and (June 21).
Solar hour angle (HRA) An angle of 15° for each hour (HR) measured from solar noon, negative for morning hours and positive for afternoon hours, i.e. $\text{HRA} = 15(\text{HR} - 12)$.
Solar noon For any given location the solar noon is the local time of day when the sun is at its highest altitude for that day, i.e. the time when the sun crosses the observer's meridian.
Solarscope A device similar to the heliodon, but having a fixed horizontal model table and a light source moveable to any solar altitude and azimuth.

Solar time The hour of the day as determined by the apparent angular motion of the sun across the sky, with solar noon as the reference point for 12 o'clock.

Sun-path diagram A graphic representation of solar altitude against solar azimuth, showing the position of the sun from sunrise to sunset for various dates of the year. Note that many different projection methods are in use.

Sunset hour angle The hour angle occurring at sunset. Note that many different definitions exist for 'sunset'; the most generally used is the geometric definition, i.e. the time when the centre of the sun's disc coincides with the extension of the observer's horizontal plane.

Zenith The highest point of the sky hemisphere, the point vertically above the observer.

Zenith angle The angular distance of the sun from the vertical, i.e. the angle subtended by a vertical line to the zenith and the line of sight to the sun, i.e. 90° minus the solar altitude angle.

23.5 RADIATION TERMS

Infrared radiation The electromagnetic radiation of wavelengths between 760 nm and approximately 1 μm.

Long-wave radiation The radiation originating from sources at terrestrial temperatures (e.g. ground and other terrestrial objects) and thus substantially all at wavelengths greater than 3 μm.

Radiant energy The energy in the form of electromagnetic waves.

Radiant flux The power emitted, transferred or received in the form of radiation.

Radiation The physical process of emission or transfer of energy in the form of electromagnetic waves.

Short-wave radiation The radiation with wavelengths less than 3 μm.

Solar energy The radiant energy originating from the sun's radiation, primarily in the wavelength region from 0.3 to 3.0 μm, or the energy made available by the reception and conversion of solar radiation.

Solar flux The radiant flux originating from the sun, measured in watts.

Solar radiation The radiation emitted by the sun. (Almost all the solar radiation incident at the earth's surface is at wavelengths less than 3 μm and is often termed short-wave radiation. The use of this term in this sense is discouraged.)

Solar spectrum The distribution by wavelength (or frequency) of electromagnetic radiation emitted from the sun.

Total (incident) radiation The radiation at all wavelengths.

Ultraviolet radiation The electromagnetic radiation with wavelengths shorter than visible light (shorter than approximately 380 nm) and longer than x-rays.

Visible radiation The radiation with wavelengths that stimulate the human optic nerves. Visible radiation lies approximately within a wavelength band from 380 to 760 nm.

23.6 RADIATION QUANTITIES

Atmospheric long-wave radiation (Synonymous with the deprecated term sky radiation) the thermal radiation emitted by gases and particles in the atmosphere.
Circumsolar radiation The radiation scattered by the atmosphere so that it appears to originate from an area of the sky immediately adjacent to the sun. It causes the solar aureole.
Diffuse (solar) radiation (Synonymous with the deprecated term sky radiation) the downward scattered and reflected solar radiation incident upon a given plane surface from a solid angle of 2π sr, excluding the sun's disc (excluding the direct radiation).
Direct (solar) radiation (Synonymous with the term beam (solar) radiation) the radiation received from a solid angle centred on the sun's disc, on a given plane. Note that normally the instruments measuring the direct radiation at normal incidence have field angles of up to 15°. See Pyrheliometer in § 23.7.
Extra-terrestrial solar radiation The solar radiation received at the outer limit of the earth's atmosphere.
Global (solar) radiation The sum of the direct and diffuse solar radiation incident upon a given plane surface from a solid angle of 2π sr. See also Pyranometer in § 23.7.
Hemispherical (solar) radiation The diffuse solar radiation received by a horizontal surface.
Insolation An imprecise term which may mean the radiation received from the sun, solar irradiation or irradiance or, in a general sense, sunlight penetration. Its use is deprecated.
Irradiance Synonymous with the deprecated terms incident solar radiation intensity, instantaneous insolation or incident radiant flux density) the power density of radiation incident on a surface, i.e. the quotient of the radiant flux incident on the surface and the area of that surface, or the rate at which radiant energy is incident on a surface, per unit area of that surface.
Irradiation (Synonymous with the deprecated term radiant exposure) the indicident energy per unit area of a surface, found by integration of irradiance over a specified time interval, often an hour or a day.
Sky radiation A deprecated term, as it can mean either atmospheric radiation or diffuse (solar) radiation.
Solar constant The solar irradiance outside the earth's atmosphere on a plane normal to the direction of this radiation, when the earth is at its mean

distance from the sun (149.5×10^6 km). The solar constant value so defined is 1353 W m^{-2}, but it is subject to revision from time to time.

23.7 RADIATION MEASUREMENT

Isohel Curve (drawn on a map) of equal sunshine duration during a given interval of time.

Isorad Curve (drawn on a map) of equal irradiation during a given interval of time.

Pyranometer An instrument for measuring the global (diffuse plus direct) solar radiation on a plane surface from a solid angle of 2π sr. The output of the instrument can be either irradiance or irradiation. It should have an approximately constant response in the wavelength range 0.3–3 μm. Note that, when radiation coming from the solid angle of the sun's disc is obscured from the instrument by a shade ring or disc, the pyranometer can be used to determine the diffuse solar radiation.

Pyrheliometer (Synonymous with the deprecated term actinometer) an instrument using a collimated detector for measuring the direct (beam) radiation received from a solid angle centred on the sun's disc, on a plane perpendicular to the axis of the solid angle. The output of the instrument can be either irradiance or irradiation. It should have an approximately constant response in the wavelength range 0.3–3.0 μm. Note that most instruments have a field (acceptance) angle of up to 15°. For the purposes of concentrating collectors the field angle should not be greater than 6°.

Pyrradiometer An instrument for measuring the total radiation (all wavelengths, solar, atmospheric, terrestrial or reflected solar) on a plane surface from a solid angle of 2π sr.

Radiometer An instrument used for measuring radiation (the output of the instrument can be either irradiance or irradiation).

Solarimeter A specific type of pyranometer based upon the Moll–Gorczynski thermopile design. Its use as a synonym for pyranometer is deprecated.

Spectral pyranometer An instrument for measuring solar radiation over restricted wavelength ranges.

Sunshine recorder An instrument which records the time interval during which solar radiation reaches sufficient intensity to cast distinct shadows. Note that the threshold value of direct irradiance of 200 W m^{-2} has been suggested.

23.8 RADIATION PROPERTIES AND PROCESSES

Absorptance (solar) The ratio of the amount of solar radiation absorbed by an element of a surface to the amount of such radiation incident on it (for terrestrial applications usually calculated for air mass 2 characteristics).

Air mass (AM) The ratio of the mass (optional thickness) of the atmosphere in the actual earth–sun path (through which the beam radiation passes) to the mass which would be traversed if the sun were directly overhead (at the zenith) and the observer at sea level.

Albedo The ratio of the radiation (radiant energy or luminous energy) reflected by a surface, to that incident on it; a generalized term of the average reflectance of a defined surface area (usually of the earth or clouds). In technical applications, its use is discouraged; the preferred term is reflectance.

Ångström (abbreviation, Å) A unit of length, 10^{-10} m or 0.1 nm (its use is discouraged; the preferred unit is the nanometre where 1 nm = 10 Å).

Attenuation The reduction in radiation flux over a given path length, due to absorption and scattering.

Emittance The ratio of the radiant energy emitted by a body to the energy emitted by a black body at the same temperature.

Reflectance The ratio of radiation reflected from a surface to the total radiation incident on that surface, which is dependent on wavelength of radiation and angle of incidence.

Reflectance, hemispherical The ratio of all the flux leaving a surface or a medium by reflection, within a solid angle of 2π sr, to the incident flux.

Scattering The interaction of radiation with matter, where its direction is changed, but the total energy and wavelength remain unaltered.

Sky temperature The atmospheric radiation received at a surface may be expressed in terms of an equivalent black-body radiation temperature, which is the sky temperature.

Index